Developments in Petroleum Science
Volume 66

Well Production Performance Analysis for Shale Gas Reservoirs

First Edition

Series Editor

Baojun Bai
Zhangxin Chen

Developments in Petroleum Science
Volume 66

Well Production Performance Analysis for Shale Gas Reservoirs

Edited by

Liehui Zhang
State Key Laboratory of Oil and Gas Reservoir Geology and Exploitation, Southwest Petroleum University

Zhangxin Chen
University of Calgary

Yu-long Zhao
State Key Laboratory of Oil and Gas Reservoir Geology and Exploitation, Southwest Petroleum University

ELSEVIER

Elsevier
Radarweg 29, PO Box 211, 1000 AE Amsterdam, Netherlands
The Boulevard, Langford Lane, Kidlington, Oxford OX5 1GB, United Kingdom
50 Hampshire Street, 5th Floor, Cambridge, MA 02139, United States

ISBN: 978-0-444-64315-5
ISSN: 0376-7361

Publisher: Candice Janco
Acquisition Editor: Amy Shapiro
Editorial Project Manager: Jennifer Horigan
Production Project Manager: Omer Mukthar
Cover Designer: Greg Harris

Typeset by SPi Global, India

Contents

Foreword vii

Preface xi

1. Shale gas reservoir characteristics and microscopic flow mechanisms 1

1.1 Introduction 1
1.2 Shale gas reservoir characteristics 2
1.3 Pore type analysis in shale gas reservoirs 6
1.4 Accumulation mechanisms in shale gas reservoirs and model description 12
1.5 Multiscale flow mechanisms in shale gas reservoirs 24
1.6 Mathematical models with various shale gas flow mechanisms 39

2. Source function derivation for gas reservoirs under different flow mechanisms 49

2.1 Introduction 49
2.2 Solutions of flow mechanism models 50
2.3 Continuous point source solutions in circular gas reservoirs 54
2.4 Continuous point source solutions in rectangular gas reservoirs 61

3. Fractured vertical wells in shale gas reservoirs without SRV 73

3.1 Introduction 73
3.2 Fractured vertical wells in circular gas reservoirs 74
3.3 Fractured vertical well in closed rectangular gas reservoirs 77
3.4 Superposition of wellbore storage and skin effects 86
3.5 Solution for production at constant bottom-hole pressure 87
3.6 A numerical inversion algorithm 88
3.7 Gas well pressure and production performance analysis 90

4. Multi-stage fractured horizontal wells in shale reservoirs without SRV 115

4.1 Introduction 115
4.2 Multi-stage fractured horizontal wells in circular gas reservoirs 116
4.3 MFHWs in rectangular gas reservoirs 121
4.4 Analysis of well bottom-hole pressure and production performance 125

5. Fractured vertical wells in circular gas reservoirs with circular SRV **151**
5.1 Introduction **151**
5.2 Continuous line source solutions in circular composite gas reservoirs **152**
5.3 Fractured vertical wells in circular composite gas reservoirs **156**
5.4 Analysis of pressure and production type curves **158**

6. Multi-stage fractured horizontal wells in circular reservoirs with SRV **177**
6.1 Introduction **177**
6.2 MFHWs in circular composite gas reservoirs **177**
6.3 Analysis of pressure and production type curves **180**

7. Fractured wells in rectangular gas reservoirs with SRV **191**
7.1 Introduction **191**
7.2 Boundary element models in composite gas reservoirs with SRV **194**
7.3 Fractured vertical wells in rectangular composite shale gas reservoirs with SRV **197**
7.4 MFHW in rectangular composite shale gas reservoirs with global SRV **210**
7.5 MFHW in shale gas reservoirs with local SRV **222**

8. Numerical simulation of MFHWs in shale gas reservoirs based on CVFEM **237**
8.1 Introduction **237**
8.2 A MFHW in a single porosity medium in a shale gas reservoir **238**
8.3 A MFHW in a dual-continuum medium in a shale gas reservoir **249**
8.4 Tri-porosity models in shale gas reservoirs **271**

9. Case studies **287**
9.1 Application of a well test analysis model **287**
9.2 Application of numerical simulation **299**

References **311**
Nomenclatures **319**
Appendices **323**
Index **369**

Foreword

Around the world, concerns about the recovery of unconventional oil and gas resources are increasing. The intensified contradiction of supply and demand, continuously decreasing reserves of conventional resources, constant advancement of development technologies in drilling and fracturing, and the urgent need to improve energy consumption worldwide are factors driving these concerns.

Coal bed methane, shale gas, tight gas, and natural gas hydrates are the most common unconventional resource types. Shale gas is a kind of biogenic or thermogenic natural gas found in organic-rich shale where interlayers are under a free and/or adsorbed state. The successful development of shale gas in North America is driving a revolution of resource development globally. According to EIA (Energy Information Administration) statistical data, the production of shale gas in the United States increased from $563.5 \times 10^8\ m^3$ in 2007 to $4294.0 \times 10^8\ m^3$ in 2015. The proportion of natural gas ranged from 8.1% to 46%. The growth of shale gas reserves has completely changed the energy structure in the United States, significantly impacting the global energy landscape.

Shale gas formations contain multi-scale storage and percolation spaces: organic nano-sized micropores, macropores, natural microfractures, hydraulic fractures and fracture networks caused by hydraulic fracturing. Shale gas can present multiple transport mechanisms during its production, including adsorption/desorption, diffusion, percolation, and slippage effects. These are additional to a special occurrence mode as the coexistence of free and adsorbed gas. Due to the synthetic action of these factors, the flow pattern of shale gas is easily distinguished from that in conventional gas reservoirs, leading to different production performance of shale gas wells. In an attempt to efficiently guide the development of shale gas reservoirs, analysis, and research methods must be established in accordance with the specific formation and flow characteristics of these reservoirs.

Many physical models of shale gas reservoirs have been developed in the past few years that describe their complex gas flow behavior and dynamical production characteristics. There is no book or monograph available that systematically interprets the dynamical production performance of shale gas wells according to their unique flow mechanisms and production processes.

This book provides an overview of the current studies related to shale gas reservoir development and production, well testing analysis methods, and numerical simulation techniques. We have built a series of complex seepage mechanisms and physical models that depict transient flow behavior while incorporating multi-scale characteristics in space. Emphasis is placed on studies of the transient pressure and production responses for wells in shale gas reservoirs. It is hoped that the book serves as a reference for researchers and engineers in the area of shale gas reservoir development and production.

The book is divided into three parts. Part I, Chapter 1, introduces the geological features and gas flow patterns in shale gas reservoirs. The corresponding seepage mechanism models are established based on these features and patterns. In Part II, Chapters 2–7, transient seepage models for fractured vertical wells and multi-stage fractured horizontal wells are built, and their production performance is studied using semi-analytical and boundary element methods. The third part of the book, Chapter 8, is focused on numerical simulation techniques in shale gas reservoirs. It features the establishment of gas–water two-phase physical and mathematical models based on the triple-media theory, the development of a numerical simulator for shale gas reservoirs based on a combination of the finite element method, the finite volume element method and a fully implicit algorithm, and an application of this simulator in production performance for multiple fractured horizontal wells.

Combining multi-scale storage spaces with various transport mechanisms, Chapter 1 establishes five microflow mechanism models that describe shale gas flow patterns under the influence of multiple scale fields. Based on the mechanism models in Chapter 1, a uniform expression of different models are obtained. The continuous point solutions for a circular or rectangular outer boundary are derived from the point source function method in Chapter 2. The physical and mathematical seepage models considering the various combinations of different well types (vertical vs horizontal), drainage areas (circular vs rectangular), seepage mechanism models (Chapter 1) and stimulated reservoir volume (SRV) shapes: circular, rectangular, or partial rectangular are all introduced in Chapters 3–7. These models are solved by the source function method and the boundary element method. Taking advantage of numerical inversion and computer programing techniques, transient pressure, and rate type curves are drawn for wells producing at a given rate or under a wellbore pressure condition, with the sensitivities of parameters analyzed further. In consideration of a stress sensitivity, the high velocity non-Darcy flow, and a variable conductivity of a fracture system, Chapter 8 develops a triple-media gas–water two-phase numerical simulator and its applications in the end, some field applications of models and methods presented in this book are analyzed.

Multiple transport mechanisms are taken into account. Gridding by unstructured tetrahedral grids and then discretizing mathematical models using the finite element–finite volume methods, a fully implicit solution algorithm for different discrete systems is established. With the help of this simulator, a variety of production cases and parametric sensitivities are analyzed for multi-stage fractured horizontal wells. Different SRV forms (circular, rectangular, or branched) and multi-well production under a well factory operation mode are considered.

We would like to express our sincere gratitude to those who supported the preparation of this book for their time, thoughts, and energy. Dr. Ruihan Zhang and Dr. Deliang Zhang prepared the numerical simulation included in Chapter 8; we want to specifically acknowledge their selfless dedication. This part is novel work and has not been published previously. We would also like to recognize all the authors whose publications are cited for their rigorous research advancements.

This work is supported jointly by the National Natural Science Foundation of PR China (Key Program) (Grant No. 51534006), the National Science Funds for Distinguished Young Scholars of PR China (Grant No. 51125019), the Natural Science Foundation of PR China (Grant Nos. 51704247, 51874251), International S&T Cooperation Program of Sichuan Province (Grant No. 2019YFH0169), Deep Marine shale gas efficient development Overseas Expertise Introduction Center for Discipline Innovation (111 Center)NSERC/AIEES/Foundation CMG Industrial Research Chair in Reservoir Simulation, and the Alberta Innovates Technology Futures (iCore) Chair in Reservoir Modelling. We express our heartfelt appreciation for their support.

Liehui Zhang, Zhangxin Chen, and Yu-long Zhao

September 2018

Preface

Around the world, concerns about the recovery of unconventional oil and gas resources are increasing. The intensified contradiction of supply and demand, continuously decreasing reserves of conventional resources, constant advancement of development technologies in drilling and fracturing, and the urgent need to improve the energy consumption worldwide are factors driving these concerns.

Coal bed methane, shale gas, tight gas, and natural gas hydrates are the most common unconventional resource types. Shale gas is a kind of biogenic or thermogenic natural gas found in organic-rich shale where gas is under a free and/or adsorbed state. The successful development of shale gas in North America is driving a revolution of resources development globally. According to EIA (Energy Information Administration) statistical data, the production of shale gas in the United States increased from $563.5 \times 10^8 m^3$ in 2007 to $5385.4 \times 10^8 m^3$ in 2017. The proportion of natural gas ranged from 8.1% to 57%. The growth of shale gas reserves has completely changed the energy structure in the United States, significantly impacting the global energy landscape.

Shale gas formations contain multiscale storage and percolation spaces: organic nanosized micropores, macropores, natural microfractures, hydraulic fractures, and fracture networks caused by hydraulic fracturing. Shale gas can present multiple transport mechanisms during its production, including adsorption/desorption, diffusion, percolation, and slippage effects. These are additional to a special occurrence mode as the coexistence of free and adsorbed gas. Due to a combined action of these factors, the flow pattern of shale gas is easily distinguished from that of gas in conventional reservoirs, leading to different production performance of shale gas wells. In an attempt to efficiently guide the development of shale gas reservoirs, analysis and research methods must be established in accordance with the specific formation and flow characteristics in these reservoirs.

Many physical models in shale gas reservoirs have been developed in the past few years that describe their complex gas flow behavior and dynamical production characteristics. There is no book or monograph available that systematically interprets the dynamical production performance of shale gas wells according to their unique flow mechanisms and production processes.

This book provides an overview of the current studies related to shale gas reservoir development and production, well testing analysis methods, and numerical simulation techniques. We have built a series of complex seepage mechanisms and physical models that depict transient flow behavior while incorporating multiscale characteristics in space. Emphasis is placed on studies of transient pressure and production responses for wells in shale gas reservoirs. It is hoped that this book could serve as a reference for researchers and engineers in the area of shale gas reservoir development and production.

The book is divided into four parts. Part I, Chapter 1, introduces the geological features and gas flow patterns in shale gas reservoirs. The corresponding seepage mechanism models are established based on these features and patterns. In Part II, Chapters 2–7, transient seepage models for fractured vertical wells and multistage fractured horizontal wells are built and their production performance is studied using semianalytical and boundary element methods. In part III, Chapter 8, is focused on numerical simulation techniques in shale gas reservoirs. It features the establishment of gas–water two-phase physical and mathematical models based on the triple-media theory; the development of a numerical simulator for shale gas reservoirs based on a combination of the finite element method, the finite volume element method, and a fully implicit algorithm; and an application of this simulator in production performance for multiple fractured horizontal wells. The final part, Chapter 9, presents case studies by using data of shale gas wells from Changnong and Weiyuan shale gas reservoirs in Sichuan Basin, China.

Combining multiscale storage spaces with various transport mechanisms, Chapter 1 establishes five microflow mechanism models that describe shale gas flow patterns under the influence of multiple scale fields. Based on the mechanism models in Chapter 1, a uniform expression of different models is obtained. The continuous point solutions for a circular or rectangular outer boundary are derived from the point source function method in Chapter 2. The physical and mathematical seepage models considering various combinations of different well types (vertical vs horizontal), drainage areas (circular vs rectangular), seepage mechanism models (Chapter 1), and stimulated reservoir volume (SRV) shapes (circular, rectangular, or partial rectangular) are all introduced in Chapters 3–7. These models are solved by the source function method and the boundary element method. Taking advantage of numerical inversion and computer programing techniques, transient pressure and rate type curves are drawn for wells producing at a given rate or under a wellbore pressure condition, with sensitivities of parameters analyzed further. In consideration of a stress sensitivity, the high-velocity non-Darcy flow, and a variable conductivity of a fracture system, Chapter 8 develops a triple-media gas–water two-phase numerical simulator and its applications. Finally, Chapter 9 presents real field applications of the models and simulators developed in the book.

Multiple transport mechanisms are taken into account. Gridding by unstructured tetrahedral grids and then discretizing mathematical models using the finite element–finite volume methods, a fully implicit solution algorithm for different discrete systems is established. With the help of the simulator developed, a variety of production cases and parametric sensitivities are analyzed for multistage fractured horizontal wells. Different SRV forms (circular, rectangular, or branched) and multiwell production under a well pad operation pattern are considered.

We would like to express our sincere gratitude to those who supported the preparation of this book for their time, thoughts, and energy. Dr. Ruihan Zhang and Dr. Deliang Zhang prepared the numerical simulation included in Chapter 8; we want to specifically acknowledge their selfless dedication. This part is novel work and has not been published previously. We would also like to recognize all the authors whose publications are cited for their rigorous research advancements.

This work is supported jointly by the National Natural Science Foundation of China (Key Program) (Grant No. 51534006), National Natural Science Foundation of China (Grant No. 51704247 and 51874251), International S&T Cooperation Program of Sichuan Province (Grant No. 2019YFH0169), Deep Marine shale gas efficient development Overseas Expertise Introduction Center for Discipline Innovation (111 Center), NSERC/Energi Simulation Industrial Research Chair in Reservoir Simulation, and the Alberta Innovates (iCore) Chair in Reservoir Modelling. We express our heartfelt appreciation for their support.

Liehui Zhang, Zhangxin Chen, and Yu-long Zhao
March 2019

Shale gas reservoir characteristics and microscopic flow mechanisms

1

Chapter Outline

1.1 Introduction 1
1.2 Shale gas reservoir characteristics 2
1.3 Pore type analysis in shale gas reservoirs 6
1.3.1 Inorganic pores 6
1.3.2 Organic intragranular pores 10
1.3.3 Natural microfractures 10
1.3.4 Hydraulic fractures 11
1.4 Accumulation mechanisms in shale gas reservoirs and model description 12
1.4.1 Free gas 15
1.4.2 Dissolved gas 15
1.4.3 Adsorbed gas 16
1.5 Multiscale flow mechanisms in shale gas reservoirs 24
1.5.1 Flow regime classification of gas flow in shale reservoirs 25
1.5.2 Gas flow in fractures and macropore systems 29
1.5.3 Gas flow in nanometer pores 30
1.5.4 Gas diffusion in nanometer pores 32
1.5.5 Desorption of adsorbed gas from surfaces of shale matrix pores 37
1.6 Mathematical models with various shale gas flow mechanisms 39
1.6.1 Microfractures + steady state adsorption/desorption and diffusion model (model 1) 39
1.6.2 Microfractures + matrix macropores + steady state adsorption/desorption and diffusion model (model 2) 41
1.6.3 Microfractures + gas adsorption/desorption + matrix Fick's diffusion model (model 3) 43
1.6.4 Microfractures + matrix macropores + gas adsorption/desorption + Fick's diffusion model in nanopores (model 4) 44
1.6.5 Microfractures + gas adsorption/desorption + Knudsen diffusion model in nanopores (model 5) 45

1.1 Introduction

Shale gas is the natural gas that exists as adsorbed or free gas in organic rich shale and its interlayers (Ahmed and Meehan, 2016). Its main component is methane, and the interlayers include laminated siltstone, slit shale, and shaly siltstone. The natural gas is generated and accumulated in the source rock, which is the typical in situ

Developments in Petroleum Science, Vol. 66. https://doi.org/10.1016/B978-0-444-64315-5.00001-2

accumulation mode in a shale gas reservoir (Zhang et al., 2004). If a shale gas reservoir is developed using the same method as for a conventional gas reservoir, gas wells generally have no or very low productivity. Its commercial development can only be realized through proper stimulation techniques (Smith and Montgomery, 2015).

As one of the important unconventional gas resources, a shale gas reservoir has completely different generation and accumulation mechanisms from a conventional gas reservoir. For a conventional gas reservoir, shale is the source rock where natural gas is generated; then it migrates and is accumulated in the reservoir pay. In this process, hydrocarbons are generated and stored in different places. The source rock, reservoir, and cap rock are independent as well as related to each other; the source rock is usually far away from the reservoir. On the other hand, for an unconventional reservoir, the source rock, reservoir, and cap rock are generally the same geobody. Shale itself is a special hydrocarbon generation and accumulation system, which has a complex microscopic pore structure and a diverse status of gas existence (Chen et al., 2009; Wu and Chen, 2016). These characteristics of a shale reservoir lead to complexity of flow mechanisms. Therefore, a complete understanding of shale reservoir characteristics and microscopic mechanisms of gas flow in such a reservoir is key to analyzing macroscopic flow mechanisms and developing the corresponding theoretical models for transient seepage flow.

1.2 Shale gas reservoir characteristics

Characteristics of a shale gas reservoir are the integration of its source rock, reservoir, and cap rock, with no obvious trap, no gas–water contact, gas existence in a continuous big area, low porosity, and low permeability. Curtis (2002) indicated that shale gas was a type of "continuous gas," and its generation could be biochemical, thermal, or a combination of these two. Such gas could exist as free gas in natural fractures and matrix pores. The gas could also exist as adsorbed gas in organic matters and on surfaces of clay mineral, or as solution gas in kerogen, asphaltene, residual water, and liquid hydrocarbons. Shale gas wells usually have no natural productivity or low productivity when using the same development method as for a conventional gas reservoir. Stimulation techniques such as hydraulic fracturing are required for commercial production (Hu et al., 2017; Smith and Montgomery, 2015). Compared to a conventional oil or gas reservoir, a shale gas reservoir has totally different geological, physical, and geomechanical characteristics, which are addressed later in this chapter.

(1) Generation and accumulation

For conventional reservoirs, gas is expelled from a source rock; it then migrates to and is accumulated in permeable formations through paths with high permeability (e.g., microfractures and faults) under the co-effects of formation static pressure, heat at a burial depth, dehydration of clay minerals, and a hydrodynamic force. Two main types of conventional reservoirs are structural and stratigraphic reservoirs. While some of the remaining gas is adsorbed on surfaces of shale matrix pores in a shale gas reservoir, most of the gas exists as free gas in matrix pores and microfractures. Therefore, the generation and accumulation of shale gas are earlier than most of other hydrocarbon reservoirs.

(2) No obvious trap

For a conventional reservoir, traps that are in favor of oil and gas accumulation are the basis for the hydrocarbon reservoir, and they determine the basic reservoir characteristics and exploration methods. However, a shale gas reservoir has no boundary; that is, shale gas does not accumulate in a trap to become a reservoir. In addition, a conventional reservoir has a certain structural background while a shale gas reservoir is not controlled or affected structurally.

(3) A variety of pore types and complex and multiscale structures

The basements of a shale play are dominated by nanopores with ultra-low permeability. According to the laboratory analysis of Javadpour et al. (2007) on 152 core samples from nine different reservoirs in North America, the permeability of 90% shale samples was less than 150×10^{-6} D, and the diameter of main flow pores was 4–200 nm. Loucks et al. (2012) analyzed the pore structure and storage space of a North America shale gas play, and indicated that micro- and nanopores were the main types of pores in the shale play, while nanopores were the most dominating ones. The diameter of nanopores was 5–800 nm, mostly around 100 nm, and the diameter of pore throats was 10–20 nm. The permeability of a typical North America shale basement is 10^{-3}–$10^{3} \times 10^{-6}$ D, and the porosity is 1%–5%. The total porosity of a Longmaxi shale play in the Sichuan basin in PR China is 2%–6%. Its gas-saturated porosity is 1.5%–2.8%, and its water-saturated porosity is 0.7%–1.2%. Therefore, a shale matrix is a tight porous medium with ultra-low porosity and permeability.

(4) Co-existence of adsorbed and free gas, and gas existence status

Free gas exists in pore space, while adsorbed gas that counts for 20%–85% of the total gas exists in organic matter. The existence status of gas includes adsorbed gas on surfaces of organic matter, free gas in inorganic intergranular pores, free gas in microfractures, free gas in hydraulic fractures, and free gas in nanopores.

(5) Abnormal high pressure of a primary shale gas reservoir

During tectonic movement, abnormal pressure increases or decreases, resulting in unpredictable formation pressure in a shale gas reservoir. Although the formation pressure is variable, most of shale plays still have the characteristics of abnormal high pressure. During the process of pyrolysis gas being massively generated, thermos chemical energy, which is the key energy for natural gas generation, converts high-density organic matters into low-density natural gas. In a relatively sealed system, a lower density leads to a volumetric expansion and pressure increase; with more and more natural gas generated, the reservoir is abnormally pressurized, which is mechanistically similar to a "pressure cooker." Under the effect of increased reservoir pressure, fractures are created along stress concentration surfaces, lithology transition surfaces, or brittle surfaces. These fractures are for gas to accumulate as free gas and lead to a commercial shale gas reservoir. The characteristics of this process are in situ gas generation, fracture creation by gas expansion, and in situ or nearby gas accumulation. In this period, most of the free gas accumulates in fractures, and the average gas saturation in the shale formation reaches a high level.

(6) Shale brittleness as an important geomechanical parameter for fracturing evaluation of hydraulic fractures

The more brittle the rock is, the easier it is for shear and induced fractures to generate during a fracturing operation. The more complex the fracture geometry is, the bigger the stimulated reservoir volume is in the reservoir and the higher production a well has. Shale brittleness is related to the components and contents of shale minerals, Young's modulus and Poisson's ratio. A lower content of clay minerals, such as kaolinite, smectite, and illite, and a higher content of brittle minerals, such as quartz, feldspar, and calcite, result in higher brittleness of rock. If the rock has a high content of clay minerals, it is more ductile and absorbs more energy during a fracturing operation. Consequently, the created fractures are mainly planar fractures, which are adverse to achieve volumetric stimulation of the reservoir rock. Therefore, a shale mineral content is important for its geomechanical characteristics. For commercial shale plays, the content of brittle minerals generally exceeds 40%, and the content of clay minerals is less than 30%. In the United States, the productive shale in main shale basins has a quartz content of 28%–52%, a carbonate content of 4%–16%, and a total brittle mineral content of 46%–60%, while, in PR China, the average content of brittle minerals is higher than 40% for marine shale, marine to continental transition carbonate shale, or continental shale. From the statistics of brittle minerals of the Longmaxi formation in the Sichuan basin, the content of brittle minerals in the Longmaxi formation is 30.75%–90.42%, averaging at 57.1%. Quartz is the most common component with a relative high percentage, averaging at 31.93%. Carbonate is the second common component with a content of 3.91%–75.99%, averaging at about 16.98%; and the content of feldspar is relatively low, around 2.39%–18.2%, averaging at 8.19%.

Young's modulus and Poisson's ratio are used to calculate brittleness of shale and can better reflect the rock failure degree under the effect of stresses and fracturing. Poisson's ratio is the index of a quartz-to-clay content ratio; the higher the quartz-to-clay content ratio is, the lower Poisson's ratio is and the higher the brittleness of the shale. A higher content of cements and a lower content of clay minerals correspond to bigger Young's modulus. Shale geomechanics and brittleness evaluation are the basis of modeling hydraulic fracturing, and they are very important to form an artificial fracture network in shale.

A comparison of reservoir properties, minerals, and rock mechanistic is made between different shale plays, as shown in Table 1.1, from which the reservoir properties of the Longmaxi shale in PR China and shale plays in the United States can be compared and analyzed. In addition, Tables 1.2 and 1.3 provide the TOC cutoffs for commercial shale gas reservoirs according to their reservoir properties, gas saturation, and rock geomechanics.

Table 1.1 Reservoir property comparison of shale plays in Sichuan basin and the United States[a]

Reservoir property	**Shale reservoir**			
Formation	**Longmaxi**	**Barnett**	**Marcellus**	**Haynesville**
Stratigraphic chronology	Silurian	Mississippian	Devonian	Upper Jurassic
Burial depth (m)	2000–4000	1950–2550	1219.2–2590.8	3200.4–4114.8
Effective thickness (m)	20–120	30.5–182.9	15.2–60.9	60.9–259.01
Formation pressure coefficient	1.3–2.32	0.96–1.16	0.93–1.56	2.00
TOC (%)	1.62	3–13 (4.5)	3–12	0.5–4
R_o (%)	2.4–3.6	1.0–1.9	0.4–1.3	0.94–2.62
Porosity (%)	1.15–5.8 (3.0)	4–5	10	8–9
Gas content (m^3/t)	0.3–5.09 (1.85)	8.49–9.91	1.69–2.83	2.83–9.34
Young's modulus (GPa)	8.6–40.9 (22.2)	39–72	30–200	30–80
Poisson's ratio	0.10–0.25 (0.18)	0.25–0.36	0.15–0.35	0.15–0.3
OGIP (10^8 m^3)	388,790.30	92,596.09	424,752.70	203,031.79
EUR (10^8 m^3)	97,126.78	12,459.41	74,190.14	71,075.28

[a] Longmaxi data from papers of Huang et al. (2012), Zeng et al. (2011) and Yao et al.. (2013a,b); the US shale parameters are from Arthur et al. (2004) and Wei's PhD thesis (2013).

Table 1.2 Evaluation standards of shale gas reservoir TOC[a]

TOC (%)	**Quality of kerogen**	**TOC (%)**	**Quality of kerogen**
<0.5	Very poor	2–4	Good
0.5–1	Poor	4–12	Very good
1–2	Average	>12	Excellent

[a] From the data of North America shale gas basin. Schlumberger pointed out that the low end of shale TOC is 2%, but this threshold will be lowered with innovation of shale gas development technologies.

Table 1.3 Key cutoff values for commercial shale gas development

Parameter	**Economic development requirement**	**Parameter**	**Economic development requirement**
GRI porosity (%)	>2	S_w (%)	<45
S_o (%)	<5	Permeability (mD)	>0.0001
TOC (%)	>2	R_o (%)	>1.1
Brittleness mineral (%)	>40	Clay mineral (%)	<30
Net thickness of organic rich shale (m)	>30–50		

1.3 Pore type analysis in shale gas reservoirs

From the previous description of shale gas reservoir properties, a shale gas reservoir, as an unconventional reservoir, has multiple types of pores with a complex structure. In recent years, using high-tech equipment like FE-SEM (field emission scanning electron microscope) and TEM (transmission electron microscopy) to observe rock slices of a shale gas reservoir, petroleum experts can identify multiple pore types according to reservoir pore occurrence and a position of rock grains; these provide important guidance on the analysis of nanopores generation and gas flow in such pores.

The typical classification methods are Slatt and O'Neal's (2011) classification and Loucks et al.'s classification (2012). Based on the pore types of Barnett and Woodford shale gas reservoirs in the United States, Slatt and Brien is divided shale reservoir pores into six types: inter-pores of clay flocs, organic pores, intra-pores of dung, intra-fossil fragment pores, intra-granular pores, and natural microfractures. Thereafter, based on Slatt and Brien's classification, Loucks et al. reclassified pore types of shale matrix and natural microfractures, and their classification includes the following four types: inter-granular pores, intra-granular pores, organic pores, and natural microfractures. The first two pore types are related to mineral grains, and the difference is that the former is developed between grains and the latter is developed in grains. The development of the third pore type is related to organic matters, and these pores are intra-pores of organic matters.

Because massive hydraulic fracturing is applied in the development of shale gas reservoirs, petroleum experts in PR China generally classify shale pores into two types: matrix pores and fracture pores. Moreover, according to pore sizes, these pores can be further divided into four categories: nano–micron inorganic pores, nano-organic intra-granular pores, micron–milli natural microfractures, and hydraulic fractures, where the first two belong to matrix pores and the latter two belong to fracture pores (Chen et al., 2009; Yao et al., 2013a,b; Zhang et al., 2004; Zou et al., 2010). Matrix pores are the main storage space for shale gas, which exists as free gas or adsorbed gas, and the gas in place in shale reservoirs directly depends on the total volume of matrix pores. Fracture pores are for not only storage of free gas but also the main factor in determining well productivity, which is very important for effective development of shale gas reservoirs. In the following sections, these four pore types are analyzed by combining with SEM (scanning electron microscope) images of rock slices from shale gas wells in the Changning area of Weiyuan, Sichuan, PR China.

1.3.1 Inorganic pores

In addition to the organics existing in shale matrix, there are also massive inorganic matters. According to their origin, inorganic pores in shale matrix can be categorized as residual primary intergranular pores, intercrystalline pores, and secondary dissolved pores formed by dissolution of unstable minerals (e.g., calcite and feldspar).

Residual primary intergranular pores are among silty grains of schistose clay (Fig. 1.1). Similar to pores in a conventional reservoir, the amount of such pores decreases with a burial depth increase.

Fig. 1.1 SEM images of shale inorganic pores—intergranular pores.

Intercrystalline pores are formed by crystallization of clay minerals, usually developed in mineral aggregates with a coarse and well-shaped crystalline form. The diameter of intercrystalline pores in a Changning shale reservoir in the Sichuan basin ranges between 10 and 500 nm, and the most common intercrystalline pores are strawberry-shaped pyrite intergranular pores formed in a deep water anoxic reducing environment (Fig. 1.2).

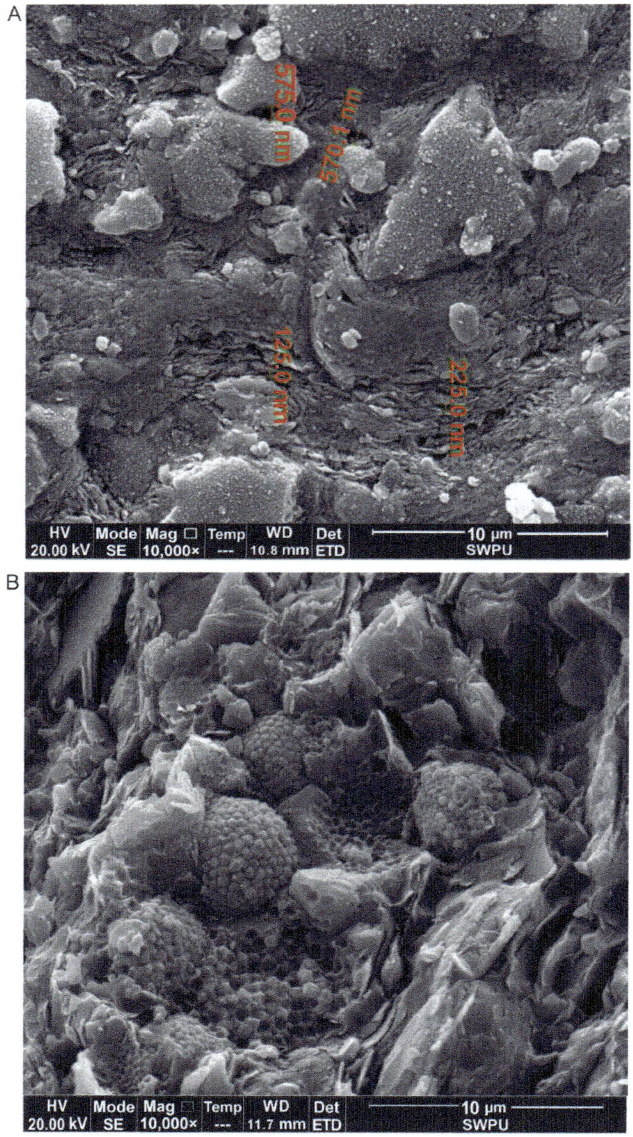

Fig. 1.2 SEM images of shale inorganic pores—intercrystalline pores.

In the Longmaxi shale of the Changning area, massive secondary dissolved pores develop (Fig. 1.3), which are formed by the solution of soluble minerals, such as feldspar and carbonates, in acidulous water mixed by air, groundwater, and decarboxylated organic acids. These pores can be divided into intragranular and intergranular-dissolved pores.

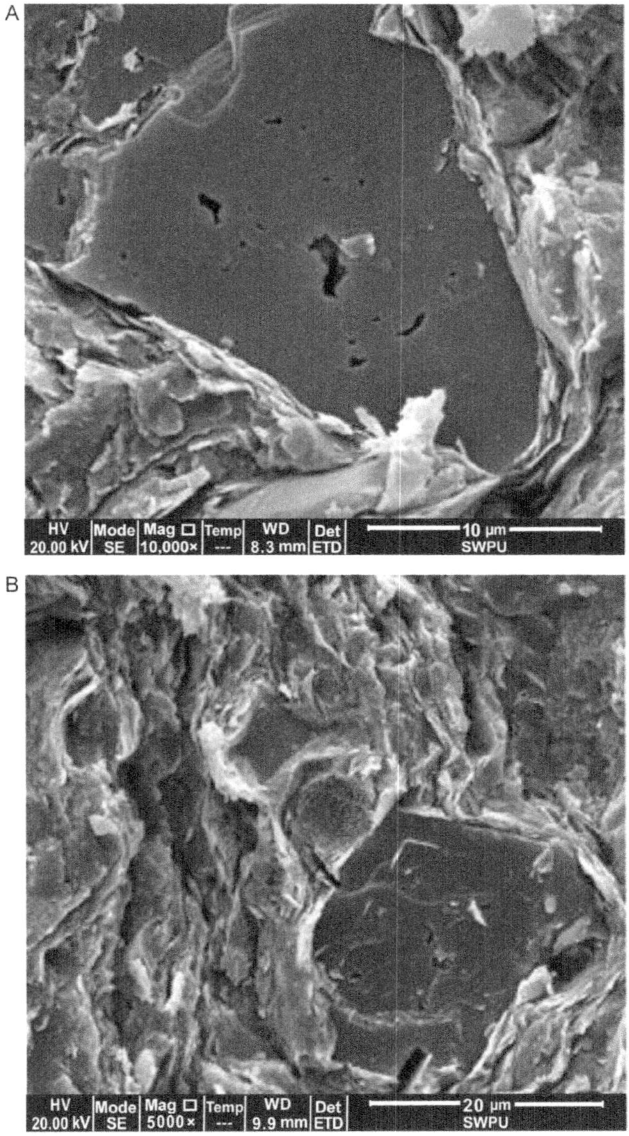

Fig. 1.3 SEM images of shale inorganic pores—intragranular dissolved pores.

Note that pores also exist between organic and inorganic matters. Although these kinds of pores are just a small portion of the total pore volume, they are the bridge connecting organic and inorganic pores and are significantly important for the storage and migration of shale gas.

1.3.2 Organic intragranular pores

In general, organic intragranular pores are only formed when R_o reaches 0.6% or above. When R_o is smaller than this value, few or no organic intragranular pores are developed. The pore development is not only controlled by the maturity of organics but also related to the type of organics (the type of kerogen). Current research results indicated that type II kerogen is more favorable for organic pores to develop than type III kerogen. The shape of organic intragranular pores is usually irregular and in a bubble shape. No connection is visible between these pores in 2D images, but the connection is discernable in 3D images (Loucks et al., 2012).

Organic pores in a shale reservoir are mainly developed during the thermal cracking of a hydrocarbon generation phase, and the sizes of organic pores range from 5 to 700 nm (Fig. 1.4). These kinds of organic nanometer pores are quite developed in a shale reservoir, providing tremendous surface areas for shale gas adsorption as well as flow paths for gas flow. For rocks with the same total pore volume, a pore superficial area is inversely proportional to the pore diameter. The smaller the pore diameter, the bigger the superficial area. Therefore, compared to micropores, nano-organic pores provide a bigger superficial area with more adsorbed gas. Organic pores in a shale reservoir are the main space for adsorbed natural gas, and about 20%–85% of shale gas is adsorbed on the surfaces of kerogen and clay minerals. Similarly, the development degree and sizes of organic pores have a very close relationship with the permeability and productivity of a gas reservoir.

1.3.3 Natural microfractures

Shale reservoirs always have massive natural microfractures, which are heterogeneously and anisotropically distributed in the reservoirs. Previous researchers have found that the width of microfractures is generally 0.01–0.5 mm. In a shale reservoir, fractures not only are the storage space for gas, but also more importantly are the connections between different types of pores. Through microfractures, different types of pores in a shale reservoir connect with each other to form a network of pores for the production of shale gas. Therefore, the scale and development level of microfractures directly impact the connectivity and permeability of a shale reservoir and determine shale gas recovery.

Fractures in shale reservoirs mainly include structural fractures, overpressured fractures, and diagenetic contraction fractures. Most microfractures in a shale reservoir develop during the organic evolution of the source rock, and they intermittently open or close with changes in the reservoir pressure. Therefore, for those shale gas reservoirs with developed microfractures, the permeability of the fracture system is stress sensitive caused by closing of the microfractures during reservoir development.

Fig. 1.5 shows structural fractures developed in a shale reservoir in Changing, the Weiyuan area in the Sichuan basin. Observable clay minerals grow in the fractures and have dissolution characteristics. Fig. 1.6 shows cleavage fractures in mineral grains in this shale reservoir. In addition, Fig. 1.7 shows intragranular fractures in mineral grains in the same shale reservoir; they have a width less than 1 μm and are possibly formed by a stress effect.

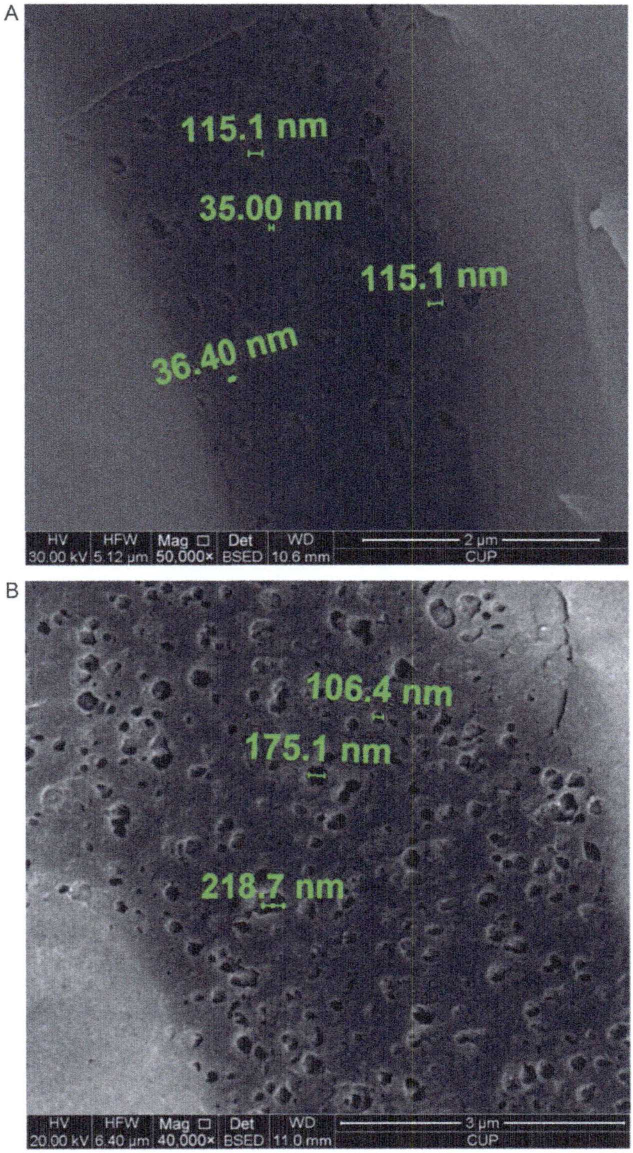

Fig. 1.4 SEM images of shale organic pores.

1.3.4 Hydraulic fractures

Due to ultra-low permeability of shale matrix (normally <1 mD and some even in nD magnitude), only a small number of wells in regions with well-developed natural fractures can be put into production directly, while more than 90% of shale gas wells must be stimulated through acidizing and/or fracturing to obtain an expected production rate.

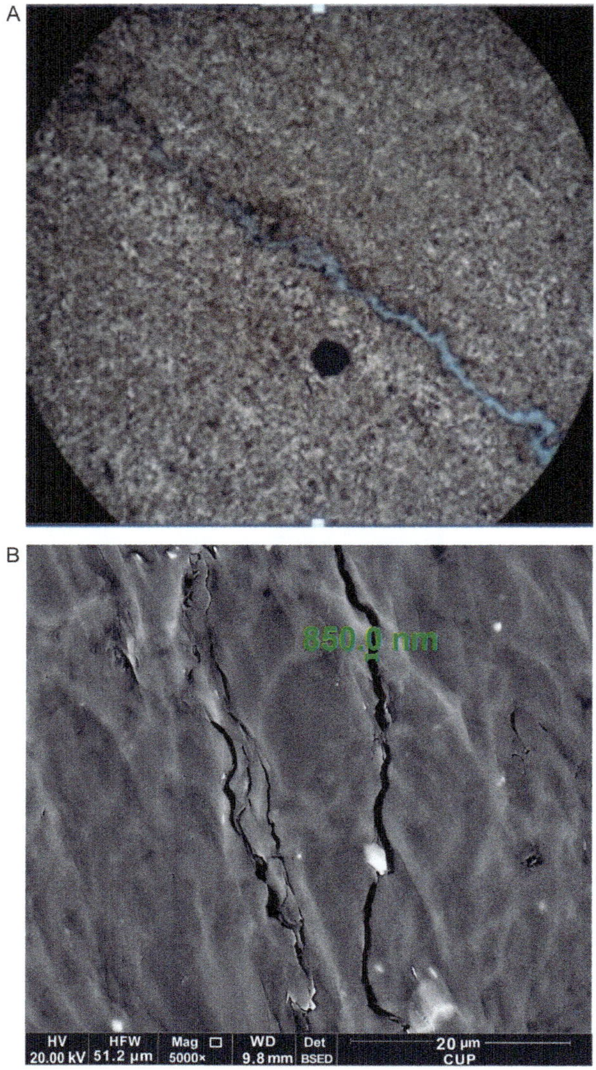

Fig. 1.5 Shale reservoir structural fractures.

Compared to natural fractures, hydraulic fractures have higher permeability and are the main gas flow paths. Furthermore, hydraulic fractures and their induced fractures can also improve the connectivity of a natural fracture system, extend the fracture network, increase migration paths, and, finally, enhance gas production, as shown in Fig. 1.8.

1.4 Accumulation mechanisms in shale gas reservoirs and model description

Compared with a conventional gas reservoir, a shale gas reservoir has obvious differences in reservoir characteristics, pore types, and accumulation mechanisms, which lead to a diverse existence status of natural gas in the shale reservoir. In addition

Fig. 1.6 Cleavage fractures in a shale reservoir of Weiyuan, Sichuan.

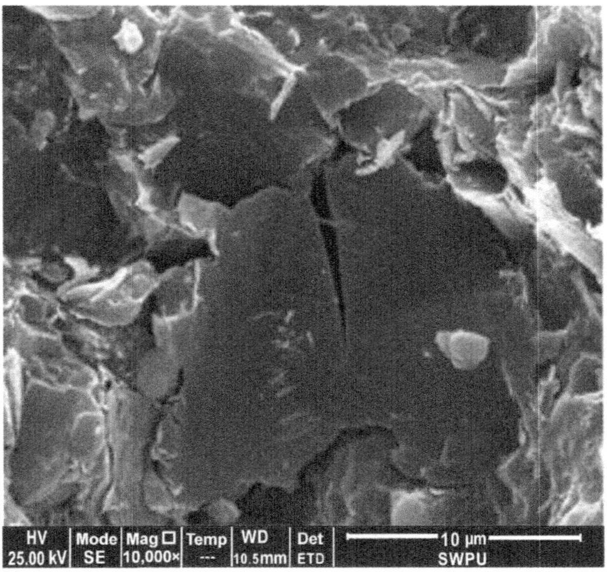

Fig. 1.7 Intragranular fractures in a shale reservoir of Weiyuan, Sichuan.

to a very small amount of natural gas dissolved in kerogen, asphaltene, liquid hydrocarbons, and formation water, most natural gas exists as free gas or adsorbed gas in shale gas reservoirs. Gas is mainly adsorbed onto the surfaces of matrix grains and organic pores, and the adsorbed gas accounts for 20%–85% of the total gas in place. Free gas mainly exists in microfractures, and inorganic and organic pores. Existence of free gas depends on the total gas content; only when there is surplus gas to dissolve and adsorb gas can it exist as free gas in a reservoir. Under different geological conditions (including temperature, pressure, kerogen type, content of organics, maturity

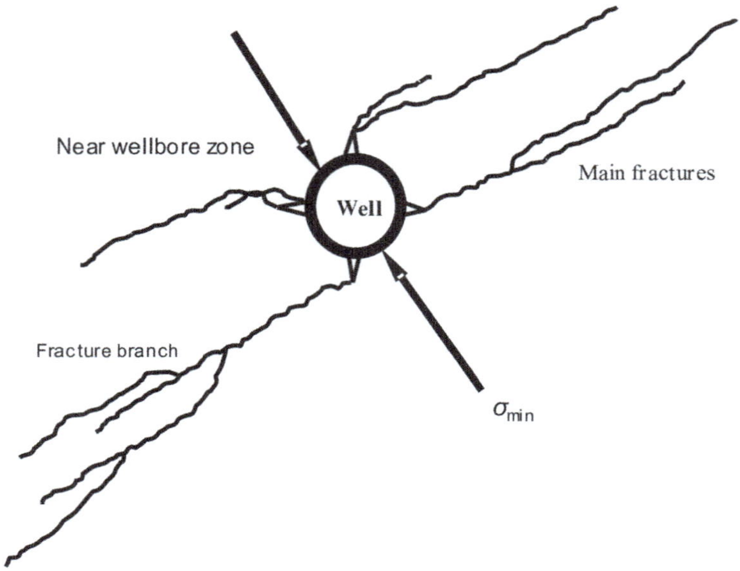

Fig. 1.8 Hydraulic fractures.

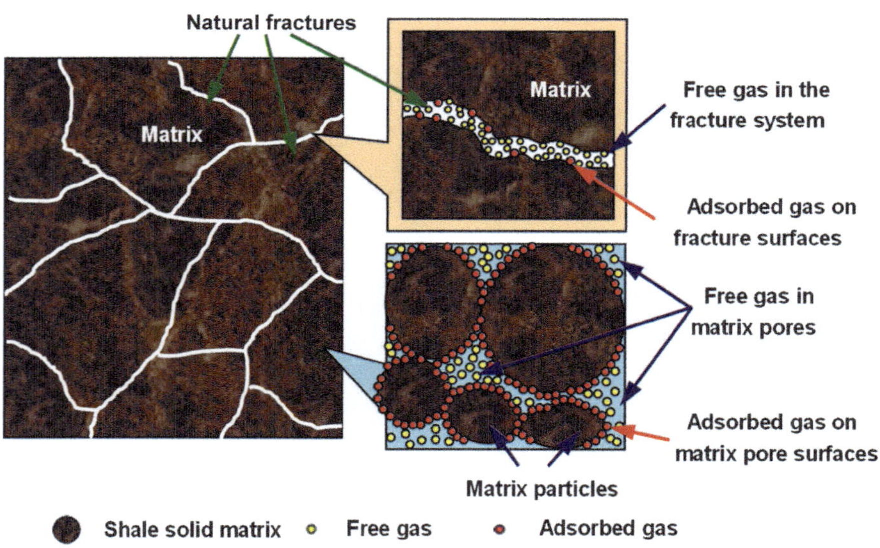

Fig. 1.9 Existence characteristics of adsorbed gas and free gas in a shale gas reservoir.

of organics, development of microfractures, and content of formation water), a relative ratio of adsorbed gas to free gas is quite different. Fig. 1.9 is a diagram showing free gas and adsorbed gas existence in a reservoir. In the next section, we will analyze quantitatively how to describe shale gas.

1.4.1 Free gas

Gas existing in a free status in organic and inorganic pores, microfractures, and artificial hydraulic fractures of a reservoir is called free gas. The amount of free gas is closely determined by a reservoir structure and a pore size. Although the flow of this kind of gas in nanopores may not follow Darcy's law of conventional gas, the status of such gas is still similar to that of conventional natural gas. It satisfies the state equation of real gas:

$$pV = ZnRT \tag{1.1}$$

where:

p—gas pressure (Pa);
V—gas volume (m^3);
T—gas absolute temperature (K);
Z—gas compressibility factor (dimensionless); for the idea gas $Z=1$;
R—gas constant [8.314 J/(mol K)];
n—number of gas moles (mol).

During the application process, the gas state equation's formula is:

$$\rho_g = \frac{pM_g}{ZRT} \tag{1.2}$$

where:

ρ_g—gas density (kg/m^3);
M_g—gas molecule mass (kg/mol).

1.4.2 Dissolved gas

Since a shale reservoir is very tight, water molecules cannot flow in nanopores, and most shale gas reservoirs produce no water and little vapor.. Based on research for conventional gas reservoirs, the dissolution of natural gas in water is very limited even though there is certain amount of clay bound water in a shale reservoir. Therefore, petroleum experts did not consider the dissolution gas in studies of shale gas in place and an unsteady state flow theory. Theoretically, Henry's law can be used to describe gas solubility in water:

$$C_b = \frac{p_b}{K_c} \tag{1.3}$$

where:

C_b—molar concentration of gas dissolved in water (mol/m^3);
p_b—gas partial pressure (Pa);
K_c—Henry constant (m^3 Pa/mol).

According to the above equation, at a certain temperature, solubility of gas in formation water is in proportion to formation pressure. Henry's constant is determined by reservoir temperature, gas components, and the total salinity of formation water. Moreover, at different temperature and salinity, solubility of shale gas is different. In general, the higher temperature is, the lower the solubility is; and the lower salinity is, the lower solubility is.

1.4.3 Adsorbed gas

Existence of organic and inorganic nanopores in a shale reservoir provides a considerable superficial area of pores in such a reservoir, which consequently becomes storage space for gas adsorption. From Curtis (2002), adsorbed gas is 20%–85% of the total shale gas in place. The amount of adsorbed gas mainly depends on three factors: shale properties (e.g., a pore structure, constitutes, and shale metamorphism); adsorbed gas properties; and environmental temperature and pressure. Figs. 1.10 and 1.11 show the relationship of TOC and the adsorbed gas amount for typical North America shale and Longmaxi shale in the Sichuan basin, respectively. According to these plots, the adsorption amount is positively correlated with TOC; that is, the amount of adsorbed gas increases with an increase in TOC.

The adsorption of shale gas at surfaces of pores is a type of physical adsorption with dynamic adsorbing and desorbing. When the amount of adsorbed gas molecules equals that of desorbed ones, the process of adsorption and desorption is in dynamic equilibrium. When shale reservoir pressure decreases, the amount of gas adsorbed is less than that of gas desorbed from pore surfaces, resulting in additional free gas. This process is called desorption. Adsorbance, V, is usually used to quantify a

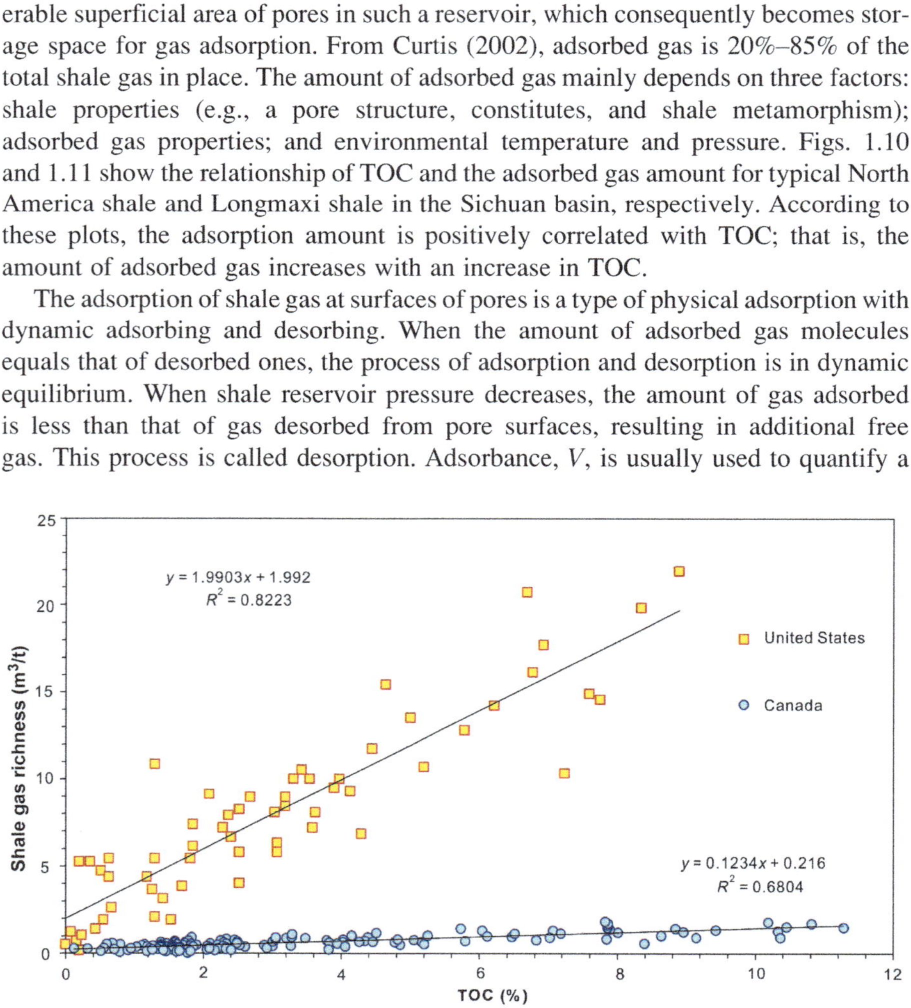

Fig. 1.10 Correlation of TOC and adsorbed gas amount in shale reservoirs in the United States and Canada.

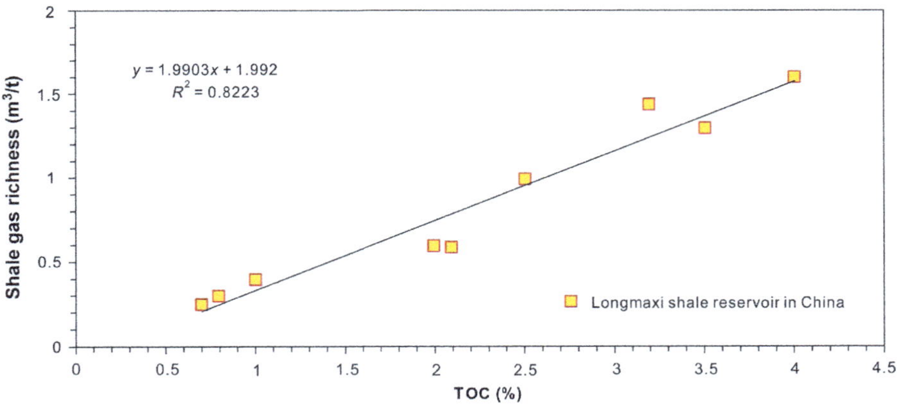

Fig. 1.11 Correlation of TOC and adsorbed gas amount in Longmaxi shale reservoirs in Sichuan basin.

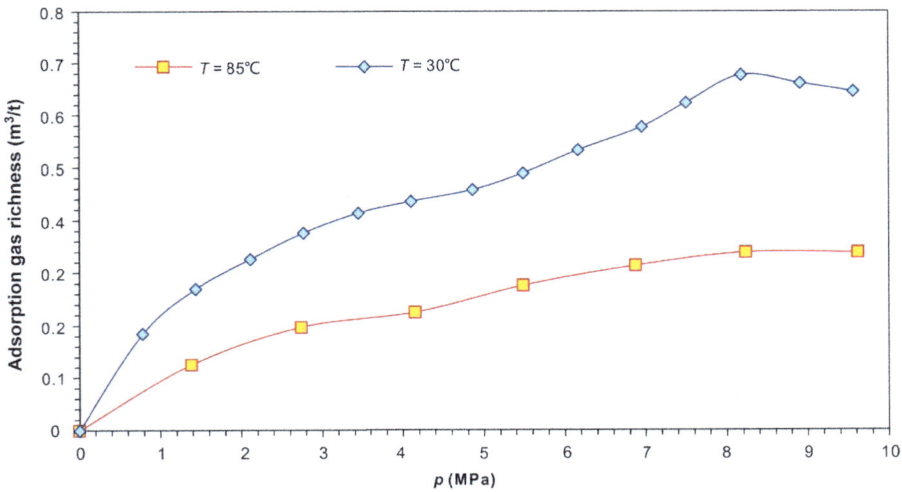

Fig. 1.12 Gas adsorption curves at different temperatures in a Longmaxi shale reservoir in Sichuan basin.

gas adsorption capacity of a shale formation, which is expressed as a gas volume, under standard conditions, adsorbed to a unit volume or unit mass of a shale reservoir. Generally speaking, adsorbance decreases with an increase in temperature and increases with an increase in pressure. Fig. 1.12 shows the relationship of gas adsorbance and pressure at different temperatures for a Longmaxi shale reservoir; as shown, adsorbance decreases significantly with an increase in temperature.

For a constant temperature, a correlation between shale gas adsorbance and pressure is called an isothermal adsorption curve, which is important for shale gas reservoir development. Isothermal adsorption curves can be used to: (1) evaluate a gas storage capacity of a shale reservoir; (2) determine the critical desorption pressure, which is the pressure for gas starting to desorb from surfaces of matrix pores; and (3) determine the adsorbed gas changes with a decrease in the formation pressure during development.

1.4.3.1 Types of isothermal adsorption curves

Adsorption phase equilibrium is an important theory to analyze the adsorption separation technology and is the basis of describing an adsorption capacity of adsorbed molecules in adsorbents and adsorptive selectivity. For research on gas–solid adsorption phase equilibrium, categorization of isothermal adsorption curves is usually the starting point, which helps in understanding different adsorption mechanisms, build up fit-for-propose theoretical models and apply these models in practice. International Union of Pure and Applied Chemistry (IUPAC) has clearly instructed in its handbook: The first step for research on an adsorption theory is to identify an appropriate type of isothermal adsorption curve, on the basis of which the nature of an adsorption process can then be analyzed. For the categorization of isothermal adsorption curves, the most common three methods being used worldwide are: BDDT's five types of isothermal adsorption curves (Brunauer et al., 1940); IUPAC's six types of isothermal adsorption curves (Rouquerol et al., 1994); and Gibbs categorization of isothermal adsorption curves (Zhou et al., 2002).

Among the above three methods, the IUPAC categorization is the most popular method for shale gas reservoirs, and, therefore, its six types of isothermal adsorption curves are analyzed later in this chapter. In 1985, based on BDDT's five types, IUPAC proposed the sixth type of physical isothermal adsorption curves (shown in Fig. 1.13), and this type of curve is widely used in the field of adsorption phase equilibrium in recent years.

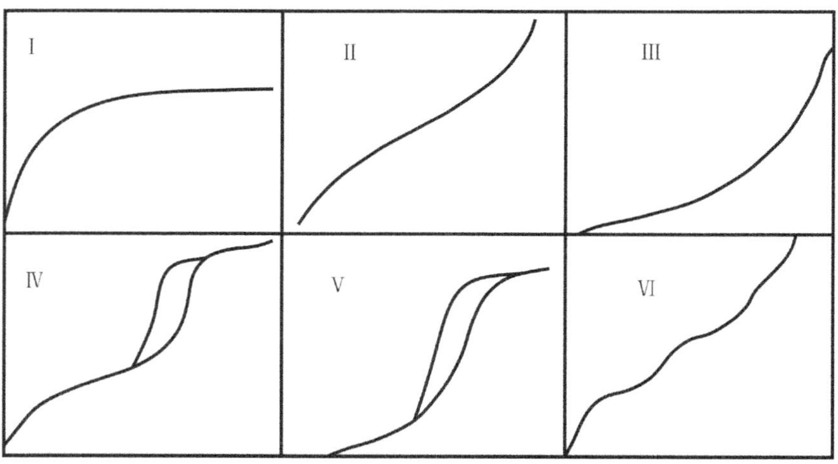

Fig. 1.13 Six types of IUPAC isothermal adsorption curves.

Type I (Langmuir isotherm) represents the adsorption status on micropore surfaces. The characteristic of this type is that gas adsorption amount has a significant increase in a relative low-pressure area. This is caused by the filling up of micropores. After that, when the adsorption pressure increases, an adsorption curve becomes horizontal or almost horizontal, which indicates the micropores being fully filled with no or little possibility of further adsorption. When pressure reaches the saturation pressure, adsorption condensation may happen. The adsorption behavior of a porous solid with a relatively small external surface (e.g., acticarbon, molecular sieve zeolite, and porous oxide) matches this type of isotherm (Fig. 1.14 shows an isothermal adsorption curve of Longmaxi shale.)

Type II (S-shape isotherm) represents adsorption on an adsorbent with big pores. In this case, the adsorbate and adsorbent have strong interaction. Type III also represents adsorption on an adsorbent with big pores, but in this case, the interaction between adsorbate molecules and adsorbent surfaces is weaker, and the interaction between the adsorbate molecules strongly impacts an isothermal adsorption curve. This type of an isothermal adsorption curve, which is uncommon, has the characteristic of concaving relative to the pressure axis. Type IV represents monolayer adsorption with capillary condensation (Fig. 1.15 illustrates the capillary condensation phenomenon). Type V represents multilayer adsorption with capillary condensation. Finally, Type VI is for multilayer adsorption happening on a non-porous adsorbent with smooth surfaces. The obvious characteristic of this isotherm type is its staircase shape during the adsorption process, which is caused by sequential adsorptions on the smooth multilayer surfaces of the non-porous adsorbent.

Introducing the capillary condensation phenomenon by IUPAC is an important supplement to the five BDDT isotherm types. The capillary condensation phenomenon also is called an adsorption hysteresis loop, or adsorption hysteresis. The non-coincidence of adsorption and desorption isothermal curves results in the hysteresis loop.

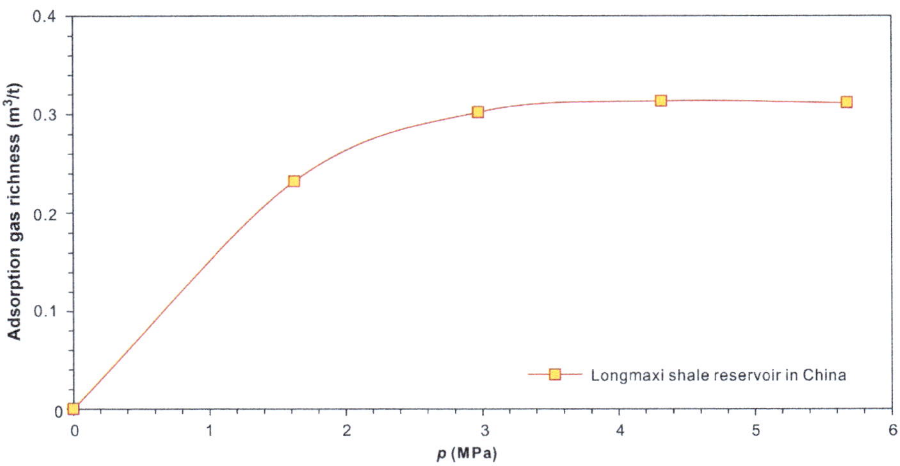

Fig. 1.14 Isothermal adsorption curve in a Longmaxi shale reservoir in Sichuan basin.

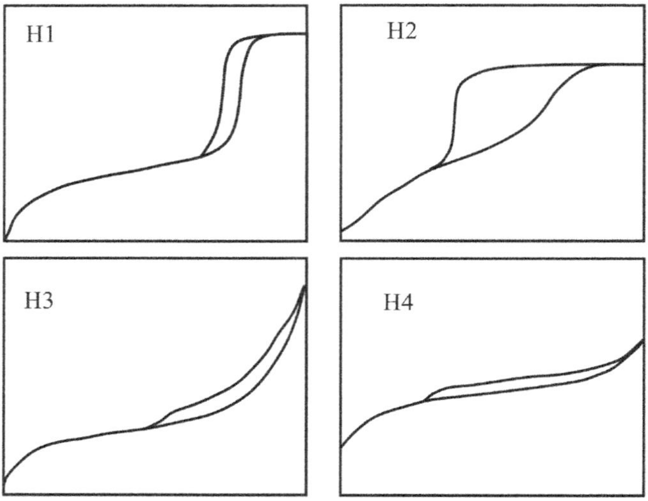

Fig. 1.15 IUPAC classification of the hysteresis loop.

This phenomenon usually occurs in adsorbents with medium size pores. IUPAC identified four types of a hysteresis loop for the adsorption hysteresis phenomenon, as shown in Fig. 1.15 (He et al., 2004).

Type H1: narrow hysteresis loop. Adsorption and desorption curves are almost vertical and parallel to each other. This type usually occurs in porous material with narrow pore diameter distribution, which is formed by conglobation or compaction.

Type H2: wide hysteresis loop. The desorption curve is much steeper than the adsorption curve. This commonly happens for porous material with relatively wide pore diameter and multiple pore types.

Type H3: no peak showing up on an adsorption curve even at high relative pressure. In this case, adsorbance monotonously increases with an increase in pressure. This usually happens for flaky material with narrow and long split-type pore structures.

Type H4: narrow hysteresis loop. Compared to H1, the adsorption and desorption curves are almost horizontal and parallel to each other.

For the above four situations, H1 and H4 are two extremes, while H2 and H3 are in between.

1.4.3.2 Theoretical adsorption model

(1) Henry adsorption model

The Henry adsorption theory is a straight line passing through the origin of adsorbance G and equilibrium adsorption pressure p. The following equation describes the linear relationship:

$$G = k_c p \tag{1.4}$$

where:

G—molar concentration of gas dissolved in the water (mol/m^3);
p—gas partial pressure (Pa);
k_c—Henry constant [mol/(m^3 Pa)].

Under the condition of low pressure, any isothermal adsorption curve is almost a straight line; therefore, Henry's law is only applicable to describing gas adsorption at low pressure.

(2) Freundlich adsorption theory

Based on Henry's equation, Freundlich proposed an adsorption model in an exponential relationship:

$$m = kp^n \tag{1.5}$$

where:

m—gas mass adsorbed by unit mass adsorbent (kg/kg);
p—gas partial pressure (Pa);
k—empirical constant (dimensionless);
n—empirical constant (dimensionless).

Note that the Freundlich equation is an empirical equation, in which the parameters have no physical meaning; however, k normally decreases with an increase in temperature, and n is between 0 and 1, which generally reflects the influence strength of pressure on adsorbance. At high pressure, $1/n$ approaches 0; so the limit of this adsorption is independent of pressure, but adsorbance is related to pressure.

(3) Langmuir adsorption theory (Langmuir, 1918)

French chemist Langmuir proposed the Langmuir isothermal equation in 1918 to analyze gas adsorption onto solid surfaces based on the molecular dynamics. The following assumptions were applied: ① the surface of an adsorbent is uniform and the adsorption of gas molecules to surfaces of solid molecules is monolayer adsorption; ② the gas adsorption process is dynamic; that is, the adsorbed molecules released by thermal motion can be back into the gaseous phase, and, reversely, the molecules in the gaseous phase can be re-adsorbed to the solid surfaces; ③ when the adsorption equilibrium is achieved, the adsorption rate equals the desorption rate; ④ the rate of gas molecule adsorption onto the solid surfaces is positively proportional to the gas partial pressure of this component; ⑤ no interaction force occurs between adsorbed gas molecules at the solid surfaces. Therefore, mass exchange can be separately considered for adsorbed gas and free gas. That is, the adsorption rate on the unit area of a rock surface (J_{ads}) is:

$$J_{ads} = k_a p(1-\theta) \tag{1.6}$$

And the desorption rate on the unit area of the rock surface (J_{des}) is:

$$J_{des} = k_d\theta \tag{1.7}$$

where:

J_{ads}—gas adsorption amount on the unit area of rock surface (m/s);
J_{des}—gas desorption amount on the unit area of rock surface (m/s);
θ—gas coverage of the porous rock surface (dimensionless);
p—gas partial pressure (Pa);
k_a—gas equilibrium adsorption rate [m/(Pa s)];
k_d—gas equilibrium desorption rate (m/s).

When dynamic equilibrium is achieved, the adsorption rate is equal to the desorption rate and then Eqs. (1.6) and (1.7) can be combined as:

$$\theta = \frac{k_a p}{k_d + k_a p} \tag{1.8}$$

Introducing the Langmuir gas adsorption equilibrium constant (b), then we see that:

$$b = \frac{k_a}{k_d} \tag{1.9}$$

Substituting Eq. (1.9) into Eq. (1.8), the latter can be transformed into the common Langmuir equation format:

$$\theta = \frac{G}{G_m} = \frac{bp}{1 + bp} \tag{1.10}$$

where:

G—gas adsorption amount under equilibrium status (m^3/m^3);
G_m—limit gas adsorption amount when gas adsorption on unit surface area is saturated (m^3/m^3);
B—Langmuir adsorption equilibrium constant (1/Pa).

According to Eq. (1.10), when the gaseous phase pressure approaches infinite ($p \to \infty$), $\theta \to 1$; that is, gas adsorption reaches saturation and equals the limit adsorption amount G_m; when the gaseous phase pressure is low, Eq. (1.10) can be transformed into Henry's equation:

$$\lim_{p \to 0}\left(\frac{G}{p}\right) = bG_m = k_c \tag{1.11}$$

Another format of the Langmuir equation can be obtained through the transformation of Eq. (1.10):

$$G = G_L \frac{p}{p_L + p} \tag{1.12}$$

where:

G_L—Langmuir adsorption volume of shale gas, representing the limit adsorbance of unit reservoir (m^3/m^3);
p_L—Langmuir pressure, which is the pressure when gas adsorbance reaches 50% of limit adsorbance (Pa).

Since pseudo-pressure is used for the analysis of gas transient flow, for convenience, the Langmuir adsorption model can be expressed in a pseudo-pressure format:

$$G = G_L = \frac{m(p)}{m(p_L) + m(p)} \tag{1.13}$$

where:

$m(p)$—gas pseudo-pressure (Pa/s);
$m(p_L)$—Langmuir pseudo-pressure (Pa/s).

Pseudo-pressure is used in the above equation, and the pseudo-pressure expression for compressible fluid is:

$$m(p) = \int_{p_o}^{p} \frac{2p}{\mu Z} \mathrm{d}p \tag{1.14}$$

where:

p_o—reference pressure; generally the standard atmospheric pressure is used as the reference pressure (Pa).

Temperature is assumed constant for the whole adsorption process of the Langmuir model. Actually, temperature can affect gas adsorption capacity. As shown in Fig. 1.12, at higher temperature, the adsorption capacity becomes weaker. Since the gas reservoir development is viewed as an isothermal process, the Langmuir equation is applicable. The Langmuir isothermal adsorption constant can be acquired from lab data matching or back calculations from the field test data.

(4) BET adsorption theory (Brunauer et al., 1938)

In 1938, Brunauer et al. proposed the assumptions based on the Langmuir model: Multilayer adsorption of gas molecules can happen on solid surfaces; there is no interaction force between layers, and the Langmuir adsorption theory is applicable for each molecule adsorption layer. These assumptions extended the monolayer adsorption theory to multilayer adsorption, and the multilayer model was named as the BET adsorption model, which can be expressed as follows:

$$\frac{G}{G_m} = \frac{b(p/p_o)}{(1 - p/p_o)[1 + (b - 1)(p/p_o)]} \tag{1.15}$$

where:

G—gas adsorption amount (m^3/m^3);
G_m—saturated gas adsorption amount at the reference temperature (m^3/m^3);
p—gas equilibrium partial pressure (Pa);
p_o—gas saturated vapor pressure at the reference temperature (Pa);
b—constant (dimensionless).

Note that the BET adsorption equation is validated for p/p_o ranging from 0.005 to 0.35. If the value is smaller than the lower limit, the model gets away from the straight line, indicating a non-uniform physical and chemical shape of a solid surface with existence of activated adsorption points. If the value is greater than the upper limit, an infinite adsorption layer most likely causes it, but an infinite solid adsorptive layer is impractical.

1.5 Multiscale flow mechanisms in shale gas reservoirs

Special accumulation features, reservoir properties, and characteristics of pore structures lead to multiscale and complex gas flow during development of shale gas reservoirs. Therefore, a systematical analysis of gas flow mechanisms in multiscale pore structures is necessary. Fig. 1.16 illustrates the microscopic flow mechanism of gas in a conventional naturally fractured dual porosity reservoir. The diagram indicates that both matrix pores and the fracture system are occupied by free gas, and some parts of the porous medium may contain residual water and even free water. For such gas reservoirs, the driving force for gas flow during development is the formation pressure. If the matrix pores are relatively small, non-Darcy's flow can happen for gas flow in these pores. In addition, some complex flow regimes, such as high velocity non-Darcy flow and gas–water two-phase flow, happen in the fracture system, which is the main flow path for gas from a reservoir to wellbore.

Compared with a conventional natural fractured reservoir, a shale gas reservoir has much more complex gas flow behavior because of the combination of source rock, a reservoir, and cap rock with nanometer pores. Therefore, how to describe the gas flow in different scale pores is always a hot research topic worldwide. Fig. 1.17 is a diagram

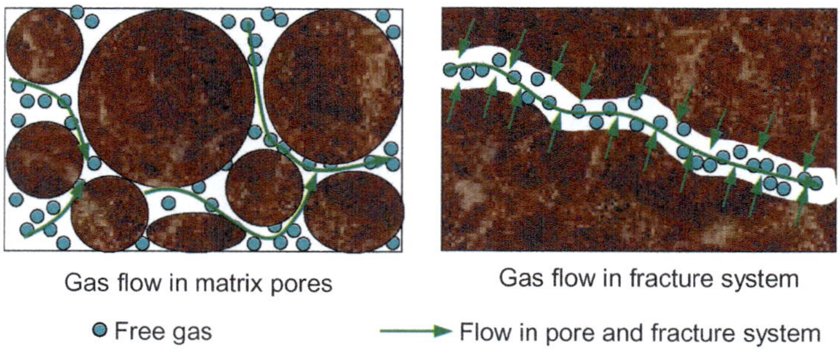

Fig. 1.16 Diagram of gas flow in a conventional dual porosity gas reservoir.

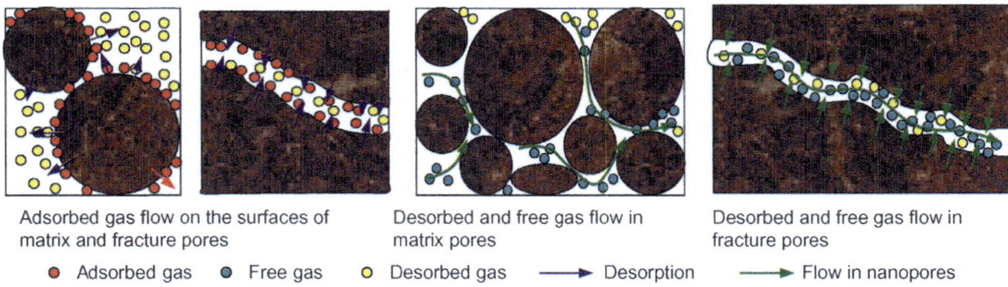

Fig. 1.17 Diagram of gas flow in a shale gas reservoir.

showing gas flow in a shale reservoir. As it is shown in this diagram, not only the shale matrix pores and microfractures are filled with free gas, but also there is a great amount of natural gas adsorbed on the surfaces of pores and microfractures. Due to the existence of adsorbed gas and nanometer pores, the gas flow mechanism in porous media becomes extremely complex. Additionally, gas flow in organic matters and matrix intragranular pores, which are extensively distributed in a shale reservoir, is different from that in inorganic pores. In general, the main flow status during the development of a shale reservoir includes: (1) gas flow in artificial and natural fractures; (2) gas flow in matrix micropores (two consideration methods: treating organic and inorganic pores differently or treating them the same way); (3) gas diffusion in the nanopores (equilibrium and non-equilibrium adsorption models); and (4) gas desorption from pore surfaces (the Langmuir isothermal adsorption equation).

1.5.1 Flow regime classification of gas flow in shale reservoirs

Mechanisms of gas migration in porous media are driven by the ratio of gas molecule motion free path to a pore radius in porous media. If the free path is much smaller than the pore radius, the probability of molecule-to-molecule collision is much higher than that of molecule-to-surface collision. In this case, gas mass transfer mainly relies on the viscous flow induced by the molecule-to-molecule collision. If the pore radius gets small enough to the same order of magnitude as the molecule motion free path, the probability of the molecule-to-surface collision becomes much higher than that of the molecule-to-molecule collision. At this time, flow is dominated by the molecule-to-surface collision.

Generally, a continuity hypothesis or molecule hypothesis can be used to model fluid flow in nanometer pores. A model with the continuity hypothesis can be used to describe a relationship between macroscopic fluid properties and spatial coordinates, which is widely used in the fluid flow. The Knudsen number is usually used to justify if fluid flow satisfies the continuity hypothesis and then determine a fluid flow regime. The Knudsen number K_n is defined as the ratio of gas molecule mean free path to the characteristic length of a porous medium (Civan, 2010):

$$K_n = \frac{\lambda}{R_h} \tag{1.16}$$

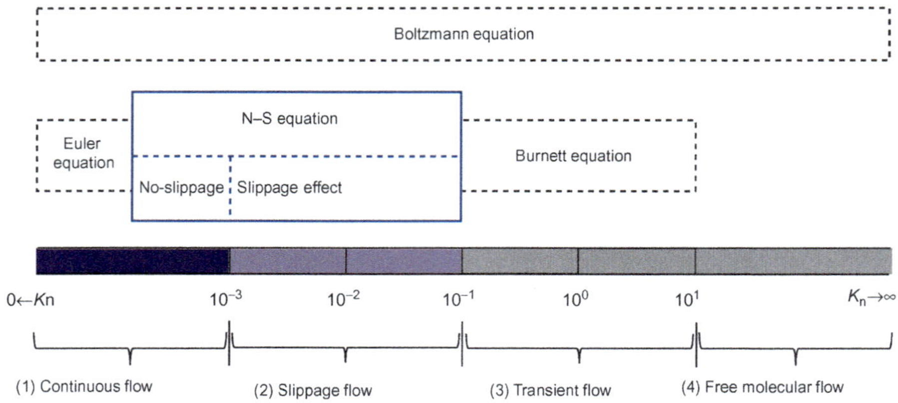

Fig. 1.18 Flow regime division by the Knudsen number and corresponding governing equations.

where:

λ—gas molecule mean free path of gas (nm);
R_h—average hydraulic radius in a pore medium (nm).

Fig. 1.18 shows the division of gas flow regimes using the Knudsen number (Roy and Raju, 2003). If the Knudsen number is close to zero, the Euler equation can be used to describe the fluid flow. For the Knudsen number < 0.001, the Navier–Stokes equation (or Darcy's flow equation) with a no-slippage boundary condition is applicable. For the Knudsen number between 0.001 and 0.1, Navier–Stokes equation with a slippage boundary condition applies. For the Knudsen number between 0.1 and 10, the flow belongs to a transition flow regime; and for the Knudsen number > 10, the Boltzmann equation with the molecule hypothesis is used to describe the fluid flow (Mohamed, 1999).

Viscous flow (continuous flow): When $K_n \leq 0.001$, gas flow in porous media is viscous flow. For a conventional gas reservoir whose pore size is in micron order, the Knudsen number is very small ($K_n < 0.001$). In other words, the gas molecule free path is negligible compared to a pore size. Therefore, the molecule-to-molecule collision dominates in molecule motion for this type of gas flow. Consequently, Darcy's law can be used to describe viscous flow with consideration of only the viscous force, without friction resistance and a slippage effect between fluids and pore surfaces.

Slippage flow: applicable for $0.001 \leq K_n \leq 0.1$. In this flow phase, the gas flow velocity on a pore surface is greater than zero, and, therefore, modifications to the viscous flow theory are required to satisfy a boundary slippage effect. When gas is in slippage flow, the collisions between gas molecules and pore surfaces become more important than the collisions between gas molecules with the molecule mean free path getting closer to the pore size. Gas flow in many tight reservoirs is in this phase for which the Klinkenberg slippage theory is applicable.

Transition flow: applicable for $0.1 \leq K_n \leq 10$.

Free gas molecule flow: applicable for $K_n > 10$.

However, to acquire an accurate Knudsen constant from Eq. (1.16), the molecule mean free path is the key. The following two methods were documented for calculations of the Knudsen constant:

Method 1 (Civan, 2010):

$$\lambda = 3.16 \times 10^9 \frac{\mu}{p} \sqrt{\frac{\pi RT}{2M_g}} \tag{1.17}$$

where:

λ—mean free path of gas molecules (nm);
μ—gas viscosity (Pa s);
R—gas constant [8.314 J/(mol K)];
T—gas absolute temperature (K);
p—gas pressure(Pa);
M_g—gas molar mass (kg/mol).

Method 2 (Javadpour et al., 2007):

$$\lambda = 10^{15} \frac{k_B T}{\sqrt{2}\pi \delta^2 p} \tag{1.18}$$

where:

k_B—Boltzmann constant (1.3805×10^{-23} J/K);
δ—collision diameter of gas molecule (nm).

The derivation of the above two methods assumes that gas is in the ideal state. However, natural gas under reservoir conditions should not be treated as ideal gas, and real gas needs to be used in calculations. Therefore, the modified Eq. (1.17) becomes:

$$\lambda = 3.16 \times 10^9 \frac{\mu Z}{p} \sqrt{\frac{\pi RT}{2M_g}} \tag{1.19}$$

where Z is the gas deviation factor (dimensionless).

Using the above three equations and the given parameters of shale gas components and temperature, a plot of the molecule mean free path under different pressures can be generated, as shown in Fig. 1.19. It can be seen from this plot that the mean free path of gas molecules calculated by the three methods is exactly the same at low pressure, while the calculated results segregate at high pressure. Since gas exhibits properties similar to liquid under a high-pressure condition, theoretically the molecule mean free path should be greater than the average molecule collision diameter and becomes constant when pressure reaches a certain level. Obviously, as shown in the plot, the calculated results by the modified equation are closer to the theoretical values, and, therefore, Eq. (1.19) is used to calculate the mean free path of gas molecules.

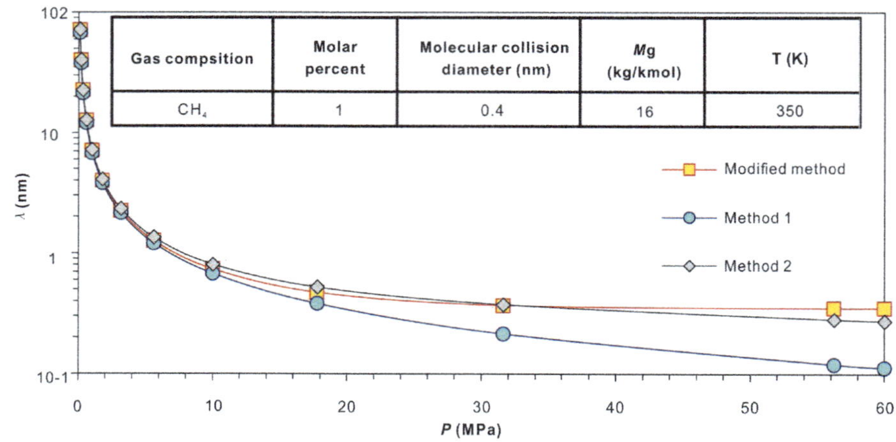

Fig. 1.19 Comparison of gas molecular mean free paths calculated by three methods.

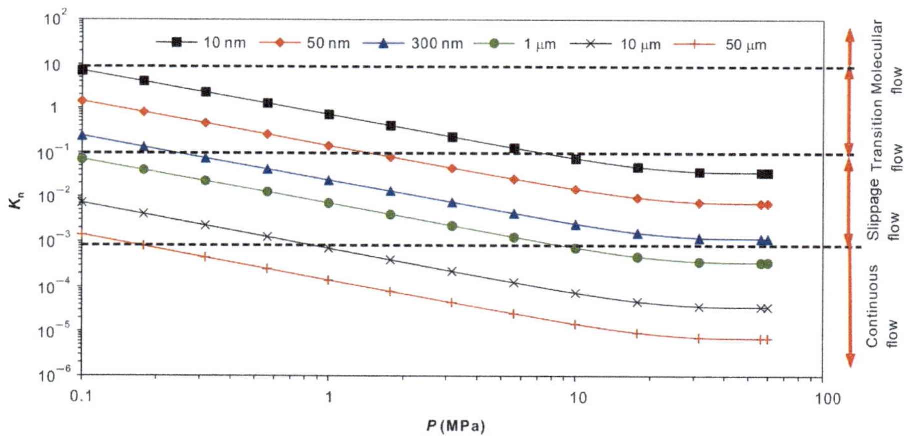

Fig. 1.20 Relationship of Knudsen number and pressure for flow in capillaries of different hydraulic diameters.

Substituting Eq. (1.19) into Eq. (1.16), a correlation between the Knudsen number and pressures for methane gas flowing in capillaries of different hydraulic diameters is shown in Fig. 1.20. According to the flow regime classification by the Knudsen number, gas flow in matrix pores is continuous flow for a conventional gas reservoir, whose pore size ranges from 1 to 200 μm. Most gas flow in a shale gas reservoir belongs to slippage flow, and part of it stays in transient flow in a low-pressure region since the pore size in the shale reservoir ranges from 1 to 200 nm. Therefore, gas flow in shale matrix pores is affected by a slippage effect.

1.5.2 Gas flow in fractures and macropore systems

Since there are massive microfractures, a certain number of macropores, and hydraulic fracture and secondary fracture networks generated around wellbores by hydraulic fracturing, which have comparatively a large pore scale, gas flow in such fractures and macropores belongs to continuous flow. According to the previous analysis, Darcy's flow equation used for a conventional gas reservoir can be used to analyze gas flow in a shale reservoir:

$$J_l = -\frac{p_l M_g}{ZRT}\frac{k_l}{\mu_g}\nabla p_l \quad (l = \text{f}, \text{mc}) \tag{1.20}$$

where:

J_l—mass velocity of gas in media l [kg/(m^2 s)];
k_l—permeability of media l (m^2);
μ_g—gas viscosity (Pa s);
p_l—pressure in media l (Pa);
∇—gradient operator,

$$\nabla = \frac{\partial}{\partial x}i + \frac{\partial}{\partial y}j + \frac{\partial}{\partial z}k;$$

f, mc—fractures and macropore media.

In Eq. (1.20), it is assumed that the reservoir permeability is uniform for both the fracture system and the macropore medium system. From the previous research on the unsteady state flow theory in shale reservoirs, stress sensitive phenomenon happens in some systems developing with microfractures and/or secondary fractures generated by hydraulic fracturing. Therefore, the permeability item in Eq. (1.20) can be treated as a function of pressure for such reservoirs:

$$k_f = k_f(p_f) \tag{1.21}$$

where p_f is the microfracture system pressure (Pa).

Empirical equations can be chosen for the permeability function of pressure. For example, the following exponential relationship can be used for stress sensitive permeability:

$$k_f = k_{f0}\exp[-\gamma(p_{f0} - p_f)] \tag{1.22}$$

where:

γ—stress sensitivity coefficient (1/Pa);
p_{f0}—reference pressure, usually original reservoir pressure (Pa);
k_{f0}—microfracture permeability at the reference pressure (m^2).

1.5.3 Gas flow in nanometer pores

Fig. 1.21 shows a physical model of gas flow in macropores and nanometer pores. As shown in the diagram, gas flow in macropores has no slippage effect due to the mean free path being much smaller than the pore hydraulic diameter. For gas flow in nanometer pores, the gas molecule mean free path is close to the size of rock pore throats. Collisions between molecules and between molecules and pore surfaces are both significant, and, therefore, the gas velocity at pore surfaces is greater than zero, and a slippage effect happens.

Klinkenberg (1941) first observed a slippage effect of oil and gas flow in porous media. According to his observation, the actual gas flow velocity under low-pressure conditions was greater than that calculated from Darcy's equation, which was caused by gas slippage at pore surfaces. Thereafter, he proposed the following equation to calculate the apparent gas phase permeability of a reservoir rock:

$$k_a = k_m\left(1 + \frac{b_k}{\overline{p}}\right) \tag{1.23}$$

where:

k_a—apparent or effective permeability (m^2);
k_m—matrix permeability (equivalent liquid permeability) (m^2);
b_k—slippage factor or coefficient, related to gas properties and pore structure (Pa);
$\overline{p}$—average pressure of core plug between outlet and inlet (Pa).

In order to bring Eq. (1.23) into flow equations, first the expression of a slippage coefficient needs to be regressed by many core lab analyses. In this regard, worldwide researchers proposed many empirical equations. Table 1.4 lists some expressions for the slippage coefficient acquired theoretically or through lab tests.

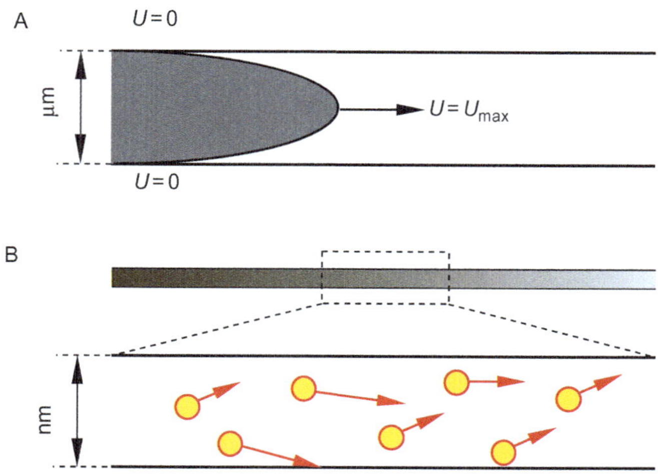

Fig. 1.21 Comparison of gas flow in macropores (A) and nanometer pores (B).

Table 1.4 Empirical expressions of a slippage coefficient

No.	b_k expressions	Author
1	$b_k = 4c\lambda p_{avg}/r$	Klinkenberg (1941)
2	$b_k = 87143(k_\infty)^{-0.33}$ (k—mD; b_k—Pa)	Jones and Owens (1980)
3	$b_k = 93638(k_\infty/\phi)^{-0.53}$ (k—mD; b_k—Pa)	Sampath and Keighin (1982)
4	$b_k = p_{avg} D_k \mu_g C_g/(a_1 k_\infty)$, $D_k = 31.54/\sqrt{M_g}(k_\infty)^{0.67}$ M—kg/kmol; k—mD; D_k—ft.^{2}/d; p—MPa; C_g—MPa^{-1}; μ_g—mPa s; $a_1 = 6.328 \times 10^{-3}$	Ertekin et al. (1986)
5	$b_k = 298853(k_\infty/\phi)^{-0.5}$ (k—mD; b_k—Pa)	Florence et al. (2007)
6	$b_k = (8\pi RT/M_g)^{0.5}\mu_g/r(2/\alpha - 1)$ R—J/mol/K; T—K; M—kg/mol; μ_g—Pa s; r—m; b_k—Pa	Javadpour (2009)
7	$b_k = \mu_g\sqrt{\pi RT\phi/(\tau M_g k_\infty)}$ R—J/mol/K; T—K; M—kg/mol; μ_g—Pa s; k—mD; b_k—Pa	Civan (2010)
8	$b_k = \frac{3\pi D_k \mu_g}{2r^2}$, $D_k = 2/3r[8RT/(M_g)]^{0.5}$ R—J/mol/K; T—K; M—kg/mol; D_k—m^2/s; r—m; μ_g—Pa s; b_k—Pa	Michel et al. (2011)

For convenience, the expressions can be simplified as follows:

$$k_a = Fk_m \tag{1.24}$$

where

$$F = 1 + \frac{b_k}{p_{avg}} \tag{1.25}$$

From the above transformation, the modification factor F is obtained, which is widely used by many petroleum experts to represent a slippage effect of gas flow in shale nanometer pores (Darabi et al., 2012; Javadpour, 2009; Ozkan et al., 2010a,b; Shabro et al., 2009, 2011a,b, 2012; Swami, 2012; Swami and Settari, 2012; Swami et al., 2013):

$$F = 1 + \left(\frac{8\pi RT}{M_g}\right)^{0.5} \frac{\mu_g}{p_{avg} r}\left(\frac{2}{\alpha} - 1\right) \tag{1.26}$$

where α is an accommodation coefficient of tangent momentum or the ratio of molecule numbers of diffusive reflection from surfaces to mirror reflection. The value of α is related to pore surface smoothness, a gas type, temperature and pressure, and it theoretically ranges from 0 to 1. If a gas slippage effect is not considered, let $\alpha = 2$ (dimensionless).

1.5.4 Gas diffusion in nanometer pores

Due to the nanometer magnitude of shale matrix pores, gas flow in such pores behaves very differently from flow in a fracture system. An assumption of molecular continuous flow and a pure application of Darcy's equation would result in a big error in representing gas flow in nanometer pores. According to previous research, shale gas flow in nanopores includes not only viscous flow but also diffusion. Especially for some ultra-tight shale reservoirs, the flow status of gas molecules involves only diffusion and no viscous flow because gas exists only in adsorption, and there is no free gas in nanometer pores.

According to the study of the existing literature, there are two models in use to describe gas diffusion in shale nanometer pores: a Fick's diffusion model (Ayala et al., 2005; Carlson and Mercer, 1991; Chawathe et al., 1996; Ertekin et al., 1986; King, 1990; Tian et al., 2014) and a Knudsen diffusion model (Darabi et al., 2012; Dehghanpour and Shirdel, 2011; Deng et al., 2014; Guo et al., 2013; Javadpour, 2009; Kuuskraa et al., 1992; Ozkan et al., 2010a,b; Schepers et al., 2009; Shabro et al., 2009, 2011a,b, 2012; Singh et al., 2013; Swami, 2012; Swami and Settari, 2012; Swami et al., 2013). In the following sections, applicable reservoir conditions and diffusion mechanisms for these two models are introduced.

Fick's diffusion. Many researchers proposed to use Fick's diffusion law to describe shale gas migration in nanopores and at surfaces of kerogen. Under the influence of a concentration difference, desorbed gas migrates from a high concentration area to a low concentration area, and diffusion stops until the gas concentration becomes uniform, as shown in Fig. 1.22. Assume that gas molecular diffusion satisfies Fick's law. Then the gas diffusion flux is:

$$J_F = -M_g D_F \cdot \nabla C_m \tag{1.27}$$

where:

J_F—Fick's mass diffusion flux (gas mass passing through unit acreage in unit time) [kg/(m^2 s)];
D_F—Fick's diffusion coefficient (m^2/s);
M_g—molar mass of gas (kg/mol);
C_m—molar mass of gas in matrix (mol/m^3).

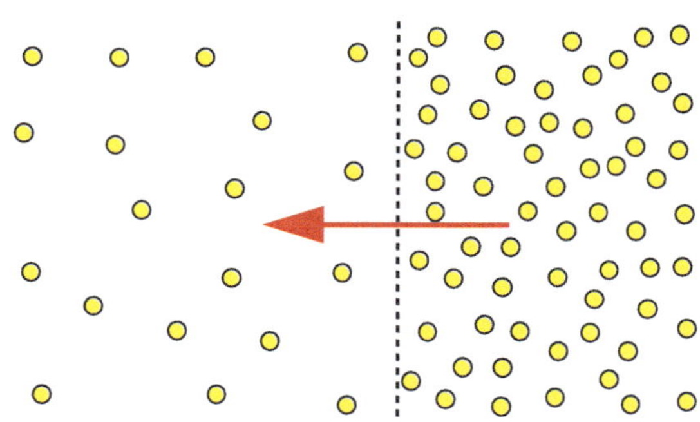

Fig. 1.22 Diagram of Fick's diffusion.

For Fick's law, when the gas concentration in matrix does not change with coordinates, meaning an equivalent concentration at any time in the matrix, it is called pseudo-steady state diffusion and can be described by Fick's First Law. Fick's Second Law is to describe unsteady state diffusion where the gas concentration in shale matrix changes with coordinates.

(1) Pseudo-steady state diffusion

According to the theory of pseudo-steady state diffusion, a change rate in gas concentration in shale matrix with time is positively proportional to the gas concentration difference between matrix and microfractures (or macropores):

$$\frac{dC_m}{dt} = D_F F_s [C_E(p) - C_m] \tag{1.28}$$

where:

C_m—gas molar concentration in the shale matrix under pseudo-steady state (mol/m^3);
C_E—gas molar concentration at the boundary between matrix and microfracture system (or macropores) (mol/m^3);
p—pore pressure in microfractures or macropores (Pa);
F_s—shape factor, which depends on the shape of matrix (refer to Table 1.5) (1/m^2);
D_F—Fick's diffusion coefficient (m^2/s);
t—time (s).

For a shale gas reservoir with volume V_b, the mass flow rate of gas diffusion from matrix to the microfracture or macropore system can be represented as:

$$q_F = M_g F_g V_b \frac{dC_m}{dt} \tag{1.29}$$

where:

q_F—Fick's mass flow rate (mass of gas passing through volume V_b in unit time) (kg/s);
F_g—geometry factor, which depends on the shape of matrix (refer to Table 1.5) (1/m^2);
V_b—shale matrix volume (m^3).

Table 1.5 Geometry and shale factor of shale matrix with different shapes

Shape of matrix	**Characteristic length**	**Shale factor F_s**	**Geometry factor F_g**
Bulk	Thickness, $2h$	2	$\frac{\pi^2}{4h^2}$
Cylinder	Radius of cylinder, R	4	$\frac{5.7832}{R^2}$
Sphere	Radius of sphere, R	6	$\frac{\pi^2}{R^2}$

(2) Unsteady state diffusion.

The pseudo-steady state diffusion introduced previously rarely happens during the development of shale gas reservoirs. In practice, gas diffusion in shale matrix is more like unsteady state diffusion. To simplify the description of a shale reservoir that has irregularly shaped matrix and microfractures, it is assumed that the matrix element is a sphere with radius R_m; each matrix element is composed of organics and clay grains, and gas flow in matrix is only diffusion (shown in Fig. 1.23). Therefore, Fick's Second Law can be applied to describe gas diffusion from matrix elements into microfractures or macro-pores.

According to the unsteady state diffusion theory, during the process of gas diffusion from matrix to its outside space, the gas concentration in the matrix C_m is a function of time and space. For a spherical matrix model, it is assumed that the change rate in gas concentration in matrix is 0, and the gas concentration at the external surface of the matrix rock is in dynamic equilibrium of pressure with the free gas in macrofractures and macropores. Therefore, the following mathematical expression can be used to describe a gas concentration change in matrix:

$$\frac{\partial C_m}{\partial t} = \frac{1}{r_m^2}\frac{\partial}{\partial r_m}\left(Dr_m^2\frac{\partial C_m}{\partial r_m}\right) \tag{1.30}$$

where:

C_m—shale gas volume concentration in matrix under unsteady state (mol/m^3);
r_m—inner diameter of sphere matrix element (m).

Before the development of a gas reservoir, the reservoir pressure (p_i) is uniform at any location, and the gas concentration at any point in matrix is related to the matrix pressure. Therefore, the initial condition of matrix unsteady state diffusion satisfies the equation:

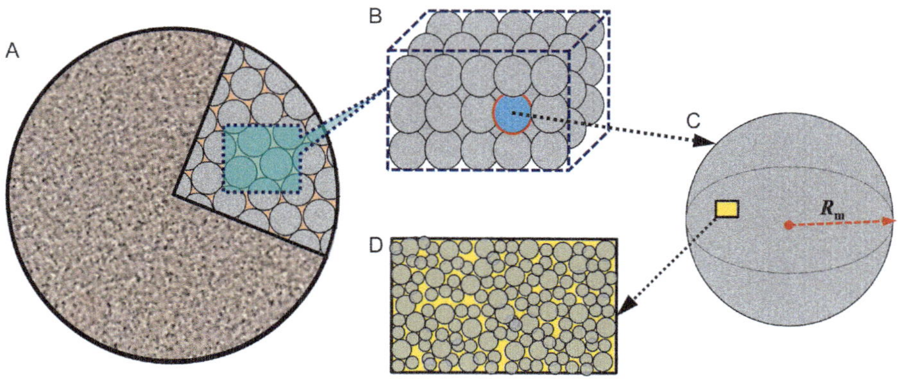

Fig. 1.23 Physical model of unsteady state Fick's diffusion.

$$C_{\mathrm{m}}(t=0, r_{\mathrm{m}})=C_{\mathrm{i}}(p_{\mathrm{i}}) \tag{1.31}$$

where:

p_i—initial reservoir pore pressure (MPa).

According to the previous assumptions, the change rate in gas concentration at the center of the matrix bulk is 0, and then the inner boundary condition can be expressed as:

$$\frac{\partial C_{\mathrm{m}}(t, r_{\mathrm{m}}=0)}{\partial r_{\mathrm{m}}}=0 \tag{1.32}$$

Since the matrix outer boundary is connected with the fracture system or macropore system, pressure at the matrix surface equals that within the fracture system. Then the outer boundary can be represented as:

$$C_{\mathrm{m}}(t, r_{\mathrm{m}}=R_{\mathrm{m}})=C_{\mathrm{m}}(p) \tag{1.33}$$

where:

p—pore pressure of microfractures or macropores, which are connected with the external surface of matrix (Pa);
R_m—radius of sphere matrix (m).

From the above model and relevant definite conditions, the distribution of gas concentration in the shale matrix can be solved. Then the following equation can be used to acquire a gas diffusion flow rate within the matrix with volume V_b:

$$q_{\mathrm{F}}=M_{\mathrm{g}}V_{\mathrm{b}}\frac{3D}{R_{\mathrm{m}}}\frac{\partial C_{\mathrm{m}}}{\partial r_{\mathrm{m}}}\bigg|_{r_{\mathrm{m}}=R_{\mathrm{m}}} \tag{1.34}$$

where:

q_F—Fick's mass flow (gas mass passing through volume V_b in unit time) (kg/s).

Previously, two scenarios of gas diffusion in shale matrix satisfying Fick's Diffusion Law have been analyzed. Although it was mentioned that the pseudo-steady state diffusion is rarely seen in practice, the calculation results from both the pseudo-steady and unsteady state models are close enough for the late time development of a shale gas reservoir according to relevant previous studies. During the early time development, the concentration change on the matrix surface is relatively significant, which means a big concentration gradient, and, therefore, the calculation results from these two models are different due to the small time gradient of average concentration in the pseudo-steady state model. As Ertekin et al. (1986) suggested, the pseudo-steady state model can be used for longer time prediction of pressure changes due to its higher calculation efficiency, while the unsteady state model is more applicable for an early time pressure analysis.

Knudsen diffusion. When the mean free path of gas molecules is close to the radius of matrix pore throats, the collisions between gas molecules and pore surface become more and more significant, and are non-negligible compared to the collisions between molecules. Therefore, for this kind of gas flow, not only the viscous flow of gas at a pore center (Darcy's flow or modified Darcy's equation for a slippage effect) but also the diffusive flow caused by the collisions between gas molecules and pore surfaces need to be considered. The viscous flow was introduced earlier, and here the diffusive flow is analyzed.

Fig. 1.24 shows physical diagrams of co-existence of viscous flow and Knudsen diffusion in nanometer pores (Guo et al., 2013). The right diagram shows the infinitesimal nanometer pore passage. These diagrams show that gas migrates in a viscous flow status along centers of pores while diffusive flow happens along pore surfaces. Therefore, the mass flux of Knudsen diffusion of gas in nanometer pores is:

$$J_K = -M_g D_k \cdot \nabla C_m \tag{1.35}$$

where:

J_K—gas mass flux of Knudsen diffusion [kg/(m^2 s)];
C_m—gas concentration in matrix pore system (mol/m^3);
D_k—Knudsen diffusion constant (m^2/s);
M_g—gas molar mass (kg/mol).

The gas concentration in Eq. (1.35) can be expressed in the following equation:

$$C_m = \frac{\rho_g}{M_g} = \frac{p_m}{ZRT} \tag{1.36}$$

where:

p_m—pore pressure of matrix (Pa);
Z—gas deviation factor (dimensionless), $Z=1$ for ideal gas;
R—gas constant [8.314 J/(mol K)];
T—reservoir temperature (K).

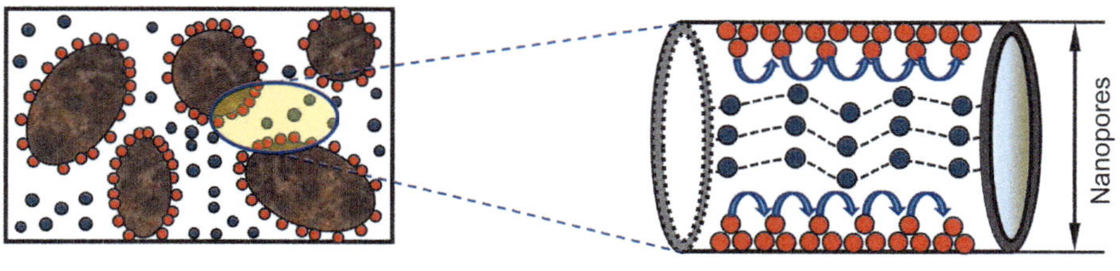

Fig. 1.24 Diagram of Knudsen diffusion and viscous flow in nanometer pores in shale matrix.

Substituting Eq. (1.36) into Eq. (1.35) yields:

$$J_{\mathrm{K}} = -\frac{M_{\mathrm{g}} D_{\mathrm{k}}}{RT} \nabla \frac{p}{Z} \tag{1.37}$$

For ideal gas, let $Z=1$, and then Eq. (1.37) becomes the same format as the expression proposed by Javadpour (2009), Civan (2010), and Darabi et al. (2012).

Similarly, Eq. (1.35) can also be transformed as:

$$J_K = -D_{\mathrm{k}} \cdot \nabla \rho_{\mathrm{g}} = -\frac{D_{\mathrm{k}} p_{\mathrm{m}} M_{\mathrm{g}} C_{\mathrm{gm}}}{ZRT} \nabla p_{\mathrm{m}} \tag{1.38}$$

where C_{gm} is the gas compressibility under the matrix pore pressure conditions (Pa^{-1}).

Then Eq. (1.38) can be written in the same format as the one proposed by Ozkan et al. (2010a,b). The gas compressibility in the above equation can be expressed as:

$$C_{\mathrm{gm}} = \frac{1}{\rho_{\mathrm{g}}} \frac{\partial \rho_{\mathrm{g}}}{\partial p_{\mathrm{m}}} = \frac{1}{p_{\mathrm{m}}} - \frac{1}{Z} \frac{\partial Z}{\partial p_{\mathrm{m}}} \tag{1.39}$$

For the Knudsen constant, there are different expressions proposed by many petroleum experts. The most frequently used ones are listed in Table 1.6.

1.5.5 Desorption of adsorbed gas from surfaces of shale matrix pores

Plenty of shale gas is adsorbed on pore surfaces and in organics. During reservoir development, with a decrease in pore pressure, adsorbed gas on pore surfaces desorbs into free gas. This is an important characteristic differentiating a shale gas reservoir from a conventional gas reservoir. Therefore, petroleum experts consider a desorption effect in their studies on shale gas reservoirs.

As described above, gas adsorption is a surface phenomenon, which is caused by the intermolecular force (van Edward force). Several adsorption models were introduced previously, and different adsorption models have different adsorption curves. According to the previous analysis, an isothermal adsorption curve generated from lab test data has the same shape as a Langmuir isothermal curve, which is frequently used to represent an adsorption–desorption process of shale gas. Under original reservoir

Table 1.6 Expressions of the Knudsen diffusion coefficient

Geometry of matrix	Dimension n	Characteristic length L_c	Geometry factor α
Layer (thickness h)	1	h	$12/h^2$
Sphere (radius r)	3	$2r$	$15/r^2$
Cylinder (radius r)	3	$2r$	$15/r^2$
Cube (length of side a)	3	a	$60/a^2$

conditions, adsorbed gas in shale is in equilibrium with the original reservoir pressure. After gas wells are completed and put in action, the reservoir pressure decreases with production, and the original equilibrium is broken. Thereafter, the adsorbed gas on surfaces of organic pores starts desorbing and converting to free gas in micropores until another equilibrium between adsorbed gas and free gas is achieved.

For emphasis, during the practice of shale gas reservoir development, when adsorbed gas starts desorbing with reservoir pressure decreasing, the amount of gas adsorption depends on initial reservoir conditions. As shown in Fig. 1.25, if gas adsorption is in saturation status under initial reservoir conditions, the amount of gas adsorption is at point B. Then gas can quickly desorb from organic surfaces with reservoir pressure decreasing. However, if gas adsorption is unsaturated, the amount of adsorbed gas is at point C. (If reservoir conditions are at point C, there is no free gas in matrix pores.) Then gas cannot desorb from organic surfaces until the reservoir pressure decreases to the pressure at point A, which is called the critical desorption pressure. The difference between the critical desorption pressure and the initial reservoir pressure determines the time at which shale gas desorption starts.

Another parameter related to adsorption phenomenon is desorption time. Under some circumstances, even when pressure decreases to the critical desorption pressure, the adsorbed gas molecules cannot depart from the surfaces of matrix grains immediately, and there is a certain delay time. The time duration between the pressure decreasing to the critical desorption pressure and the real happening of gas molecule desorption is called desorption time. However, for the sake of simplification, an instantaneous desorption model is commonly used, which represents immediate desorption once pressure decreases to the critical point.

Based on the above analysis, the Langmuir isothermal adsorption equation is expressed as:

$$G = G_{\mathrm{L}} \frac{p_{\mathrm{m}}}{p_{\mathrm{L}} + p_{\mathrm{m}}} \tag{1.40}$$

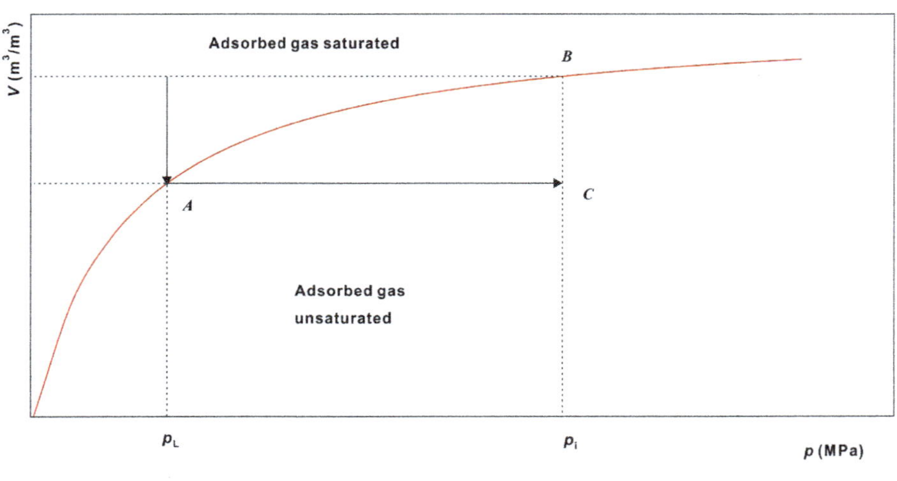

Fig. 1.25 Shale gas isothermal adsorption curve and desorption process diagram.

Then the mass flow rate of desorption from a shale reservoir with volume V_b in unit time is:

$$q_{\text{des}} = \rho_{\text{gsc}} V_{\text{b}} \frac{\partial G}{\partial t} \tag{1.41}$$

where:

q_{des}—mass flow rate of gas desorption from reservoir with volume V_b (kg/s);
ρ_{gsc}—shale gas density at standard conditions (kg/m^3).

Substitute Eq. (1.40) into Eq. (1.41) and rearrange to obtain:

$$q_{\text{des}} = \rho_{\text{gsc}} V_{\text{b}} G_{\text{L}} \frac{p_{\text{L}}}{(p_{\text{L}} + p_{\text{m}})^2} \frac{\partial p_{\text{m}}}{\partial t} \tag{1.42}$$

If the pseudo-pressure is used to describe the Langmuir equation, the above equation can be transformed as:

$$q_{\text{des}} = \rho_{\text{gsc}} V_{\text{b}} G_{\text{L}} \frac{m(p_{\text{L}})}{[m(p_{\text{L}}) + m(p_{\text{m}})]^2} \frac{\partial m(p_{\text{m}})}{\partial t} \tag{1.43}$$

Assume that shale gas desorption happens instantaneously, and all desorbed gas goes into free gas. Then Eq. (1.42) or (1.43) can be directly introduced into the continuity equation of the matrix or fracture system, resulting in the steady state adsorption–desorption model mentioned above. Many petroleum experts used this method to analyze the unsteady state flow model for shale gas reservoirs (Bumb and McKee, 1988; Clarkson et al., 2007; Civan, 2010; Gao et al., 1994; Lane et al., 1989; Wang et al., 2013; Xu et al., 2013).

1.6 Mathematical models with various shale gas flow mechanisms

In the previous section, various microscopic flow mechanisms and different description methods for gas flow in multiscale pores of a shale gas reservoir were introduced individually. For a specific analysis, different investigators considered different factors and methods. Based on the existing research achievements, the following integrated microscopic flow models are introduced and used for the analysis of an unsteady state flow theory for wells in the later macroscopic flow models.

1.6.1 Microfractures + steady state adsorption/desorption and diffusion model (model 1)

The following assumptions are used in the model: The shale gas reservoir is composed of only microfractures and matrix. Adsorption and inter-porosity flow within the matrix are negligible. Adsorbed gas only exists on the matrix surface, and the

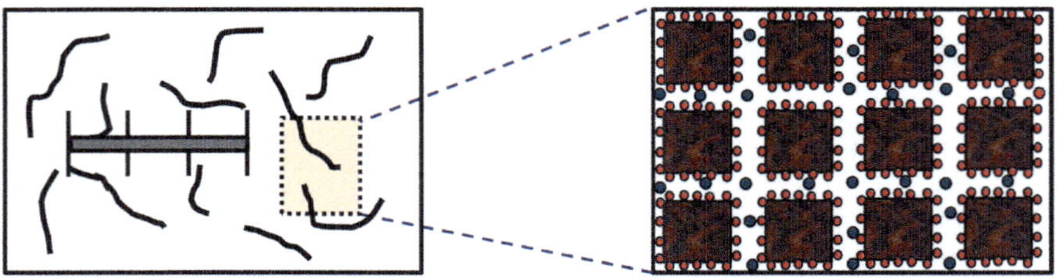

Fig. 1.26 Physical process of pseudo-steady state desorption and diffusion from matrix to fractures.

microfractures are not only the storage space for free gas but also the main passages for desorbed gas flowing into hydraulic fractures or wellbores. The physical model is shown in Fig. 1.26 (Bumb and McKee, 1988; Clarkson et al., 2007; Civan, 2010; Gao et al., 1994; Lane et al., 1989; Wang et al., 2013; Xu et al., 2013).

In the radial coordinates, the continuity equation of shale gas flow in fractures is:

$$\frac{1}{r}\frac{\partial}{\partial r}\left(\frac{k_{\mathrm{f}}}{\mu_{\mathrm{g}}}\rho_{\mathrm{g}}r\frac{\partial p_{\mathrm{f}}}{\partial r}\right)=\frac{\partial(\phi_{\mathrm{f}}\rho_{\mathrm{g}})}{\partial t}+q_{\mathrm{des}} \tag{1.44}$$

where:

p_{f}—pressure of microfracture system (Pa);
ρ_g—gas density at the given conditions (kg/m^3);
ϕ_f—porosity of microfracture system (dimensionless);
k_{f}—permeability of microfracture system (m^2);
q_{des}—gas mass by steady state desorption from unit reservoir volume [kg/(m^3 s)].

Since the corresponding point source functions of various flow models will be derived in the following chapters of this book, the spherical coordinates are used for the following continuity equation, and Eq. (1.44) is rewritten as:

$$\frac{1}{r^2}\frac{\partial}{\partial r}\left(\frac{k_{\mathrm{f}}}{\mu_{\mathrm{g}}}\rho_{\mathrm{g}}r^2\frac{\partial p_{\mathrm{f}}}{\partial r}\right)=\frac{\partial(\phi_{\mathrm{f}}\rho_{\mathrm{g}})}{\partial t}+q_{\mathrm{des}} \tag{1.45}$$

For steady state desorption and diffusion, according to the Langmuir isothermal adsorption equation, Eq. (1.42) becomes:

$$q_{\mathrm{des}}=\rho_{\mathrm{gsc}}(1-\phi_{\mathrm{f}})G_{\mathrm{L}}\frac{p_{\mathrm{L}}}{(p_{\mathrm{L}}+p_{\mathrm{f}})^2}\frac{\partial p_{\mathrm{f}}}{\partial t} \tag{1.46}$$

where ρ_{gc} is the gas density at the standard condition (kg/m^3).

Consequently, the integrated mathematical model considering shale gas pseudo-steady state desorption can be obtained through combining Eqs. (1.45) and (1.46).

This model has been widely applied in the unsteady state flow theory of CBM (coal bed methane) and shale gas reservoirs.

1.6.2 Microfractures + matrix macropores + steady state adsorption/desorption and diffusion model (model 2)

When there are a number of macropores developing in the shale reservoir matrix, ignoring gas contained and flowing in such pores can result in a certain error. Therefore, some scholars proposed the conceptual model of triple porosity. That is, the adsorbed gas at the matrix surface goes into macropores instead of microfractures after desorption and then flows into microfractures from the macropores (as shown in Fig. 1.27) (Song, 2010; Zhao et al., 2013).

In the spherical coordinates, the continuity equation of shale gas flow in fractures is:

$$\frac{1}{r^2}\frac{\partial}{\partial r}\left(\frac{k_{\mathrm{f}}}{\mu_{\mathrm{g}}}\rho_{\mathrm{g}}r^2\frac{\partial p_{\mathrm{f}}}{\partial r}\right)+q_{\mathrm{m}}=\frac{\partial(\phi_{\mathrm{f}}\rho_{\mathrm{g}})}{\partial t} \tag{1.47}$$

where q_{m} is the gas mass flow from the matrix macropores to microfractures in unit reservoir volume [kg/(m^3 s)].

1.6.2.1 Transient interporosity flow model

When gas flow from matrix to fractures is transient interporosity flow, the continuity equation of gas flow in the matrix can be expressed as:

$$\frac{1}{r_{\mathrm{m}}^2}\frac{\partial}{\partial r_{\mathrm{m}}}\left(\frac{k_{\mathrm{m}}}{\mu_{\mathrm{g}}}\rho_{\mathrm{g}}r_{\mathrm{m}}^2\frac{\partial p_{\mathrm{m}}}{\partial r_{\mathrm{m}}}\right)=\frac{\partial(\phi_{\mathrm{m}}\rho_{\mathrm{g}})}{\partial t}+q_{\mathrm{des}} \tag{1.48}$$

where $\emptyset_{\mathrm{m}}$ is the porosity of the matrix (dimensionless). The initialization condition of the matrix system is:

$$p_{\mathrm{m}}(t=0,r_{\mathrm{m}})=p_{\mathrm{i}} \tag{1.49}$$

where p_{i} is the initial reservoir pressure (Pa).

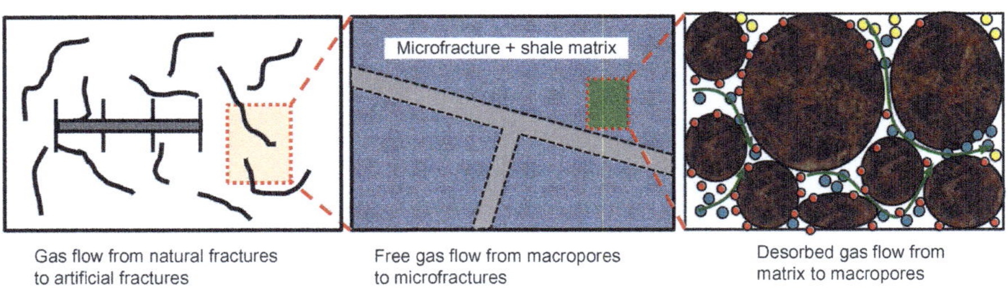

Fig. 1.27 Pseudo-steady state desorption and diffusion of gas flow from matrix to fractures.

Due to the symmetry of gas flow in the matrix macropores, the inner boundary condition is:

$$\frac{\partial p_m}{\partial r_m}(t, r_m = 0) = 0 \tag{1.50}$$

Since the external surfaces of the spherical matrix elements are connected with the fracture system, the matrix external boundary pressure equals the pressure of the fracture system:

$$p_m(t, r_m = R_m) = p_f \tag{1.51}$$

where R_m is the radius of each spherical matrix element (m).

The interporosity flow rate q_m can be expressed as:

$$q_m = -\frac{3\rho_g}{R_m}\frac{k_m}{\mu_g}\frac{\partial p_m}{\partial r_m}\bigg|_{r_m = R_m} \tag{1.52}$$

where ρ_{gf} is the gas density in the fracture conditions (m).

For the gas desorption rate q_{des}, according to the Langmuir isothermal Eq. (1.42), there is:

$$q_{des} = \rho_g(1 - \phi_f - \phi_m)G_L\frac{p_L}{(p_L + p_m)^2}\frac{\partial p_m}{\partial t} \tag{1.53}$$

Combine and solve Eqs. (1.48) and (1.51), and then substitute the calculated macropore pressures into Eq. (1.52) to obtain the interporosity flow rate, which can be used to solve for the pressure of the fracture system through the diffusion equation.

1.6.2.2 Pseudo-steady state flow

When gas flow from macropores to microfractures is pseudo-steady state flow, the interporosity flow in macropores can be represented as:

$$-q_m = \frac{\partial(\phi_m\rho_g)}{\partial t} + q_{des} \tag{1.54}$$

Assuming that the flow from matrix to the fracture system is pseudo-steady state flow, there is:

$$q_m = \frac{\alpha k_m}{\mu_g}(\rho_{gm}p_m - \rho_{gf}p_f) \tag{1.55}$$

where:

α—matrix shape factor, expressed in the next equation (1/m^2);
ρ_{gm}—gas density in the matrix pressure condition (kg/m^3);
ρ_{gf}—gas density in the fracture pressure condition (kg/m^3).

$$\alpha = \frac{4n(n+2)}{L_c} \tag{1.56}$$

where:

n—geometrical dimension of matrix, as shown in Table 1.7 (dimensionless);
L_c—characteristic length of matrix, as shown in Table 1.7 (m).

Substitute Eqs. (1.55) and (1.53) into Eq. (1.54) and combine the resulting equation with Eq. (1.48); then the integrated mathematical model for triple porosity transient interporosity flow for pseudo-steady state diffusion can be obtained.

1.6.3 Microfractures + gas adsorption/desorption + matrix Fick's diffusion model (model 3)

Gas diffusion was assumed to be steady state for the two models introduced above. Now, the Fick diffusion theory is introduced to describe shale gas desorption and diffusion. Compared to the previous microfractures+ steady state diffusion model (model 1), the difference of this model uses Fick diffusion instead of steady state diffusion. The physical desorption and diffusion process is illustrated in Fig. 1.28.

Table 1.7 Shape factors of a typical matrix block

Geometry of matrix	Dimension n	Characteristic length L_c	Geometry factor α
Layer (thickness h)	1	h	$12/h^2$
Sphere (radius r)	3	$2r$	$15/r^2$
Cylinder (radius r)	3	$2r$	$15/r^2$
Cube (length of side a)	3	a	$60/a^2$

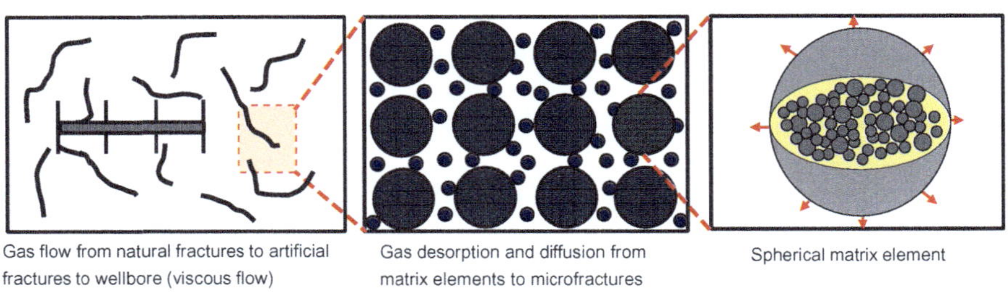

Fig. 1.28 Physical model of matrix desorption and Fick's diffusion.

In the spherical coordinates, the continuity equation of shale gas flow in fractures is:

$$\frac{1}{r^2}\frac{\partial}{\partial r}\left(\frac{k_{\mathrm{f}}}{\mu_{\mathrm{g}}}\rho_{\mathrm{g}}r^2\frac{\partial p_{\mathrm{f}}}{\partial r}\right)=\frac{\partial(\phi_{\mathrm{f}}\rho_{\mathrm{g}})}{\partial t}+q_{\mathrm{F}} \tag{1.57}$$

Gas diffusion from matrix to fractures in unit reservoir volume is:

$$q_{\mathrm{F}}=M_{\mathrm{g}}(1-\phi_{\mathrm{f}})\frac{dC_{\mathrm{m}}}{\mathrm{d}t} \tag{1.58}$$

For pseudo-steady state and unsteady state gas diffusion from matrix to fractures, there is:

$$\frac{\mathrm{d}C_{\mathrm{m}}}{\mathrm{d}t}=\begin{cases}\left.\dfrac{3D_{\mathrm{F}}}{R_{\mathrm{m}}}\dfrac{\partial C_{\mathrm{m}}}{\partial r_{\mathrm{m}}}\right|_{r_{\mathrm{m}}=R_{\mathrm{m}}} & \text{unsteady-state diffusion}\\ \dfrac{6D_{\mathrm{F}}\pi^2}{R_{\mathrm{m}}^2}\left[C_{\mathrm{E}}(p_{\mathrm{f}})-C_{\mathrm{m}}\right] & \text{pseudo-steady-state diffusion}\end{cases} \tag{1.59}$$

where:

D_{F}—Fick diffusion coefficient (m^2/s);
R_{m}—radius of a spherical matrix element (m);
C_{E}—gas molar concentration when the gas adsorption at a matrix surface is in equilibrium with free gas in microfractures (mol/m^3).

1.6.4 Microfractures + matrix macropores + gas adsorption/desorption + Fick's diffusion model in nanopores (model 4)

Similar to the steady state diffusion model, if macropores are well developed in shale matrix, desorbed gas is assumed to flow into macropores and then to microfractures. The physical process is shown in Fig. 1.29.

In the spherical coordinates, the continuity equation of shale gas flow in fractures is:

$$\frac{1}{r^2}\frac{\partial}{\partial r}\left(\frac{k_{\mathrm{f}}}{\mu_{\mathrm{g}}}\rho_{\mathrm{g}}r^2\frac{\partial p_{\mathrm{f}}}{\partial r}\right)+q_{\mathrm{m}}=\frac{\partial(\phi_{\mathrm{f}}\rho_{\mathrm{g}})}{\partial t} \tag{1.60}$$

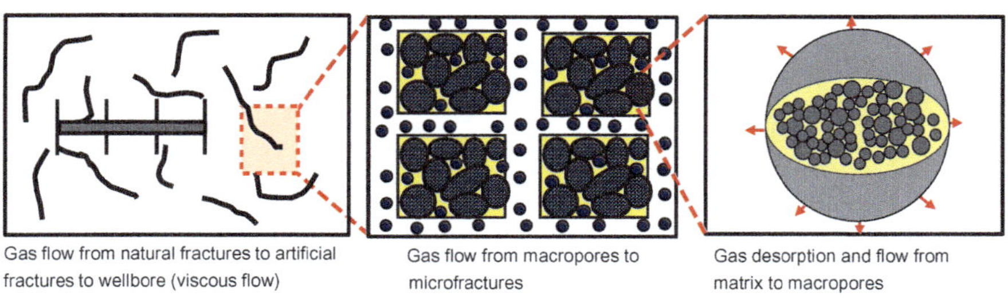

Fig. 1.29 Diagram of matrix desorbed gas flow from macropores to microfractures.

(1) For transient interporosity flow from matrix to fractures, the continuity equation for flow in matrix is:

$$\frac{1}{r_\mathrm{m}^2}\frac{\partial}{\partial r_\mathrm{m}}\left(\frac{k_\mathrm{m}}{\mu_\mathrm{g}}\rho_\mathrm{g}r_\mathrm{m}^2\frac{\partial p_\mathrm{m}}{\partial r_\mathrm{m}}\right)=\frac{\partial(\phi_\mathrm{m}\rho_\mathrm{g})}{\partial t}+q_\mathrm{F} \tag{1.61}$$

For unsteady state flow from macropores to fractures, the flow rate q_m is expressed as:

$$q_\mathrm{m}=-\frac{3\rho_\mathrm{g}}{R_\mathrm{mac}}\frac{k_\mathrm{m}}{\mu_\mathrm{g}}\frac{\partial p_\mathrm{m}}{\partial r_\mathrm{m}}\bigg|_{r_\mathrm{m}=R_\mathrm{m}} \tag{1.62}$$

where R_mac is the radius of a spherical matrix element (m).

The initial, inner boundary and outer boundary conditions are the same as in Eqs. (1.49)–(1.51) for the macropore matrix under transient interporosity flow. In addition, the flow rate has the same format as in Eq. (1.52), and q_F is the same as in Eq. (1.58).

(2) For pseudo-steady state flow from macropores to microfractures, the continuity equation for flow in macropores is:

$$-q_\mathrm{m}=\frac{\partial(\phi_\mathrm{m}\rho_\mathrm{g})}{\partial t}+q_\mathrm{F} \tag{1.63}$$

The flow rate is expressed by:

$$q_\mathrm{m}=\frac{\alpha k_\mathrm{m}}{\mu_\mathrm{g}}(\rho_\mathrm{gm}p_\mathrm{m}-\rho_\mathrm{gf}p_\mathrm{f}) \tag{1.64}$$

The model of matrix desorption and diffusion is the same as that described in Section 1.5.3, and the model of flow from macropores to microfractures is the same as that in Section 1.5.2. Now, q_F is expressed by:

$$q_\mathrm{F}=M_\mathrm{g}(1-\phi_\mathrm{f}-\phi_\mathrm{m})\frac{\mathrm{d}C_\mathrm{m}}{\mathrm{d}t} \tag{1.65}$$

1.6.5 Microfractures + gas adsorption/desorption + Knudsen diffusion model in nanopores (model 5)

In recent years, with studies on reservoir structure and flow mechanisms of shale gas reservoirs getting deeper and deeper, it is observed that a Knudsen diffusion model can be used to represent gas flow in pores in nanometer magnitude. Most researchers thought that gas flow in shale nanopores is a kind of complex flow under combined influence of multiple mechanisms, such as a slippage effect, Darcy flow, Knudsen diffusion and adsorption–desorption (Civan, 2010; Darabi et al., 2012; Javadpour, 2009; Swami, 2012; Swami et al., 2013). According to their research methods, the physical model of microscopic gas flow in shale reservoirs is shown in Fig. 1.30.

Since gas flow in a microfracture system satisfies Darcy's law, the continuity equation is similar to that in the previous models and can be expressed in the spherical coordinates by:

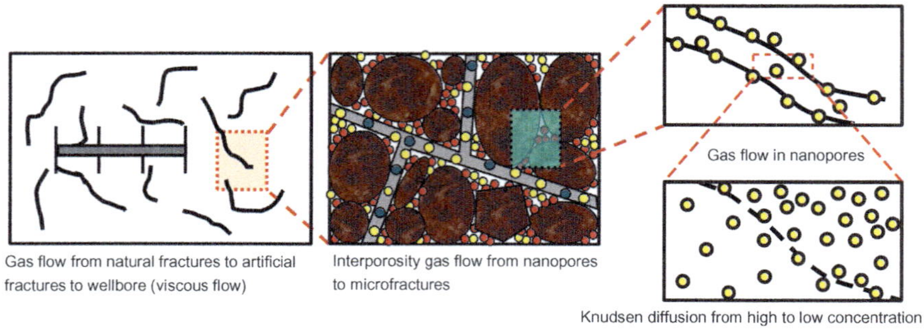

Fig. 1.30 Physical model of gas Knudsen diffusion in nanopores.

$$\frac{1}{r^2}\frac{\partial}{\partial r}\left(\frac{k_f}{\mu_g}\rho_g r^2 \frac{\partial p_f}{\partial r}\right) + q_m = \frac{\partial(\phi_f \rho_g)}{\partial t} \tag{1.66}$$

Since both are Darcy's flow controlled by a pressure difference and Knudsen diffusion caused by a concentration difference in matrix nanometer pores. According to Section 1.5.3, the gas mass flux induced by the pressure difference in nanopores is:

$$J_{mp} = -\frac{\rho_{gm} k_m}{\mu}\frac{\partial p_m}{\partial r_m} \tag{1.67}$$

According to Eq. (1.38), the mass flux caused by a concentration difference is:

$$J_K = -\frac{D_k p_m M_g C_{gm}}{ZRT}\frac{\partial p_m}{\partial r_m} \tag{1.68}$$

Then the total mass flux by the co-effect of pressure and concentration differences is:

$$J = J_{mp} + J_K = -\frac{\rho_g}{\mu}\left(k_m + D_k \mu C_{gm}\right)\frac{\partial p_m}{\partial r_m} \tag{1.69}$$

In order to simplify the analysis, based on the format of Eq. (1.69) and Darcy's equation, an apparent permeability k_{app} is introduced to represent a matrix permeability change affected by both pressure and Knudsen diffusion. It is expressed by:

$$k_{app} = k_m + D_k \mu C_{gm} \tag{1.70}$$

(1) For transient interporosity flow from matrix to fractures, the continuity equation of flow in matrix is:

$$\frac{1}{r_m^2}\frac{\partial}{\partial r_m}\left(\frac{k_{app}}{\mu_g}\rho_g r_m^2 \frac{\partial p_m}{\partial r_m}\right) = \frac{\partial(\phi_m \rho_g)}{\partial t} + q_{des} \tag{1.71}$$

In addition, the interporosity flow rate q_{m} is:

$$q_{\mathrm{m}} = -\frac{3\rho_{\mathrm{g}}}{R_{\mathrm{m}}}\frac{k_{\mathrm{app}}}{\mu_{\mathrm{g}}}\frac{\partial p_{\mathrm{m}}}{\partial r_{\mathrm{m}}}\bigg|_{r_{\mathrm{m}}=R_{\mathrm{m}}} \tag{1.72}$$

For the unsteady state interporosity flow model, the definite solution condition for a spherical matrix element is the same as in the previous transient interporosity model. Moreover, the desorbed gas item in Eq. (1.71) has the same format as that in the previous desorption model:

$$q_{\mathrm{des}} = \rho_{\mathrm{gsc}}(1-\phi_{\mathrm{f}})G_{\mathrm{L}}\frac{p_{\mathrm{L}}}{(p_{\mathrm{L}}+p_{\mathrm{m}})^2}\frac{\partial p_{\mathrm{m}}}{\partial t} \tag{1.73}$$

(2) For the pseudo-steady state flow from macropores to microfractures, the continuity equation of flow in macropores is:

$$-q_{\mathrm{m}} = \frac{\partial(\phi_{\mathrm{m}}\rho_{\mathrm{g}})}{\partial t} + q_{\mathrm{des}} \tag{1.74}$$

Furthermore, for the pseudo-steady state interporosity flow model, the equation of gas flow from nanopores to microfractures is:

$$q_{\mathrm{m}} = \frac{\alpha k_{\mathrm{app}}}{\mu}\left(\rho_{\mathrm{gm}}p_{\mathrm{m}} - \rho_{\mathrm{gf}}p_{\mathrm{f}}\right) \tag{1.75}$$

Through the above analysis, the integrated mathematical model in a shale gas reservoir for gas flow in nanopores under the co-effect of pressure and concentration difference is established.

Source function derivation for gas reservoirs under different flow mechanisms

2

Chapter Outline

2.1 Introduction 49
2.2 Solutions of flow mechanism models 50
2.2.1 Microfractures + steady state adsorption/desorption and diffusion model (model 1) 50
2.2.2 Microfractures + matrix macropores + steady state adsorption/desorption and diffusion model (model 2) 51
2.2.3 Microfractures + gas adsorption/desorption + matrix Fick's diffusion model (model 3) 51
2.2.4 Microfractures + matrix macropores + gas adsorption/desorption + Fick's diffusion model in nanopores (model 4) 52
2.2.5 Microfractures + gas adsorption/desorption + Knudsen diffusion model in nanopores (model 5) 54
2.3 Continuous point source solutions in circular gas reservoirs 54
2.3.1 Mathematical models 55
2.3.2 Model solutions 58
2.3.3 Point source solutions in various outer boundary conditions 60
2.4 Continuous point source solutions in rectangular gas reservoirs 61
2.4.1 Derivation of a continuous point source solution 62
2.4.2 Computational remarks 69

2.1 Introduction

In the previous chapter, multi-scale microscopic flow models in shale gas reservoirs considering various complex flow mechanisms have been derived. These models are the basis for the study of the transient flow theory for fractured wells in the shale reservoirs. According to the research on the transient flow theory for fractured wells (Aguilera, 1995; Zhang et al., 2004), two popular methods for this theory are frequently used: a source function and an orthogonal transformation. The latter one is often used for linear flow, and the former is the main method to consider for interference among fractures or interference between wellbores and fractures. In this chapter, the microscopic flow models established in the previous chapter are first solved based on the basic oil and gas diffusivity equations. Then the continuous point source

Developments in Petroleum Science, Vol. 66. https://doi.org/10.1016/B978-0-444-64315-5.00002-4

solutions are derived based on the source function theory, together with applications of mathematic methods including the Delta function, Laplace transformation, orthogonal transformation, and superposition principle (Carslaw and Jaeger, 1959).

For consistency, the isothermal adsorption curves and variables in this chapter are in SI units. In addition, the descriptions of most variables are the same as those in the previous chapter.

2.2 Solutions of flow mechanism models

During the development of shale gas reservoirs, microfractures are the main paths for gas transporting from a reservoir location to natural fractures and wellbores. Therefore, the flow models for these microfractures are key to solve for a point source function. The models for different flow mechanisms in a microfracture system have different parameter groups. In this chapter, pseudo differential pressure equations in the different mechanism models in spherical coordinates in a Laplace space for the fracture system are established. The detailed derivation and solutions of these mechanism models are given in Appendix A.

The general flow equation in a microfracture system under different flow mechanisms is:

$$\frac{1}{r_D^2}\frac{\partial}{\partial r_D}\left(r_D^2\frac{\partial \Delta\overline{m}_f}{\partial r_D}\right)=f(s)\Delta\overline{m}_f \tag{2.1}$$

where:

$f(s)$—the parameter group for different flow mechanism models, dimensionless;
s—Laplace variable, dimensionless.

Expressions for $f(s)$ and s in different flow mechanism models are shown in the next sections.

2.2.1 Microfractures + steady state adsorption/desorption and diffusion model (model 1)

For flow mechanisms in microfractures and steady state adsorption/desorption, the model can be easily solved by introducing an additional compressibility coefficient for adsorption/desorption. The parameter group $f(s)$ is expressed by:

$$f(s)=(1+\omega)s \tag{2.2}$$

and the variables in this model are:

$$c_d=\frac{2Tp_{sc}}{\phi_f\mu_{gi}T_{sc}}\frac{(1-\phi_f)G_L m(p_L)}{[m(p_L)+m(p_f)]^2},\quad \omega=\frac{c_d}{c_{gi}},\quad r_D=\frac{r}{L_{ref}},\quad t_D=\frac{k_f t}{\phi_f\mu_{gi}c_{fgi}L_{ref}^2}$$

2.2.2 Microfractures + matrix macropores + steady state adsorption/desorption and diffusion model (model 2)

With consideration of macroscopic pores in the matrix as a medium for gas to flow through, the model turns into a tri-porosity model. Because there are both transient and pseudo-steady state flow of gas from matrix pores to the microfracture system, the following variables are defined for the derivation convenience:

$$r_D = \frac{r}{L_{ref}},\ t_D = \frac{k_f t}{\left(\phi_m c_{mgi} + \phi_f c_{fgi}\right)\mu_{gi} L_{ref}^2},\ \omega_f = \frac{\phi_f c_{fgi}}{\phi_m c_{mgi} + \phi_f c_{fgi}},\ \omega_d = \frac{\phi_m c_d}{\phi_m c_{mgi} + \phi_f c_{fgi}},\ \lambda = \alpha \frac{k_m}{k_f} L_{ref}^2,$$

$$\Delta m_f = m(p_i) - m(p_f),\quad \Delta m_m = m(p_i) - m(p_m),\quad c_d = \frac{2Tp_{sc}}{\phi_m \mu_{gi} T_{sc}} \frac{(1-\phi_f-\phi_m) G_L m(p_L)}{[m(p_L)+m(p_m)]^2}$$

(1) For transient flow, $f(s)$ is:

$$f(s) = \omega_f s + \frac{\lambda}{5}\left[\sqrt{\frac{15(1-\omega_f+\omega_d)s}{\lambda}} \coth\sqrt{\frac{15(1-\omega_f+\omega_d)s}{\lambda}} - 1\right] \tag{2.3}$$

where $r_{mD} = \frac{r_m}{R_m}$, $\alpha = \frac{15}{R_m^2}$, and $g = \frac{15(1-\omega_f+\omega_d)s}{\lambda}$.

(2) For pseudo-steady state flow, $f(s)$ is:

$$f(s) = \frac{\lambda(1+\omega_d) + \omega_f(1-\omega_f+\omega_d)s}{\lambda + (1-\omega_f+\omega_d)s} s \tag{2.4}$$

2.2.3 Microfractures + gas adsorption/desorption + matrix Fick's diffusion model (model 3)

When the diffusion of the adsorbed gas from matrix to microfractures satisfies Fick's diffusion law, the model is similar to that used for transient flow in CBM (cold bed methane). The variables are defined:

$$r_{mD} = \frac{r_m}{R_m},\quad t_D = \frac{k_f t}{N L_{ref}^2},\quad \omega = \frac{\phi_f \mu_{gi} c_{fgi}}{\Lambda},\quad \Lambda = \begin{cases} \phi_f \mu_{gi} c_{fgi} + \dfrac{6k_f h}{q_{sc}} & \text{unsteady state diffusion} \\ \phi_f \mu_{gi} c_{fgi} + \dfrac{2k_f h}{q_{sc}} & \text{pseudo-steady state diffusion} \end{cases},$$

$$\lambda = \begin{cases} \dfrac{k_f \tau}{\Lambda L_{ref}^2} & \text{unsteady state diffusion} \\ \dfrac{k_f \tau}{6\Lambda L_{ref}^2} & \text{pseudo-steady state diffusion} \end{cases},\quad \tau = \begin{cases} \dfrac{R_m^2}{D_F} & \text{unsteady state diffusion} \\ \dfrac{R_m^2}{\pi^2 D_F} & \text{pseudo-steady state diffusion} \end{cases}$$

The dimensionless adsorption concentration and coefficient are defined as:

$$C_{mD} = C_m(p_m) - C_m(p_i), \quad C_{ED} = C_E(p_f) - C_m(p_i) \tag{2.5}$$

$$\sigma = \frac{G_L m(p_L)}{[m(p_L) + m(p_f)][m(p_L) + m(p_i)]} \frac{q_{sc} p_{sc} T}{k_f h T_{sc}} \tag{2.6}$$

(1) For unsteady state diffusion, $f(s)$ is expressed as:

$$f(s) = \omega s + \frac{(1-\omega)(1-\phi_f)\sigma}{\lambda}\left[\sqrt{\lambda s}\coth\left(\sqrt{\lambda s}\right) - 1\right] \tag{2.7}$$

(2) For pseudo-steady state diffusion, $f(s)$ is expressed as:

$$f(s) = \omega s + \frac{\sigma(1-\omega)(1-\phi_f)s}{\lambda s + 1} \tag{2.8}$$

2.2.4 Microfractures + matrix macropores + gas adsorption/desorption + Fick's diffusion model in nanopores (model 4)

When considering the effect of matrix macropores on gas flow, desorbed gas diffuses into macropores through matrix surfaces following Fick's diffusion law and then flows into microfractures under a pressure gradient. Because diffusion is in an unsteady or steady state and the flow from macropores to microfractures is in transient or pseudo-steady state, there are four combinations for such flow models, which are unsteady state diffusion + transient interporosity flow, unsteady state diffusion + pseudo-steady state interporosity flow, steady state diffusion + transient interporosity flow and steady state diffusion + pseudo-steady state interporosity flow. Their flow models are shown below.

(1) For steady state diffusion + transient interporosity flow, the variables are defined by:

$$\lambda_{mf} = \alpha\frac{k_m}{k_f}L_{ref}^2, \ r_D = \frac{r}{L_{ref}}, \ r_{mD} = \frac{r_m}{R_{mac}}, \ t_D = \frac{k_f t}{\Lambda L_{ref}^2}, \ \omega_f = \frac{\phi_f \mu_{gi} c_{fgi}}{\Lambda}, \ \omega_m = \frac{\phi_m \mu_{gi} c_{mgi}}{\Lambda},$$

$$\Lambda = \phi_m \mu_{gi} c_{mgi} + \phi_f \mu_{gi} c_{fgi} + \frac{2k_f h R_{mac}^2}{q_{sc}\ L_{ref}^2}, \ \lambda = \frac{k_m \tau}{6\Lambda L_{ref}^2}, \ \tau = \frac{R_m^2}{\pi^2 D_F}, \ \theta_{mf} = \frac{k_f}{k_m}$$

The adsorption/desorption coefficient is expressed as:

$$\sigma = \frac{G_L m(p_L)}{[m(p_L) + m(p_f)][m(p_L) + m(p_i)]} \frac{q_{sc} p_{sc} T}{k_f h T_{sc}} \tag{2.9}$$

For this model, $f(s)$ is:

$$f(s) = \omega_f s + \frac{\lambda_{mf}}{5}\left[\sqrt{g(s)}\coth\left(\sqrt{g(s)}\right) - 1\right] \tag{2.10}$$

where $g(s) = \left(\frac{15\omega_m}{\lambda_{mf}} + \frac{\beta\sigma\theta_{mf}}{\lambda\theta_{mf}s + 1}\right)s$ and $\beta = (1-\omega_f-\omega_m)(1-\phi_f-\phi_m)$.

(2) For unsteady state diffusion + transient interporosity flow, the variables are defines by:

$$\lambda_{\mathrm{mf}}=\alpha\frac{k_{\mathrm{m}}}{k_{\mathrm{f}}}L_{\mathrm{ref}}^{2},\ r_{D}=\frac{r}{L_{\mathrm{ref}}},\ r_{\mathrm{mD}}=\frac{r_{\mathrm{m}}}{R_{\mathrm{mac}}},\ t_{\mathrm{D}}=\frac{k_{\mathrm{f}}t}{\Lambda L_{\mathrm{ref}}^{2}},\ \omega_{\mathrm{f}}=\frac{\phi_{\mathrm{f}}\mu_{\mathrm{gi}}c_{\mathrm{fgi}}}{\Lambda},\ \omega_{\mathrm{m}}=\frac{\phi_{\mathrm{m}}\mu_{\mathrm{gi}}c_{\mathrm{mgi}}}{\Lambda},$$
$$\Lambda=\phi_{\mathrm{m}}\mu_{\mathrm{gi}}c_{\mathrm{mgi}}+\phi_{\mathrm{f}}\mu_{\mathrm{gi}}c_{\mathrm{fgi}}+\frac{6k_{\mathrm{f}}hR_{\mathrm{mac}}^{2}}{q_{\mathrm{sc}}\ L_{\mathrm{ref}}^{2}},\ \lambda=\frac{k_{\mathrm{m}}\tau}{\Lambda L_{\mathrm{ref}}^{2}},\ \tau=\frac{R_{\mathrm{m}}^{2}}{D_{\mathrm{F}}},\ \theta_{\mathrm{mf}}=\frac{k_{\mathrm{f}}}{k_{\mathrm{m}}}$$

The expression of $f(s)$ is:

$$f(s)=\omega_{\mathrm{f}}s+\frac{\lambda_{\mathrm{mf}}}{5}\left[\sqrt{g(s)}\coth\left(\sqrt{g(s)}\right)-1\right] \tag{2.11}$$

where $g(s)=\left(\frac{15\omega_{\mathrm{m}}}{\lambda_{\mathrm{mf}}}+\frac{\beta}{\lambda}\sigma\left[\sqrt{\lambda\theta_{\mathrm{mf}}s}\coth\left(\sqrt{\lambda\theta_{\mathrm{mf}}s}\right)-1\right]\right)s$ and $\beta=(1-\omega_{\mathrm{f}}-\omega_{\mathrm{m}})(1-\phi_{\mathrm{f}}-\phi_{\mathrm{m}})$.

(3) For steady state diffusion + pseudo-steady state interporosity flow, the variables are defined by:

$$\lambda_{\mathrm{mf}}=\alpha\frac{k_{\mathrm{m}}}{k_{\mathrm{f}}}L_{\mathrm{ref}}^{2},\ r_{D}=\frac{r}{L_{\mathrm{ref}}},\ t_{\mathrm{D}}=\frac{k_{\mathrm{f}}t}{\Lambda L_{\mathrm{ref}}^{2}},\ \omega_{\mathrm{f}}=\frac{\phi_{\mathrm{f}}\mu_{\mathrm{gi}}c_{\mathrm{fgi}}}{\Lambda},\ \omega_{\mathrm{m}}=\frac{\phi_{\mathrm{m}}\mu_{\mathrm{gi}}c_{\mathrm{mgi}}}{\Lambda},$$
$$\Lambda=\phi_{\mathrm{m}}\mu_{\mathrm{gi}}c_{\mathrm{mgi}}+\phi_{\mathrm{f}}\mu_{\mathrm{gi}}c_{\mathrm{fgi}}+\frac{2k_{\mathrm{f}}h}{q_{\mathrm{sc}}},\ \lambda=\frac{k_{\mathrm{f}}\tau}{6\Lambda L_{\mathrm{ref}}^{2}},\ \tau=\frac{R_{\mathrm{m}}^{2}}{\pi^{2}D_{\mathrm{F}}}$$

Then $f(s)$ is:

$$f(s)=\omega_{\mathrm{f}}s+\frac{\lambda_{\mathrm{mf}}g(s)}{g(s)+\lambda_{\mathrm{mf}}} \tag{2.12}$$

where $g(s)=\omega_{\mathrm{m}}s+\frac{\sigma\beta s}{\lambda s+1}$ and $\beta=(1-\omega_{\mathrm{f}}-\omega_{\mathrm{m}})(1-\phi_{\mathrm{f}}-\phi_{\mathrm{m}})$.

(4) For unsteady state diffusion + pseudo-steady state interporosity flow, the variables are defined by:

$$\lambda_{\mathrm{mf}}=\alpha\frac{k_{\mathrm{m}}}{k_{\mathrm{f}}}L_{\mathrm{ref}}^{2},\ r_{D}=\frac{r}{L_{\mathrm{ref}}},\ t_{\mathrm{D}}=\frac{k_{\mathrm{f}}t}{\Lambda L_{\mathrm{ref}}^{2}},\ \omega_{\mathrm{f}}=\frac{\phi_{\mathrm{f}}\mu_{\mathrm{gi}}c_{\mathrm{fgi}}}{\Lambda},\ \omega_{\mathrm{m}}=\frac{\phi_{\mathrm{m}}\mu_{\mathrm{gi}}c_{\mathrm{mgi}}}{\Lambda},$$
$$\Lambda=\phi_{\mathrm{m}}\mu_{\mathrm{gi}}c_{\mathrm{mgi}}+\phi_{\mathrm{f}}\mu_{\mathrm{gi}}c_{\mathrm{fgi}}+\frac{6k_{\mathrm{f}}h}{q_{\mathrm{sc}}},\ \lambda=\frac{k_{\mathrm{f}}\tau}{\Lambda L_{\mathrm{ref}}^{2}},\ \tau=\frac{R_{\mathrm{m}}^{2}}{D_{\mathrm{F}}}$$

Then $f(s)$ is:

$$f(s)=\omega_{\mathrm{f}}s+\frac{\lambda_{\mathrm{mf}}g(s)}{g(s)+\lambda_{\mathrm{mf}}} \tag{2.13}$$

where $g(s)=\omega_{\mathrm{m}}s+\frac{\sigma\beta}{\lambda}\left[\sqrt{\lambda s}\coth\left(\sqrt{\lambda s}\right)-1\right]$ and $\beta=(1-\omega_{\mathrm{f}}-\omega_{\mathrm{m}})(1-\phi_{\mathrm{f}}-\phi_{\mathrm{m}})$.

2.2.5 Microfractures + gas adsorption/desorption + Knudsen diffusion model in nanopores (model 5)

When considering both Darcy's flow and the Knudsen diffusion of gas in nanopores, an apparent permeability is introduced to represent matrix permeability. Because both transient and pseudo-steady state interporosity flow exist for gas flow from nanopores to microfractures, $f(s)$ for microfracture flow equations is introduced for these two flow statuses.

(1) For transient interporostiy flow, the compressibility coefficient for adsorption/desorption is defined by:

$$c_{\mathrm{d}}=\frac{2Tp_{\mathrm{sc}}}{\phi_{\mathrm{m}}\mu_{\mathrm{gi}}T_{\mathrm{sc}}}\frac{(1-\phi_{\mathrm{f}}-\phi_{\mathrm{m}})G_{\mathrm{L}}m(p_{\mathrm{L}})}{[m(p_{\mathrm{L}})+m(p_{\mathrm{m}})]^{2}} \tag{2.14}$$

In addition, the variables are defined by:

$$r_{\mathrm{mD}}=\frac{r_{\mathrm{m}}}{R_{\mathrm{m}}},\ r_{\mathrm{D}}=\frac{r}{L_{\mathrm{ref}}},\ t_{\mathrm{D}}=\frac{k_{\mathrm{f}}t}{\left(\phi_{\mathrm{m}}c_{\mathrm{mgi}}+\phi_{\mathrm{f}}c_{\mathrm{fgi}}\right)\mu_{\mathrm{gi}}L_{\mathrm{ref}}^{2}},\ \omega_{\mathrm{f}}=\frac{\phi_{\mathrm{f}}c_{\mathrm{fgi}}}{\phi_{\mathrm{m}}c_{\mathrm{mgi}}+\phi_{\mathrm{f}}c_{\mathrm{fgi}}},\ \omega_{\mathrm{d}}=\frac{\phi_{\mathrm{m}}c_{\mathrm{d}}}{\phi_{\mathrm{m}}c_{\mathrm{mgi}}+\phi_{\mathrm{f}}c_{\mathrm{fgi}}},$$

$$\lambda=\alpha\frac{k_{\mathrm{m}}}{k_{\mathrm{f}}}L_{\mathrm{ref}}^{2},\ \theta_{\mathrm{ka\text{-}m}}=\frac{k_{\mathrm{app}}}{k_{\mathrm{m}}}$$

According to the derivation in Appendix A, $f(s)$ is expressed as:

$$f(s)=\omega_{\mathrm{f}}s+\frac{\lambda\theta_{\mathrm{ka\text{-}m}}}{5}\left[\sqrt{g}\coth\left(\sqrt{g}\right)-1\right] \tag{2.15}$$

where $g=\frac{15(1-\omega_{\mathrm{f}}+\omega_{\mathrm{d}})s}{\lambda\theta_{\mathrm{ka\text{-}m}}}$.

(2) For pseudo-steady state interporosity flow, the dimensionless variables are defined by:

$$r_{\mathrm{D}}=\frac{r}{L_{\mathrm{ref}}},\ t_{\mathrm{D}}=\frac{k_{\mathrm{f}}t}{\left(\phi_{\mathrm{m}}c_{\mathrm{mgi}}+\phi_{\mathrm{f}}c_{\mathrm{fgi}}\right)\mu_{\mathrm{gi}}L_{\mathrm{ref}}^{2}},\ \omega_{\mathrm{f}}=\frac{\phi_{\mathrm{f}}c_{\mathrm{fgi}}}{\phi_{\mathrm{m}}c_{\mathrm{mgi}}+\phi_{\mathrm{f}}c_{\mathrm{fgi}}},\ \omega_{\mathrm{d}}=\frac{\phi_{\mathrm{m}}c_{\mathrm{d}}}{\phi_{\mathrm{m}}c_{\mathrm{mgi}}+\phi_{\mathrm{f}}c_{\mathrm{fgi}}},\ \lambda=\alpha\frac{k_{\mathrm{m}}}{k_{\mathrm{f}}}L_{\mathrm{ref}}^{2},$$

$$\theta_{\mathrm{ka\text{-}m}}=\frac{k_{\mathrm{app}}}{k_{\mathrm{m}}},\ c_{\mathrm{d}}=\frac{2Tp_{\mathrm{sc}}}{\phi_{\mathrm{m}}\mu_{\mathrm{gi}}T_{\mathrm{sc}}}\frac{(1-\phi_{\mathrm{f}}-\phi_{\mathrm{m}})G_{\mathrm{L}}m(p_{\mathrm{L}})}{[m(p_{\mathrm{L}})+m(p_{\mathrm{m}})]^{2}}$$

Therefore, $f(s)$ is expressed as:

$$f(s)=\frac{\lambda\theta_{\mathrm{ka\text{-}m}}(1+\omega_{\mathrm{d}})+\omega_{\mathrm{f}}(1-\omega_{\mathrm{f}}+\omega_{\mathrm{d}})s}{\lambda\theta_{\mathrm{ka-m}}+(1-\omega_{\mathrm{f}}+\omega_{\mathrm{d}})s}s \tag{2.16}$$

2.3 Continuous point source solutions in circular gas reservoirs

In the previous section, the dimensionless pressure expressions in microfractures under different flow mechanisms are derived and given in spherical coordinates for shale gas reservoirs. In the following sections, these mechanism models are used to derive continuous point source solutions for gas reservoirs and shale gas reservoirs with a circular boundary.

2.3.1 Mathematical models

In this part, a continuous point source solution for any point in an anisotropic circular gas reservoir with their sealed upper and lower boundaries is derived. The physical model is shown in Fig. 2.1. The assumptions for this model are: (a) the gas reservoir is horizontally homogeneous, with thickness h and initial pressure p_i; (b) the horizontal and vertical permeabilities of the gas reservoir are k_{fh} and k_{fz}; (c) an infinitely small cylindrical source–sink exists at z_w from the lower boundary, and the radius and height of the infinitesimal cylinder are r and ε, respectively; (d) the intensity of the cylindrical source-sink is q_{scin}; (e) gas flow obeys Darcy's law, and the capillary pressure and gravity are ignorable.

According to the basic mass conservation theory, the continuity equation for gas flow is:

$$\frac{1}{r}\frac{\partial(r\rho_g v_r)}{\partial r}+\frac{\partial(\rho_g v_z)}{\partial z}=-\frac{\partial(\rho_g \phi_f)}{\partial t} \tag{2.17}$$

where:

ρ_g—gas density in the fracture system (kg/m^3);
v_r—gas radial flow velocity (m/s);
v_z—gas vertical flow velocity (m/s);
ϕ_f—dimensionless porosity of the fracture system;
r/z—radial and vertical coordinates (m);
t—production time (s) .

Based on Darcy's law, the equations of motion in different directions are:

$$v_r=-\frac{k_{fh}}{\mu_g}\frac{\partial p_f}{\partial r},\quad v_z=-\frac{k_{fz}}{\mu_g}\frac{\partial p_f}{\partial r} \tag{2.18}$$

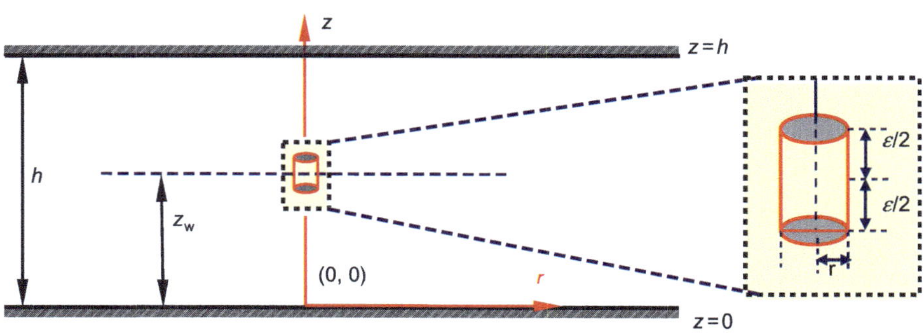

Fig. 2.1 A continuous point source in a gas reservoir with sealed upper and lower boundaries.

where:

k_{fh}—horizontal permeability of the fracture system (m^2);
k_{fz}—vertical permeability of the fracture system (m^2);
p_{f}—fracture system pressure (Pa);
μ_g—gas viscosity (Pa s).

According to the real gas equation of state, the gas density is:

$$\rho_{\mathrm{g}} = \frac{p_{\mathrm{f}} M_{\mathrm{g}}}{ZRT} \tag{2.19}$$

where:

M_{g}—gas molar mass (kg/mol);
T—gas absolute temperature (K);
Z—gas deviation factor (dimensionless), for ideal gas, $\mathrm{Z}=1$;
R—gas constant [8.314 J/(mol K)].

Based on the definition of rock compressibility, the relationship between the fracture porosity and the fracture system pressure is:

$$c_{\phi_{\mathrm{f}}} = \frac{1}{\phi_{\mathrm{f}}} \frac{\partial \phi_{\mathrm{f}}}{\partial p_{\mathrm{f}}} \tag{2.20}$$

where $c_{\phi_{\mathrm{f}}}$ is the compressibility of fracture pores. Since the gas compressibility is far higher than the formation compressibility, the rock porosity of a gas reservoir is usually treated as a constant (Pa^{-1}).

Because it is assumed that gas flows into the cylinder from surfaces and the cylinder unit is infinitely small, it can be treated as a continuous point sink and its inner boundary expression is:

$$\lim_{\varepsilon \to 0} \left[\lim_{r \to 0} \int_{z_{\mathrm{w}}-\varepsilon/2}^{z_{\mathrm{w}}+\varepsilon/2} \frac{2\pi r}{B_{\mathrm{g}}} \frac{k_{\mathrm{fh}}}{\mu_{\mathrm{g}}} \frac{\partial p_{\mathrm{f}}}{\partial r} \mathrm{d}z_{\mathrm{w}} \right] = \begin{cases} 0, & |z - z_{\mathrm{w}}| > \varepsilon/2 \\ q_{\mathrm{scins}}, & |z - z_{\mathrm{w}}| < \varepsilon/2 \end{cases} \tag{2.21}$$

where the gas volume factor B_{g} is:

$$B_{\mathrm{g}} = \frac{p_{\mathrm{sc}} ZT}{p_{\mathrm{f}} T_{\mathrm{sc}}} \tag{2.22}$$

For the sealed upper and lower boundaries:

$$\left.\frac{\partial p_{\mathrm{f}}}{\partial z}\right|_{z=0} = 0, \quad \left.\frac{\partial p_{\mathrm{f}}}{\partial z}\right|_{z=h} = 0 \tag{2.23}$$

For the outer boundary:

$$\left. r \frac{\partial p_{\mathrm{f}}}{\partial r}\right|_{r=r_{\mathrm{e}}} = \vartheta p_{\mathrm{f}}|_{r=r_{\mathrm{e}}} \tag{2.24}$$

The pseudo pressure is introduced and expressed as follows:

$$m(p_f) = \int_{p_o}^{p_f} \frac{2p}{Z\mu_g} dp \tag{2.25}$$

Introducing the gas density, volume factor, and pseudo pressure into the fracture flow equation, we have:

$$\frac{1}{r}\frac{\partial}{\partial r}\left(r\frac{\partial m(p_f)}{\partial r}\right) + \frac{k_{fz}}{k_{fh}}\frac{\partial^2 m(p_f)}{\partial z^2} = \frac{\phi_f \mu_g c_{fg}}{k_{fh}}\frac{\partial m(p_f)}{\partial t} \tag{2.26}$$

$$\lim_{\varepsilon \to 0}\left[\lim_{r \to 0}\int_{z_w - \varepsilon/2}^{z_w + \varepsilon/2} r\frac{\partial m(p_f)}{\partial r} dz_w\right] = \begin{cases} 0, & |z - z_w| > \varepsilon/2 \\ \dfrac{p_{sc}T}{\pi k_{fh} T_{sc} h} q_{scins}, & |z - z_w| < \varepsilon/2 \end{cases} \tag{2.27}$$

In the above continuity equations, the gas viscosity μ_g and compressibility c_{fg} are functions of pressure. To obtain analytical solutions for these equations, both viscosity and compressibility are assumed as their initial values, and then $\mu_g = \mu_{gi}$ and $c_{fg} = c_{fgi}$.

Define the following dimensionless variables:

$$r_D = \frac{r}{L_{ref}},\ r_{eD} = \frac{r_e}{L_{ref}},\ z_D = \frac{z}{h},\ h_D = \frac{h}{L_{ref}}\sqrt{\frac{k_{fh}}{k_{fz}}},\ t_D = \frac{k_{fh} t}{\phi_f \mu_{gi} c_{fgi} L_{ref}^2}$$

Introducing these variables into Eqs. (2.26) and (2.27), then the fracture flow equation becomes:

$$\frac{1}{r_D}\frac{\partial}{\partial r_D}\left(r_D\frac{\partial \Delta m_f}{\partial r_D}\right) + \frac{1}{h_D^2}\frac{\partial^2 \Delta m_f}{\partial z_D^2} = \frac{\partial \Delta m_f}{\partial t_D} \tag{2.28}$$

The inner boundary condition becomes:

$$\lim_{\varepsilon \to 0}\left[\lim_{r_D \to 0}\int_{z_{wD} - \varepsilon/2}^{z_{wD} + \varepsilon/2} r_D\frac{\partial \Delta m_f}{\partial r_D} dz_{wD}\right] = \begin{cases} 0, & |z_D - z_{wD}| > \varepsilon/2 \\ -\dfrac{p_{sc}T}{\pi k_{fh} h T_{sc}} q_{scins}, & |z_D - z_{wD}| < \varepsilon/2 \end{cases} \tag{2.29}$$

The upper and lower boundary conditions become:

$$\left.\frac{\partial \Delta m_f}{\partial z_D}\right|_{z_D=0} = 0,\ \left.\frac{\partial \Delta m_f}{\partial z_D}\right|_{z_D=1} = 0 \tag{2.30}$$

Finally, for the outer boundary condition:

$$\left.r_D\frac{\partial \Delta m_f}{\partial r_D}\right|_{r_D=r_{eD}} = \vartheta \Delta m_f|_{r_D=r_{eD}} \tag{2.31}$$

2.3.2 Model solutions

2.3.2.1 Laplace transformation

Take the following Laplace transformation for the above model:

$$\Delta\overline{m}_{\mathrm{f}} = \int_0^{\infty} \Delta m_{\mathrm{f}} \cdot e^{st_{\mathrm{D}}}\, \mathrm{d}t_{\mathrm{D}} \tag{2.32}$$

Then the continuity equation of the natural fracture system becomes:

$$\frac{1}{r_{\mathrm{D}}}\frac{\partial}{\partial r_{\mathrm{D}}}\left(r_{\mathrm{D}}\frac{\partial \Delta\overline{m}_{\mathrm{f}}}{\partial r_{\mathrm{D}}}\right) + \frac{1}{h_{\mathrm{D}}^2}\frac{\partial^2 \Delta\overline{m}_{\mathrm{f}}}{\partial z_{\mathrm{D}}^2} = s\Delta\overline{m}_{\mathrm{f}} \tag{2.33}$$

To be combined with the transient flow equation of the shale fracture system, the Laplace variable s in the above expression can be replaced with $f(s)$. Then we see that:

$$\frac{1}{r_{\mathrm{D}}}\frac{\partial}{\partial r_{\mathrm{D}}}\left(r_{\mathrm{D}}\frac{\partial \Delta\overline{m}_{\mathrm{f}}}{\partial r_{\mathrm{D}}}\right) + \frac{1}{h_{\mathrm{D}}^2}\frac{\partial^2 \Delta\overline{m}_{\mathrm{f}}}{\partial z_{\mathrm{D}}^2} = f(s)\Delta\overline{m}_{\mathrm{f}} \tag{2.34}$$

$$\lim_{\varepsilon\to 0}\left[\lim_{r_{\mathrm{D}}\to 0}\int_{z_{\mathrm{wD}}-\varepsilon/2}^{z_{\mathrm{wD}}+\varepsilon/2} r_{\mathrm{D}}\frac{\partial \Delta\overline{m}_{\mathrm{f}}}{\partial r_{\mathrm{D}}}\mathrm{d}z_{\mathrm{wD}}\right] = \begin{cases} 0, & |z_{\mathrm{D}}-z_{\mathrm{wD}}| > \varepsilon/2 \\ -\dfrac{p_{\mathrm{sc}}T}{\pi k_{\mathrm{fh}} h T_{\mathrm{sc}}}\overline{q}_{\mathrm{scins}}, & |z_{\mathrm{D}}-z_{\mathrm{wD}}| < \varepsilon/2 \end{cases} \tag{2.35}$$

$$\left.\frac{\partial \Delta\overline{m}_{\mathrm{f}}}{\partial z_{\mathrm{D}}}\right|_{z_{\mathrm{D}}=0} = 0, \quad \left.\frac{\partial \Delta\overline{m}_{\mathrm{f}}}{\partial z_{\mathrm{D}}}\right|_{z_{\mathrm{D}}=1} = 0 \tag{2.36}$$

$$\left. r_{\mathrm{D}}\frac{\partial \Delta\overline{m}_{\mathrm{f}}}{\partial r_{\mathrm{D}}}\right|_{r_{\mathrm{D}}=r_{\mathrm{eD}}} = \vartheta \Delta\overline{m}_{\mathrm{f}}\big|_{r_{\mathrm{D}}=r_{\mathrm{eD}}} \tag{2.37}$$

2.3.2.2 Orthogonal transformation *(Carslaw and Jaeger, 1959)*

According to the upper and lower boundary Eq. (2.36), the following Fourier finite cosine transformation can be done for Eqs. (2.34) and (2.37):

$$\text{Forward transformation: } \Delta\tilde{\overline{m}}_{\mathrm{f}} = \int_0^1 \Delta\overline{m}_{\mathrm{f}}\cos(n\pi z_{\mathrm{D}})\mathrm{d}z_{\mathrm{D}} \tag{2.38}$$

$$\frac{\text{Inverse transformation: } \Delta\overline{m}_{\mathrm{f}} = \sum_{n=0}^{\infty}\Delta\tilde{\overline{m}}_{\mathrm{f}}\cos(n\pi z_{\mathrm{D}})}{N(n)} \tag{2.39}$$

where $N(n)$ is expressed by $N(n) = \int_0^1 \cos^2(n\pi z_{\mathrm{D}})\mathrm{d}z_{\mathrm{D}} = \begin{cases} 1 & n=0 \\ 1/2 & n=1,2,\ldots \end{cases}$.

The model transformed through the Fourier forward cosine transformation is:

$$\frac{\partial^2 \Delta \tilde{\overline{m}}_{\mathrm{f}}}{\partial r_{\mathrm{D}}^2} + \frac{1}{r_{\mathrm{D}}}\frac{\partial \Delta \tilde{\overline{m}}_{\mathrm{f}}}{\partial r_{\mathrm{D}}} - \left[f(s) + \frac{(n\pi)^2}{h_{\mathrm{D}}^2} \right] \Delta \tilde{\overline{m}}_{\mathrm{f}} = 0 \tag{2.40}$$

$$\lim_{r_{\mathrm{D}}\to 0} r_{\mathrm{D}} \frac{\partial \Delta \tilde{\overline{m}}_{\mathrm{f}}}{\partial r_{\mathrm{D}}} = -\frac{p_{\mathrm{sc}} T \overline{q}_{\mathrm{scins}}}{\pi k_{\mathrm{fh}} h T_{\mathrm{sc}}} \cos(n\pi z_{\mathrm{wD}}) \tag{2.41}$$

$$r_{\mathrm{D}} \left. \frac{\partial \Delta \tilde{\overline{m}}_{\mathrm{f}}}{\partial r_{\mathrm{D}}} \right|_{r_{\mathrm{D}} = r_{\mathrm{eD}}} = \vartheta \Delta \tilde{\overline{m}}_{\mathrm{f}} \Big|_{r_{\mathrm{D}} = r_{\mathrm{eD}}} \tag{2.42}$$

According to Eq. (2.40) and the Bessel equation, the general solution to Eq. (2.40) is:

$$\Delta \tilde{\overline{m}}_{\mathrm{f}} = A_n I_0(\xi_n r_{\mathrm{D}}) + B_n K_0(\xi_n r_{\mathrm{D}}) \tag{2.43}$$

where $\xi_n = \sqrt{f(s) + \frac{(n\pi)^2}{h_{\mathrm{D}}^2}}$, $(n = 0, 1, 2, 3, \cdots)$.

Introducing solution (2.43) into the inner boundary Eq. (2.41) yields:

$$B_n = -\frac{p_{\mathrm{sc}} T \overline{q}_{\mathrm{scins}}}{\pi k_{\mathrm{fh}} h T_{\mathrm{sc}}} \cos(n\pi z_{\mathrm{wD}}) \tag{2.44}$$

Thereafter, introducing Eq. (2.44) into Eq. (2.43) and then combining it with Eq. (2.42) yields:

$$A_n r_{\mathrm{eD}} \xi_n I_1(\xi_n r_{\mathrm{eD}}) - B_n r_{\mathrm{eD}} \xi_n K_1(\xi_n r_{\mathrm{eD}}) = A_n \vartheta I_0(\xi_n r_{\mathrm{eD}}) + B_n \vartheta K_0(\xi_n r_{\mathrm{eD}}) \tag{2.45}$$

By solving the above equation, we have:

$$C_n = \frac{A_n}{B_n} = \frac{r_{\mathrm{eD}} \xi_n K_1(\xi_n r_{\mathrm{eD}}) + \vartheta K_0(\xi_n r_{\mathrm{eD}})}{r_{\mathrm{eD}} \xi_n I_1(\xi_n r_{\mathrm{eD}}) - \vartheta I_0(\xi_n r_{\mathrm{eD}})} \tag{2.46}$$

Substituting the above equation into the general solution, we see that:

$$\Delta \tilde{\overline{m}}_{\mathrm{f}} = -\frac{p_{\mathrm{sc}} T \overline{q}_{\mathrm{scins}}}{\pi k_{\mathrm{fh}} h T_{\mathrm{sc}}} \left[I_0(\xi_n r_{\mathrm{D}}) + C_n K_0(\xi_n r_{\mathrm{D}}) \right] \cos(n\pi z_{\mathrm{wD}}) \tag{2.47}$$

Transforming the above expression through the Fourier inverse cosine transformation, it then becomes:

$$\Delta \overline{m}_{\mathrm{f}} = \frac{p_{\mathrm{sc}} T \overline{q}_{\mathrm{scins}}}{\pi k_{\mathrm{fh}} h T_{\mathrm{sc}}} \left\{ \begin{array}{l} K_0(\xi_0 r_{\mathrm{D}}) + C_n I_0(\xi_0 r_{\mathrm{D}}) + \\ 2\sum_{n=1}^{\infty} \left[K_0(\xi_n r_{\mathrm{D}}) + C_n I_0(\xi_n r_{\mathrm{D}}) \right] \cos(n\pi z_{\mathrm{wD}}) \cos(n\pi z_{\mathrm{D}}) \end{array} \right\} \tag{2.48}$$

where $r_{\mathrm{D}} = \sqrt{(x_{\mathrm{D}} - x_{\mathrm{wD}})^2 + (y_{\mathrm{D}} - y_{\mathrm{wD}})^2}$.

2.3.3 Point source solutions in various outer boundary conditions

In the previous part, the models for the mixed outer boundary have been solved. However, the outer boundary of a gas reservoir is usually one of three conditions: infinite, constant pressure, or closed. In the following introduction, the continuous point source solutions for these three boundary conditions are analyzed.

2.3.3.1 Infinite outer boundary

When the outer boundary of a gas reservoir is infinite, it can be represented as:

$$\lim_{r_{\mathrm{D}}\to 0} \left.\frac{\partial \Delta m_{\mathrm{f}}}{\partial r_{\mathrm{D}}}\right|_{r_{\mathrm{D}}=r_{\mathrm{eD}}} = 0 \tag{2.49}$$

Then we see that:

$$A_n = 0 \tag{2.50}$$

Therefore, we have:

$$C_n = 0 \tag{2.51}$$

Introducing Eq. (2.51) into Eq. (2.48), the continuous point source solution for a gas reservoir with the closed upper and lower boundaries and infinite outer boundary is expressed as (Ozkan and Raghavan, 1991a,b, 1994):

$$\Delta \overline{m}_{\mathrm{f}} = \frac{p_{\mathrm{sc}} T \overline{q}_{\mathrm{scins}}}{\pi k_{\mathrm{fh}} h T_{\mathrm{sc}}} \left[K_0(\xi_0 r_{\mathrm{D}}) + 2\sum_{n=1}^{\infty} K_0(\xi_n r_{\mathrm{D}}) \cos(n\pi z_{\mathrm{wD}}) \cos(n\pi z_{\mathrm{D}}) \right] \tag{2.52}$$

Generally, the continuous point source intensity q_{scins} is constant; therefore:

$$\Delta \overline{m}_{\mathrm{f}} = \frac{p_{\mathrm{sc}} T q_{\mathrm{scins}}}{\pi k_{\mathrm{fh}} h T_{\mathrm{sc}} s} \left[K_0(\xi_0 r_{\mathrm{D}}) + 2\sum_{n=1}^{\infty} K_0(\xi_n r_{\mathrm{D}}) \cos(n\pi z_{\mathrm{wD}}) \cos(n\pi z_{\mathrm{D}}) \right] \tag{2.53}$$

2.3.3.2 Constant pressure outer boundary

For a constant pressure outer boundary:

$$\Delta \overline{m}_{\mathrm{f}}\big|_{r_{\mathrm{D}}=r_{\mathrm{eD}}} = 0 \tag{2.54}$$

Then the following equation is valid:

$$\vartheta = \infty \tag{2.55}$$

Introducing Eq. (2.55) into Eq. (2.46), we have:

$$C_n = \lim_{\vartheta\to\infty} \frac{r_{\mathrm{eD}}\xi_n K_1(\xi_n r_{\mathrm{eD}}) + \vartheta K_0(\xi_n r_{\mathrm{eD}})}{r_{\mathrm{eD}}\xi_n I_1(\xi_n r_{\mathrm{eD}}) - \vartheta I_0(\xi_n r_{\mathrm{eD}})} = -\frac{K_0(\xi_n r_{\mathrm{eD}})}{I_0(\xi_n r_{\mathrm{eD}})} \tag{2.56}$$

Thus the continuous point source solution for the constant pressure outer boundary is:

$$\Delta\overline{m}_{\mathrm{f}}=\frac{p_{\mathrm{sc}}Tq_{\mathrm{scins}}}{\pi k_{\mathrm{fh}}hT_{\mathrm{sc}}s}\left\{\begin{array}{l}K_0(\xi_0 r_{\mathrm{D}})-\dfrac{K_0(\xi_0 r_{\mathrm{eD}})}{I_0(\xi_0 r_{\mathrm{eD}})}I_0(\xi_0 r_{\mathrm{D}})+\\ 2\displaystyle\sum_{n=1}^{\infty}\left[K_0(\xi_n r_{\mathrm{D}})-\dfrac{K_0(\xi_n r_{\mathrm{eD}})}{I_0(\xi_n r_{\mathrm{eD}})}I_0(\xi_n r_{\mathrm{D}})\right]\cos(n\pi z_{\mathrm{wD}})\cos(n\pi z_{\mathrm{D}})\end{array}\right\} \tag{2.57}$$

2.3.3.3 Closed outer boundary

When there is no flow through the outer boundary, there is:

$$\vartheta=0 \tag{2.58}$$

Substituting Eq. (2.58) into Eq. (2.46), we see that:

$$C_n=\frac{K_1(\xi_n r_{\mathrm{eD}})}{I_1(\xi_n r_{\mathrm{eD}})} \tag{2.59}$$

Then the continuous point source solution for the closed outer boundary is:

$$\Delta\overline{m}_{\mathrm{f}}=\frac{p_{\mathrm{sc}}Tq_{\mathrm{scins}}}{\pi k_{\mathrm{fh}}hT_{\mathrm{sc}}s}\left\{\begin{array}{l}K_0(\xi_0 r_{\mathrm{D}})+\dfrac{K_1(\xi_0 r_{\mathrm{eD}})}{I_1(\xi_0 r_{\mathrm{eD}})}I_0(\xi_0 r_{\mathrm{D}})+\\ 2\displaystyle\sum_{n=1}^{\infty}\left[K_0(\xi_n r_{\mathrm{D}})+\dfrac{K_1(\xi_n r_{\mathrm{eD}})}{I_1(\xi_n r_{\mathrm{eD}})}I_0(\xi_n r_{\mathrm{D}})\right]\cos(n\pi z_{\mathrm{wD}})\cos(n\pi z_{\mathrm{D}})\end{array}\right\} \tag{2.60}$$

2.4 Continuous point source solutions in rectangular gas reservoirs

The main development method for shale gas reservoirs includes horizontal wells with massive hydraulic fractures; in the early exploration and appraisal stage, hydraulically fractured vertical wells are used. For a hydraulic multiple-staged fractured horizontal well (MFHW), due to its long well length and comparatively short half fracture length, the pressure distribution around its wellbore during production is more like an ellipse. In this situation, a circular boundary assumption is not valid, and, therefore, a rectangular boundary is applied for the analysis of MFHWs. Solving a mathematical model constructed for a rectangular gas reservoir causes massive computation efforts. To simplify the problem, the pressure potential for a continuous source sink at any point in the rectangular reservoir is analyzed by the point source function.

2.4.1 Derivation of a continuous point source solution

Before solving for the source/sink functions for reservoirs with different shapes, pressure for a continuous point source at any point in a rectangular gas reservoir is derived. Fig. 2.2 shows a continuous point source and its physical image model at any point of a rectangular gas reservoir with a closed outer boundary. The superposition principle is used to solve this model.

According to the superposition principle, the instantaneous point source solution at the observation point M_D' generated by the unit source/sink at point M_D is (Newman, 1936; Ozkan and Raghavan, 1991a,b, 1994):

$$\overline{G}\left(M_D, M_D', s\right) = \sum_{k=-\infty}^{+\infty} \sum_{m=-\infty}^{+\infty} \sum_{n=-\infty}^{+\infty} \left(S_{1,1,1} + S_{2,1,1} + S_{1,2,1} + S_{1,1,2} + S_{2,2,1} + S_{1,2,2} + S_{2,1,2} + S_{2,2,2}\right) \tag{2.61}$$

In the above equation, $S_{i,j,\chi}$ represents a pressure response at any point of the gas reservoir caused by each image well due to the closed boundary:

$$S_{i,j,\chi} = \frac{\exp\left[-\sqrt{f(s)}\sqrt{(\tilde{x}_{Di} - 2kx_{eD})^2 + \left(\tilde{y}_{Dj} - 2my_{eD}\right)^2 + (\tilde{z}_{Dk} - 2nh_D)^2}\right]}{\sqrt{(\tilde{x}_{Di} - 2kx_{eD})^2 + \left(\tilde{y}_{Dj} - 2my_{eD}\right)^2 + \left(\tilde{z}_{D\chi} - 2nh_D\right)^2}} \quad (i, j, \chi = 1, 2) \tag{2.62}$$

where:

$$\tilde{x}_{D1} = x_D - x_{wD} \tag{2.63}$$

$$\tilde{x}_{D2} = x_D - x_{wD} \tag{2.64}$$

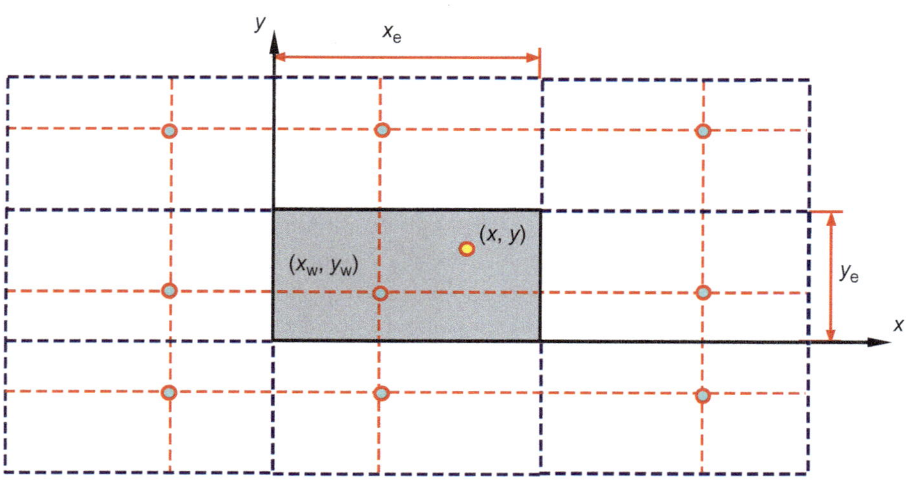

Fig. 2.2 Physical image model of a point source in a rectangular gas reservoir with closed boundary.

$$\tilde{y}_{D1} = y_D - y_{wD} \tag{2.65}$$

$$\tilde{y}_{D2} = y_D - y_{wD} \tag{2.66}$$

$$\tilde{z}_{D1} = z_D - z_{wD} \tag{2.67}$$

$$\tilde{z}_{D2} = z_D - z_{wD} \tag{2.68}$$

Note that the units in this chapter are the SI units. For an explanation of repeated symbols, refer to previous sections.

For the derivation of the continuous point source function, the following expressions are used:

$$u = f(s) \tag{2.69}$$

$$\zeta_D = \frac{\zeta}{L_{ref}} \sqrt{\frac{k}{k_\zeta}} \quad (\zeta = x, y, z) \tag{2.70}$$

$$h_D = \frac{h}{L_{ref}} \sqrt{\frac{k}{k_z}} \tag{2.71}$$

where:

x, y, z—coordinates in a Cartesian coordinate system (m);
h—effective thickness of the gas reservoir (m);
L_{ref}—reference length (m);
k—effective permeability of the gas reservoir (m^2);
k_z—vertical permeability of the gas reservoir (m^2);
s—Laplace variable, dimensionless.

Eq. (2.61) can be simplified as follows:

$$TS = \sum_{k=-\infty}^{+\infty} \sum_{m=-\infty}^{+\infty} \sum_{n=-\infty}^{+\infty} S \tag{2.72}$$

where:

$$S = \frac{\exp\left[-\sqrt{u}\sqrt{(\tilde{x}_D - 2kx_{eD})^2 + (\tilde{y}_D - 2my_{eD})^2 + (\tilde{z}_D - 2nh_D)^2}\right]}{\sqrt{(\tilde{x}_D - 2kx_{eD})^2 + (\tilde{y}_D - 2my_{eD})^2 + (\tilde{z}_D - 2nh_D)^2}} \tag{2.73}$$

The sum can be expressed as:

$$\sum_{n=-\infty}^{+\infty} \frac{\exp\left[-\sqrt{v}\sqrt{a^2 + (\xi - 2n\xi_e)^2}\right]}{\sqrt{a^2 + (\xi - 2n\xi_e)^2}} = \frac{1}{\xi_e}\left[K_0(a\sqrt{v}) + 2\sum_{n=1}^{\infty} K_0\left(a\sqrt{v + \frac{n^2\pi^2}{\xi_e^2}}\right)\cos\left(n\pi\frac{\xi}{\xi_e}\right)\right] \tag{2.74}$$

Comparing Eq. (2.73) and Eq. (2.74), the following definitions are applied:

$$a^2 = (\tilde{x}_D - 2kx_{eD})^2 + (\tilde{y}_D - 2my_{eD})^2 \tag{2.75}$$

$$\xi_e = h_D \tag{2.76}$$

Then the sum of the infinite series in Eq. (2.72) can be simplified to:

$$TS = \sum_{k=-\infty}^{+\infty} \sum_{m=-\infty}^{+\infty} \frac{1}{h_D} \left[K_0\left(a\sqrt{u}\right) + 2\sum_{n=1}^{\infty} K_0\left(a\sqrt{u + \frac{n^2\pi^2}{h_D^2}} \right) \cos\left(n\pi \frac{\tilde{z}_D}{h_D} \right) \right] \tag{2.77}$$

For the Bessel function, there is:

$$K_0(z) = \frac{1}{2}\int_0^{\infty} \exp\left(-\xi - \frac{z^2}{4\xi} \right) \frac{d\xi}{\xi} \tag{2.78}$$

Combining with Eq. (2.75), the Bessel term in Eq. (2.77) can be transformed to:

$$\begin{aligned} K_0\left(a\sqrt{u}\right) &= \frac{1}{2}\int_0^{\infty} \exp\left(-\xi - \frac{a^2 u}{4\xi} \right) \frac{d\xi}{\xi} \\ &= \frac{1}{2}\int_0^{\infty} \exp(-\xi) \exp\left(-\frac{(\tilde{x}_D - 2kx_{eD})^2 u}{4\xi} \right) \exp\left(-\frac{(\tilde{y}_D - 2my_{eD})^2 u}{4\xi} \right) \frac{d\xi}{\xi} \end{aligned} \tag{2.79}$$

In the same manner, the following equation is also valid:

$$\begin{aligned} K_0\left(a\sqrt{u + \frac{n^2\pi^2}{h_D^2}} \right) &= \frac{1}{2}\int_0^{\infty} \exp(-\xi) \times \exp\left(-\frac{(\tilde{x}_D - 2kx_{eD})^2 \left(u + n^2\pi^2/h_D^2\right)}{4\xi} \right) \\ &\quad \times \exp\left(-\frac{(\tilde{y}_D - 2my_{eD})^2 \left(u + n^2\pi^2/h_D^2\right)}{4\xi} \right) \frac{d\xi}{\xi} \end{aligned} \tag{2.80}$$

Introducing Eqs. (2.79) and (2.80) into Eq. (2.77), there is:

$$\begin{aligned} TS = \frac{1}{2h_D} \Bigg\{ & \int_0^{\infty} \exp(-\xi) \times \sum_{k=-\infty}^{+\infty} \exp\left(-\frac{(\tilde{x}_D - 2kx_{eD})^2 u}{4\xi} \right) \times \sum_{m=-\infty}^{+\infty} \exp\left(-\frac{(\tilde{y}_D - 2my_{eD})^2 u}{4\xi} \right) \frac{d\xi}{\xi} \\ & + 2\sum_{n=1}^{\infty} \cos\left(n\pi \frac{\tilde{z}_D}{h_D} \right) \times \int_0^{\infty} \exp(-\xi) \times \sum_{k=-\infty}^{+\infty} \exp\left(-\frac{(\tilde{x}_D - 2kx_{eD})^2 \left(u + n^2\pi^2/h_D^2\right)}{4\xi} \right) \\ & \sum_{m=-\infty}^{+\infty} \exp\left(-\frac{(\tilde{y}_D - 2my_{eD})^2 \left(u + n^2\pi^2/h_D^2\right)}{4\xi} \right) \frac{d\xi}{\xi} \Bigg\} \end{aligned} \tag{2.81}$$

According to the following Poisson summation formula:

$$\sum_{n=-\infty}^{+\infty} \exp\left[-\frac{(\xi - 2n\xi_e)^2}{4\tau} \right] = \frac{\sqrt{\pi\tau}}{\xi_e} \left[1 + 2\sum_{n=1}^{+\infty} \exp\left(-\frac{n^2\pi^2\tau}{\xi_e^2} \right) \cos\left(n\pi \frac{\xi}{\xi_e} \right) \right] \tag{2.82}$$

The sum of the infinite series in Eq. (2.81) can be simplified. For the first term:

$$\sum_{k=-\infty}^{+\infty} \exp\left(-\frac{(\tilde{x}_D - 2kx_{eD})^2 u}{4\xi}\right) = \frac{\sqrt{\pi\xi}}{x_{eD}\sqrt{u}}\left[1 + 2\sum_{k=1}^{+\infty} \exp\left(-\frac{k^2\pi^2\xi}{ux_{eD}^2}\right)\cos\left(k\pi\frac{\tilde{x}_D}{x_{eD}}\right)\right] \tag{2.83}$$

For the second term:

$$\sum_{m=-\infty}^{+\infty} \exp\left(-\frac{(\tilde{y}_D - 2my_{eD})^2 u}{4\xi}\right) = \frac{\sqrt{\pi\xi}}{y_{eD}\sqrt{u}}\left[1 + 2\sum_{m=1}^{+\infty} \exp\left(-\frac{m^2\pi^2\xi}{uy_{eD}^2}\right)\cos\left(m\pi\frac{\tilde{y}_D}{y_{eD}}\right)\right] \tag{2.84}$$

For the third term:

$$\begin{aligned}&\sum_{k=-\infty}^{+\infty} \exp\left(-\frac{(\tilde{x}_D - 2kx_{eD})^2 \left(u + n^2\pi^2/h_D^2\right)}{4\xi}\right) = \\ &\frac{\sqrt{\pi\xi}}{x_{eD}\sqrt{u + n^2\pi^2/h_D^2}}\left[1 + 2\sum_{k=1}^{+\infty} \exp\left(-\frac{k^2\pi^2\xi}{(u + n^2\pi^2/h_D^2)x_{eD}^2}\right)\cos\left(k\pi\frac{\tilde{x}_D}{x_{eD}}\right)\right]\end{aligned} \tag{2.85}$$

Furthermore, for the fourth term:

$$\begin{aligned}&\sum_{m=-\infty}^{+\infty} \exp\left(-\frac{(\tilde{y}_D - 2my_{eD})^2 \left(u + n^2\pi^2/h_D^2\right)}{4\xi}\right) = \\ &\frac{\sqrt{\pi\xi}}{y_{eD}\sqrt{u + n^2\pi^2/h_D^2}}\left[1 + 2\sum_{m=1}^{+\infty} \exp\left(-\frac{m^2\pi^2\xi}{(u + n^2\pi^2/h_D^2)y_{eD}^2}\right)\cos\left(m\pi\frac{\tilde{y}_D}{y_{eD}}\right)\right]\end{aligned} \tag{2.86}$$

Then Eq. (2.81) becomes:

$$\begin{aligned}&\int_0^\infty \exp(-\xi) \times \sum_{k=-\infty}^{+\infty} \exp\left(-\frac{(\tilde{x}_D - 2kx_{eD})^2 u}{4\xi}\right) \times \sum_{m=-\infty}^{+\infty} \exp\left(-\frac{(\tilde{y}_D - 2my_{eD})^2 u}{4\xi}\right)\frac{d\xi}{\xi} \\ &= \int_0^\infty \exp(-\xi) \times \frac{\pi}{x_{eD}y_{eD}u}\left[1 + 2\sum_{k=1}^{+\infty} \exp\left(-\frac{k^2\pi^2\xi}{ux_{eD}^2}\right)\cos\left(k\pi\frac{\tilde{x}_D}{x_{eD}}\right)\right] \\ &\quad \left[1 + 2\sum_{m=1}^{+\infty} \exp\left(-\frac{m^2\pi^2\xi}{uy_{eD}^2}\right)\cos\left(m\pi\frac{\tilde{y}_D}{y_{eD}}\right)\right]d\xi \\ &= \frac{\pi}{x_{eD}y_{eD}u}\left[\begin{aligned}&1 + 2\sum_{k=1}^{+\infty}\frac{ux_{eD}^2}{ux_{eD}^2 + k^2\pi^2}\cos\left(k\pi\frac{\tilde{x}_D}{x_{eD}}\right) + 2\sum_{m=1}^{+\infty}\frac{uy_{eD}^2}{uy_{eD}^2 + m^2\pi^2}\cos\left(m\pi\frac{\tilde{y}_D}{y_{eD}}\right) + \\ &4\sum_{k=1}^{+\infty}\sum_{m=1}^{+\infty}\left(1 + \frac{m^2\pi^2}{uy_{eD}^2} + \frac{k^2\pi^2}{ux_{eD}^2}\right)^{-1}\cos\left(k\pi\frac{\tilde{x}_D}{x_{eD}}\right)\cos\left(m\pi\frac{\tilde{y}_D}{y_{eD}}\right)\end{aligned}\right]\end{aligned} \tag{2.87}$$

Applying the following relationship:

$$\sum_{k=1}^{\infty}\frac{\cos(k\pi x)}{k^2+a^2}=\frac{\pi}{2a}\frac{\cosh[a\pi(1-x)]}{\sinh(a\pi)}-\frac{1}{2a^2}\quad[0\le x\le 2\pi] \tag{2.88}$$

Expression (2.87) can be further simplified to:

$$\begin{aligned}\sum_{k=1}^{+\infty}\frac{ux_{\mathrm{eD}}^2}{ux_{\mathrm{eD}}^2+k^2\pi^2}\cos\left(k\pi\frac{\tilde{x}_\mathrm{D}}{x_{\mathrm{eD}}}\right)&=\frac{ux_{\mathrm{eD}}^2}{\pi^2}\sum_{k=1}^{+\infty}\frac{1}{\frac{ux_{\mathrm{eD}}^2}{\pi^2}+k^2}\cos\left(k\pi\frac{\tilde{x}_\mathrm{D}}{x_{\mathrm{eD}}}\right)\\&=\frac{\sqrt{u}x_{\mathrm{eD}}}{2}\frac{\cosh[\sqrt{u}(x_{\mathrm{eD}}-|\tilde{x}_\mathrm{D}|)]}{\sinh(\sqrt{u}x_{\mathrm{eD}})}-\frac{1}{2}\end{aligned} \tag{2.89}$$

In the same manner, there is:

$$\sum_{m=1}^{+\infty}\frac{uy_{\mathrm{eD}}^2}{uy_{\mathrm{eD}}^2+m^2\pi^2}\cos\left(m\pi\frac{\tilde{y}_\mathrm{D}}{y_{\mathrm{eD}}}\right)=\frac{\sqrt{u}y_{\mathrm{eD}}}{2}\frac{\cosh[\sqrt{u}(y_{\mathrm{eD}}-|\tilde{y}_\mathrm{D}|)]}{\sinh(\sqrt{u}y_{\mathrm{eD}})}-\frac{1}{2} \tag{2.90}$$

For the sum of the infinite series in the right-hand side of Eq. (2.87), there is:

$$\begin{aligned}&\sum_{k=1}^{+\infty}\sum_{m=1}^{+\infty}\left(1+\frac{m^2\pi^2}{uy_{\mathrm{eD}}^2}+\frac{k^2\pi^2}{ux_{\mathrm{eD}}^2}\right)^{-1}\cos\left(k\pi\frac{\tilde{x}_\mathrm{D}}{x_{\mathrm{eD}}}\right)\cos\left(m\pi\frac{\tilde{y}_\mathrm{D}}{y_{\mathrm{eD}}}\right)\\&=\frac{uy_{\mathrm{eD}}^2}{\pi^2}\sum_{k=1}^{+\infty}\sum_{m=1}^{+\infty}\left(\frac{uy_{\mathrm{eD}}^2}{\pi^2}+\frac{k^2y_{\mathrm{eD}}^2}{x_{\mathrm{eD}}^2}+m^2\right)^{-1}\cos\left(m\pi\frac{\tilde{y}_\mathrm{D}}{y_{\mathrm{eD}}}\right)\cos\left(k\pi\frac{\tilde{x}_\mathrm{D}}{x_{\mathrm{eD}}}\right)\\&=\frac{uy_{\mathrm{eD}}^2}{\pi^2}\sum_{k=1}^{+\infty}\left(\sum_{m=1}^{+\infty}\frac{\cos\left(m\pi\frac{\tilde{y}_\mathrm{D}}{y_{\mathrm{eD}}}\right)}{\left(\frac{uy_{\mathrm{eD}}^2}{\pi^2}+\frac{k^2y_{\mathrm{eD}}^2}{x_{\mathrm{eD}}^2}\right)+m^2}\right)\cos\left(k\pi\frac{\tilde{x}_\mathrm{D}}{x_{\mathrm{eD}}}\right)\end{aligned} \tag{2.91}$$

According to Eq. (2.88), the sum in the right-hand side of Eq. (2.91) can be expressed by:

$$\begin{aligned}\sum_{m=1}^{+\infty}\left(\frac{\cos\left(m\pi\frac{\tilde{y}_\mathrm{D}}{y_{\mathrm{eD}}}\right)}{\left(\frac{uy_{\mathrm{eD}}^2}{\pi^2}+\frac{k^2y_{\mathrm{eD}}^2}{x_{\mathrm{eD}}^2}\right)+m^2}\right)&=\frac{\pi^2}{2y_{\mathrm{eD}}\sqrt{u+\frac{\pi^2k^2}{x_{\mathrm{eD}}^2}}}\frac{\cosh\left[\sqrt{u+\frac{k^2\pi^2}{x_{\mathrm{eD}}^2}}(y_{\mathrm{eD}}-\tilde{y}_\mathrm{D})\right]}{\sinh\left(\sqrt{u+\frac{k^2\pi^2}{x_{\mathrm{eD}}^2}}y_{\mathrm{eD}}\right)}\\&-\frac{1}{2\left(\frac{uy_{\mathrm{eD}}^2}{\pi^2}+\frac{k^2y_{\mathrm{eD}}^2}{x_{\mathrm{eD}}^2}\right)}\end{aligned} \tag{2.92}$$

Define the variable:

$$\varepsilon_k = u + \frac{\pi^2 k^2}{x_{eD}^2} \tag{2.93}$$

Then introducing Eq. (2.92) into Eq. (2.91) gives:

$$\sum_{k=1}^{+\infty}\sum_{m=1}^{+8}\left(1+\frac{m^2\pi^2}{uy_{eD}^2}+\frac{k^2\pi^2}{ux_{eD}^2}\right)^{-1}\cos\left(k\pi\frac{\tilde{x}_D}{x_{eD}}\right)\cos\left(m\pi\frac{\tilde{y}_D}{y_{eD}}\right)=\frac{uy_{eD}^2}{\pi^2}\times$$
$$\sum_{k=1}^{+\infty}\left(\frac{\pi^2}{2y_{eD}\sqrt{\varepsilon_k}}\frac{\cosh\left[\sqrt{\varepsilon_k}(y_{eD}-\tilde{y}_D)\right]}{\sinh\left(\sqrt{\varepsilon_k}y_{eD}\right)}-\frac{1}{2\left(\frac{uy_{eD}^2}{\pi^2}+\frac{k^2y_{eD}^2}{x_{eD}^2}\right)}\right)\cos\left(k\pi\frac{\tilde{x}_D}{x_{eD}}\right) \tag{2.94}$$

Introducing Eqs. (2.94), (2.89) and (2.90) into Eq. (2.87), there is:

$$\frac{\pi}{x_{eD}y_{eD}u}\left[\begin{array}{l}1+2\sum_{k=1}^{+\infty}\frac{ux_{eD}^2}{ux_{eD}^2+k^2\pi^2}\cos\left(k\pi\frac{\tilde{x}_D}{x_{eD}}\right)+2\sum_{m=1}^{+\infty}\frac{uy_{eD}^2}{uy_{eD}^2+m^2\pi^2}\cos\left(m\pi\frac{\tilde{y}_D}{y_{eD}}\right)\\ 4\sum_{k-1}^{+\infty}\sum_{m=1}^{+\infty}\left(1+\frac{m^2\pi^2}{uy_{eD}^2}+\frac{k^2\pi^2}{ux_{eD}^2}\right)^{-1}\cos\left(k\pi\frac{\tilde{x}_D}{x_{eD}}\right)\cos\left(m\pi\frac{\tilde{y}_D}{y_{eD}}\right)\end{array}\right]$$
$$=\frac{\pi}{x_{eD}y_{eD}u}\left[\begin{array}{l}\sqrt{u}x_{eD}\frac{\cosh\left[\sqrt{u}(x_{eD}-\tilde{x}_D)\right]}{\sinh\left(\sqrt{u}x_{eD}\right)}+\sqrt{u}y_{eD}\frac{\cosh\left[\sqrt{u}(y_{eD}-\tilde{y}_D)\right]}{\sinh\left(\sqrt{u}y_{eD}\right)}-1\\ +2\sum_{k=1}^{+\infty}\frac{uy_{eD}}{\sqrt{\varepsilon_k}}\frac{\cosh\left[\sqrt{\varepsilon_k}(y_{eD}-\tilde{y}_D)\right]}{\sinh\left(\sqrt{\varepsilon_k}y_{eD}\right)}\cos\left(k\pi\frac{\tilde{x}_D}{x_{eD}}\right)\\ -\frac{uy_{eD}^2}{\pi^2}\sum_{k=1}^{+\infty}\frac{2}{\left(\frac{uy_{eD}^2}{\pi^2}+\frac{k^2y_{eD}^2}{x_{eD}^2}\right)}\cos\left(k\pi\frac{\tilde{x}_D}{x_{eD}}\right)\end{array}\right] \tag{2.95}$$

Simplifying the last term in the right-hand side of Eq. (2.95) based on Eq. (2.88), there is:

$$\frac{uy_{eD}^2}{\pi^2}\frac{x_{eD}^2}{y_{eD}^2}\sum_{k=1}^{+\infty}\frac{2}{\left(\frac{ux_{eD}^2}{\pi^2}+k^2\right)}\cos\left(k\pi\frac{\tilde{x}_D}{x_{eD}}\right)=\sqrt{u}x_{eD}\frac{\cosh\left[\sqrt{u}(x_{eD}-\tilde{x}_D)\right]}{\sinh\left(\sqrt{u}x_{eD}\right)}-1 \tag{2.96}$$

By introducing Eq. (2.96) into Eq. (2.95) and combining it with Eq. (2.87), there is:

$$\int_0^{\infty} \exp(-\xi) \times \sum_{k=-\infty}^{+\infty} \exp\left(-\frac{(\tilde{x}_{\mathrm{D}}-2kx_{\mathrm{eD}})^2 u}{4\xi}\right) \times \sum_{m=-\infty}^{+\infty} \exp\left(-\frac{(\tilde{y}_{\mathrm{D}}-2my_{\mathrm{eD}})^2 u}{4\xi}\right) \frac{\mathrm{d}\xi}{\xi}$$
$$= \frac{\pi}{x_{\mathrm{eD}}}\left[\frac{\cosh[\sqrt{u}(y_{\mathrm{eD}}-\tilde{y}_{\mathrm{D}})]}{\sqrt{u}\sinh(\sqrt{u}y_{\mathrm{eD}})} + 2\sum_{k=1}^{+\infty}\frac{\cosh\left[\sqrt{\varepsilon_k}(y_{\mathrm{eD}}-\tilde{y}_{\mathrm{D}})\right]}{\sqrt{\varepsilon_k}\sinh\left(\sqrt{\varepsilon_k}y_{\mathrm{eD}}\right)}\cos\left(k\pi\frac{\tilde{x}_{\mathrm{D}}}{x_{\mathrm{eD}}}\right)\right] \tag{2.97}$$

Define the following variables:

$$\varepsilon_n = u + \frac{n^2\pi^2}{h_{\mathrm{D}}^2} \tag{2.98}$$

$$\varepsilon_{k,n} = u + \frac{n^2\pi^2}{h_{\mathrm{D}}^2} + \frac{\pi^2 k^2}{x_{\mathrm{eD}}^2} \tag{2.99}$$

Similarly, the second term in the right-hand side of Eq. (2.81) can be simplified to:

$$\int_0^{\infty} \exp(-\xi) \times \sum_{k=-\infty}^{+\infty} \exp\left(-\frac{(\tilde{x}_{\mathrm{D}}-2kx_{\mathrm{eD}})^2 \left(u+n^2\pi^2/h_{\mathrm{D}}^2\right)}{4\xi}\right)$$
$$\times \sum_{m=-\infty}^{+\infty} \exp\left(-\frac{(\tilde{y}_{\mathrm{D}}-2my_{\mathrm{eD}})^2 \left(u+n^2\pi^2/h_{\mathrm{D}}^2\right)}{4\xi}\right) \frac{\mathrm{d}\xi}{\xi}$$
$$= \frac{\pi}{x_{\mathrm{eD}}}\left[\frac{\cosh\left[\sqrt{\varepsilon_n}(y_{\mathrm{eD}}-\tilde{y}_{\mathrm{D}})\right]}{\sqrt{\varepsilon_n}\sinh\left(\sqrt{\varepsilon_n}y_{\mathrm{eD}}\right)} + 2\sum_{k=1}^{+\infty}\cos\left(k\pi\frac{\tilde{x}_{\mathrm{D}}}{x_{\mathrm{eD}}}\right)\frac{\cosh\left[\sqrt{\varepsilon_{k,n}}(y_{\mathrm{eD}}-\tilde{y}_{\mathrm{D}})\right]}{\sqrt{\varepsilon_{k,n}}\sinh\left(\sqrt{\varepsilon_{k,n}}y_{\mathrm{eD}}\right)}\right] \tag{2.100}$$

Then substituting Eqs. (2.97) and (2.100) into Eq. (2.81) yields:

$$\sum_{k=-\infty}^{+\infty}\sum_{m=-\infty}^{+\infty}\sum_{n=-\infty}^{+\infty} S = \frac{\pi}{2h_{\mathrm{D}}x_{\mathrm{eD}}} \times \left\{\frac{\cosh[\sqrt{u}(y_{\mathrm{eD}}-\tilde{y}_{\mathrm{D}})]}{\sqrt{u}\sinh(\sqrt{u}y_{\mathrm{eD}})}\right.$$
$$+2\sum_{k=1}^{+\infty}\frac{\cosh\left[\sqrt{\varepsilon_k}(y_{\mathrm{eD}}-\tilde{y}_{\mathrm{D}})\right]}{\sqrt{\varepsilon_k}\sinh\left(\sqrt{\varepsilon_k}y_{\mathrm{eD}}\right)}\cos\left(\frac{k\pi\tilde{x}_{\mathrm{D}}}{x_{\mathrm{eD}}}\right) + 2\sum_{n=1}^{\infty}\cos\left(\frac{n\pi\tilde{z}_{\mathrm{D}}}{h_{\mathrm{D}}}\right)\left[\frac{\cosh\left[\sqrt{\varepsilon_n}(y_{\mathrm{eD}}-\tilde{y}_{\mathrm{D}})\right]}{\sqrt{\varepsilon_n}\sinh\left(\sqrt{\varepsilon_n}y_{\mathrm{eD}}\right)}\right.$$
$$\left.\left.+2\sum_{k=1}^{+\infty}\cos\left(k\pi\frac{\tilde{x}_{\mathrm{D}}}{x_{\mathrm{eD}}}\right)\frac{\cosh\left[\sqrt{\varepsilon_{k,n}}(y_{\mathrm{eD}}-\tilde{y}_{\mathrm{D}})\right]}{\sqrt{\varepsilon_{k,n}}\sinh\left(\sqrt{\varepsilon_{k,n}}y_{\mathrm{eD}}\right)}\right]\right\} \tag{2.101}$$

Introducing Eq. (2.101) into Eq. (2.61), the continuous point source solution for a rectangular gas reservoir with the closed boundary can be simplified and expressed by:

$$\overline{G}(M_{\rm D},M'_{\rm D},s)=\frac{2\pi}{h_{\rm D}x_{\rm eD}}\left\{\frac{\cosh(\sqrt{u}\tilde{y}_{\rm D1})+\cosh(\sqrt{u}\tilde{y}_{\rm D2})}{\sqrt{u}\sinh(\sqrt{u}y_{\rm eD})}\right.$$
$$+2\sum_{k=1}^{+\infty}\cos\left(k\pi\frac{x_{\rm D}}{x_{\rm eD}}\right)\cos\left(k\pi\frac{x_{\rm wD}}{x_{\rm eD}}\right)\frac{\cosh(\sqrt{\varepsilon_k}\tilde{y}_{\rm D1})+\cosh(\sqrt{\varepsilon_k}\tilde{y}_{\rm D2})}{\sqrt{\varepsilon_k}\sinh(\sqrt{\varepsilon_k}y_{\rm eD})}$$
$$+2\sum_{n=1}^{\infty}\cos\left(n\pi\frac{z_{\rm D}}{h_{\rm D}}\right)\cos\left(n\pi\frac{z_{\rm wD}}{h_{\rm D}}\right)\times\left[\frac{\cosh[\sqrt{\varepsilon_n}(y_{\rm eD}-\tilde{y}_{\rm D})]}{\sqrt{\varepsilon_n}\sinh(\sqrt{\varepsilon_n}y_{\rm eD})}\right.$$
$$\left.\left.+2\sum_{k=1}^{+\infty}\cos\left(k\pi\frac{x_{\rm D}}{x_{\rm eD}}\right)\cos\left(k\pi\frac{x_{\rm wD}}{x_{\rm eD}}\right)\frac{\cosh\left(\sqrt{\varepsilon_{k,n}}\tilde{y}_{\rm D1}\right)+\cosh\left(\sqrt{\varepsilon_{k,n}}\tilde{y}_{\rm D2}\right)}{\sqrt{\varepsilon_{k,n}}\sinh\left(\sqrt{\varepsilon_{k,n}}y_{\rm eD}\right)}\right]\right\} \tag{2.102}$$

where:

$$\tilde{y}_{\rm D1}=y_{\rm eD}-|y_{\rm D}-y_{\rm wD}| \tag{2.103}$$

$$\tilde{y}_{\rm D2}=y_{\rm eD}-|y_{\rm D}+y_{\rm wD}| \tag{2.104}$$

Then the pressure drop in the rectangular reservoir generated by the intensity $q_{\rm scins}$ of the continuous point source is:

$$\Delta\overline{m}_{\rm f}=\frac{p_{\rm sc}T}{T_{\rm sc}}\frac{q_{\rm scins}}{2\pi k_{\rm f}L_{\rm ref}s}\overline{G}(M_{\rm D},M'_{\rm D},s) \tag{2.105}$$

Introducing Eq. (2.102) into Eq. (2.105), we see that:

$$\Delta\overline{m}_{\rm f}=\frac{p_{\rm sc}T}{T_{\rm sc}}\frac{q_{\rm scins}}{\pi k_{\rm f}L_{\rm ref}h_{\rm D}s}\frac{\pi}{x_{\rm eD}}\left\{\frac{\cosh(\sqrt{u}\tilde{y}_{\rm D1})+\cosh(\sqrt{u}\tilde{y}_{\rm D2})}{\sqrt{u}\sinh(\sqrt{u}y_{\rm eD})}+2\sum_{k=1}^{+\infty}\cos\left(\frac{k\pi x_{\rm D}}{x_{\rm eD}}\right)\cos\left(\frac{k\pi x_{\rm wD}}{x_{\rm eD}}\right)\right.$$
$$\times\frac{\cosh\left(\sqrt{\varepsilon_k}\tilde{y}_{\rm D1}\right)+\cosh\left(\sqrt{\varepsilon_k}\tilde{y}_{\rm D2}\right)}{\sqrt{\varepsilon_k}\sinh\left(\sqrt{\varepsilon_k}y_{\rm eD}\right)}+2\sum_{n=1}^{\infty}\cos\left(\frac{n\pi z_{\rm D}}{h_{\rm D}}\right)\cos\left(\frac{n\pi z_{\rm wD}}{h_{\rm D}}\right)\left[\frac{\cosh[\sqrt{\varepsilon_n}(y_{\rm eD}-\tilde{y}_{\rm D})]}{\sqrt{\varepsilon_n}\sinh(\sqrt{\varepsilon_n}y_{\rm eD})}\right.$$
$$\left.\left.+2\sum_{k=1}^{+\infty}\cos\left(k\pi\frac{x_{\rm D}}{x_{\rm eD}}\right)\cos\left(k\pi\frac{x_{\rm wD}}{x_{\rm eD}}\right)\frac{\cosh\left(\sqrt{\varepsilon_{k,n}}\tilde{y}_{\rm D1}\right)+\cosh\left(\sqrt{\varepsilon_{k,n}}\tilde{y}_{\rm D2}\right)}{\sqrt{\varepsilon_{k,n}}\sinh\left(\sqrt{\varepsilon_{k,n}}y_{\rm eD}\right)}\right]\right\} \tag{2.106}$$

The above expression is the pressure drop at any point of the reservoir generated by the constant intensity $q_{\rm scins}$ in the rectangular reservoir.

2.4.2 Computational remarks

In the preceding sections, the continuous point source solution for a point source producing at a constant intensity $q_{\rm scin}$ in a rectangular gas reservoir is derived. However, the solution of its equation can greatly impact computational speed and accuracy.

Therefore, a further analysis on the algorithm is continued in this section (Ozkan and Raghavan, 1991a,b, 1994):

(1) Computation of hyperbolic function $\frac{\cosh(\sqrt{u}\tilde{y}_{D1})+\cosh(\sqrt{u}\tilde{y}_{D2})}{\sinh(\sqrt{u}y_{eD})}$.

According to the expression of the hyperbolic function, the above expression can be spread out:

$$\begin{aligned}\frac{\cosh\left[\sqrt{u}\left(y_{eD}-y'_D\right)\right]}{\sinh\left(\sqrt{u}y_{eD}\right)} &= \frac{\exp\left[\sqrt{u}\left(y_{eD}-y'_D\right)\right]+\exp\left[-\sqrt{u}\left(y_{eD}-y'_D\right)\right]}{\exp\left(\sqrt{u}y_{eD}\right)-\exp\left(-\sqrt{u}y_{eD}\right)}\\ &= \frac{\exp\left[-\sqrt{u}y'_D\right]+\exp\left[-\sqrt{u}\left(2y_{eD}-y'_D\right)\right]}{1-\exp\left(-2\sqrt{u}y_{eD}\right)}\end{aligned} \tag{2.107}$$

For the denominator in Eq. (2.107), when $\sqrt{u}y_{eD}$ is very small, the denominator $1-\exp\left(-2\sqrt{u}y_{eD}\right)\to 0$; this would impact the rate of convergence. To overcome this problem, the following expression can be applied:

$$\frac{1}{1-\exp\left(-2\sqrt{u}y_{eD}\right)} = 1+\sum_{m=1}^{\infty}\exp\left(-2m\sqrt{u}y_{eD}\right) \tag{2.108}$$

Then Eq. (2.107) becomes

$$\frac{\cosh\left(\sqrt{u}\tilde{y}_{D1}\right)+\cosh\left(\sqrt{u}\tilde{y}_{D2}\right)}{\sinh\left(\sqrt{u}y_{eD}\right)} = \left\{e^{-\sqrt{u}\left(y_D+y_{wD}\right)}+e^{-\sqrt{u}\left(y_{eD}+\tilde{y}_{D1}\right)}+e^{-\sqrt{u}\left(y_{eD}+\tilde{y}_{D2}\right)}+e^{-\sqrt{u}\left|y_D+y_{wD}\right|}\right\}\left[1+\sum_{m=1}^{\infty}e^{-2m\sqrt{u}y_{eD}}\right] \tag{2.109}$$

Reforming the above expression results in:

$$\begin{aligned}\frac{\cosh\left(\sqrt{u}\tilde{y}_{D1}\right)+\cosh\left(\sqrt{u}\tilde{y}_{D2}\right)}{\sinh\left(\sqrt{u}y_{eD}\right)} &= e^{-\sqrt{u}\left|y_D-y_{wD}\right|}+e^{-\sqrt{u}\left|y_D-y_{wD}\right|}\sum_{m=1}^{\infty}e^{-2m\sqrt{u}y_{eD}}\\ &+\left\{e^{-\sqrt{u}\left(y_D+y_{wD}\right)}+e^{-\sqrt{u}\left(y_{eD}+\tilde{y}_{D2}\right)}+e^{-\sqrt{u}\left(y_{eD}+\tilde{y}_{D1}\right)}\right\}\\ &\left[1+\sum_{m=1}^{\infty}e^{-2m\sqrt{u}y_{eD}}\right]\end{aligned} \tag{2.110}$$

(2) Computation of series $\sum_{n=1}^{+\infty}\cos(n\pi z)\cos(n\pi z_w)\frac{e^{-\sqrt{u+(n\pi/h_D)^2+a^2}\left|y_D-y_{wD}\right|}}{\sqrt{u+(n\pi/h_D)^2+a^2}}$.

Applying a product-to-sum transformation to the trigonometric functions in the above series, there is:

$$\sum_{n=1}^{\infty}\cos(n\pi z)\cos(n\pi z_w)\frac{e^{-\sqrt{u+(n\pi/h_D)^2+a^2}|y_D-y_{wD}|}}{\sqrt{u+(n\pi/h_D)^2+a^2}}$$

$$=\frac{1}{2}\left[\sum_{n=1}^{\infty}\cos[n\pi(z-z_w)]\frac{e^{-\sqrt{u+(n\pi/h_D)^2+a^2}|y_D-y_{wD}|}}{\sqrt{u+(n\pi/h_D)^2+a^2}}\right.$$

$$\left.+\sum_{n=1}^{\infty}\cos[n\pi(z+z_w)]\frac{e^{-\sqrt{u+(n\pi/h_D)^2+a^2}|y_D-y_{wD}|}}{\sqrt{u+(n\pi/h_D)^2+a^2}}\right] \tag{2.111}$$

Assuming that the parameter group u is the expression of function F regarding variable τ in the Laplace space, then we see that:

$$\mathbf{L}[F(\tau)]=\sum_{n=1}^{\infty}\frac{\cos[n\pi(z-z_w)]}{\sqrt{u+(n\pi/h_D)^2+a^2}}e^{-\sqrt{u+(n\pi/h_D)^2+a^2}|y_D-y_{wD}|} \tag{2.112}$$

According to the displacement property of the Laplace transform:

$$\mathbf{L}[e^{-at}f(t)]=F(s+a) \tag{2.113}$$

There is the following equation:

$$\mathbf{L}^{-1}\left[\frac{e^{-\sqrt{u+(n\pi/h_D)^2+a^2}|y_D-y_{wD}|}}{\sqrt{u+(n\pi/h_D)^2+a^2}}\right]=e^{-\left[(n\pi/h_D)^2+a^2\right]\tau}\mathbf{L}^{-1}\left[\frac{e^{-\sqrt{u}|y_D-y_{wD}|}}{\sqrt{u}}\right] \tag{2.114}$$

Looking up in the Laplace transform table, there is:

$$\mathbf{L}\left[e^{-x^2/4\kappa t}\sqrt{\frac{\kappa}{\pi t}}\right]=\frac{e^{-\sqrt{s/\kappa}x}}{\sqrt{s/\kappa}} \tag{2.115}$$

Introducing Eq. (2.115) into Eq. (2.114) yields:

$$\mathbf{L}^{-1}\left[\frac{e^{-\sqrt{u+(n\pi/h_D)^2+a^2}|y_D-y_{wD}|}}{\sqrt{u+(n\pi/h_D)^2+a^2}}\right]=\frac{1}{\sqrt{\pi\tau}}e^{-\left[(n\pi/h_D)^2+a^2\right]\tau}e^{-(y_D-y_{wD})^2/4\tau} \tag{2.116}$$

Substituting Eq. (2.116) into Eq. (2.112) gives:

$$F(\tau)=\frac{1}{\sqrt{\pi\tau}}\exp\left(-\frac{(y_{\mathrm{D}}-y_{\mathrm{wD}})^2}{4\tau}\right)\exp\left(-a^2\tau\right)\sum_{n=1}^{\infty}\cos\left[(z-z_{\mathrm{w}})\right]\exp\left(-\frac{n^2\pi^2}{h_{\mathrm{D}}^2}\tau\right) \tag{2.117}$$

According to the Poisson formula in Eq. (2.82), we see that:

$$\sum_{n=1}^{\infty}\cos\left[n\pi(z-z_{\mathrm{w}})\right]\exp\left(-\frac{n^2\pi^2\tau}{h_{\mathrm{D}}^2}\right)=\frac{h_{\mathrm{D}}}{2\sqrt{\pi\tau}}\sum_{n=-\infty}^{+\infty}\exp\left[-\frac{((z-z_{\mathrm{w}}-2n)h_{\mathrm{D}})^2}{4\tau}\right]-\frac{1}{2} \tag{2.118}$$

Eq. (2.117) can be reformed as:

$$\mathrm{F}(\tau)=\frac{h_{\mathrm{D}}\exp\left(-a^2\tau\right)}{2\pi\tau}\sum_{n=-\infty}^{+\infty}\exp\left[-\frac{(z-z_{\mathrm{w}}-2n)^2h_{\mathrm{D}}^2+(y_{\mathrm{D}}-y_{\mathrm{wD}})^2}{4\tau}\right]-\frac{\exp\left(-\frac{(y_{\mathrm{D}}-y_{\mathrm{wD}})^2}{4\tau}\right)\exp\left(-a^2\tau\right)}{2\sqrt{\pi\tau}} \tag{2.119}$$

According to the following Laplace transform equation:

$$\mathbf{L}\left[\frac{\exp\left(-\frac{a^2}{4\kappa\tau}\right)}{2\tau}\right]=K_0\left(a\sqrt{u/\kappa}\right) \tag{2.120}$$

Combining Eqs. (2.115), (2.120), and (2.113), the expression of Eq. (2.119) in the Laplace space is:

$$\mathbf{L}[F(\tau)]=\frac{h_{\mathrm{D}}}{\pi}\sum_{n=-\infty}^{+\infty}K_0\left[\sqrt{(z-z_{\mathrm{w}}-2n)^2h_{\mathrm{D}}^2+(y_{\mathrm{D}}-y_{\mathrm{wD}})^2}\sqrt{u+a^2}\right]-\frac{\exp\left(-\sqrt{u+a^2}\left|y_{\mathrm{D}}-y_{\mathrm{wD}}\right|\right)}{2\sqrt{u+a^2}} \tag{2.121}$$

Substituting Eqs. (2.112) and (2.121) into Eq. (2.111), the resulting equation becomes:

$$\sum_{n=1}^{\infty}\cos(n\pi z)\cos(n\pi z_{\mathrm{w}})\frac{e^{-\sqrt{u+(n\pi/h_{\mathrm{D}})^2+a^2}\left|y_{\mathrm{D}}-y_{\mathrm{wD}}\right|}}{\sqrt{u+(n\pi/h_{\mathrm{D}})^2+a^2}}=$$
$$\frac{h_{\mathrm{D}}}{2\pi}\sum_{n=-\infty}^{+\infty}\left\{K_0\left[\sqrt{(z-z_{\mathrm{w}}-2n)^2h_{\mathrm{D}}^2+(y_{\mathrm{D}}-y_{\mathrm{wD}})^2}\sqrt{u+a^2}\right]+K_0\left[\sqrt{(z+z_{\mathrm{w}}-2n)^2h_{\mathrm{D}}^2+(y_{\mathrm{D}}-y_{\mathrm{wD}})^2}\sqrt{u+a^2}\right]\right\}-\frac{\exp\left(-\sqrt{u+a^2}\left|y_{\mathrm{D}}-y_{\mathrm{wD}}\right|\right)}{2\sqrt{u+a^2}} \tag{2.122}$$

Fractured vertical wells in shale gas reservoirs without SRV

3

Chapter Outline

3.1 Introduction 73
3.2 Fractured vertical wells in circular gas reservoirs 74
3.2.1 Physical models 74
3.2.2 Bottom-hole pressure expressions 75
3.3 Fractured vertical well in closed rectangular gas reservoirs 77
3.3.1 Continuous line source solutions in closed rectangular gas reservoirs 78
3.3.2 Bottom-hole pressure responses of fully penetrated fractured wells 82
3.4 Superposition of wellbore storage and skin effects 86
3.5 Solution for production at constant bottom-hole pressure 87
3.6 A numerical inversion algorithm 88
3.7 Gas well pressure and production performance analysis 90
3.7.1 Microfractures + steady state adsorption/desorption and diffusion 90
3.7.2 Microfractures + matrix macropores + steady state adsorption/desorption and diffusion 96
3.7.3 Microfractures + gas adsorption/desorption + Fick's diffusion 102
3.7.4 Microfractures + matrix macropores + gas adsorption/desorption + nanopore Fick's diffusion 107
3.7.5 Microfractures + gas adsorption/desorption + nanopore Knudsen diffusion 111

3.1 Introduction

During the early appraisal and development phases of shale gas reservoirs, in order to evaluate reservoir properties and gas resources, vertical wells are usually drilled and completed with hydraulic fracturing techniques to improve well productivity. According to the research and statistics available, these hydraulic fractures mostly extend vertically when the reservoir burial depth is greater than 700 m. Burial depths of shale gas reservoirs in North America range from 450 to 2300 m, while those in PR China are even deeper, ranging from 1500 to 4000 m. Therefore, vertical hydraulic fractures are common, and fractures mentioned in this chapter without special mention are vertical fractures.

In this chapter, transient flow mechanisms in vertical shale gas wells are discussed on the basis of the continuous point source solutions for circular and rectangular reservoirs derived in Chapter 2. Well test type curves of wells at constant production rates and production decline curves of wells at constant bottom-hole pressure under different flow

Developments in Petroleum Science, Vol. 66. https://doi.org/10.1016/B978-0-444-64315-5.00003-6

mechanisms are generated and analyzed through techniques of numerical inversion and programming. Two models studied are: vertical wells with infinite conductivity fractures in circular reservoirs and vertical wells with infinite conductivity fractures in rectangular reservoirs.

3.2 Fractured vertical wells in circular gas reservoirs

In this section, transient flow models for fractured vertical wells in shale gas reservoirs are discussed based on the continuous point source functions in circular gas reservoirs. Two physical models for fully penetrated fractured and partially penetrated fractured vertical wells are introduced, which correspond to the models with infinite conductivity and finite conductivity fractures, respectively. For a fracture shape, there are a bi-wing fracture model and an asymmetric bi-wing fracture model. The transient flow mechanisms in these models for conventional oil and gas reservoirs have been studied and discussed by many scholars. But few of them have discussed flow mechanisms for shale gas reservoirs regarding various complex flow mechanisms and multi-scale effects. Shale permeability is ultra-low, and even the microscopic fracture systems in a reservoir are much less permeable than artificial hydraulic fractures. In addition, shale gas wells have very low productivity. Therefore, hydraulic fractures in shale reservoirs can be considered to have an infinite conductivity. In this section, a well test analysis and a decline curve analysis for fractured shale gas wells with infinite conductivity fractures are introduced based on the five flow mechanism models studied in Chapter 1.

3.2.1 Physical models

Fig. 3.1 shows the physical model of a fully penetrated fractured well in a cylindrical shale gas reservoir with its sealed upper and lower boundaries. The assumptions for this model are: The reservoir thickness is h; the half fracture length of a symmetric bi-wing fracture is x_f; the radius of the outer boundary is r_e; the well is producing

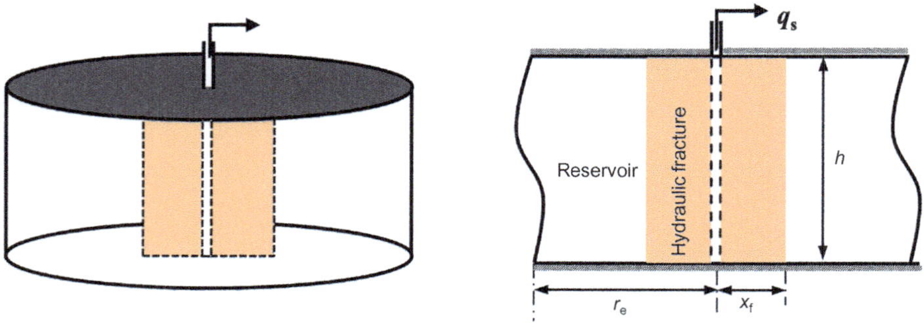

Fig. 3.1 A fully penetrated fractured well in a cylindrical shale gas reservoir.

at a constant rate q_{sc} or constant bottom-hole pressure p_{wf}; the fracture is of infinite conductivity; and the assumptions for the shale reservoir are the same as in the corresponding mechanism model.

3.2.2 Bottom-hole pressure expressions

For the well producing at a constant rate, the gas volume is evenly distributed along the fracture surface. Therefore, its bottom-hole pressure can be expressed by the integral of the continuous point source solutions derived in Section 2.3.2 for circular gas reservoirs with different boundary conditions. Because the constant pressure boundary condition is rarely seen for shale gas reservoirs, the following analysis mainly focuses on infinite and closed boundaries.

For an infinite shale gas reservoir, integrating in z_w Eq. (2.53) from 0 to h and then x_w from $-x_f$ to x_f, there is:

$$\Delta\overline{m}_f = \frac{p_{sc}Tq_{scins}}{\pi k_{fh}hT_{sc}s}\int_0^h\int_{-x_f}^{x_f}\left[K_0(\xi_0 r_D) + 2\sum_{n=1}^{\infty}K_0(\xi_n r_D)\cos(n\pi z_{wD})\cos(n\pi z_D)\right]dx_w dz_w \tag{3.1}$$

where

$$\xi_n = \sqrt{f(s) + \frac{(n\pi)^2}{h_D^2}},\quad (n = 0, 1, 2, 3, \ldots)\ \text{ and }\ r_D = \sqrt{(x_D - x_{wD})^2 + (y_D - y_{wD})^2}$$

We simplify the above equation to:

$$\Delta\overline{m}_f = \frac{p_{sc}Tq_{scins}h}{\pi k_{fh}hT_{sc}s}\int_{-x_f}^{x_f}K_0[\xi_0 r_D]dx_w \tag{3.2}$$

For a closed shale gas reservoir, the same methodology is applied:

$$\Delta\overline{m}_f = \frac{p_{sc}Tq_{scins}h}{\pi k_{fh}hT_{sc}s}\int_{-x_f}^{x_f}\left[K_0(\xi_0 r_D) + \frac{K_1(\xi_0 r_{eD})}{I_1(\xi_0 r_{eD})}I_0(\xi_0 r_D)\right]dx_w \tag{3.3}$$

For a fully penetrated fractured vertical well with an evenly distributed production volume along the fracture, there is the following relationship between q_{sc} and q_{scins}:

$$q_{sc} = 2hx_f q_{scins} \tag{3.4}$$

Define a dimensionless pressure by:

$$m_{\mathrm{fD}}=\frac{\pi k_{\mathrm{fh}}hT_{\mathrm{sc}}}{p_{\mathrm{sc}}Tq_{\mathrm{sc}}}\Delta m_{\mathrm{f}} \tag{3.5}$$

Substituting Eqs. (3.4) and (3.5) into Eqs. (3.2) and (3.3) and then converting the integration variable to a dimensionless format by applying the dimensionless length defined before, the dimensionless pressure solution can be obtained for a multi-scale fractured vertical well under infinite and closed boundary conditions in a shale gas reservoir considering various complex flow mechanisms. For an infinite boundary condition:

$$\overline{m}_{\mathrm{fD}}=\frac{1}{2s}\int_{-1}^{1}K_0[\xi_0 r_{\mathrm{D}}]\mathrm{d}x_{\mathrm{wD}} \tag{3.6}$$

For a closed boundary condition:

$$\overline{m}_{\mathrm{fD}}=\frac{1}{2s}\int_{-1}^{1}\left[K_0(\xi_0 r_{\mathrm{D}})+\frac{K_1(\xi_0 r_{\mathrm{eD}})}{I_1(\xi_0 r_{\mathrm{eD}})}I_0(\xi_0 r_{\mathrm{D}})\right]\mathrm{d}x_{\mathrm{wD}} \tag{3.7}$$

The pressure response expression at any point in the reservoir for a fractured vertical well producing at a constant rate have been obtained. But in reality, the bottom-hole pressure is critical to know for a well test analysis. Moreover, the pressure distributions are totally different for fractures with an evenly distributed flow rate than an infinite conductivity. Therefore, to analyze the dynamical pressure changes of infinite conductivity fractures through the above solutions, the pressure solution at a special point ($x_{\mathrm{D}}=0.732$) in fractures with evenly distributed flow can be found to represent approximately the bottom-hole pressure of a fractured vertical well with infinite conductivity fractures according to the research results of Gringarten et al. (1973). This method is the main method to analyze transient flow of wells with infinite conductivity fractures. Then, in Eqs. (3.6) and (3.7), assuming $x_{\mathrm{D}}=0.732$ and $y_{\mathrm{D}}=0$, the dimensionless bottom-hole pressure can be obtained. Because the coordinates are along the fracture surfaces, there is $y_{\mathrm{wD}}=0$ for the source/sink term. For an infinite boundary condition:

$$\overline{m}_{\mathrm{wD}}=\frac{1}{2s}\int_{-1}^{1}K_0(\xi_0|x_{\mathrm{D}}-\alpha|)\mathrm{d}\alpha \tag{3.8}$$

For a closed boundary condition:

$$\overline{m}_{\mathrm{wD}}=\frac{1}{2s}\int_{-1}^{1}\left[K_0(\xi_0|x_{\mathrm{D}}-\alpha|)+\frac{K_1(\xi_0 r_{\mathrm{eD}})}{I_1(\xi_0 r_{\mathrm{eD}})}I_0(\xi_0|x_{\mathrm{D}}-\alpha|)\right]\mathrm{d}\alpha \tag{3.9}$$

According to the properties of the Bessel function, when $\alpha \to x_D$, the value of $K_0(|x_D - \alpha|\sqrt{s})$ approaches zero. Therefore, the values close to this point are infinite when numerical integration is applied, and errors could result in this integration. In addition, the bottom-hole pressure must be constrained at the boundary so it is not appropriate to acquire the bottom-hole pressure through direct integration of the above equation. Therefore, an appropriate transformation is applied when calculating the dimensionless bottom-hole pressure. For an infinite boundary condition:

$$\overline{m}_{wD} = \frac{1}{2s\xi_0}\left[\int_0^{\xi_0(1+x_D)} K_0(\alpha)d\alpha + \int_0^{\xi_0(1-x_D)} K_0(\alpha)d\alpha\right] \tag{3.10}$$

For a closed boundary condition:

$$\overline{m}_{wD} = \frac{1}{2s\xi_0}\left[\begin{array}{l}\int_0^{\xi_0(1+x_D)} K_0(\alpha)d\alpha + \int_0^{\xi_0(1-x_D)} K_0(\alpha)d\alpha + \\ \frac{K_1(\xi_0 r_{eD})}{I_1(\xi_0 r_{eD})}\left(\int_0^{\xi_0(1+x_D)} I_0(\alpha)d\alpha + \int_0^{\xi_0(1-x_D)} I_0(\alpha)d\alpha\right)\end{array}\right] \tag{3.11}$$

3.3 Fractured vertical well in closed rectangular gas reservoirs

In the previous section, a transient flow theory is analyzed for a fractured vertical well in a circular gas reservoir. However, for fractured wells in low permeability oil and gas reservoirs, a pressure contour distribution is closer to a rectangle due to ultra-low permeability, a limited well drainage area, and an extended half fracture length by massive hydraulic fracturing. For such a scenario, a trilinear model was brought up by scholars in earlier days to analyze flow of fractured wells with finite conductivity fractures. Although this kind of model can well describe linear flow stages of fractures in a reservoir, it cannot represent well a boundary response and elliptical flow commonly seen in fractured wells. In this section, the point source function methodology is applied to analyze the transient flow theory of a fractured well in a rectangular shale gas reservoir. First, a continuous line source expression is derived by the continuous point source solution acquired previously. Then a dimensionless bottom-hole pseudo pressure is acquired by integration of the continuous line source solution along a fracture surface. Finally, well test analysis type curves and transient production decline curves for a fractured vertical well in a shale gas reservoir are analyzed by a numerical inversion method.

3.3.1 Continuous line source solutions in closed rectangular gas reservoirs

Fig. 3.2 shows the physical model of a continuous line source in a closed rectangular gas reservoir. The line source is assumed to be perpendicular to the upper and lower boundaries. Other assumptions are: The intensity of the line source is q_{scL}; the height of the line source equals the reservoir thickness h; the coordinates of any point on the line source is (x_w, y_w, z_w). Then the continuous line source solution can be acquired by integration along the line source. Because the height of the line source equals the reservoir thickness and is parallel to the z axis, the integration of Eq. (2.106) along the line source is:

$$\Delta\overline{m}_f = \frac{p_{sc}T}{T_{sc}}\frac{q_{scins}}{2\pi k_f L_{ref} s}\int_0^h \overline{G}\left(x_D, y_D, z_D, x'_D, y'_D, z'_D\right)dz' \tag{3.12}$$

Additionally, because the line source is producing at a constant rate q_{scL}, there is the following relationship with the continuous point source intensity q_{scins}:

$$q_{scL} = q_{scins}h \tag{3.13}$$

Introducing Eqs. (3.13) and (2.102) into Eq. (3.12), the pseudo pressure difference in the Laplace space generated by a continuous line source in a closed rectangular gas reservoir is:

$$\begin{aligned}\Delta\overline{m}_f = &\frac{p_{sc}T}{T_{sc}}\frac{q_{scL}}{\pi k_f L_{ref} h_D s x_{eD}}\frac{\pi}{}\left\{\frac{\cosh\left(\sqrt{u}\tilde{y}_{D1}\right)+\cosh\left(\sqrt{u}\tilde{y}_{D2}\right)}{\sqrt{u}\sinh\left(\sqrt{u}y_{eD}\right)}\right.\\ &\left.+2\sum_{k=1}^{+\infty}\cos\left(k\pi\frac{x_D}{x_{eD}}\right)\cos\left(k\pi\frac{x_{wD}}{x_{eD}}\right)\frac{\cosh\left(\sqrt{\varepsilon_k}\tilde{y}_{D1}\right)+\cosh\left(\sqrt{\varepsilon_k}\tilde{y}_{D2}\right)}{\sqrt{\varepsilon_k}\sinh\left(\sqrt{\varepsilon_k}y_{eD}\right)}\right\}\end{aligned} \tag{3.14}$$

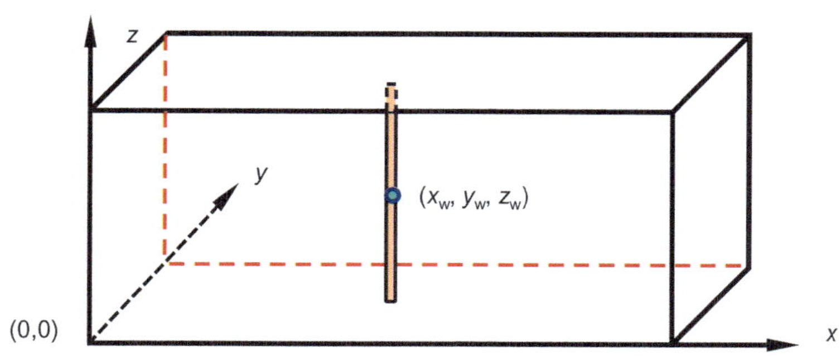

Fig. 3.2 A continuous line source in a closed rectangular gas reservoir.

The above expression is the continuous line source solution in a rectangular gas reservoir. By substituting the well location coordinates into this expression, the transient flow theory of a vertical well in a rectangular gas reservoir can be analyzed. To differentiate the contributions of different items in the above equation to the early and late stages of the curve, the items in the brackets can be expressed as follows:

$$\overline{S}_{\mathrm{VW}}=\overline{S}_{\mathrm{Vb1}}+\overline{S}_{\mathrm{V2}}+\overline{S}_{\mathrm{Vb3}}=\frac{\cosh\left(\sqrt{u}\tilde{y}_{\mathrm{D1}}\right)+\cosh\left(\sqrt{u}\tilde{y}_{\mathrm{D2}}\right)}{\sqrt{u}\sinh\left(\sqrt{u}y_{\mathrm{eD}}\right)}$$
$$+2\sum_{k=1}^{+\infty}\cos\left(k\pi\frac{x_{\mathrm{D}}}{x_{\mathrm{eD}}}\right)\cos\left(k\pi\frac{x_{\mathrm{wD}}}{x_{\mathrm{eD}}}\right)\frac{\cosh\left(\sqrt{\varepsilon_k}\tilde{y}_{\mathrm{D1}}\right)+\cosh\left(\sqrt{\varepsilon_k}\tilde{y}_{\mathrm{D2}}\right)}{\sqrt{\varepsilon_k}\sinh\left(\sqrt{\varepsilon_k}y_{\mathrm{eD}}\right)} \tag{3.15}$$

where

$$\overline{S}_{\mathrm{Vb1}}=\frac{\cosh\left(\sqrt{u}\tilde{y}_{\mathrm{D1}}\right)+\cosh\left(\sqrt{u}\tilde{y}_{\mathrm{D2}}\right)}{\sqrt{u}\sinh\left(\sqrt{u}y_{\mathrm{eD}}\right)}$$
$$=\frac{1}{\sqrt{u}}\left[\begin{array}{l}e^{-\sqrt{u}(y_{\mathrm{D}}+y_{\mathrm{wD}})}+e^{-\sqrt{u}\left(y_{\mathrm{eD}}+\tilde{y}_{\mathrm{D1}}\right)}+\\ e^{-\sqrt{u}\left(y_{\mathrm{eD}}+\tilde{y}_{\mathrm{D2}}\right)}+e^{-\sqrt{u}\left|y_{\mathrm{D}}-y_{\mathrm{wD}}\right|}\end{array}\right]\left[1+\sum_{m=1}^{\infty}e^{-2m\sqrt{u}y_{\mathrm{eD}}}\right] \tag{3.16}$$

It is noted that the convergent solutions in Eq. (3.16) can be acquired for both the early and late stages. However, when u and y_{eD} are very small, the rate of convergence can be greatly slowed down to affect the computational speed. Therefore, for the late stage of pressure propagation (u is very small), when y_{eD} is very small (this happens for a MFHW model in rectangular oil and gas reservoirs) and the following relationship is valid. Eq. (3.16) can be directly used to calculate $\overline{S}_{\mathrm{Vb1}}$ according to the research results of Medeiros et al. (2007, 2010):

$$\sqrt{u}y_{\mathrm{eD}}>5\times10^{-4} \tag{3.17}$$

When Eq. (3.17) is invalid, there is the next relationship according to Eq. (2.88):

$$\sum_{k=1}^{\infty}\frac{\cos(kx)}{k^2+a^2}=\frac{\pi}{2a}\frac{\cosh\left[a(\pi-x)\right]}{\sinh(a\pi)}-\frac{1}{2a^2}\quad[0\le x\le 2\pi] \tag{3.18}$$

By using Eq. (3.18), there is:

$$\frac{\cosh\left[\sqrt{u}\left(y_{\mathrm{eD}}-\left|y_{\mathrm{D}}-y_{\mathrm{wD}}\right|\right)\right]}{\sqrt{u}\sinh\left(\sqrt{u}y_{\mathrm{eD}}\right)}=\frac{\cosh\left[\sqrt{u}\frac{y_{\mathrm{eD}}}{\pi}\left(\pi-\pi\frac{\left|y_{\mathrm{D}}-y_{\mathrm{wD}}\right|}{y_{\mathrm{eD}}}\right)\right]}{\sqrt{u}\sinh\left(\sqrt{u}\frac{y_{\mathrm{eD}}}{\pi}\pi\right)}$$
$$=\frac{2}{y_{\mathrm{eD}}}\sum_{k=1}^{\infty}\frac{\cos\left(k\pi\frac{\left|y_{\mathrm{D}}-y_{\mathrm{wD}}\right|}{y_{\mathrm{eD}}}\right)}{k^2\frac{\pi^2}{y_{\mathrm{eD}}^2}+u}+\frac{1}{uy_{\mathrm{eD}}} \tag{3.19}$$

Similarly,

$$\frac{\cosh\left[\sqrt{u}(y_{eD}-(y_D+y_{wD}))\right]}{\sqrt{u}\sinh\left(\sqrt{u}y_{eD}\right)}=\frac{2}{y_{eD}}\sum_{k=1}^{\infty}\frac{\cos\left(k\pi\frac{y_D+y_{wD}}{y_{eD}}\right)}{k^2\frac{\pi^2}{y_{eD}^2}+u}+\frac{1}{uy_{eD}} \tag{3.20}$$

Then, substituting Eqs. (3.19) and (3.20) into Eq. (3.16), $\overline{S}_{Vb1}$ for $\sqrt{u}y_{eD}\leq 5\times10^{-4}$ can be expressed by:

$$\overline{S}_{Vb1}=\frac{2}{y_{eD}}\sum_{k=1}^{\infty}\frac{\cos\left(k\pi\frac{|y_D-y_{wD}|}{y_{eD}}\right)+\cos\left(k\pi\frac{y_D+y_{wD}}{y_{eD}}\right)}{k^2\frac{\pi^2}{y_{eD}^2}+u}+\frac{2}{uy_{eD}} \tag{3.21}$$

Summing the infinite series term in Eq. (3.21) by ignoring the effect of u for $u<k^2\frac{\pi^2}{y_{eD}^2}$, there is:

$$\frac{\cos\left(k\pi\frac{|y_D-y_{wD}|}{y_{eD}}\right)+\cos\left(k\pi\frac{y_D+y_{wD}}{y_{eD}}\right)}{k^2\frac{\pi^2}{y_{eD}^2}+u}\cong\frac{y_{eD}^2}{\pi^2}\frac{\cos\left(k\pi\frac{|y_D-y_{wD}|}{y_{eD}}\right)+\cos\left(k\pi\frac{y_D+y_{wD}}{y_{eD}}\right)}{k^2} \tag{3.22}$$

According to the relationship:

$$\sum_{k=1}^{\infty}\frac{\cos(kx)}{k^2}=\frac{\pi^2}{6}-\frac{\pi x}{2}+\frac{x^2}{4},\quad(0\leq x\leq 2\pi) \tag{3.23}$$

and substituting Eqs. (3.22) and (3.23) into Eq. (3.21), the $\overline{S}_{Vb1}$ expression for $\sqrt{u}y_{eD}\leq 5\times10^{-4}$ and $\frac{uy_{eD}}{k\pi}\leq 0.01$ can be written as:

$$\overline{S}_{Vb1}=\frac{2}{uy_{eD}}+2y_{eD}\left[\frac{1}{3}-\left(\frac{y_D+y_{wD}+|y_D-y_{wD}|}{2y_{eD}}\right)+\frac{y_D^2+y_{wD}^2}{2y_{eD}^2}\right] \tag{3.24}$$

In the calculation process, k in $uy_{ed}/k\pi\leq 0.01$ is a variable. If $uy_{ed}/k\pi\leq 0.01$ is valid with $k=1$, then the reliability of the simplified Eq. (3.24) can be ensured. Therefore, Eq. (3.24) is applicable when: $\sqrt{u}y_{eD}\leq 5\times10^{-4}$ and $uy_{eD}\leq 0.01\pi$, and Eq. (3.16) applies in the rest of the conditions.

Due to existence of the infinite series summation in $\overline{S}_{v2}$ and $\overline{S}_{Vb3}$ in Eq. (3.15), different methods need to be used to improve a convergence rate for calculations of the early and late stages.

(1) For the early stage (s is big enough and $\sqrt{u}y_{eD} > 5\times10^{-4}$),

$$\overline{S}_{V2} = 2\sum_{k=1}^{+\infty}\cos\left(k\pi\frac{x_D}{x_{eD}}\right)\cos\left(k\pi\frac{x_{wD}}{x_{eD}}\right)\frac{e^{-\sqrt{\varepsilon_k}|y_D-y_{wD}|}}{\sqrt{\varepsilon_k}} \tag{3.25}$$

$$\begin{aligned}\overline{S}_{Vb3} = 2\sum_{k=1}^{+\infty}\cos\left(k\pi\frac{x_D}{x_{eD}}\right)\cos\left(k\pi\frac{x_{wD}}{x_{eD}}\right)&\left\{\frac{e^{-\sqrt{\varepsilon_k}|y_D-y_{wD}|}}{\sqrt{\varepsilon_k}}\sum_{m=1}^{\infty}e^{-2m\sqrt{\varepsilon_k}y_{eD}}\right.\\ &\left.+\frac{e^{-\sqrt{\varepsilon_k}(y_D+y_{wD})}+e^{-\sqrt{\varepsilon_k}(y_{eD}+\tilde{y}_{D1})}+e^{-\sqrt{\varepsilon_k}(y_{eD}+\tilde{y}_{D2})}}{\sqrt{\varepsilon_k}}\left[1+\sum_{m=1}^{\infty}e^{-2m\sqrt{\varepsilon_k}y_{eD}}\right]\right\}\end{aligned} \tag{3.26}$$

Eq. (3.25) can be simplified, based on Eq. (2.122), to:

$$\overline{S}_{V2} = \frac{x_{eD}}{\pi}\left\{\begin{array}{l}K_0\left[\sqrt{(x_D-x_{wD}-2nx_{eD})^2+(y_D-y_{wD})^2}\sqrt{u}\right]+\\ K_0\left[\sqrt{(x_D+x_{wD}-2nx_{eD})^2+(y_D-y_{wD})^2}\sqrt{u}\right]\end{array}\right\}-\frac{\exp(-\sqrt{u}|y_D-y_{wD}|)}{\sqrt{u}} \tag{3.27}$$

Simplify the summation of the infinite series in the above equation to obtain:

$$\overline{S}_{V2} = \overline{S}_{Vb2} + \overline{S}_{inf} \tag{3.28}$$

where:

$$\begin{aligned}\overline{S}_{Vb2} = &\frac{x_{eD}}{\pi}\sum_{n=1}^{+\infty}\left\{K_0\left[\sqrt{(x_D-x_{wD}-2nx_{eD})^2+(y_D-y_{wD})^2}\sqrt{u}\right]\right.\\ &+K_0\left[\sqrt{(x_D+x_{wD}-2nx_{eD})^2+(y_D-y_{wD})^2}\sqrt{u}\right]\\ &+K_0\left[\sqrt{(x_D-x_{wD}+2nx_{eD})^2+(y_D-y_{wD})^2}\sqrt{u}\right]\\ &\left.+K_0\left[\sqrt{(x_D+x_{wD}+2nx_{eD})^2+(y_D-y_{wD})^2}\sqrt{u}\right]\right\}\\ &+\frac{x_{eD}}{\pi}K_0\left[\sqrt{(x_D+x_{wD})^2+(y_D-y_{wD})^2}\sqrt{u}\right]-\frac{\exp(-\sqrt{u}|y_D-y_{wD}|)}{\sqrt{u}}\end{aligned} \tag{3.29}$$

$$\overline{S}_{inf} = \frac{x_{eD}}{\pi}K_0\left[\sqrt{(x_D-x_{wD})^2+(y_D-y_{wD})^2}\sqrt{u}\right] \tag{3.30}$$

(2) For the late stage (s is very small and $\sqrt{u}y_{eD} \le 5\times10^{-4}$), according to the research results of Medeiros et al. (2007, 2008, 2010), the following method can be used to improve the computational convergence rate:

$$\overline{S}_{\mathrm{Vb2}}=2\sum_{k=1}^{+\infty}\frac{\cos\left(k\pi\frac{x_{\mathrm{D}}}{x_{\mathrm{eD}}}\right)\cos\left(k\pi\frac{x_{\mathrm{wD}}}{x_{\mathrm{eD}}}\right)}{\sqrt{\varepsilon_k}}\left\{\left[e^{-\sqrt{\varepsilon_k}(y_{\mathrm{D}}+y_{\mathrm{wD}})}+e^{-\sqrt{\varepsilon_k}\left(y_{\mathrm{eD}}+\tilde{y}_{\mathrm{D1}}\right)}\right.\right.$$
$$\left.\left.+e^{-\sqrt{\varepsilon_k}\left(y_{\mathrm{eD}}+\tilde{y}_{\mathrm{D2}}\right)}+e^{-\sqrt{\varepsilon_k}|y_{\mathrm{D}}-y_{\mathrm{wD}}|}\right]\left(1+\sum_{m=1}^{\infty}e^{-2m\sqrt{\varepsilon_k}y_{\mathrm{eD}}}\right)\right\} \tag{3.31}$$

$$\overline{S}_{\mathrm{Vb3}}=0 \tag{3.32}$$

$$\overline{S}_{\mathrm{inf}}=0 \tag{3.33}$$

If the following dimensionless pressure is defined:

$$m_{\mathrm{fD}}=\frac{\pi k_{\mathrm{f}}hT_{\mathrm{sc}}}{q_{\mathrm{scL}}p_{\mathrm{sc}}T}\Delta m_{\mathrm{f}} \tag{3.34}$$

then the dimensionless pseudo pressure for a vertical well at the center of a rectangular gas reservoir can be expressed as:

$$\overline{m}_{\mathrm{fD}}=\frac{\pi}{x_{\mathrm{eD}}s}\left(\overline{S}_{\mathrm{Vb1}}+\overline{S}_{\mathrm{Vb2}}+\overline{S}_{\mathrm{Vb3}}+\overline{S}_{\mathrm{inf}}\right) \tag{3.35}$$

Although vertical wells have not been analyzed in the previous discussions, the above analysis is meaningful to improve computational convergence rates and accuracy for fractured wells in the following discussions.

3.3.2 Bottom-hole pressure responses of fully penetrated fractured wells

In the previous section, a fully penetrated continuous line source in a rectangular gas reservoir is analyzed, and the factors affecting a computational speed are discussed. In this section, the bottom-hole pressure response of a fully penetrated fractured well with infinite conductivity fractures will be analyzed based on the continuous line source solution introduced previously.

Fig. 3.3 shows the physical model of a fully penetrated fractured well in a closed rectangular shale gas reservoir. The assumptions for this model are: x_{e}, y_{e} and h for the reservoir length, width and height, respectively; infinite conductivity fractures with their height equal to the reservoir thickness; symmetric bi-wing rectangular fractures with half-length x_f and negligible fracture width. The yellow line in this figure represents a line source at any location of the fracture surface. Then the dimensionless pseudo pressure solution at any point in the rectangular gas reservoir generated by a fractured well producing at a constant rate can be acquired by integration of the line source along the x axis, which is the fracture surface.

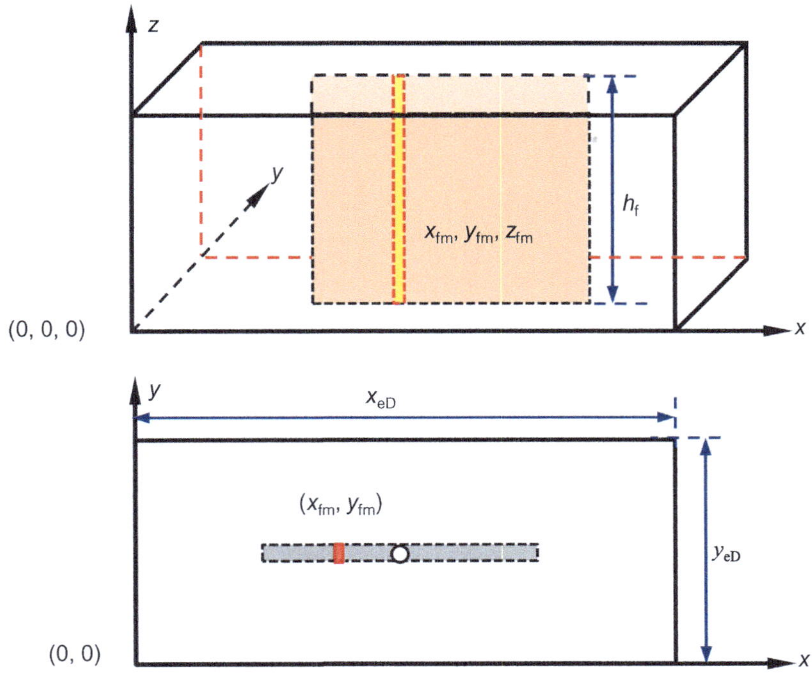

Fig. 3.3 A fractured well in a closed rectangular shale gas reservoir.

$$\overline{m}_{\mathrm{f}} = \int_{x_{\mathrm{fm}}-x_{\mathrm{f}}}^{x_{\mathrm{fm}}+x_{\mathrm{f}}} \overline{G}_{\mathrm{L}}\left(x_{\mathrm{D}}, y_{\mathrm{D}}, x'_{\mathrm{D}}, y'_{\mathrm{D}}\right) \mathrm{d}x' \tag{3.36}$$

In the above equation, $\overline{G}_{\mathrm{L}}$ is the continuous line source solution with intensity q_{scL} in the rectangular gas reservoir, and its expression is given by Eq. (3.14). Then the above equation can be written, by introducing Eq. (3.14), as:

$$\begin{aligned}\overline{S}_{FS} = {} & C2x_{\mathrm{f}} \frac{\cosh\left(\sqrt{u}\tilde{y}_{\mathrm{D1}}\right) + \cosh\left(\sqrt{u}\tilde{y}_{\mathrm{D2}}\right)}{\sqrt{u}\sinh\left(\sqrt{u}y_{\mathrm{eD}}\right)} \\ & + 2Cz_{\mathrm{f}} \sum_{k=1}^{+\infty} \frac{\cosh\left(\sqrt{\varepsilon_k}\tilde{y}_{\mathrm{D1}}\right) + \cosh\left(\sqrt{\varepsilon_k}\tilde{y}_{\mathrm{D2}}\right)}{\sqrt{\varepsilon_k}\sinh\left(\sqrt{\varepsilon_k}y_{\mathrm{eD}}\right)} \cos\left(k\pi \frac{x_{\mathrm{D}}}{x_{\mathrm{eD}}}\right) \int_{x_{\mathrm{fm}}-x_{\mathrm{f}}}^{x_{\mathrm{fm}}+x_{\mathrm{f}}} \cos\left(k\pi \frac{x_{\mathrm{wD}}}{x_{\mathrm{eD}}}\right) \mathrm{d}x_{\mathrm{w}}\end{aligned} \tag{3.37}$$

where

$$C = \frac{p_{\mathrm{sc}}T}{T_{\mathrm{sc}}} \frac{q_{\mathrm{scL}}}{\pi k_{\mathrm{f}} L_{\mathrm{ref}} h_{\mathrm{D}} s x_{\mathrm{eD}} s} \frac{\pi}{}$$

There is the relationship between the line source intensity and well production rate for a fractured well:

$$q_{\mathrm{sc}}=2q_{\mathrm{scL}}x_{\mathrm{f}} \tag{3.38}$$

Using the expression of the dimensionless pseudo pressure in the same format as in Eq. (3.34), the dimensionless pseudo pressure at any point in the rectangular shale gas reservoir generated by a fully penetrated fractured well can be expressed as:

$$\overline{m}_{\mathrm{fD}}=\frac{\pi}{x_{\mathrm{eD}}s}\left\{\frac{\cosh\left(\sqrt{u}\tilde{y}_{\mathrm{D1}}\right)+\cosh\left(\sqrt{u}\tilde{y}_{\mathrm{D2}}\right)}{\sqrt{u}\sinh\left(\sqrt{u}y_{\mathrm{eD}}\right)}+\frac{L_{\mathrm{ref}}}{x_{\mathrm{f}}}\sum_{k=1}^{+\infty}\int_{x_{\mathrm{fmD}}-x_{\mathrm{fD}}}^{x_{\mathrm{fmD}}+x_{\mathrm{fD}}}\cos\left(k\pi\frac{x_{\mathrm{wD}}}{x_{\mathrm{eD}}}\right)\mathrm{d}x_{\mathrm{wD}}\right.$$
$$\left.\times\cos\left(k\pi\frac{x_{\mathrm{D}}}{x_{\mathrm{eD}}}\right)\frac{\cosh\left(\sqrt{\varepsilon_k}\tilde{y}_{\mathrm{D1}}\right)+\cosh\left(\sqrt{\varepsilon_k}\tilde{y}_{\mathrm{D2}}\right)}{\sqrt{\varepsilon_k}\sinh\left(\sqrt{\varepsilon_k}y_{\mathrm{eD}}\right)}\right\} \tag{3.39}$$

where

$$x_{\mathrm{fmD}}=\frac{x_{\mathrm{fm}}}{L_{\mathrm{ref}}}\sqrt{\frac{k}{k_x}}$$

The items in the brackets in Eq. (3.39) can be expressed by the following format, the same as that for the line source of a vertical well:

$$\overline{S}_{\mathrm{VW}}=\overline{S}_{\mathrm{Vb1}}+\overline{S}_{\mathrm{V2}}+\overline{S}_{\mathrm{Vb3}} \tag{3.40}$$

where the expression of $\overline{S}_{\mathrm{Vb1}}$ is the same as in Eq. (3.16). In addition, $\overline{S}_{\mathrm{V2}}$ and $\overline{S}_{\mathrm{Vb3}}$ can be expressed, respectively, as:

$$\overline{S}_{\mathrm{V2}}=\frac{2}{x_{\mathrm{fD}}}\int_{x_{\mathrm{fmD}}-x_{\mathrm{fD}}}^{x_{\mathrm{fmD}}+x_{\mathrm{fD}}}\sum_{k=1}^{+\infty}\cos\left(k\pi\frac{x_{\mathrm{D}}}{x_{\mathrm{eD}}}\right)\cos\left(k\pi\frac{x_{\mathrm{wD}}}{x_{\mathrm{eD}}}\right)\frac{e^{-\sqrt{\varepsilon_k}|y_{\mathrm{D}}-y_{\mathrm{wD}}|}}{\sqrt{\varepsilon_k}}\mathrm{d}x_{\mathrm{wD}} \tag{3.41}$$

$$\overline{S}_{\mathrm{Vb3}}=\frac{2}{x_{\mathrm{fD}}}\int_{x_{\mathrm{fmD}}-x_{\mathrm{fD}}}^{x_{\mathrm{fmD}}+x_{\mathrm{fD}}}\sum_{k=1}^{+\infty}\cos\left(k\pi\frac{x_{\mathrm{D}}}{x_{\mathrm{eD}}}\right)\cos\left(k\pi\frac{x_{\mathrm{wD}}}{x_{\mathrm{eD}}}\right)\left\{\frac{e^{-\sqrt{\varepsilon_k}|y_{\mathrm{D}}-y_{\mathrm{wD}}|}}{\sqrt{\varepsilon_k}}\sum_{m=1}^{\infty}e^{-2m\sqrt{\varepsilon_k}y_{\mathrm{eD}}}\right.$$
$$\left.+\frac{e^{-\sqrt{\varepsilon_k}(y_{\mathrm{D}}-y_{\mathrm{wD}})}+e^{-\sqrt{\varepsilon_k}\left(y_{\mathrm{eD}}+\tilde{y}_{\mathrm{D1}}\right)}+e^{-\sqrt{\varepsilon_k}\left(y_{\mathrm{eD}}+\tilde{y}_{\mathrm{D2}}\right)}}{\sqrt{\varepsilon_k}}\left[1+\sum_{m=1}^{\infty}e^{-2m\sqrt{\varepsilon_k}y_{\mathrm{eD}}}\right]\right\}\mathrm{d}x_{\mathrm{wD}} \tag{3.42}$$

Rearranging Eq. (3.41) results in:

$$\overline{S}_{V2} = \overline{S}_{Vb2} + \overline{S}_{inf} \tag{3.43}$$

where:

$$\begin{aligned}\overline{S}_{Vb2} = {} & \frac{x_{eD}}{\pi}\sum_{n=1}^{+\infty}\int_{-1}^{1}\left\{K_0\left[\sqrt{(x_D - x_{fmD} - 2nx_{eD} - \alpha)^2 + (y_D - y_{wD})^2}\sqrt{u}\right]\right.\\ & + K_0\left[\sqrt{(x_D + x_{fmD} - 2nx_{eD} + \alpha)^2 + (y_D - y_{wD})^2}\sqrt{u}\right]\\ & + K_0\left[\sqrt{(x_D - x_{fmD} + 2nx_{eD} - \alpha)^2 + (y_D - y_{wD})^2}\sqrt{u}\right]\\ & \left. + K_0\left[\sqrt{(x_D + x_{fmD} + 2nx_{eD} + \alpha)^2 + (y_D - y_{wD})^2}\sqrt{u}\right]\right\}d\alpha\\ & + \frac{x_{eD}}{\pi}\int_{-1}^{1}K_0\left[\sqrt{(x_D + x_{fmD} + \alpha)^2 + (y_D - y_{wD})^2}\sqrt{u}\right]d\alpha - \frac{2\exp(-\sqrt{u}|y_D - y_{wD}|)}{\sqrt{u}}\end{aligned} \tag{3.44}$$

$$\overline{S}_{inf} = \frac{x_{eD}}{\pi}\int_{-1}^{1}K_0\left[\sqrt{(x_D - x_{fmD} - \alpha)^2 + (y_D - y_{wD})^2}\sqrt{u}\right]d\alpha \tag{3.45}$$

According to the integration of trigonometric functions:

$$\begin{aligned}\int_{x_{fmD}-x_{fX}}^{x_{fmD}+x_{fD}}\cos\left(k\pi\frac{x_{wD}}{x_{eD}}\right)dx_{wD} &= \frac{x_{eD}}{k\pi}\sin\left(k\pi\frac{x_{wD}}{x_{eD}}\right)\Bigg|_{x_{fmD}-x_{fD}}^{x_{fmD}+x_{fD}}\\ &= 2\frac{x_{eD}}{\pi}\frac{1}{k}\cos\left(k\pi\frac{x_{fmD}}{x_{eD}}\right)\sin\left(k\pi\frac{x_{fX}}{x_{eD}}\right)\end{aligned} \tag{3.46}$$

and substituting Eq. (3.46) into Eq. (3.42), we see that:

$$\begin{aligned}\overline{S}_{Vb3} = {} & \frac{4x_{eD}}{\pi}\sum_{k=1}^{+\infty}\frac{\cos\left(k\pi\frac{x_D}{x_{eD}}\right)\cos\left(k\pi\frac{x_{fmD}}{x_{eD}}\right)\sin\left(k\pi\frac{1}{x_{eD}}\right)}{k}\left\{\frac{e^{-\sqrt{\varepsilon_k}|y_D + y_{wD}|}}{\sqrt{\varepsilon_k}}\sum_{m=1}^{\infty}e^{-2m\sqrt{\varepsilon_k}y_{eD}}\right.\\ & \left. + \frac{e^{-\sqrt{\varepsilon_k}(y_D + y_{wD})} + e^{-\sqrt{\varepsilon_k}(y_{eD} + \tilde{y}_{D1})} + e^{-\sqrt{\varepsilon_k}(y_{eD} + \tilde{y}_{D2})}}{\sqrt{\varepsilon_k}}\left[1 + \sum_{m=1}^{\infty}e^{-2m\sqrt{\varepsilon_k}y_{eD}}\right]\right\}\end{aligned} \tag{3.47}$$

3.4 Superposition of wellbore storage and skin effects

In the previous discussions, wellbore storage and skin effects have not been considered when deriving the point source solutions and well test analysis models for fractured wells by using these point source solutions. Because a pressure gauge usually locates at a certain distance above the middle depth of a reservoir, the wellbore storage effect is comparatively obvious during shut-in and open-well tests. In addition, a fracturing and completion fluid can cause certain damage to a formation during drilling, completion, and hydraulic fracturing. Furthermore, this damage is not negligible for well test and production decline analyses, and there is a convergence skin for fractured wells. In fact, the skin is an additional dimensionless pressure drop caused by the formation damage. Therefore, the dimensionless pseudo pressure in a gas reservoir considering the skin effect is:

$$m_{\mathrm{wD2}} = m_{\mathrm{wD1}} + S_{\mathrm{kin}} \tag{3.48}$$

where:

S_{kin}—skin factor (dimensionless):
m_{wD1}—dimensionless pseudo pressure difference excluding the skin effect;
m_{wD2}—dimensionless pseudo pressure difference including the skin effect.

The skin factor is defined by:

$$S_{\mathrm{kin}} = \frac{\pi k_{\mathrm{fh}} h T_{\mathrm{sc}}}{p_{\mathrm{sc}} T q_{\mathrm{sc}}} \Delta m_{\mathrm{s}} \tag{3.49}$$

where Δm_{s} is the additional pseudo pressure drop caused by the skin effect (Pa/s).

For the wellbore storage effect, according to the superposition principle proposed by Everdinggen and Hurst (1949), the following convolution equation can be used to calculate the dimensionless bottom-hole pressure drop for a well producing at a constant rate at $t=0$:

$$m_{\mathrm{wD3}} = \int_0^{t_{\mathrm{D}}} q_{\mathrm{D}}(\tau) \frac{\partial m_{\mathrm{wD2}}}{\partial t_{\mathrm{D}}} (t_{\mathrm{D}} - \tau) \mathrm{d}\tau \tag{3.50}$$

where:

m_{wD3}—dimensionless bottom-hole pressure drop counting for the wellbore storage and skin effects;
$q_{\mathrm{D}}(t_{\mathrm{D}})$—dimensionless well production rate, $q_{\mathrm{D}} = q/q_{\mathrm{sc}}$.

The relationship between a subsurface rate and a surface rate at standard conditions can be represented by a wellbore storage coefficient and bottom-hole pressure:

$$q_{\mathrm{sf}} - q_{\mathrm{sc}} = \frac{C}{B_{\mathrm{g}}} \frac{\partial p_{\mathrm{wf}}}{\partial t} \tag{3.51}$$

where:

q_{sf}—subsurface production rate at wellbore or fracture surface conditions under standard conditions (m^3/s);
q_{sc}—surface production rate under standard conditions (m^3/s);
C—wellbore and fracture system storage coefficient (m^3/Pa);
B_g—gas volume factor at bottom-hole pressure (sm^3/m^3);
p_{wf}—flowing bottom-hole pressure (Pa).

To be consistent with the expression of dimensionless time defined for the mechanism models mentioned earlier, the dimensionless time can be written as:

$$t_D = \frac{k_{fh} t}{\Lambda L_{ref}^2} \tag{3.52}$$

where Λ is a parameter group, different for different mechanism models, referring to Section 2.2.

Substituting the dimensionless variables, time, production rate, volume factor, and pseudo pressure, into Eq. (3.51), we obtain:

$$q_D = 1 - C_D \frac{\partial m_{wD2}}{\partial t_D} \tag{3.53}$$

where the dimensionless wellbore storage coefficients is expressed as:

$$C_D = \frac{C \mu_{gi}}{2\pi h \Lambda L_{ref}^2} \tag{3.54}$$

Introducing Eqs. (3.48) and (3.53) into Eq. (3.50) for the Laplace transformation, the dimensionless bottom-hole pseudo pressure with consideration of the wellbore storage and skin effects is:

$$\overline{m}_{wD3} = \frac{s\overline{m}_{wD1} + S_{kin}}{s + C_D s^2 (s\overline{m}_{wD1} + S_{kin})} \tag{3.55}$$

For convenience, the dimensionless pressure drop without consideration of wellbore storage and skin is represented by $\overline{m}_{wDN}$, and that with consideration of these effects is represented by $\overline{m}_{wD}$.

3.5 Solution for production at constant bottom-hole pressure

The point source solutions and models introduced previously are based on the assumption of wells producing at constant rates. When a gas well is producing at constant bottom-hole pressure, according to the research results of Everdingen and Hurst (1949) and Ozkan and Raghavan (1991a,b), there is the following relationship between the dimensionless pressure and production rate:

$$\overline{q}_{\mathrm{D}} = \frac{1}{\overline{m}_{\mathrm{fD}} \times s^2} \tag{3.56}$$

where the definitions of the dimensionless pseudo-pressure and rate are:

$$m_{\mathrm{fD}} = \frac{\pi k_{\mathrm{fh}} h T_{\mathrm{sc}}}{p_{\mathrm{sc}} T q_{\mathrm{sc}}} \Delta m_{\mathrm{f}} \tag{3.57}$$

$$q_{\mathrm{D}} = \frac{q_{\mathrm{sc}} p_{\mathrm{sc}} T}{\pi k_{\mathrm{fh}} h T_{\mathrm{sc}} (m_{\mathrm{fi}} - m_{\mathrm{wf}})} \tag{3.58}$$

To solve the transient flow model for gas wells producing at constant pressure, the pseudo pressure needs be defined as:

$$m_{\mathrm{fD}} = \frac{\Delta m_{\mathrm{f}}}{m_i - m_{\mathrm{wf}}} \tag{3.59}$$

Note that if the ad-/desorption of shale gas satisfies the Fick diffusion law in a mechanism model, the following parameters must be redefined for constant bottom-hole pressure conditions to give the model dimension:

$$\sigma = \frac{G_{\mathrm{L}} m(p_{\mathrm{L}})}{[m(p_{\mathrm{L}}) + m(p_{\mathrm{f}})][m(p_{\mathrm{L}}) + m(p_{\mathrm{i}})]} (m_i - m_{\mathrm{wf}}) \tag{3.60}$$

$$\Lambda = \begin{cases} X + \dfrac{6 p_{\mathrm{sc}} T}{T_{\mathrm{sc}} (m_i - m_{\mathrm{wf}})} & \text{unsteady-state diffusion} \\ X + \dfrac{2 p_{\mathrm{sc}} T}{T_{\mathrm{sc}} (m_i - m_{\mathrm{wf}})} & \text{steady-state diffusion} \end{cases} \tag{3.61}$$

where, for Model 3, the parameter group in Eq. (3.61) is defined by:

$$X = \phi_{\mathrm{f}} \mu_{\mathrm{gi}} c_{\mathrm{fgi}} \tag{3.62}$$

And, for Model 4, it is defined by:

$$X = \phi_{\mathrm{m}} \mu_{\mathrm{gi}} c_{\mathrm{mgi}} + \phi_{\mathrm{f}} \mu_{\mathrm{gi}} c_{\mathrm{fgi}} \tag{3.63}$$

3.6 A numerical inversion algorithm

All the solutions derived previously are in a Laplace space. To plot their type curves, they need be converted to a real space. Since the dimensionless bottom-hole pressure is in a complex format, it is difficult to get a primitive function through the Laplace inverse transformation or contour integral. In engineering applications, what people usually care about are the characteristics of the pressure type curves in different flow stages, and the real data can be explained through matching with the type curves. The difficulty in inversion through a contour integral and the complexity of image functions result in a difficulty in acquisition of their primitive functions. This constrains an application of the Laplace transformation in the engineering field. For a long time, people had been seeking for a satisfactory numerical inversion method to avoid shortages of analytical inversion. Until the 1970s, various numerical inversion algorithms were developed and accelerated the application of the Laplace transformation in engineering. In petroleum engineering, there are two main methods in numerical inversion; one is the Stehfest inversion algorithm (Stehfest,1970), which is based on a function probability density theory, and the other one is the Crump inversion algorithm, which is based on the Fourier series theory. Due to its simplicity and programming friendliness, the Stehfest inversion algorithm is the most common one in engineering applications.

For the Stehfest numerical inversion algorithm, assuming that $F(s)$ is the Laplace transformation of a time function $f(t)$, then:

$$F(s) = \mathbf{L}[f(t)] \tag{3.64}$$

Therefore, the Stehfest numerical inversion algorithm to invert a solution from the Laplace space to the real space solution is:

$$f(t) = \frac{\ln 2}{t} \sum_{j=1}^{N} V_j F(s_j) \tag{3.65}$$

where

$$s_j = \frac{\ln 2}{t} j$$

$$V_j = (-1)^{N/2+j} \sum_{k=\left[\frac{j+1}{2}\right]}^{\min(j,N/2)} \frac{k^{N/2}(2k)!}{(N/2-k)!k!(k-1)!(j-k)!(2k-j)!} \tag{3.66}$$

According to the above inversion method, the image function $f(t)$ can be calculated by Eq. (3.65) when N and t are known. It is worthy to know that N must be an even number, and its value can greatly affect the calculation accuracy. The N value can be decided for different types of functions through practical calculations. In most cases, the N values of 8, 10, and 12 are appropriate. The bigger N is, the less accurate is the calculation.

3.7 Gas well pressure and production performance analysis

Expressions of transient pressure responses of a fully penetrated fractured well with infinite conductivity fractures in circular and rectangular shale gas reservoirs have been obtained for various complex flow mechanisms in the previous sections. Moreover, the wellbore storage and skin effects have also been considered by the Duhamel and superposition principles. Thereafter, the pressure type curves and the dimensionless production decline curves for constant producing bottom-hole pressure have been calculated by numerical inversion and computational programming. Finally, the production performance of a gas well producing at constant bottom-hole pressure have been analyzed for given reservoir properties. In the following sections, the relationships of the dimensionless pseudo pressure, dimensionless production rate, production rate and cumulative production with time for a fractured gas well flowing under different mechanisms are discussed.

3.7.1 Microfractures + steady state adsorption/desorption and diffusion

Table 3.1 lists the basic properties of a reservoir pay, gas, and a fractured well for the mechanism Model 1. The dimensionless parameters can be calculated by inserting these values into this model. Then the dimensionless pseudo pressure at a given dimensionless time is solved by using these dimensionless parameters. Thereafter, the dimensionless production rate at the given dimensionless time can be acquired according to the relationship between the dimensionless pseudo pressure for constant rate production and the dimensionless rate for constant pressure production of a gas well.

Table 3.1 Shale reservoir properties

Reservoir properties	Value	Reservoir properties	Value
Initial reservoir pressure, p_i (MPa)	25	Reservoir temperature, T (K)	320
Reservoir thickness, h (m)	60	Fracture half length, x_f (m)	100
Specific gas gravity, γ_g (fraction)	0.65	Production rate, q_{sc} (m^3/d)	1×10^4
Gas compressibility at reservoir conditions, C_{gi} (MPa^{-1})	0.02	Bottom-hole pressure, p_{wf} (MPa)	1
Microfracture system permeability, k_f (mD)	0.01	Microfracture system porosity, $\varnothing_f$ (fraction)	0.02
Skin, S_{kin} (dimensionless)	0.1	Dimensionless wellbore storage coefficient, C_D	10^{-3}
Langmuir pressure, P_L (MPa)	4	Langmuir volume, G_L (m^3/m^3)	10
Outer boundary radius of the closed circular gas reservoir, r_e (m)	1000		

In the end, the gas well production curves can be generated by converting the dimensionless production rate and time into the real production rate and time.

Fig. 3.4 shows the well test type curves of fractured conventional gas wells in circular infinite and closed outer boundary gas reservoirs without consideration of gas ad-/desorption. Based on the curve characteristics, the gas flow in the reservoirs can be classified into the following stages:

Flow stage 1: wellbore storage and transition flow periods. During the wellbore storage period, the dimensionless pseudo pressure and pseudo pressure derivative overlay each other as a line with their slope equal to 1 on a log–log plot. Following the wellbore storage period, the transition flow exhibits as a hump whose scale and duration depend on the dimensionless wellbore storage coefficient C_D and the skin effect S_{kin}.

Flow stage 2: formation linear flow. This period corresponds to the period of linear flow from the fracture system in the shale gas reservoir to the hydraulic fracture surfaces. The slope of the pseudo pressure derivative curve equals 0.5.

Flow stage 3: fracture elliptical flow. This flow period is viewed, by most scholars, as the formation linear flow or the transition from formation linear flow to radial flow. However, still some scholars consider it as the elliptical flow period, whose pseudo pressure derivative curve exhibits as a line with a slope of 0.36 according to the elliptical flow equation.

Flow stage 4: formation radial flow. On a log–log plot, the pseudo pressure derivative of this flow period is a horizontal line with a value of 0.5.

Flow stage 5: boundary dominated flow. For a reservoir with a closed outer boundary, when pressure propagates to the boundary, the pseudo pressure and its derivative curves overlay each other and exhibit as lines with a slope of 1.

According to the above analysis, the flow stages of fractured wells in conventional and shale gas reservoirs are exactly the same for this mechanism model. To differentiate them, the well test type curves for the conventional gas reservoir without consideration of ad-/desorption are marked as dotted lines in Fig. 3.4. Due to the ad-/desorption

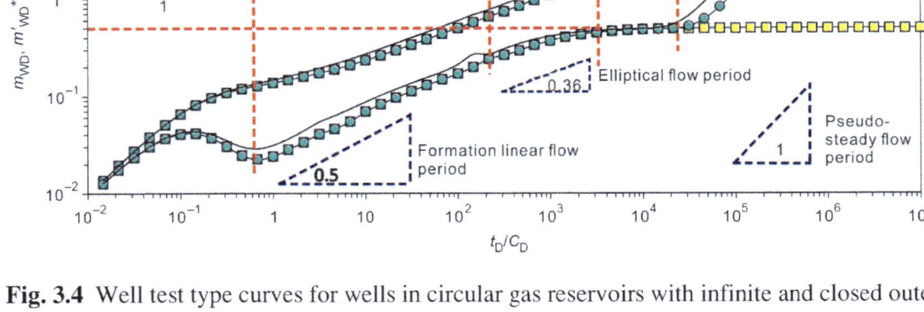

Fig. 3.4 Well test type curves for wells in circular gas reservoirs with infinite and closed outer boundaries.

characteristics of the shale gas reservoir, the adsorbed gas desorbs when reservoir pressure decreases. Therefore, when the pressure propagation reaches the reservoir boundary, there is plenty of desorbed gas in this reservoir. When the wells are producing at a constant rate, the shale gas reservoir has a lower pressure drop than the conventional gas reservoir, which is shown in the plot as the curves of the shale gas reservoir are above those of the conventional gas reservoir. Because the total compressibility includes ad-/desorption compressibility for the shale gas reservoir, it is higher than that of the conventional gas reservoir. The resulted low conductivity of the shale gas reservoir delays the boundary response time when the boundary conditions are the same for both types of reservoirs.

Fig. 3.5 shows the well test type curves of gas transient flow for different reservoir drainages. It can be seen that the size of the reservoir only affects the late boundary response time and has no impact on other flow stages.

Fig. 3.6 shows the type curves for different equivalent ad-/desorption levels of compressibility. It can be seen that the higher the compressibility is, the more gas supply to the microfracture system and the smaller pressure drop is required for the same production rate.

Fig. 3.7 shows the production rate and cumulative production vs time for fractured gas wells with different half fracture lengths producing at constant pressure. The longer the fracture is, the higher the initial production rate but the higher the decline. After entering a stable production period, the wells with different x_f all produce at about $0.8–1.6 \times 10^4 m^3$. However, after 1000 day's production, due to a higher gas volume being produced at an earlier stage, the shorter the fracture is, the lower the production rate. The EURs (estimated ultimate reserves) are the same for wells with different x_f.

Fig. 3.8 illustrates the effect of an adsorption volume on gas well performance. Since gas ad-/desorption is a steady state process, the bigger the adsorption volume is, the higher the well production rate. According to the study in this chapter,

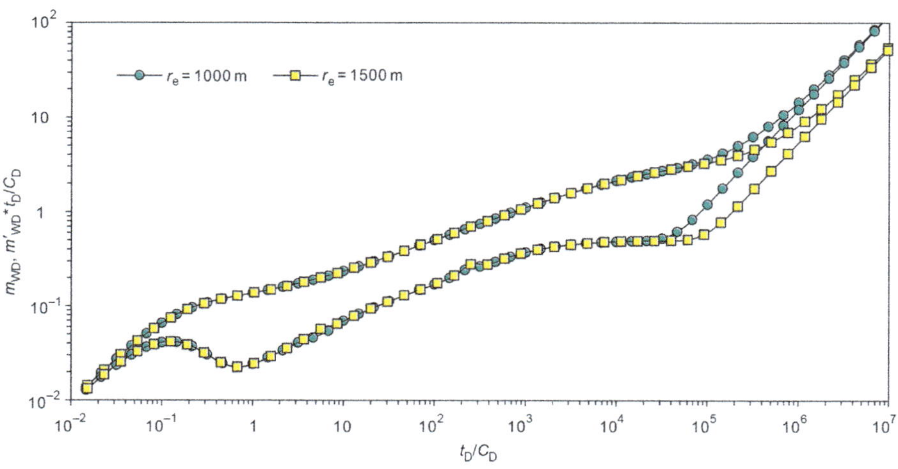

Fig. 3.5 Effect of reservoir drainage on well test type curves.

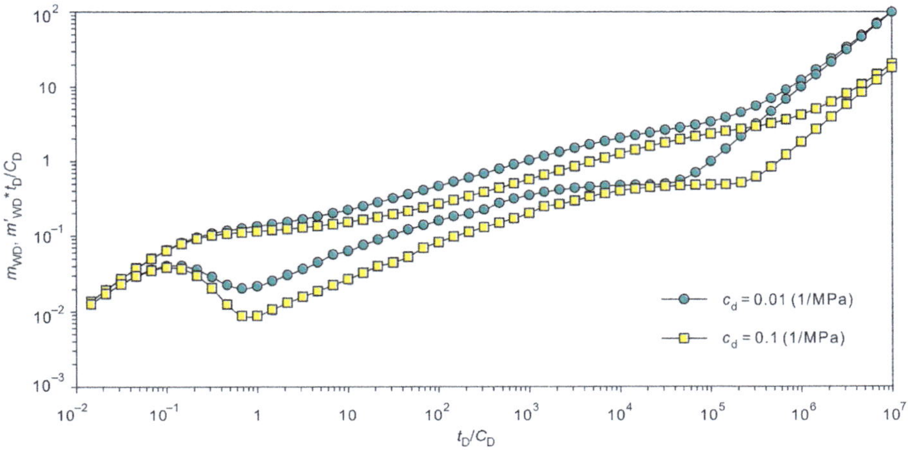

Fig. 3.6 Effect of equivalent compressibility on well test type curves.

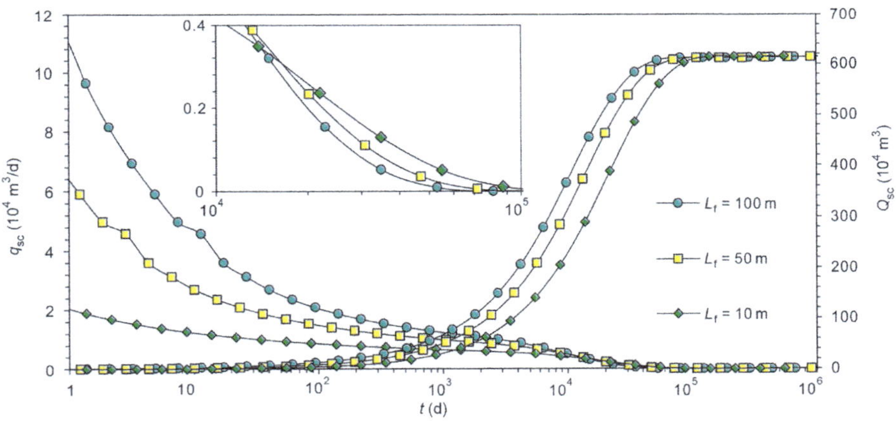

Fig. 3.7 Effect of fracture half-length on production rate and cumulative production.

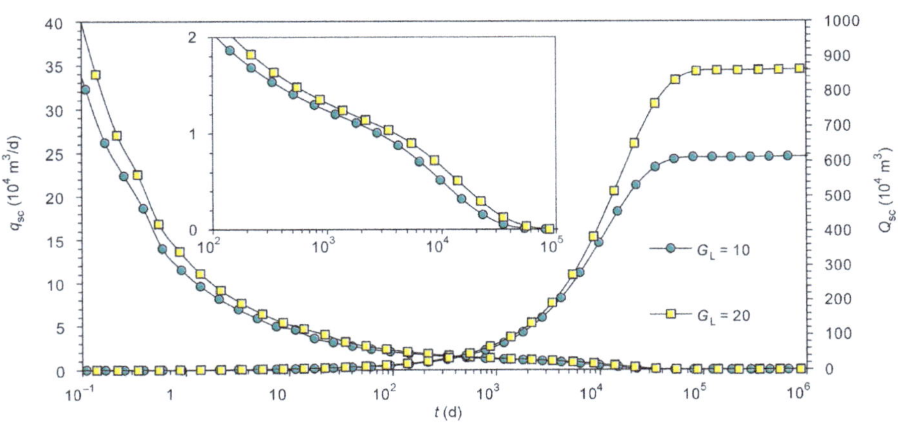

Fig. 3.8 Effect of gas adsorption on production rate and cumulative production.

for the matrix adsorption of 10 and $20\,m^3/m^3$, the difference of the well production rates is about $0.2 \times 10^4\,m^3$, and the longer the hydraulic fracture, the bigger the difference. Therefore, the production rate and cumulative production of a shale gas reservoir greatly rely on gas content in shale. For economic shale gas development, an evaluation of a high gas content area is very important.

The transient flow theory for a fractured gas well with infinite conductivity fractures in a closed rectangular reservoir can be used to generate well test type curves and production decline curves. In Fig. 3.9, there are type curves for closed circular and rectangular gas reservoirs with the same reservoir and gas properties. From this plot, it can be observed that the shape of a reservoir mainly influences the lasting time of the radial flow period and the characteristics of a boundary response. When the ratio of the rectangle height to width is relatively high, there is a linear flow period when pressure propagates to the closer boundary. This period exhibits as a line with a slope of 0.5 on the pseudo pressure derivative curve.

Fig. 3.10 shows the effect of the reservoir width on the well test type curves. For the same reservoir length, the shorter the reservoir width is, the earlier the pressure reaches the boundary and the longer the linear flow period lasts.

Figs. 3.11 and 3.12, respectively, show the effect of fracture orientations on the well test type curves and production decline curves. It can be seen that the difference of the well test type curves when a fracture extends along the reservoir width and length only shows up slightly at the beginning of the linear flow period, which has little effect on the well test analysis. As for the production decline curves, they overlay each other before pressure propagates to the closer boundary; then the production of the gas well whose fracture extends along the rectangular length is obviously higher than that of the well whose fracture extends along the width. But in the end, the cumulative production at the well abandonment is the same and has nothing to do with the fracture orientation.

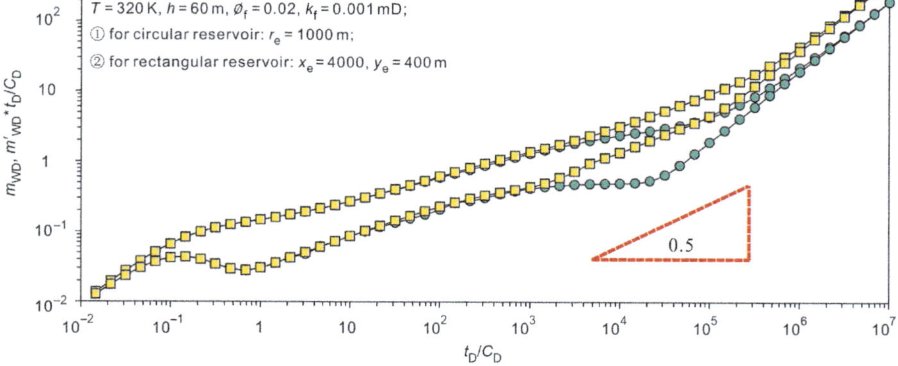

Fig. 3.9 Well test type curve comparison of rectangular and circular gas reservoirs.

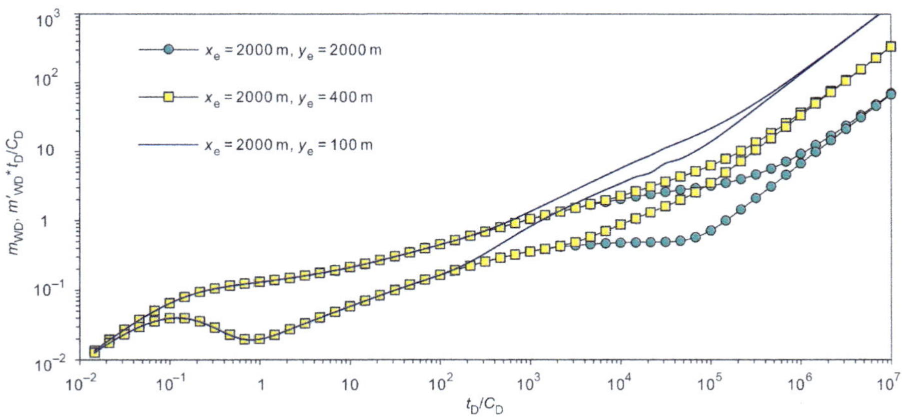

Fig. 3.10 Effect of reservoir width on well test type curves.

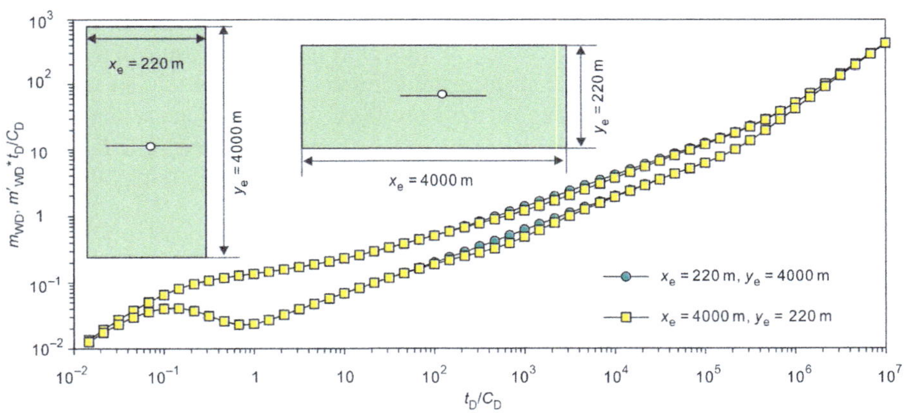

Fig. 3.11 Effect of reservoir L/W (length/width) ratio on well test type curves.

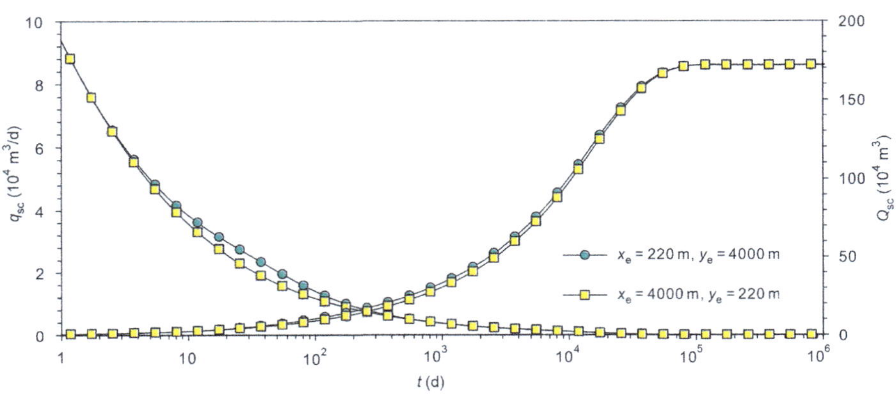

Fig. 3.12 Effect of reservoir L/W ratio on gas production rate and cumulative production.

3.7.2 Microfractures + matrix macropores + steady state adsorption/desorption and diffusion

The reservoir and gas properties for the mechanism Model 2 are listed in Table 3.2. By inputting these parameters into this mechanism model, the well test type curves and production decline curves can be generated through numerical inversion and computer programming.

Fig. 3.13 shows the well test type curves of fractured gas wells in the shale gas reservoir with different boundary conditions based on the triple-porosity model with consideration of shale gas steady state ad-/desorption and diffusion. Compared to the mechanism Model 1, for pseudo steady state flow, Model 2 has a concave part corresponding to the flow from macropores to the microfracture system, and other flow stages exhibit the same characteristics as Model 1. The pseudo pressure derivative curve is at 0.5 for the late time radial flow under the infinite reservoir condition, while, under the closed outer boundary condition, the pseudo pressure and its derivative curves go up with a slope of 1.

For transient interporosity flow from macropores to the fracture system, the type curves are shown in Fig. 3.14. According to this plot, the interporosity flow is very sensitive to the microfracture system pressure compared to the pseudo steady state flow, resulting in earlier happening of the flow from the macropore system to the microfracture system. Then the sharp concave part of the pseudo pressure derivative curve becomes flat and the lowest point corresponds to a value of 0.25. When the

Table 3.2 Shale reservoir properties

Reservoir properties	Value	Reservoir properties	Value
Microfracture permeability, k_f (mD)	0.01	Microfracture porosity, $\varnothing_f$	0.02
Matrix permeability, k_m (mD)	0.0001	Matrix porosity, $\varnothing_m$	0.12

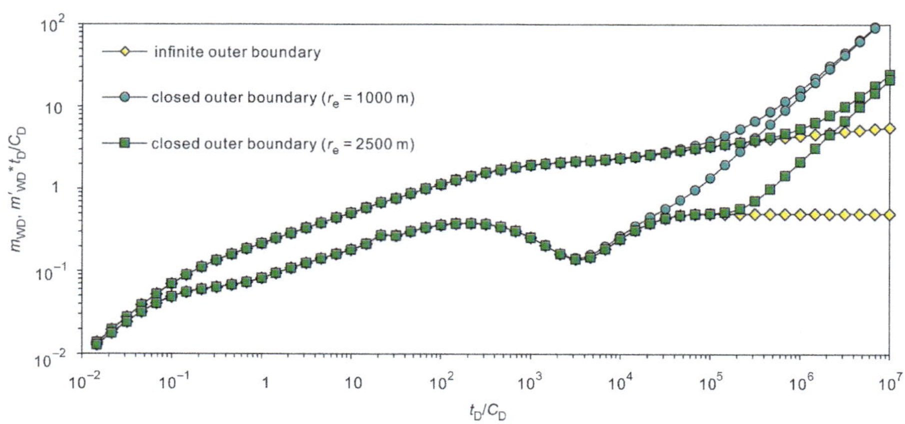

Fig. 3.13 Well test type curves for different outer boundary conditions.

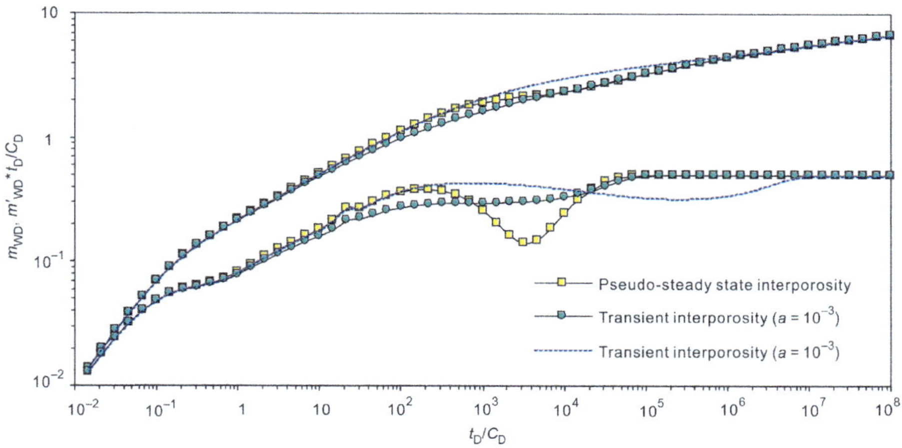

Fig. 3.14 Effect of macropore to microfracture interporosity flow status on well test type curves.

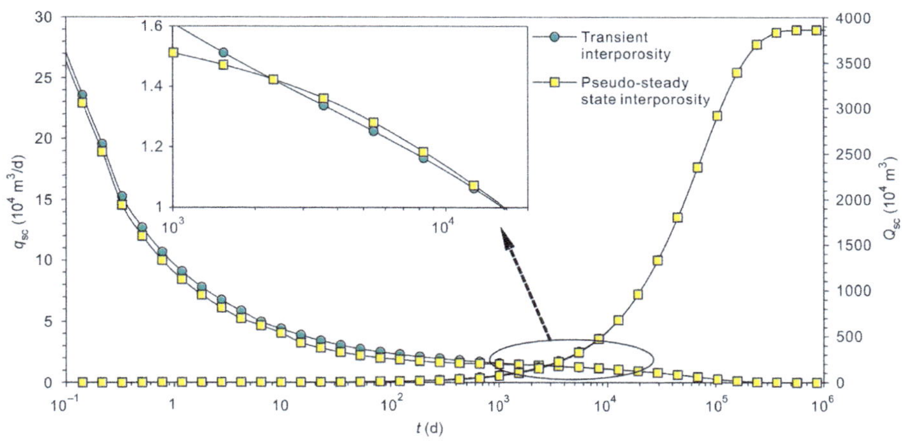

Fig. 3.15 Production and cumulative production vs time for pseudo steady state and transient interporosity flow.

interporosity flow coefficient is relatively big, only if the flow rate is equivalent to the well production rate, we can see the radial flow in the microfracture system; otherwise, it is difficult to see it. Therefore, the well shut-in time should be as long as possible to acquire reliable radial flow period data for a well test analysis of such gas reservoirs.

The production and cumulative production vs time of a well producing at constant bottom-hole pressure for different interporosity flow status are shown in Fig. 3.15. It can be seen that transient flow is more sensitive to the microfracture system pressure and the flow from the macropores to the fracture system happens earlier. Therefore, the well production is higher under transient flow than pseudo steady state flow. Consequently, the production declines faster due to a faster drop of the reservoir pressure, and after a certain period of production, the production under the pseudo steady state exceeds that under the transient state until abandonment of the gas well.

Fig. 3.16 shows the effect of the fracture system storativity ratio on the well test type curves. It can be seen from this plot that the ratio mainly affects the timing and scale of the concave curve for the pseudo steady state interporosity flow. The smaller the storativity ratio is, the earlier the interporosity flow happens, and the deeper the concave is.

Figs. 3.17 and 3.18 illustrate the influence of the microfracture system permeability on the well test type curves and the production curves, respectively. The higher the

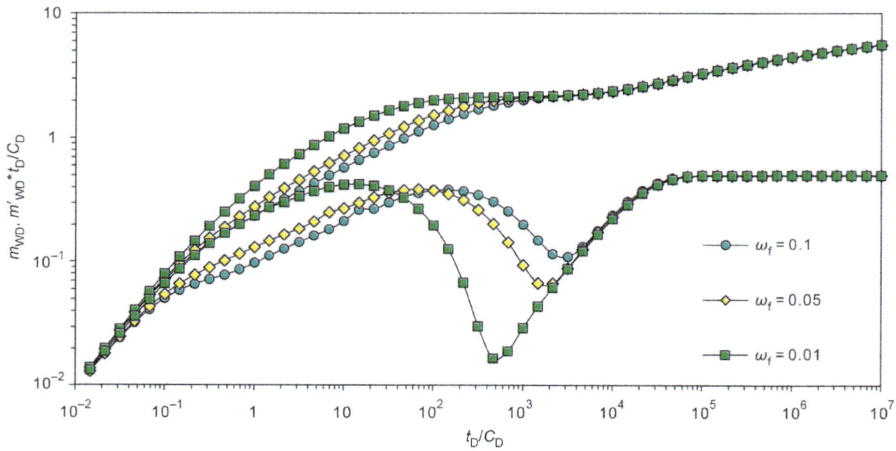

Fig. 3.16 Effect of fracture system storativity ratio on well test type curves.

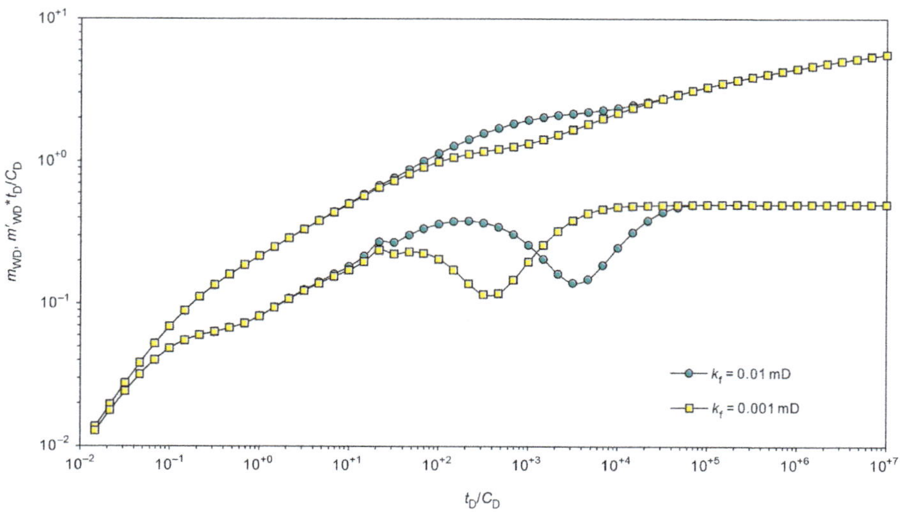

Fig. 3.17 Effect of fracture system permeability k_f on well test type curves.

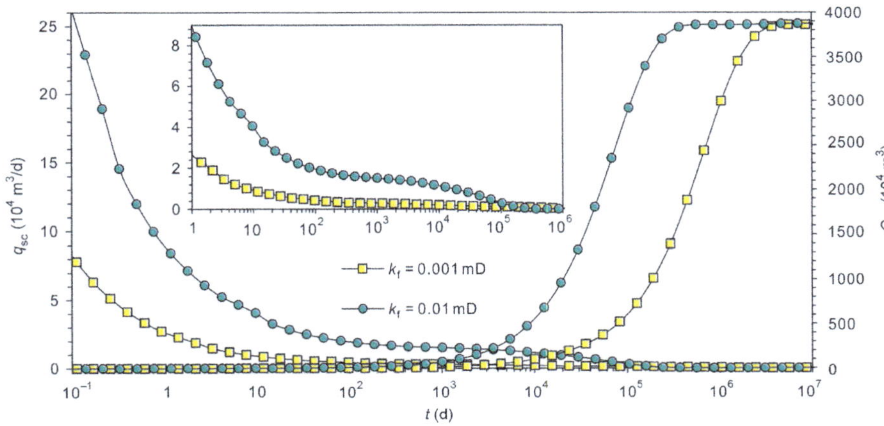

Fig. 3.18 Effect of microfracture system permeability on well production rate and cumulative production.

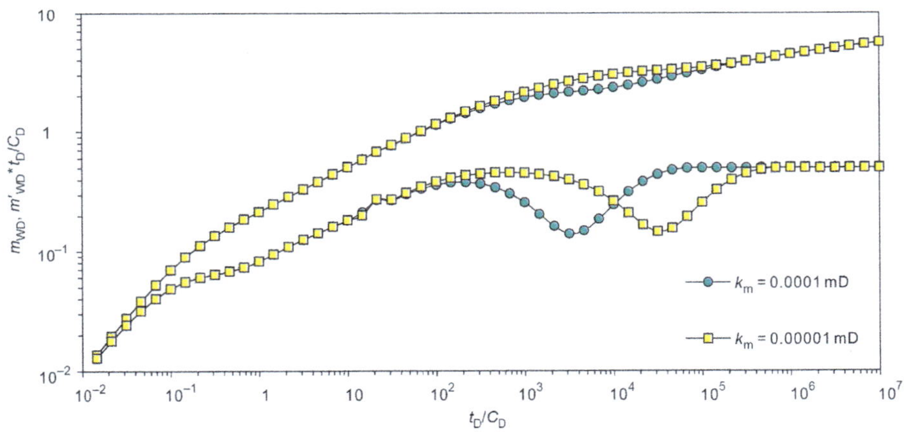

Fig. 3.19 Effect of matrix permeability k_m on well test type curves.

system permeability is, the earlier the interporosity flow happens, and the higher the well production is. This is because the fracture system permeability has no effect on the storativity ratios of both the fracture system and the adsorbed gas, and only has an effect on the interporosity flow coefficient of the flow from macropores to the fracture system. According to the definition of the interporosity flow coefficient, the higher the fracture permeability is, the smaller the coefficient is, and the later the interporosity flow happens. Additionally, the microfracture system permeability is positively correlative to the well production rate based on Darcy's law. The higher the permeability is, the higher the production rate is.

Figs. 3.19 and 3.20 show the influence of the matrix permeability on the well test type curves and production curves. According to the analysis of the fracture system

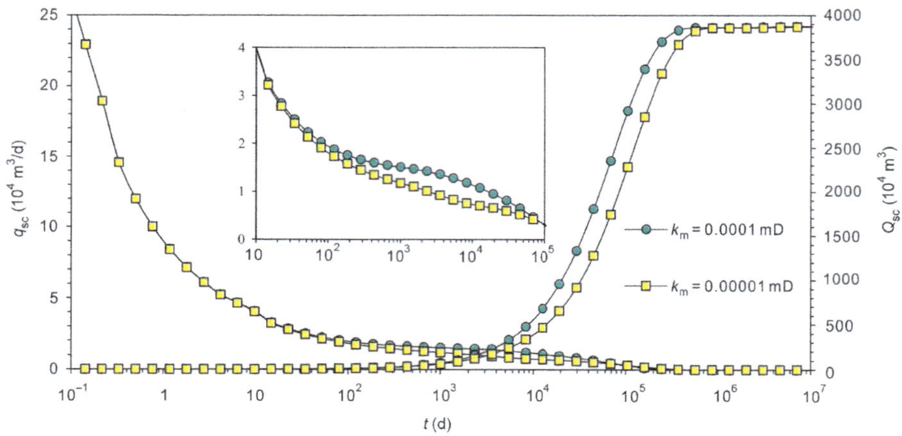

Fig. 3.20 Effect of matrix permeability k_m on well production rate and cumulative production.

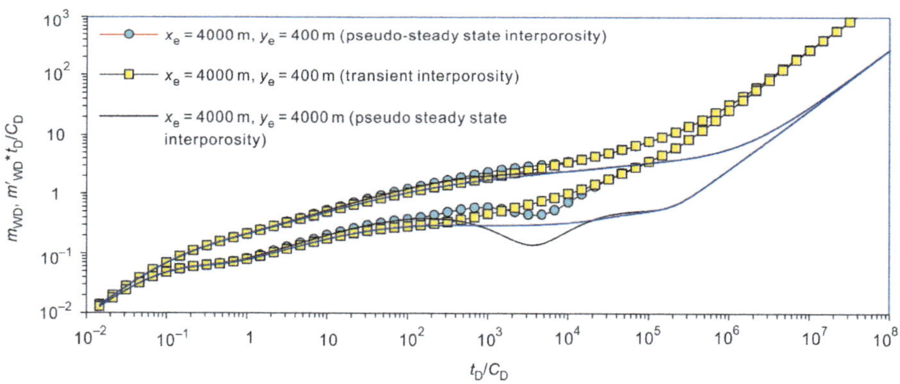

Fig. 3.21 Effect of interporosity flow status on well test type curves.

effect on these curves, the higher the matrix permeability is, the higher the interporosity flow coefficient is, and the earlier the interporosity flow (dig on the derivative curve) happens. For the effect of the matrix permeability, since the early time gas supply to the wellbore is mainly from the free gas in the microfracture system, the matrix permeability has little effect on the initial production rate. After the interporosity flow between the matrix macropores and the microfracture system happens, the bigger the interporosity flow coefficient is, the bigger supply volume from the matrix to the fracture system and the higher the well production rate. Similarly, with a faster pressure drop in the matrix with higher permeability, the production of less permeable matrix exceeds that of more permeable matrix after a certain period of production.

For wells in a closed rectangular gas reservoir, due to the short time period for pressure to propagate to the boundary, the concave part for the interporosity flow is hard to observe (shown in Fig. 3.21). For big L/W values, the linear flow period can be identified on the type curves, which is not observable on the type curves for a closed circular gas reservoir. For a fracture system without a very small storativity ratio, the interporosity flow concave part is not obvious on the pseudo pressure derivative curve. Especially for shale gas reservoirs, microfractures develop with poor connectivity, which results in even later time happening of the interporosity flow. There is less of a chance to observe the theoretical concave part on the pseudo pressure derivative curve (Fig. 3.22).

Fig. 3.23 shows an effect of the adsorption storativity ratio on gas production. From this plot, it can be seen that the initial gas production is the same for different ω_d values; this is because the early production is dominated by the fracture system permeability. However, after the interporosity flow from the matrix to the microfracture

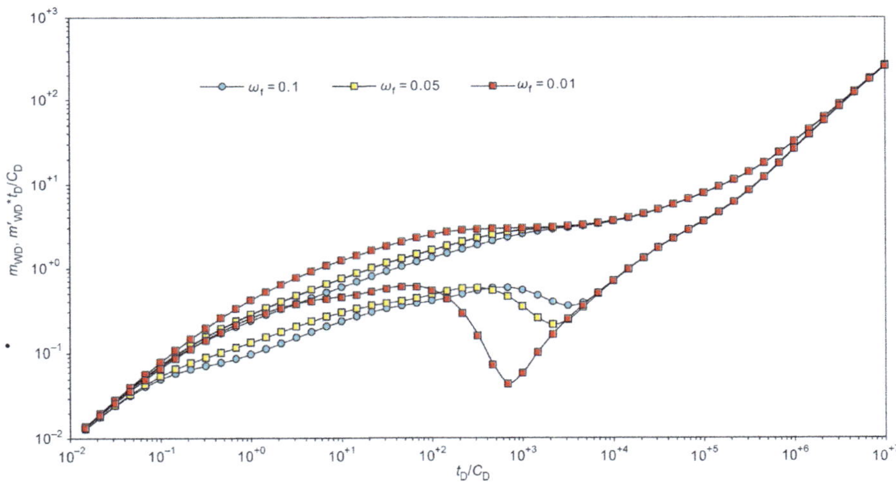

Fig. 3.22 Effect of fracture storativity ratio on well test type curves.

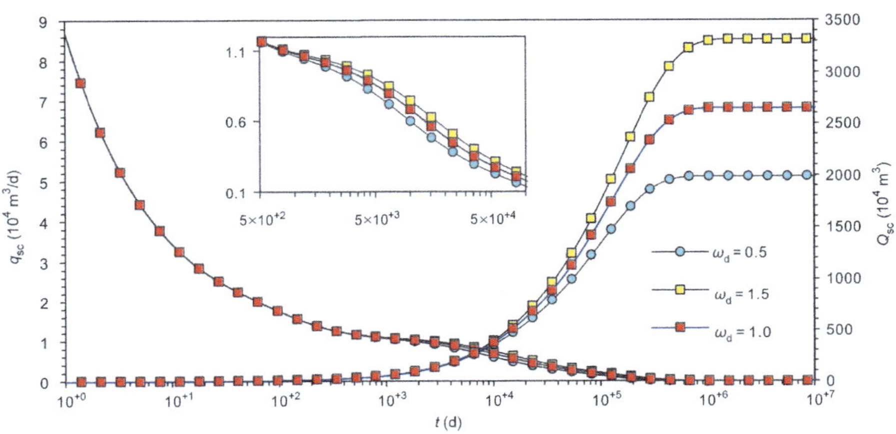

Fig. 3.23 Effect of equivalent adsorption and desorption compressibility coefficient on well production.

system happens, the well production is higher for bigger ω_d, which represents a bigger amount of gas adsorption in the macropore system. In addition, the original gas in place (OGIP) is higher for bigger ω_d, and for the same amount of pressure drop in the reservoir, the cumulative production is also higher for bigger ω_d.

3.7.3 Microfractures + gas adsorption/desorption + Fick's diffusion

The reservoir properties used for the mechanism Model 3 are listed in Table 3.3. The well test type curves and production curves are generated by these parameters.

Figs. 3.24 and 3.25 show, respectively, the well test type curves and production curves for the adsorbed gas desorption and diffusion to the microfracture system under Fick's unsteady and pseudo steady state diffusion laws. For the early production stage of the shale gas reservoir, the produced gas is mainly free gas in the fracture system and a small amount of desorbed gas; therefore, the type curves are similar to those in conventional fractured wells. However, due to the small amount of desorbed gas flow into the microfracture system, the pseudo pressure drop required for the same production rate is smaller than that in a conventional reservoir. Also, for transient and pseudo steady state desorption, the desorbed gas amount of the latter state is bigger than that of the former one, and thus the pseudo pressure drop required for the same production rate is higher for unsteady state desorption.

Table 3.3 Shale reservoir properties.

Reservoir properties	Value	Reservoir properties	Value
Microfracture permeability, k_f (mD)	0.01	Microfracture porosity, $\varnothing_f$	0.02
Constant production rate, q_{sc} (m^3/d)	1×10^4	$(R_m)^2/D_F$	2×10^6

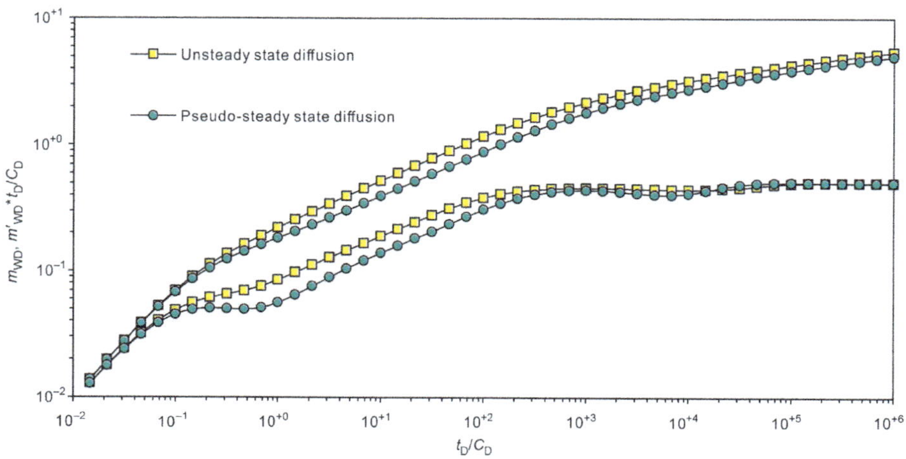

Fig. 3.24 Well test type curves of pseudo steady state and unsteady state diffusion.

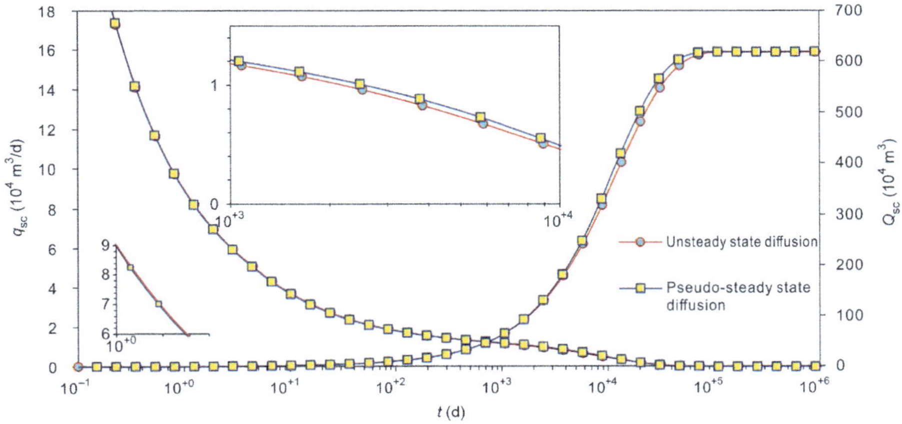

Fig. 3.25 Well production rate and cumulative production vs time for pseudo steady state and unsteady state diffusion.

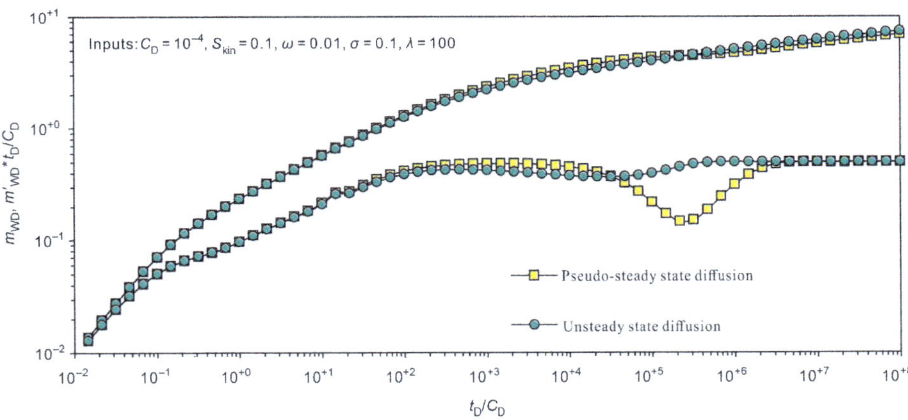

Fig. 3.26 Well test type curves for theoretical dimensionless parameters in pseudo steady state and unsteady state.

When a well is producing at constant bottom-hole pressure (Fig. 3.25), the production rate is the same no matter if diffusion is pseudo steady or unsteady because the early time gas supply is mainly from free gas in the microfracture system. But with production time extending, the production is dominated by the desorbed gas. During this period, according to the previous analysis, the production rate of pseudo steady state diffusion is higher than that of unsteady state diffusion, but the cumulative production at the well abandonment is the same. To clearly illustrate the type curve characteristics of pseudo steady and unsteady state diffusion, theoretical parameters (their values are not realistic for real reservoirs) are given to generate the type curves, as shown in Fig. 3.26.

Figs. 3.27 and 3.28 show the effect of the interporosity flow coefficient and fracture storativity ratio on the well test type curves. It can be seen from these plots that the interporosity flow coefficient λ mainly influences the desorption and diffusion time of adsorbed gas; the bigger the λ is, the earlier the diffusion (concave on the derivative curves) happens, and the fracture storativity ratio ω affects the shape of the curves of both the interporosity flow and early time linear flow. The smaller the ω is, the deeper and wider the concave part on the curves is and the higher the pseudo pressure and its derivative curves are, and vice versa.

The influence of an adsorbed gas amount on the well test type curves and production curves is shown in Figs. 3.29 and 3.30. From these plots, it can be seen that the bigger the volume of adsorbed gas the matrix has, the deeper the concave part of the pseudo pressure derivative is. When the well is producing at constant pressure, a higher volume of

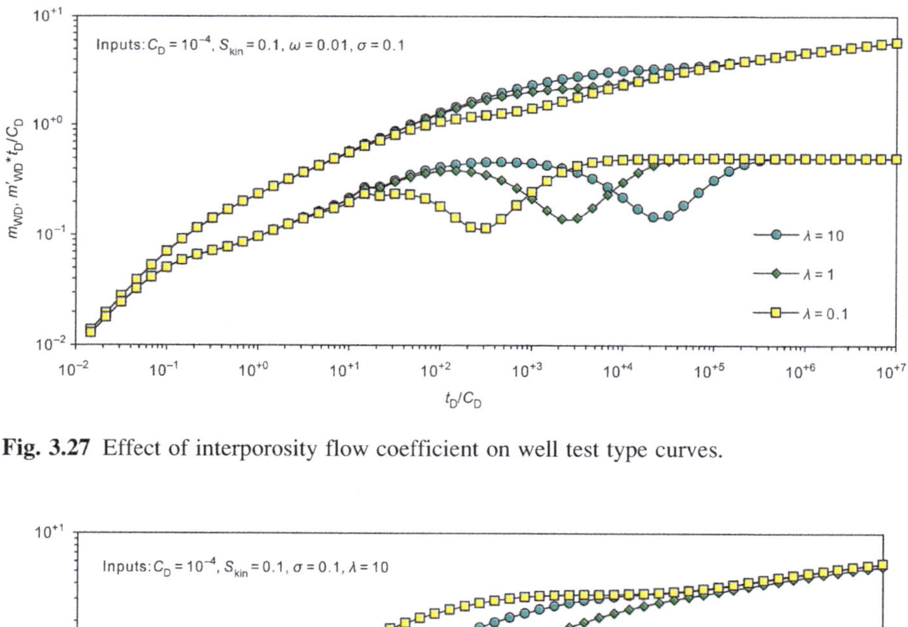

Fig. 3.27 Effect of interporosity flow coefficient on well test type curves.

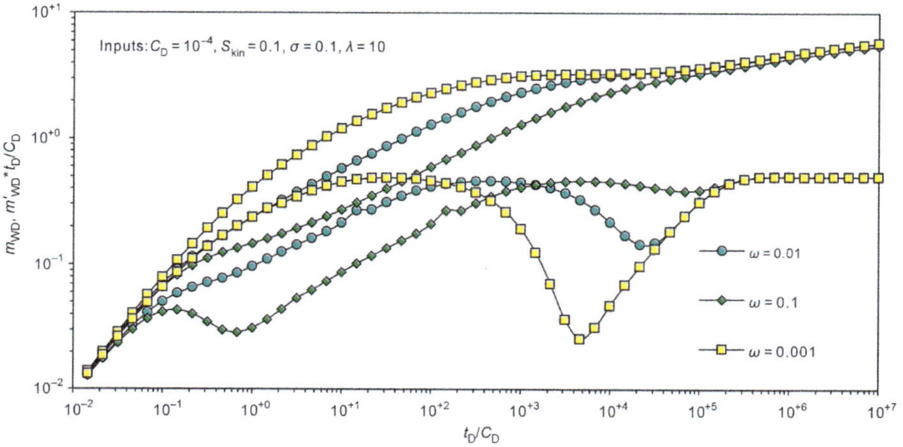

Fig. 3.28 Effect of storativity ratio on well test type curves.

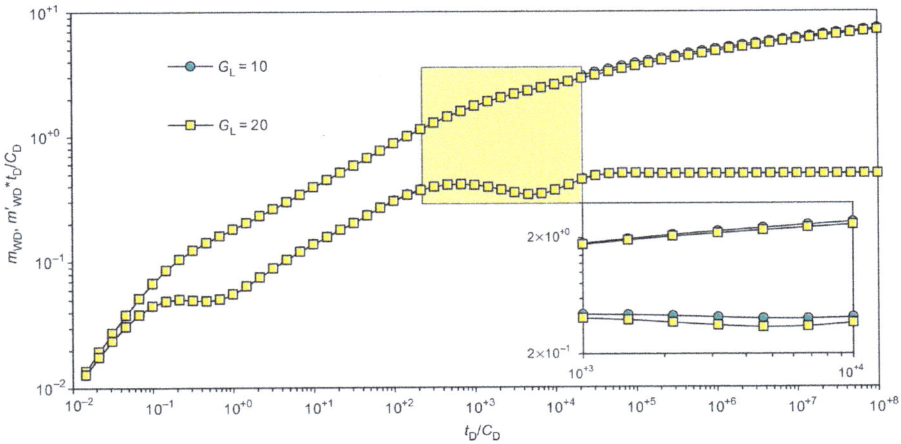

Fig. 3.29 Effect of adsorbed gas volume on well test type curves.

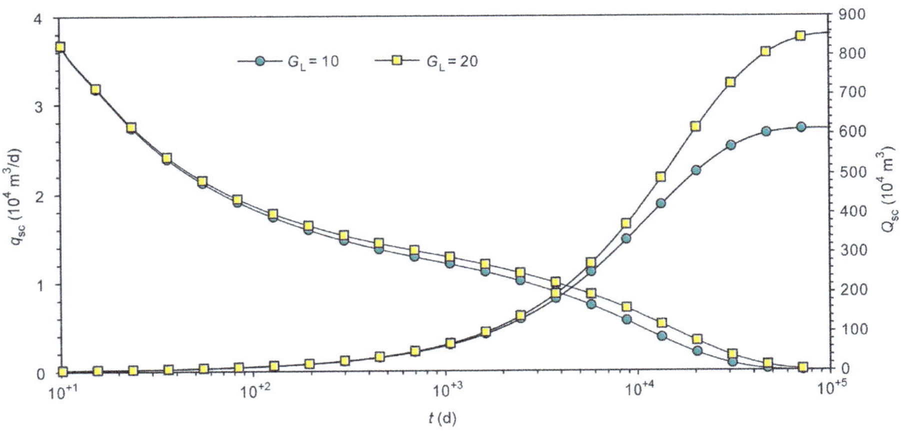

Fig. 3.30 Effect of adsorption volume on production rate and cumulative production.

the adsorbed gas means a bigger amount of diffusion gas under the same pressure difference, which results in a higher production rate of the gas well. Additionally, a higher volume of the adsorbed gas also means bigger OGIP in the well-controlled area, which results in higher cumulative production at well abandonment.

The well test type curves shown in Fig. 3.31 are for different initial reservoir pressures without consideration of the wellbore storage and skin effects. For typical shale gas reservoirs, the Langmuir isothermal adsorption pressure is relatively low. Thus the higher the initial formation pressure is, the smaller the adsorption–desorption constant σ is and the shallower the concave part on the pseudo pressure derivative curve is, and vice versa. The difference of the curves for the early time linear flow period in Fig. 3.31 is caused by a different input of gas viscosity and compressibility for

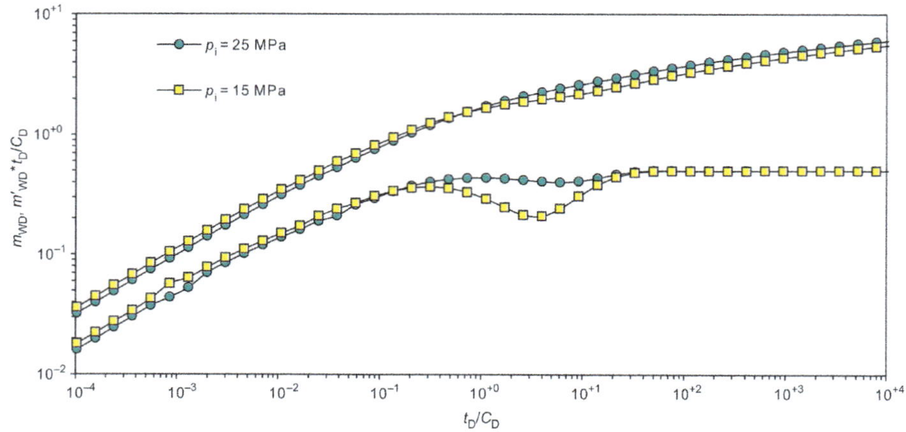

Fig. 3.31 Well test type curves at different initial reservoir pressures.

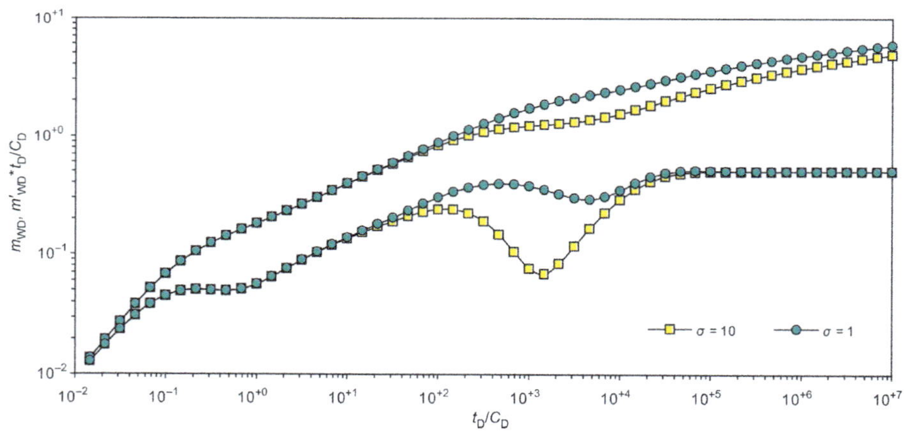

Fig. 3.32 Effect of ad-/desorption constant on well test type curves.

different initial reservoir conditions. Theoretically, if only changing σ, the curves should overlay each other during the early linear flow period (as shown in Fig. 3.32).

For rectangular gas reservoirs, due to the short distance between the fractured well and the reservoir boundary, it is almost impossible to observe the interporosity flow concave curve for the given parameters in this section. In addition, the formation radial flow quickly transfers into the linear flow period (as shown in Fig. 3.33) caused by the short distance to the reservoir boundary. Fig. 3.34 shows the effect of a reservoir width on the type curves. It can be seen that the reservoir width mainly influences the start of the late time linear flow. The wider the reservoir is, the later the linear flow starts, and the shorter time it lasts. In addition, the wider the reservoir is, the bigger the flow area is, and the smaller the pressure drop required for the same production rate, which exhibits at a lower position of the pseudo pressure and its derivative curves in the plot.

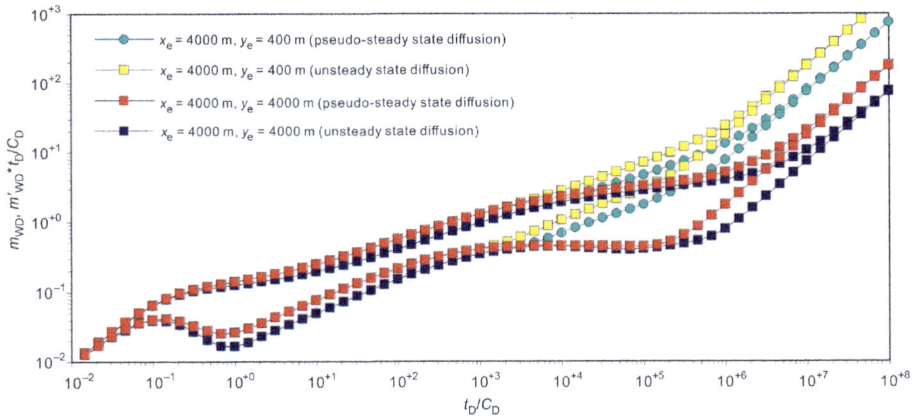

Fig. 3.33 Well test type curves in different diffusion status.

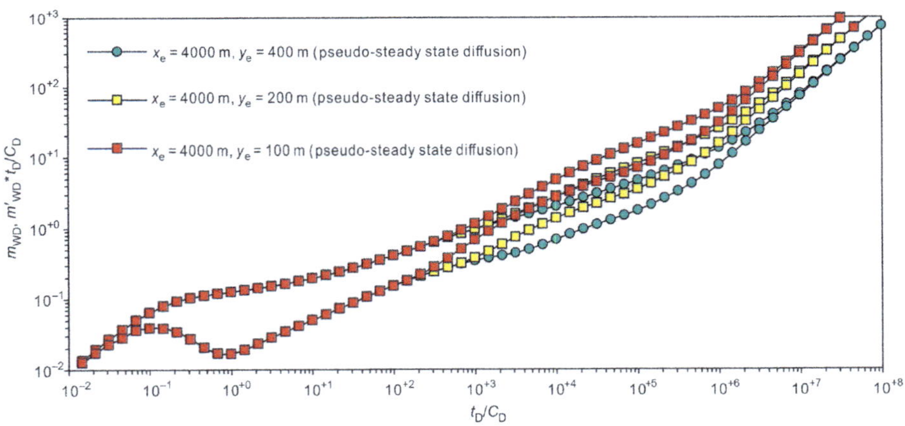

Fig. 3.34 Effect of reservoir width on well test type curves.

3.7.4 Microfractures + matrix macropores + gas adsorption/desorption + nanopore Fick's diffusion

Table 3.4 lists the reservoir parameters used for the mechanism Model 4. The well test type curves and production curves can be generated based on these values and Model 4 equations.

Figs. 3.35 and 3.36 show the well test type curves for the pseudo steady state and transient interporosity flow under different diffusion models. Theoretically, there should be two concave parts for the flow from macropores to the fracture system and the desorbed gas diffusion from matrix surfaces to macropores, respectively. However, the two types of flow happen almost at the same time based on the parameters in Table 3.4. According to the analysis of the mechanism Model 3, the concave

Table 3.4 Shale reservoir properties

Reservoir properties	Value	Reservoir properties	Value
Microfracture system perm, k_f (mD)	0.01	Microfracture system porosity, $\varnothing_f$	0.02
Matrix macropore perm, k_m (mD)	0.0001	Matrix macropore porosity, $\varnothing_m$	0.12
Constant production rate, q_{sc} (m^3/d)	1×10^4	$(R_m)^2/D_F$	2×10^6
Interporosity flow shape factor, α ($1/m^2$)	10^{-3}		

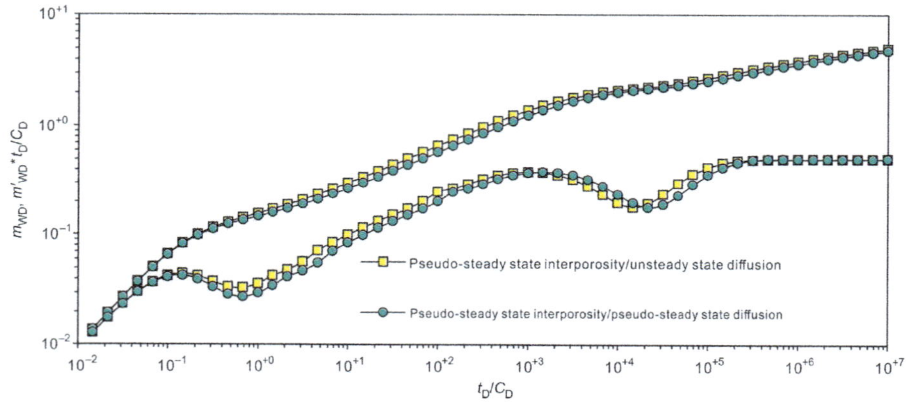

Fig. 3.35 Well test type curves for pseudo steady state interporosity flow under different diffusion models.

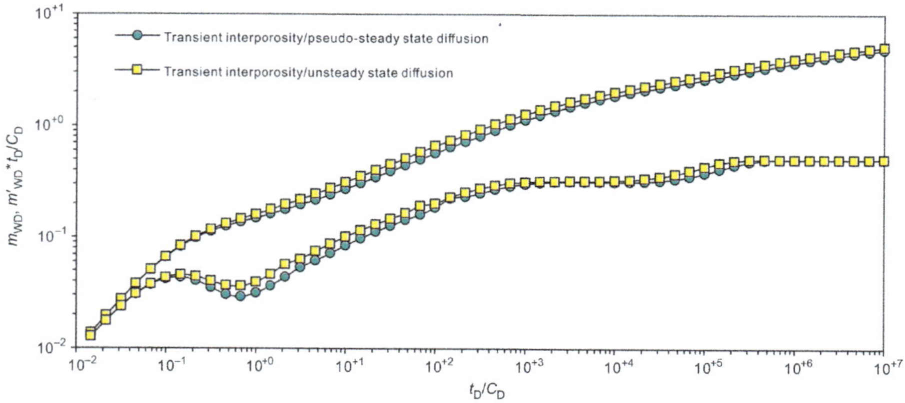

Fig. 3.36 Well test type curves for unsteady state interporosity flow under different diffusion models.

part for diffusive interporosity flow is not clearly visible. Therefore, the two sets of curves for different diffusion models, respectively, in Figs. 3.35 and 3.36, have no major differences. For the unsteady state diffusion, the pseudo pressure and its derivative curves for formation linear flow are higher compared to the pseudo steady state diffusion.

To theoretically identify the two concave parts on the type curves of the mechanism Model 4, the type curves for different diffusion coefficients without consideration of the wellbore storage and skin effects are plotted in Fig. 3.37. The bigger the diffusion coefficient is, the earlier the diffusion happens. For this mechanism model, small diffusion coefficients lead to coincidence of the macropores to the microfracture system interporosity flow and the diffusion, which exhibits as only one wider and deeper concave part observable on the curves. Only when the diffusion coefficient is big enough, the happening time of the two types of flow can be differentiated by the two concave parts on the pseudo pressure derivative curves.

For a fractured well in a closed rectangular gas reservoir, the continuous point source solution can be used to generate the well test type curves, which are shown in Figs. 3.38 and 3.39. There is only one concave part for the interporosity flow observable on the curves due to a boundary response effect. Also, because the flow happens during the late time linear flow period, there is a concave part (pseudo steady state flow) or a deviation from the linear flow line with a slope of 0.5 (unsteady state flow) showing up on the pseudo pressure derivative curves.

For a well producing at constant bottom-hole pressure in a rectangular gas reservoir, the production rate and cumulative production vs time curves are shown in Fig. 3.40 for combinations of pseudo steady state diffusion and different macropores to the microfracture system flow status. From this plot, it can be seen that the early time production of unsteady state interporosity flow is slightly higher than that of the pseudo steady state, and reverses in the late time. However, the two-flow status results in equivalent cumulative production at the well abandonment pressure.

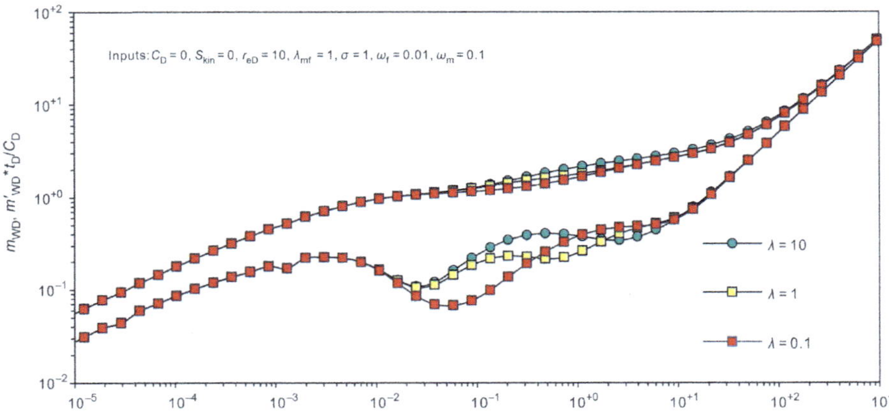

Fig. 3.37 Effect of interporosity flow coefficient on well test type curves.

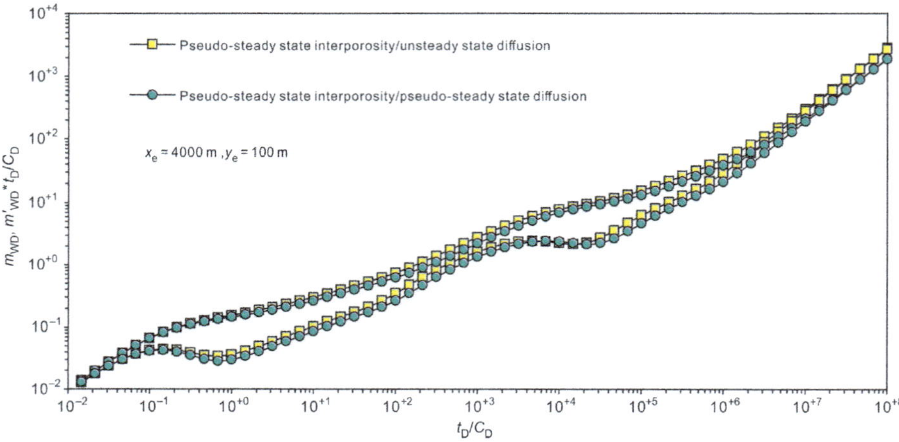

Fig. 3.38 Well test type curves for pseudo steady state interporosity flow under different diffusion models in a rectangular gas reservoir.

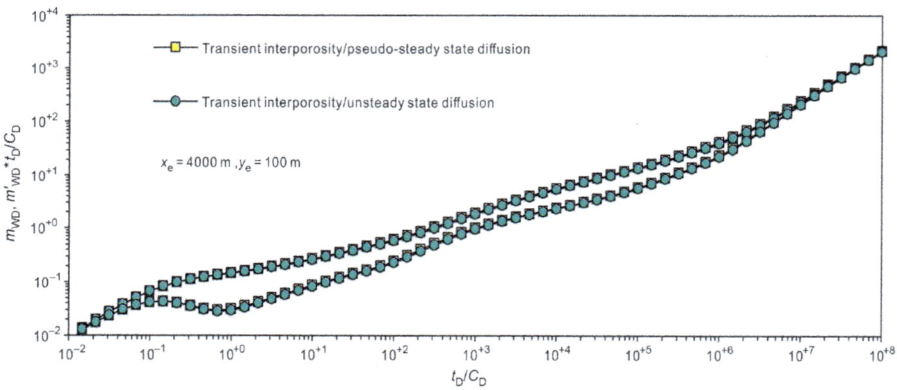

Fig. 3.39 Well test type curves for unsteady state interporosity flow under different diffusion models in a rectangular gas reservoir.

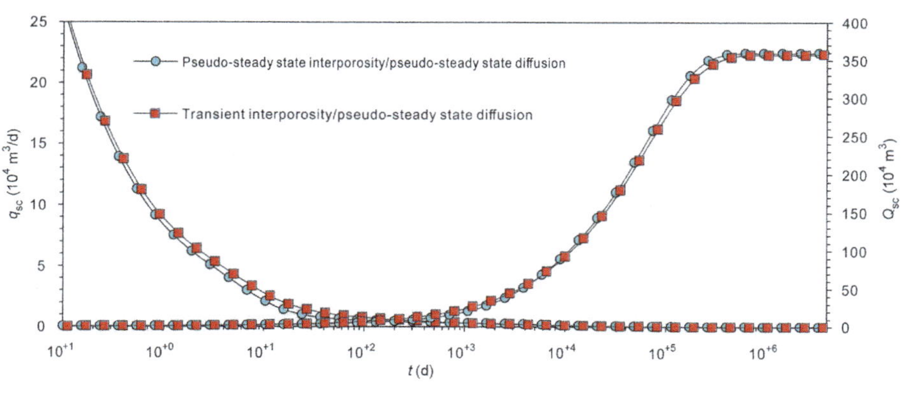

Fig. 3.40 Well production rate and cumulative production curves for pseudo steady state diffusion with different interporosity flow models in a rectangular gas reservoir.

3.7.5 Microfractures + gas adsorption/desorption + nanopore Knudsen diffusion

Table 3.5 lists the shale gas reservoir properties for the mechanism Model 5 and the calculated dimensionless variables. They are input into the model solutions to generate the well test type curves and production curves.

The transient well test type curves shown in Fig. 3.41 are for a fractured vertical well in a shale gas reservoir considering multiple flow mechanisms of Darcy's law, Klinkenberg's effect, and Knudsen diffusion. The shape of the type curves from the mechanism Model 5 is similar to that from Model 2. The difference is that the concave part in Fig. 3.41 is determined by not only interporosity flow from nanopores to microfractures but also other effects including gas slippage and the Knudsen diffusion caused by a concentration gradient. In fact, these mechanisms only affect the interporosity flow from matrix nanopores to the microfracture system.

Figs. 3.42 and 3.43 show, respectively, the effect of the slippage factor (F_f) on well test type curves and production curves. According to the definition of the slippage factor, $F_f = 1$ means no slippage effect, and the bigger the number is, the more severe the slippage effect is. Because the slippage effect actually increases the interporosity flow capability in the matrix nanopores and, consequently, increases the flow capability from the matrix to the microfractures, the concave part for the interporosity flow shows up

Table 3.5 Shale reservoir properties

Reservoir properties	Value	Reservoir properties	Value
Microfracture system perm, k_f (mD)	0.01	Microfracture system porosity, $\varnothing_f$	0.02
Matrix perm, k_m (mD)	0.0001	Matrix porosity, $\varnothing_m$	0.12
Interporosity flow shape factor, α (1/m^2)	10^{-3}	Knudsen diffusion coefficient, D_k (1/m^2)	10^{-6}
Average matrix pore radius, r_n (m)	2×10^{-9}	α value in Eq. (1.26)	0.8

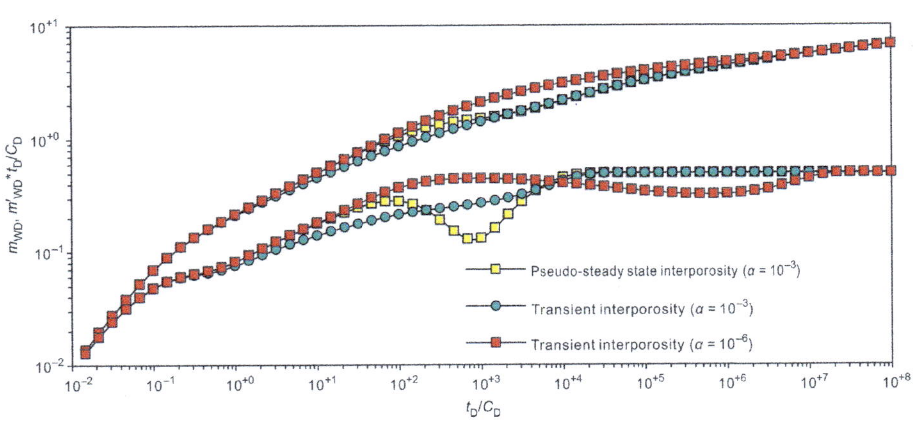

Fig. 3.41 Well test type curves for pseudo steady state and unsteady state interporosity flow.

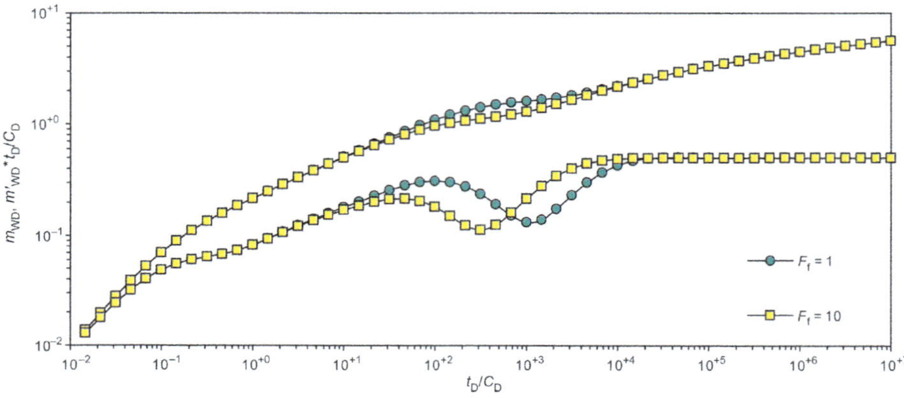

Fig. 3.42 Effect of slippage factor on well test type curves.

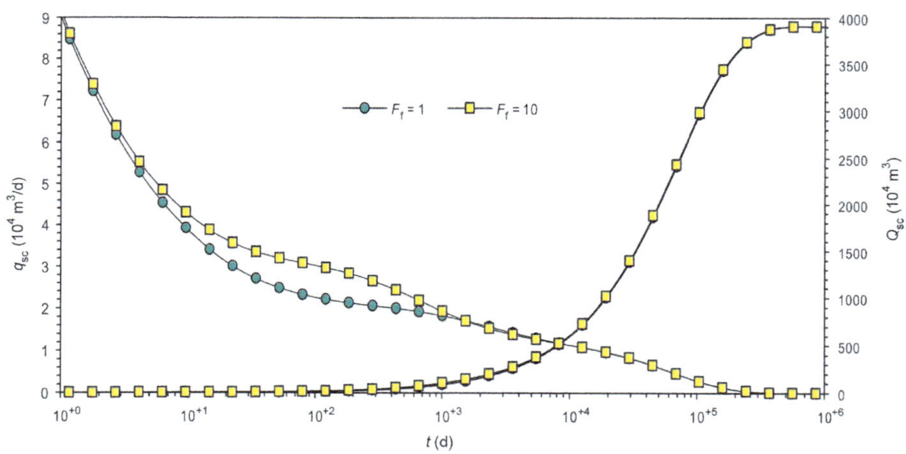

Fig. 3.43 Effect of slippage factor on production rate and cumulative production.

earlier on the derivative curves. In addition, the production rate also increases as the interporosity flow coefficient gets bigger. However, higher production leads to a quicker formation pressure drop and poorer later time production. Therefore, with production time extending, the production rate from wells with a lower F_f can gradually catch up and exceed that from wells with a higher F_f, resulting in the same cumulative production.

Figs. 3.44 and 3.45 show the effect of the Knudsen diffusion coefficient on well test type curves and production curves, respectively. Similar to the effect of the slippage factor, the matrix interporosity flow capacity is higher due to the Knudsen diffusion effect, and the concave part shows up earlier on the curves. Moreover, the early time production rate is positively related to the diffusion coefficient D_k, but the cumulative production at the well abandonment is the same.

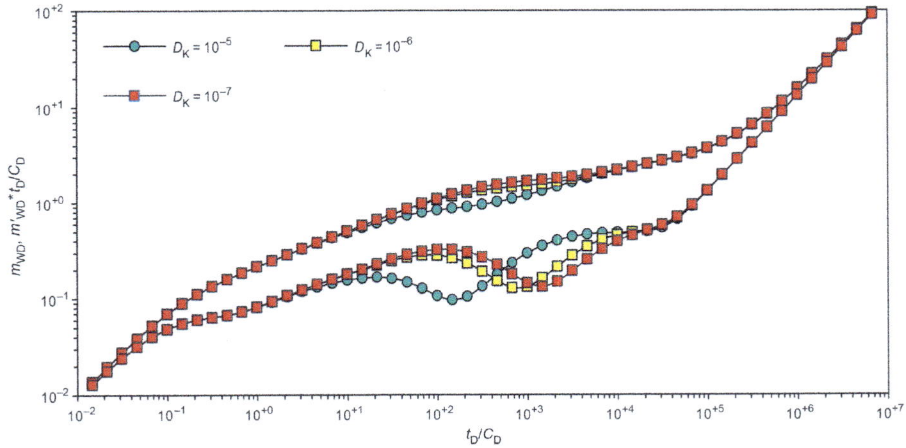

Fig. 3.44 Effect of Knudsen diffusion coefficient on well test type curves.

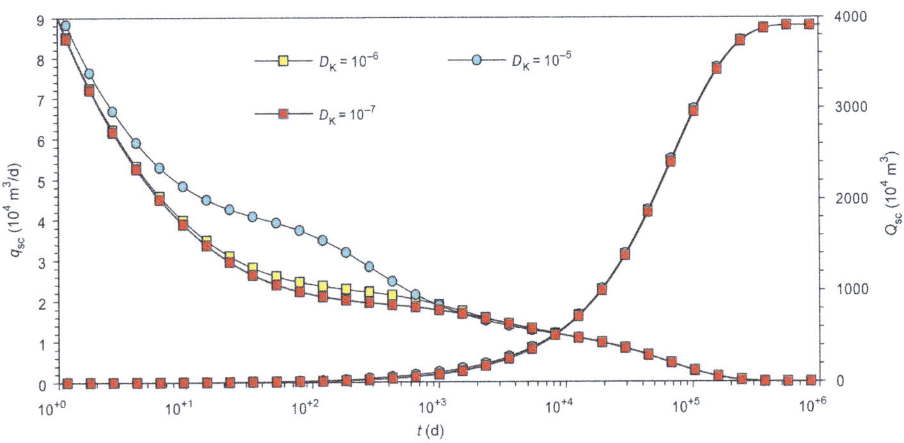

Fig. 3.45 Effect of Knudsen diffusion coefficient on production rate and cumulative production.

On the plots of well test type curves and production decline curves for the mechanism Model 5, there is only one late time linear flow period showing up on the curves. Also, due to the short time period for pressure to propagate to the reservoir boundary, the shape and position of the interporosity flow concave part are both influenced by the well distance to the reservoir boundary. For the well production rate, it is slightly higher under transient interporosity flow than pseudo steady state flow, and the difference gets smaller and reversed after a certain period of production, finally resulting in the same cumulative production, as shown in Figs. 3.46–3.48.

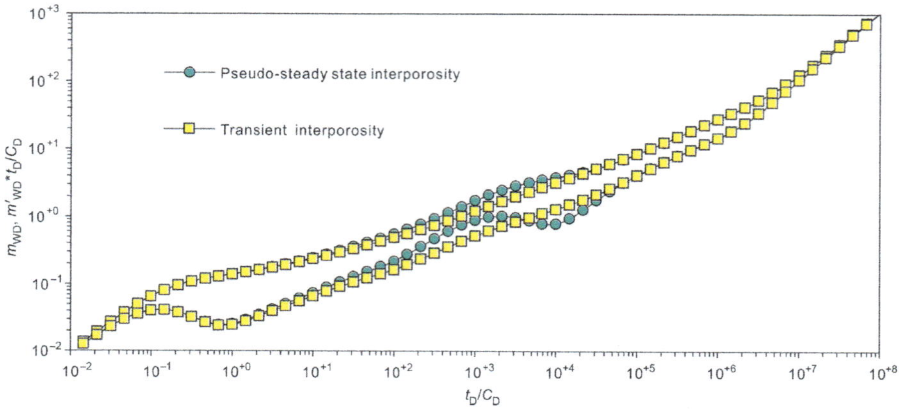

Fig. 3.46 Well test type curves for different interporosity flow models in a rectangular gas reservoir.

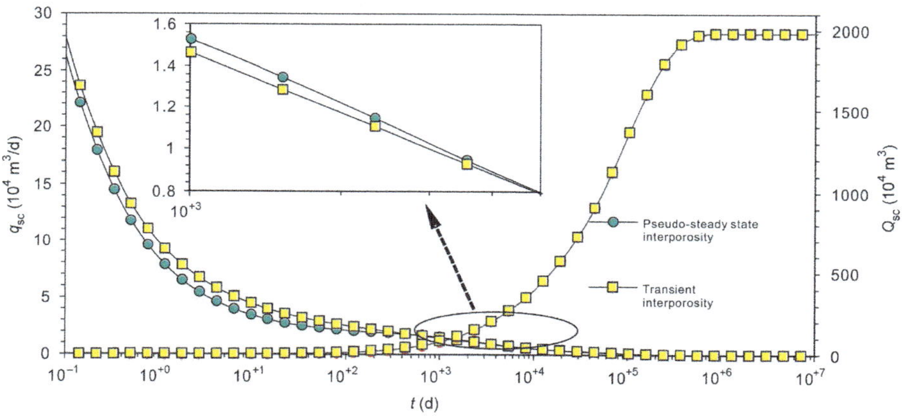

Fig. 3.47 Well production rate and cumulative production vs time for different interporosity flow models in a rectangular gas reservoir.

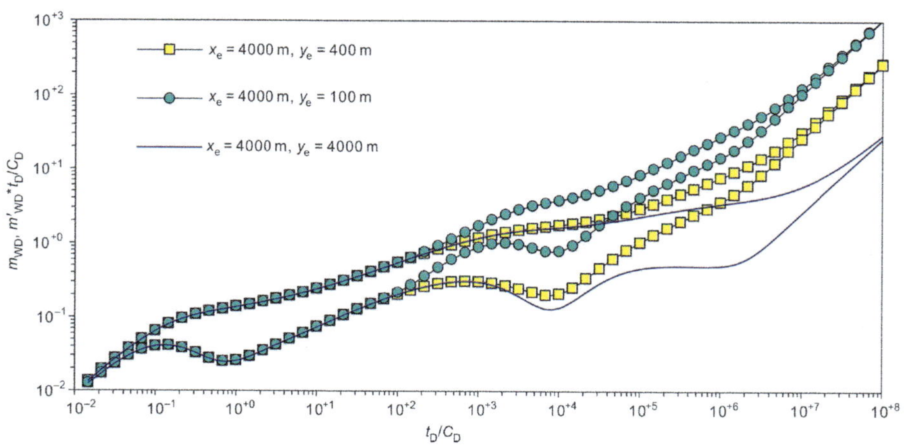

Fig. 3.48 Well test type curves in rectangular gas reservoirs with different dimensions.

Multi-stage fractured horizontal wells in shale reservoirs without SRV

4

Chapter Outline

4.1 Introduction 115
4.2 Multi-stage fractured horizontal wells in circular gas reservoirs 116
4.2.1 Physical model 116
4.2.2 Mathematical model and solutions 116
4.3 MFHWs in rectangular gas reservoirs 121
4.3.1 Physical model 121
4.3.2 Mathematical model and solutions 122
4.4 Analysis of well bottom-hole pressure and production performance 125
4.4.1 Microfractures + steady state adsorption/desorption and diffusion 125
4.4.2 Microfracture + matrix macropores + steady state adsorption/desorption and diffusion 130
4.4.3 Microfractures + gas adsorption/desorption + Fick's diffusion 136
4.4.4 Microfractures + matrix macropores + gas adsorption/desorption + Fick's diffusion in nanopores 142
4.4.5 Microfractures + gas adsorption/desorption + nanopore Knudsen diffusion 146

4.1 Introduction

Although shale gas reservoirs are widely distributed around the world with considerable amounts of gas resources, they had not been extensively and effectively developed for a long period of time. This is mainly due to their super low permeability. Massive hydraulic fractures must be achieved in wells to obtain economic production. Higher gas production is from bigger well drainage areas, which requires more fractures and longer fracture length. This makes horizontal wells with multi-stage fractures desirable, which can enlarge larger well drainage areas compared to vertical wells, and be applied in shale gas development. However, constrained by fracturing techniques and operation conditions, multi-stage fractured horizontal wells (MFHWs) had not been widely implemented until the early 21st century when multi-stage fracturing was successfully applied in shale gas development in North America (Chen and Raghavan, 1996; Horne and Temeng, 1995; Wan and Aziz, 1999, 2002).

On the basis of the point source solutions introduced in the previous two chapters, the transient flow theory of MFHWs in shale gas reservoirs is investigated in this

Developments in Petroleum Science, Vol. 66. https://doi.org/10.1016/B978-0-444-64315-5.00004-8

chapter. Numerical inversion and computer programming are used to generate well test type curves and production type curves, whose influence factors are analyzed.

4.2 Multi-stage fractured horizontal wells in circular gas reservoirs

4.2.1 Physical model

Fig. 4.1 shows the physical model of a MFHW in a gas reservoir with a circular boundary. To facilitate the subsequent analysis, the following assumptions are made:

1. The reservoir is horizontal and homogenous with thickness h and sealed upper and lower boundaries. The rest of the outer boundary is infinite or closed.
2. A horizontal well with length L located in the middle of the gas reservoir, and the pressure drop for fluid flow in the wellbore and the flow from the reservoir to the horizontal wellbore are ignorable.
3. There are M hydraulic fractures with infinite conductivity and negligible fracture width.
4. These fractures are evenly or unevenly distributed along the wellbore, and are symmetrical or unsymmetrical bi-wing fractures.
5. Gas flows from the reservoir to the wellbore through the fractures. The flow in the reservoir is isothermal, and the gravity and capillary pressure are negligible.

4.2.2 Mathematical model and solutions

We set up a coordinate system as shown in Fig. 4.1 (the y-axis along the horizontal wellbore, the x-axis along the fracture surface and the z-axis perpendicular to the xy plane). The flow in such a horizontal well with multi-stage fractures is very complex due to the interference between the fractures and an uneven flow velocity along the fracture surfaces. The method used to acquire the bottom-hole pressure for a vertical well is not applicable any more. Fracture discretization and potential superposition are combined to solve such problems. First, we divide the fractures into discrete

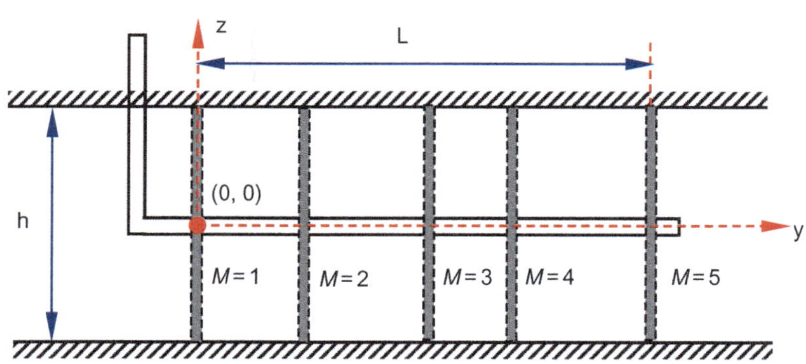

Fig. 4.1 A MFHW in a gas reservoir.

units and assume a unique flow rate for each fracture unit; the flow rates for different units can be different. Second, each fracture unit is treated as an observation point, and the pressure response at each point can be acquired by a potential superposition method. Therefore, the same number of equations are built for these observation points. Finally, based on the conditions of each unit pressure equal to the bottom-hole pressure and the sum of the unit rates equal to the well rate, the equations of these fracture units can be integrated to obtain the bottom-hole pressure of the MFHW and rate distribution along the fracture surfaces.

Because the fractures are assumed fully open, the continuous line source solution at any point of the gas reservoir can be acquired by integration of the continuous point source solution along the vertical direction z_w. We assume that the intensity of the continuous line source is q_{scL}, and then, for the gas reservoir with an infinite outer boundary, there is:

$$\Delta \overline{m}_{fL} = \frac{p_{sc}Tq_{scins}}{\pi k_{fh} h T_{sc} s} \int_0^h \left[K_0(\xi_0 r_D) + 2\sum_{n=1}^{\infty} K_0(\xi_n r_D)\cos(n\pi z_{wD})\cos(n\pi z_D) dz_w \right] \tag{4.1}$$

where $\xi_n = \sqrt{f(s) + \frac{(n\pi)^2}{h_D^2}}, (n=0, 1, 2, 3, \cdots)$ and $r_D = \sqrt{(x_D - x_{wD})^2 + (y_D - y_{wD})^2}$.

The relationship between the continuous point source and the line source intensity is:

$$q_{scL} = q_{scins} h \tag{4.2}$$

The dimensionless distance z_{wD} is defined as:

$$z_{wD} = \frac{z_w}{h} \tag{4.3}$$

Substituting Eqs. (4.2) and (4.3) into Eq. (4.1), the continuous line source solution for the infinite circular shale gas reservoir is:

$$\Delta \overline{m}_{fL} = \frac{p_{sc}Tq_{scL}}{\pi k_{fh} h T_{sc} s} K_0(\xi_0 r_D) \tag{4.4}$$

Similarly, the continuous line source solution for a closed circular shale gas reservoir is:

$$\Delta \overline{m}_{fL} = \frac{p_{sc}Tq_{scL}}{\pi k_{fh} h T_{sc} s} \left[K_0(\xi_0 r_D) + \frac{K_1(\xi_0 r_{eD})}{I_1(\xi_0 r_{eD})} I_0(\xi_0 r_D) \right] \tag{4.5}$$

The partition of fracture grids, grid coordinates' determination, and the solution of these units by superposition principle are introduced below. According to the model assumptions, the number of fractures is M, and the effective horizontal well length is L equal to the distance between the fractures at the two ends. We divide each fracture into $2N$ units and assume that the middle points of the fracture units are pressure observation points (nodes). Then the discrete grid units for a MFHW are shown in Fig. 4.2.

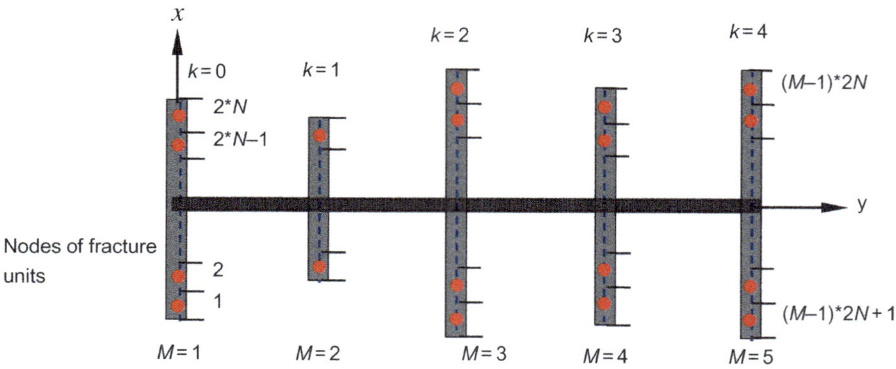

Fig. 4.2 Discrete grid units for a MFHW.

According to the coordinate system set up above, the node coordinates of each fracture unit can be represented for the negative and positive x axis as follows:

For the nodes along the negative x axis:

$$\begin{cases} \hat{x}_i = -\left[N-(i-k^*2N)+\frac{1}{2}\right]\Delta L_{\mathrm{fL}i}, & 1 \le i-k^*2N \le N \\ \hat{y}_i = y_{k+1} \end{cases} \tag{4.6}$$

For the nodes along the positive x axis:

$$\begin{cases} \hat{x}_i = \left[(i-k^*2N)-N-\frac{1}{2}\right]\Delta L_{\mathrm{fL}i}, & N+1 \le i-k^*2N \le 2N \\ \hat{y}_i = y_{k+1} \end{cases} \tag{4.7}$$

where $\hat{x}_i$ and $\hat{y}_i$ are the x and y values of the node of fracture unit i (m), respectively; $\Delta L_{\mathrm{fL}i}$ is the length of the ith fracture unit (m).

Assuming that each half fracture is evenly divided into N units, the length of the ith fracture unit is represented by:

$$\Delta L_{\mathrm{fLi}} = \frac{L_{\mathrm{fL(k+1)}}}{N} \tag{4.8}$$

Because it is assumed that the rate is evenly distributed among these discrete fracture units, the pseudo pressure drop generated by the ith node at any point (x, y) in the reservoir can be acquired by integration of Eq. (4.4) or (4.5) along the ith node unit:

$$\Delta\overline{m}_{\mathrm{fi}} = \int_{\Gamma} \Delta\overline{m}_{\mathrm{fL}}\,\mathrm{d}l \tag{4.9}$$

The fractures discussed in this chapter are assumed parallel to the x axis, and, therefore, the above integration can be converted into integration along the x axis. For an infinite reservoir, substituting Eq. (4.4) and the coordinates of the ith node into Eq. (4.9), there is:

$$\Delta\overline{m}_{\mathrm{fi}} = \int_{\hat{x}_i-\frac{\Delta L_{\mathrm{f}i}}{2}}^{\hat{x}_i+\frac{\Delta L_{\mathrm{f}i}}{2}} \frac{p_{\mathrm{sc}}Tq_{\mathrm{scL}i}}{\pi k_{\mathrm{fh}}hT_{\mathrm{sc}}s}\mathrm{K}_0\left(\xi_0\sqrt{(x_{\mathrm{D}}-x_{\mathrm{wD}i})^2+(y_{\mathrm{D}}-y_{\mathrm{wD}i})^2}\right)\mathrm{d}x_{\mathrm{w}} \tag{4.10}$$

Assuming that the rate of the ith discrete unit is $q_{\mathrm{sc}i}$, there is the following relationship between the line rate and the unit rate based on the assumption of an evenly distributed rate:

$$q_{\mathrm{sc}i} = q_{\mathrm{scL}i}\Delta L_{\mathrm{f}i} \tag{4.11}$$

We define the following dimensionless rate and bottom-hole pressure for a discrete unit:

$$q_{\mathrm{D}i} = \frac{q_{\mathrm{sc}i}}{q_{\mathrm{sc}}} \tag{4.12}$$

$$m_{\mathrm{fD}} = \frac{\pi k_{\mathrm{fh}}hT_{\mathrm{sc}}}{p_{\mathrm{sc}}Tq_{\mathrm{sc}}}\Delta m_{\mathrm{f}} \tag{4.13}$$

Substituting Eqs. (4.12) and (4.13) into Eq. (4.10) yields:

$$\Delta\overline{m}_{\mathrm{fD}i} = \frac{q_{\mathrm{D}i}}{s\Delta L_{\mathrm{fD}i}}\int_{-\frac{\Delta L_{\mathrm{fD}i}}{2}}^{\frac{\Delta L_{\mathrm{fD}i}}{2}} \mathrm{K}_0\left(\xi_0\sqrt{(x_{\mathrm{D}}-\hat{x}_{\mathrm{D}i}-\alpha)^2+(y_{\mathrm{D}}-y_{\mathrm{wD}i})^2}\right)\mathrm{d}\alpha \tag{4.14}$$

where $\Delta L_{\mathrm{fD}i}$ is the dimensionless length of the ith fracture unit, the same definition as x_{D}.

According to the potential superposition, the pressure response at any point (x, y) in the reservoir generated by the $M \times 2N$ discrete fracture units of M fractures can be expressed as:

$$\overline{m}_{\mathrm{D}}(x_{\mathrm{D}}, y_{\mathrm{D}}) = \sum_{i=1}^{M*2N} \overline{m}_{\mathrm{fD}i}(x_{\mathrm{D}}, y_{\mathrm{D}}) \tag{4.15}$$

We take the node $(x_{\mathrm{D}j}, y_{\mathrm{D}j})$ $(1 \le j \le M*2N)$ of any discrete fracture unit as the observation point. Then the pseudo pressure drop caused by all units at node j is:

$$\overline{m}_{\mathrm{D}}(x_{\mathrm{D}j}, y_{\mathrm{D}j}) = \sum_{i=1}^{M*2N} \overline{m}_{\mathrm{fD}i}(x_{\mathrm{D}j}, y_{\mathrm{D}j}) \tag{4.16}$$

Based on the assumption of infinitely conductive fractures, which means no pressure drop in a fracture and no pressure drop in the wellbore, the pseudo pressure at any node is identical and equals the flowing bottom-hole pressure. Then we see that:

$$\overline{m}_{\mathrm{wD}} = \overline{m}_{\mathrm{D}}(x_{\mathrm{D}j}, y_{\mathrm{D}j}) = \sum_{i=1}^{M*2N} \overline{m}_{\mathrm{fD}i}(x_{\mathrm{D}j}, y_{\mathrm{D}j}) \tag{4.17}$$

In addition, the flow rates of all fracture units sum up to the horizontal well production rate q_{sc}; that is:

$$\sum_{i=1}^{M*2N}\left[q_{scLi}\Delta L_{fi}\right]=q_{sc} \tag{4.18}$$

Combining Eqs. (4.12) and (4.18), we have:

$$\sum_{i=1}^{M*2N} q_{Di}=1 \tag{4.19}$$

According to Eq. (4.17), there are $M\times 2N$ linear equations for all nodes. With Eq. (4.19), the number of the total equations is $M\times 2N+1$, which is identical to the number of the total variables, the bottom-hole pseudo pressure m_{wD}, and the dimensionless rates at all discrete fracture units q_{Di} ($i=1, 2, 3,\ldots, M\times 2N$). The equations and variables can be expressed in matrix form as:

$$\begin{bmatrix} A_{1,1} & \cdots & A_{1,k} & \cdots & A_{1,2N*M} & -1 \\ \cdots & \cdots & \cdots & \cdots & \cdots & -1 \\ A_{k,1} & \cdots & A_{k,k} & \cdot\cdot & A_{k,2N*M} & -1 \\ \cdots & \cdots & \cdots & \cdots & \cdots & -1 \\ A_{2N*M,1} & \cdots & A_{2N*M,k} & \cdots & A_{2N*M,2N*M} & -1 \\ 1 & \cdots & 1 & \cdots & 1 & 0 \end{bmatrix}\begin{bmatrix} q_{D1} \\ q_{D2} \\ \cdot \\ \cdot \\ q_{DM*2N} \\ \overline{m}_{wD} \end{bmatrix}=\begin{bmatrix} 0 \\ 0 \\ \cdots \\ \cdots \\ 0 \\ 1 \end{bmatrix} \tag{4.20}$$

where $A_{i,j}$ can be expressed, according to Eq. (4.14), as:

$$A_{j,i}=\frac{1}{s\Delta L_{fDi}}\int_{-\frac{\Delta L_{fDi}}{2}}^{\frac{\Delta L_{fDi}}{2}} K_0\left(\xi_0\sqrt{\left(x_{Dj}-x_{Di}-\alpha\right)^2+\left(y_{Dj}-y_{wDi}\right)^2}\right)d\alpha \tag{4.21}$$

The dimensionless production rate and flow rate derived above are for a MFHW in an infinite shale gas reservoir. For a closed outer boundary, the procedure and coefficient matrix are the same except that the expression of $A_{i,j}$ becomes:

$$A_{j,i}=\frac{1}{s\Delta L_{fDi}}\left[\int_{-\frac{\Delta L_{fDi}}{2}}^{\frac{\Delta L_{fDi}}{2}} K_0(\xi_0 r_D)+\frac{K_1(\xi_0 r_{eD})}{I_1(\xi_0 r_{eD})}I_0(\xi_0 r_D)\right]d\alpha \tag{4.22}$$

where $r_D=\sqrt{\left(x_{Dj}-x_{Di}-\alpha\right)^2+\left(y_{Dj}-y_{wDi}\right)^2}$.

The integral in Eq. (4.22) can be calculated by numerical integration, such as an adaptive algorithm, a Legendre Gauss integral and a variable step-length Runge–Kutta method (Zwillinger, 1996). When an observation point and a sink unit are on the same fracture, then $y_{Dj} = y_{Dwi}$. In such a case, the following algorithm can greatly speed up the integration of the coefficient matrix. Let H_0 represent function $I_0()$ or $K_0()$; then there is:

$$\int_{-\frac{\Delta L_{fDi}}{2}}^{\frac{\Delta L_{fDi}}{2}} H_0\left(\xi_0 \left|x_{Dj} - x_{Di} - \alpha\right|\right) d\alpha = \frac{1}{\xi_0} \begin{cases} \int_0^{\beta_1} H_0(\alpha) d\alpha - \int_0^{\beta_2} H_0(\alpha) d\alpha, & i > j \\ 2\int_0^{\beta_1} H_0(\alpha) d\alpha, & i = j \\ \int_0^{-\beta_2} H_0(\alpha) d\alpha - \int_0^{-\beta_1} H_0(\alpha) d\alpha, & i < j \end{cases} \tag{4.23}$$

where $\beta_1 = \xi_0\left(x_{Dj} - x_{Di} + \frac{\Delta L_{fDi}}{2}\right)$ and $\beta_2 = \xi_0\left(x_{Dj} - x_{Di} + \frac{\Delta L_{fDi}}{2}\right)$.

After solving for the coefficient matrix by numerical integration, the dimensionless pressure and production rate can be acquired by solving the system of the matrix equations. For such a dense matrix, a direct solution method for linear algebraic equations (e.g., Gaussian elimination) instead of an iterative solution method (e.g., Jacobi and Gauss–Seidel iterative methods) can be used (Zwillinger, 1996).

4.3 MFHWs in rectangular gas reservoirs

With development of drilling and fracturing techniques and equipment, longer well length and more fracture stages are achieved to improve well productivity. Due to the ultra-low permeability of shale, a well factory mode is applied for shale gas development. The well spacing is small relative to horizontal well length. Therefore, the assumption of a circular reservoir boundary is not applicable and could lead to data misinterpretation or even a wrong analysis, especially for an analysis of a boundary response. More and more investigators brought up the application of a linear flow model for such a problem; however, a linear flow model cannot represent interference between fractures and the flow from a reservoir to fracture tips. To better describe a transient flow model for MFHWs, the point source function methodology is used in this chapter to analyze transient flow for a horizontal well with multi-stage fractures in a closed rectangular gas reservoir, and well test and production decline type curves for different mechanism models are analyzed (Zerzar, et al., 2004).

4.3.1 Physical model

Fig. 4.3 shows the physical model of a MFHW in a closed rectangular gas reservoir. To simplify the subsequent analysis, the following assumptions are made:

1. The gas reservoir is homogeneous and anisotropic with horizontal and vertical permeability k_h and k_z, and the length, width, and height of the reservoir are x_e, y_e and h, respectively.

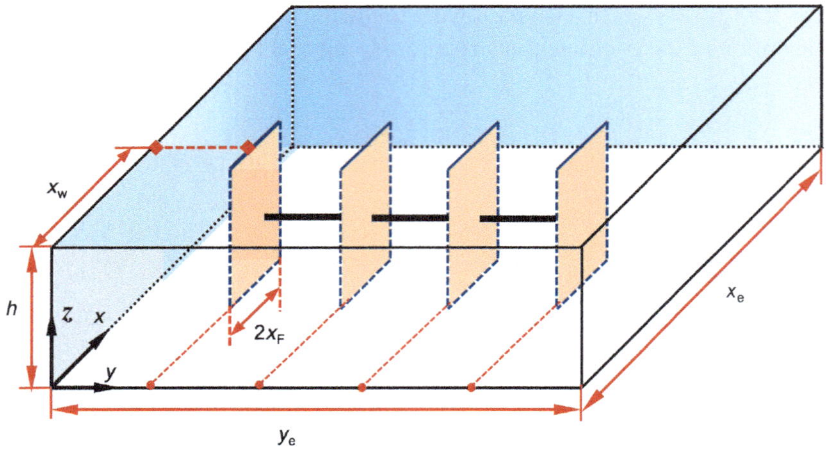

Fig. 4.3 A MFHW in a rectangular gas reservoir.

2. A MFHW locates in the middle of the reservoir and parallel to the reservoir boundary. The well length is L. All fractures are symmetrical and perpendicular to the wellbore; the fracture half length of the ith fracture is x_{fi}; and the fractures are fully open with y_i as the coordinate in the y direction.
3. The well is perforated at the joints of the fractures and wellbore, which means that the fluid flows from the reservoir to the fractures and then to the wellbore. There is no pressure drop in the infinite conductivity fractures and horizontal wellbore.
4. Each fracture is evenly divided into $2N$ units with an identical flow rate.

4.3.2 Mathematical model and solutions

Similar to the construction and solution of the mathematical model in a circular gas reservoir, the model for a MFHW in a closed rectangular gas reservoir can be constructed and solved by discretization of the fractures and the potential superposition principle (Fig. 4.4).

According to the solution of a fully penetrated continuous line source in a rectangular gas reservoir, the pressure drop generated by the ith fracture unit at any reservoir location (x_D, y_D) can be acquired by the integration of the continuous line source solution Eq. (3.14) along the infinitesimal direction:

$$\Delta\overline{m}_{fi}(x_D, y_D) = \int_{\Gamma} \Delta\overline{m}_{fL}(x_D, y_D, x_{wDi}, y_{wDi}, s)dl \tag{4.24}$$

For the fractures in this model, which are parallel to the x-axis, the line integral in the above equation can be converted into:

$$\Delta\overline{m}_{fi}(x_D, y_D) = \int_{x_{mi}-\Delta L_{fi}/2}^{x_{mi}+\Delta L_{fi}/2} \Delta\overline{m}_{fL}(x_D, y_D, x_{wDi}, y_{wDi}, s)dx_w \tag{4.25}$$

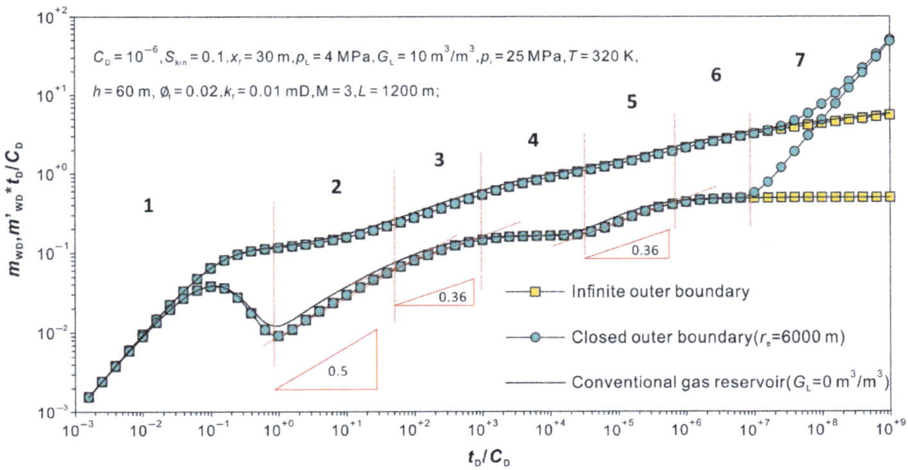

Fig. 4.4 Discrete fracture units on an *x-y* plane.

Substituting Eq. (3.14) into Eq. (4.25) yields:

$$\Delta \overline{m}_{fi} = \frac{p_{sc}T}{T_{sc}} \frac{q_{scLi}}{\pi k_f L_{ref} h_D s} \frac{\pi}{x_{eD}} \left\{ \Delta L_{fi} \frac{\cosh(\sqrt{u}\tilde{y}_{D1}) + \cosh(\sqrt{u}\tilde{y}_{D2})}{\sqrt{u}\sinh(\sqrt{u}y_{eD})} \right.$$
$$\left. + 2L_{ref} \int_{-\Delta L_{fDi}/2}^{+\Delta L_{fDi}/2} \sum_{k=1}^{+\infty} \left(k\pi \frac{x_D}{x_{eD}}\right) \cos\left(k\pi \frac{x_{mDi} + \alpha}{x_{eD}}\right) \frac{\cosh\left(\sqrt{\varepsilon_k}\tilde{y}_{D1}\right) + \cosh\left(\sqrt{\varepsilon_k}\tilde{y}_{D2}\right)}{\sqrt{\varepsilon_k}\sinh\left(\sqrt{\varepsilon_k}y_{eD}\right)} \right\} d\alpha \tag{4.26}$$

We define the following dimensionless production rate and pressure:

$$q_{Di} = \frac{q_{scLi}\Delta L_{fi}}{q_{sc}} \tag{4.27}$$

$$m_{fD} = \frac{\pi k_{fh} h T_{sc}}{p_{sc} T q_{sc}} \Delta m_f \tag{4.28}$$

Introducing Eqs. (4.27) and (4.28) into Eq. (4.26), the dimensionless pressure drop generated by the *i*th unit at any reservoir location can be expressed as:

$$\overline{m}_{fDi} = \frac{q_{Di}}{s} \frac{\pi}{x_{eD}} \left\{ \frac{\cosh(\sqrt{u}\tilde{y}_{D1}) + \cosh(\sqrt{u}\tilde{y}_{D2})}{\sqrt{u}\sinh(\sqrt{u}y_{eD})} \right.$$
$$\left. + \frac{2}{\Delta L_{fDi}} \int_{-\Delta L_{fDi}/2}^{\Delta L_{fDi}/2} \sum_{k=1}^{+\infty} \cos\left(k\pi \frac{x_D}{x_{eD}}\right) \cos\left(k\pi \frac{x_{mDi} + \alpha}{x_{eD}}\right) \frac{\cosh\left(\sqrt{\varepsilon_k}\tilde{y}_{D1}\right) + \cosh\left(\sqrt{\varepsilon_k}\tilde{y}_{D2}\right)}{\sqrt{\varepsilon_k}\sinh\left(\sqrt{\varepsilon_k}y_{eD}\right)} \right\} d\alpha \tag{4.29}$$

Because the horizontal well is composed of many discrete fracture units, the pressure at any point in the reservoir is the superposition of the pressure drops caused by all these units at the same point. For the observation point (x_{Dj}, y_{Dj}), based on the superposition principle, there is:

$$\overline{m}_D\left(x_{Dj}, y_{Dj}\right) = \sum_{i=1}^{M*2N} q_{Di} A_{ji}\left(x_{Di}, y_{Di}, x_{Dj}, y_{Dj}, s\right) \tag{4.30}$$

where A_{ji} is expressed as:

$$A_{ji} = \left(\overline{S}_{\text{Dinf}} + \overline{S}_{\text{Db1}} + \overline{S}_{\text{Db2}} + \overline{S}_{\text{Db3}}\right)_{ji} \tag{4.31}$$

$$\left(\overline{S}_{\text{Dinf}}\right)_{ji} = \frac{1}{s\Delta L_{fDi}} \int_{-\Delta L_{fDi}/2}^{\Delta L_{fDi}/2} \mathrm{K}_0\left[\sqrt{\left(x_{\text{Di}} - x_{\text{Dj}} + \alpha\right)^2 \left(y_{\text{Di}} - y_{\text{Dj}}\right)^2}\sqrt{u}\right] \mathrm{d}\alpha \tag{4.32}$$

$$\begin{aligned}\left(\overline{S}_{\text{Db1}}\right)_{ji} = {} & \frac{\pi}{x_{eD} s\sqrt{u}}\left\{\exp\left[-\sqrt{u}\left(y_{\text{Dj}} + y_{\text{Di}}\right)\right] + \exp\left[-\sqrt{u}\left[2y_{eD} - \left(y_{\text{Dj}} + y_{\text{Di}}\right)\right]\right]\right. \\ & \left. + \exp\left[-\sqrt{u}\left(2y_{eD} - \left|y_{\text{Di}} - y_{\text{Dj}}\right|\right)\right] + \exp\left[-\sqrt{u}\left|y_{\text{Di}} - y_{\text{Dj}}\right|\right]\right\}\left[1 + \sum_{m=1}^{\infty}\left(-2m\sqrt{u}y_{eD}\right)\right]\end{aligned} \tag{4.33}$$

$$\begin{aligned}\left(\overline{S}_{\text{Db2}}\right)_{ji} = {} & \frac{1}{s\Delta L_{fDi}} \int_{-\Delta L_{fDi}/2}^{\Delta L_{fDi}/2} \mathrm{K}_0\left[\sqrt{\left(x_{\text{Di}} - x_{\text{Dj}} + \alpha\right)^2 + \Delta y_{\text{D}}^2}\sqrt{u}\right] \mathrm{d}\alpha \\ & + \frac{1}{s\Delta L_{fDi}} \sum_{n=1}^{+\infty} \int_{-\Delta L_{fDi}/2}^{\Delta L_{fDi}/2} \left\{\mathrm{K}_0\left[\sqrt{\left(x_{\text{Di}} - x_{\text{Dj}} - 2nx_{eD} + \alpha\right)^2 + \Delta y_{\text{D}}^2}\sqrt{u}\right]\right. \\ & + \mathrm{K}_0\left[\sqrt{\left(x_{\text{Di}} + x_{\text{Dj}} - 2nx_{eD} + \alpha\right)^2 + \Delta y_{\text{D}}^2}\sqrt{u}\right] + \mathrm{K}_0\left[\sqrt{\left(x_{\text{Di}} - x_{\text{Dj}} - 2nx_{eD} + \alpha\right)^2 + \Delta y_{\text{D}}^2}\sqrt{u}\right] \\ & \left. + \mathrm{K}_0\left[\sqrt{\left(x_{\text{Di}} + x_{\text{Dj}} + 2nx_{eD} + \alpha\right)^2 + \Delta y_{\text{D}}^2}\sqrt{u}\right]\right\} d\alpha - \frac{\pi}{x_{eD} s} \frac{\exp\left(-\sqrt{u}\left|\Delta y_{\text{D}}\right|\right)}{\sqrt{u}}\end{aligned} \tag{4.34}$$

$$\begin{aligned}\left(\overline{S}_{\text{Db3}}\right)_{ji} = {} & \frac{4}{s\Delta L_{fDi}} \sum_{k=1}^{+\infty} \frac{\cos\left(k\pi\frac{x_{\text{Dj}}}{x_{eD}}\right)\cos\left(k\pi\frac{x_{\text{Di}}}{x_{eD}}\right)\sin\left(k\pi\frac{x_{f\text{Di}}}{x_{eD}}\right)}{k} \left\{\frac{e^{-\sqrt{\varepsilon_k}\left|y_{\text{Di}} - y_{\text{Dj}}\right|}}{\sqrt{\varepsilon_k}} \sum_{m=1}^{\infty} e^{-2m\sqrt{\varepsilon_k}y_{eD}}\right. \\ & \left. + \frac{e^{-\sqrt{\varepsilon_k}\left(y_{\text{Di}} + y_{\text{Dj}}\right)} + e^{-\sqrt{\varepsilon_k}\left(2y_{eD} - \left|y_{\text{Di}} - y_{\text{Dj}}\right|\right)} + e^{-\sqrt{\varepsilon_k}\left[2y_{eD} - \left(y_{\text{Dj}} + y_{\text{Dj}}\right)\right]}}{\sqrt{\varepsilon_k}}\left[1 + \sum_{m=1}^{\infty} e^{-2m\sqrt{\varepsilon_k}y_{eD}}\right]\right\}\end{aligned} \tag{4.35}$$

Since the fractures are assumed to be of infinite conductivity, the pressure at any point along the fractures is identical to the bottom-hole pressure:

$$\overline{m}_{\mathrm{wD}} = \overline{m}_{\mathrm{D}}\left(\hat{x}_{\mathrm{D}j}, \hat{y}_{\mathrm{D}j}\right) \tag{4.36}$$

Also, on the basis of material balance, the sum of the unit production rates is equal to the well production rate:

$$\sum_{i=1}^{\mathrm{M*2N}} q_{\mathrm{D}i} = 1 \tag{4.37}$$

Integrating Eqs. (4.30), (4.36), and (4.37) and using the similar solution method to that for a circular gas reservoir to build the linear equation system, the resulting coefficient matrix system is:

$$\begin{bmatrix} A_{1,1} & \cdots & A_{1,k} & \cdots & A_{1,2N*M} & -1 \\ \cdots & \cdots & \cdots & \cdots & \cdots & -1 \\ A_{k,1} & \cdots & A_{k,k} & \cdots & A_{k,2N*M} & -1 \\ \cdots & \cdots & \cdots & \cdots & \cdots & -1 \\ A_{2N*M,1} & \cdots & A_{2N*M,k} & \cdots & A_{2N*M,2N*M} & -1 \\ 1 & \cdots & 1 & \cdots & 1 & 0 \end{bmatrix} \begin{bmatrix} q_{\mathrm{D1}} \\ q_{\mathrm{D2}} \\ \cdot \\ \cdot \\ q_{\mathrm{D}M*2N} \\ \overline{m}_{\mathrm{wD}} \end{bmatrix} = \begin{bmatrix} 0 \\ 0 \\ \cdots \\ \cdots \\ 0 \\ 1 \end{bmatrix} \tag{4.38}$$

The coefficient matrix can be obtained through numerical integration, and this system can be solved by Gaussian elimination, for example.

4.4 Analysis of well bottom-hole pressure and production performance

4.4.1 *Microfractures + steady state adsorption/desorption and diffusion*

Inserting the parameters in Table 4.1 into the semi-analytical solutions of bottom-hole pressure for a MFHW producing at a constant rate in circular and rectangular gas reservoirs, the dimensionless bottom-hole pressure response curves, and production and cumulative production of the well producing at constant bottom-hole pressure can be acquired. Fig. 4.5 shows the comparison of well test type curves for a multi-stage horizontal well using the mechanism Model 1 in a shale gas reservoir and a conventional gas reservoir (without consideration of adsorption, desorption, and diffusion). According to the pseudo pressure log–log curves, the transient flow in a MFHW by using the mechanism Model 1 can be divided into the following stages:

Flow stage 1: the wellbore storage period followed by the transition period. During the wellbore storage period, the pseudo pressure and its derivative curves overlay each other and their slope equals 1. Since pressure has not propagated into the reservoir during this period, the curves for the shale reservoir and conventional reservoir are the same.

Flow stage 2: the early time linear flow perpendicular to the hydraulic fractures. After the wellbore effect terminates, the free gas in the microfracture system flows to the fractures

Table 4.1 Shale reservoir properties

Reservoir properties	Value	Reservoir properties	Value
Initial reservoir pressure, p_i (MPa)	25	Reservoir temperature, T (K)	320
Reservoir thickness, h (m)	60	Fracture half length, x_f (m)	30
Specific gas gravity, γ_g	0.65	Well production rate, q_{sc} (m^3/d)	1×10^4
Gas compressibility at initial conditions, C_{gi} (MPa^{-1})	0.02	Bottom-hole pressure, p_{wf} (MPa)	1
Microfracture permeability, k_f (mD)	0.01	Microfracture porosity, $\varnothing_f$ (fraction)	0.02
Skin, S_{kin} (dimensionless)	0.1	Dimensionless wellbore storage coefficient, C_D	10^{-6}
Langmuir pressure, P_L (MPa)	4	Langmuir volume, G_L (m^3/m^3)	10
Number of hydraulic pressures, M	3	Effective horizontal well length, L (m)	1200
Outer boundary radius of closed circular reservoir, r_e(m)	6000		

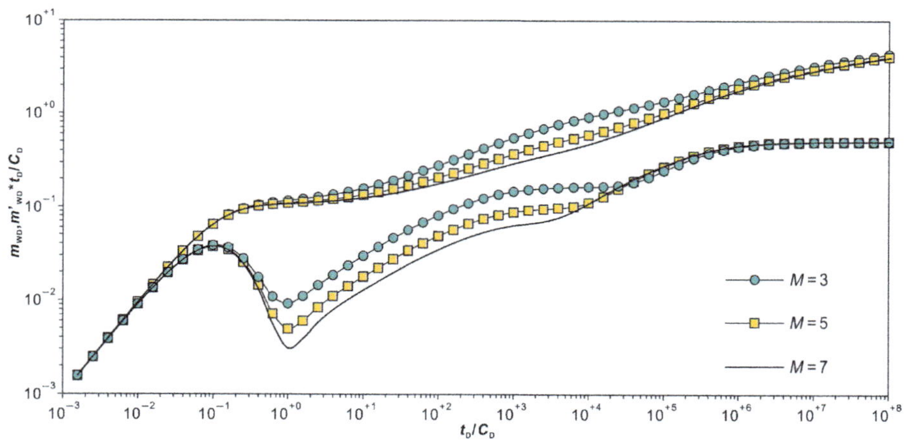

Fig. 4.5 Well test type curves of MFHW in a circular gas reservoir with different boundary conditions.

perpendicular to the fracture surfaces. During this period, there is no interference between these fractures, and the slope of the pseudo pressure derivative is 0.5.

Flow stage 3: the early time elliptical flow. Due to the long fracture half length, there is elliptical flow with the two fracture ends as its centers, and the slope of the pseudo pressure derivative is 0.36. When the fracture half length is relatively small or fracture spacing is relatively small, this stage will be covered by the interference between the fractures.

Flow stage 4: the early time radial flow. If the fracture spacing is big and the half length is small, radial flow happens and surrounds each fracture before interference happens. During this period, the pseudo pressure derivative exhibits the pseudo radial flow characteristics of a single fracture, and becomes a horizontal line at a value of $1/(2M)$.

Flow stage 5: the elliptical flow in the natural fracture system. It is also viewed as the late time linear flow by some scholars. However, on the pressure contour map of a MFHW, there is usually elliptical flow surrounding the horizontal wellbore after the fracture early time radial flow, which is not absolutely linear flow. The pseudo pressure derivative curve has a slope close to 0.36; and, therefore, it is viewed as the elliptical flow in the natural fracture system in this chapter.

Flow stage 6: the radial flow in the natural fracture system. During this period, the pseudo pressure derivative is horizontal at 0.5.

Flow stage 7: the boundary dominated flow. For a gas reservoir with a closed boundary, the boundary dominated flow happens when pressure propagates to the boundary. During this period, the pseudo pressure and its derivative overlay each other as a line with slope 1.

Figs. 4.6 and 4.7 show the effect of the fracture number M on the well test type and production decline curves. From these plots, it can be seen that M influences the pseudo pressure during the early and middle time. The more fractures there are, the lower the early and middle time pseudo pressure and its derivative curves are on their plots. This is because the area for flow from the reservoir to the fractures is increased with an increase in the fracture number, and a less pressure drop is required for the same production rate. During the elliptical flow period in the natural fracture system, although the pseudo pressure derivative curves overlay, the differential pseudo pressure is different. After the pressure propagates further into the reservoir, all curves overlay for different M, which confirms that the late radial flow represents the flow characteristics of a further reservoir region. For the well production rate, the bigger the M is, the higher the production rate is. But the production difference for different M gets smaller with pressure propagation into the reservoir. After 10 years' production, the difference in the production rate for $M=3$ and $M=7$ is

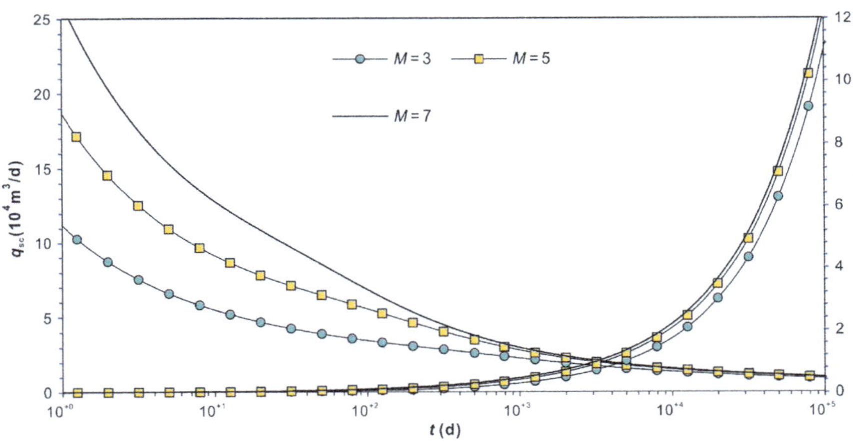

Fig. 4.6 Effect of fracture number on well test type curves.

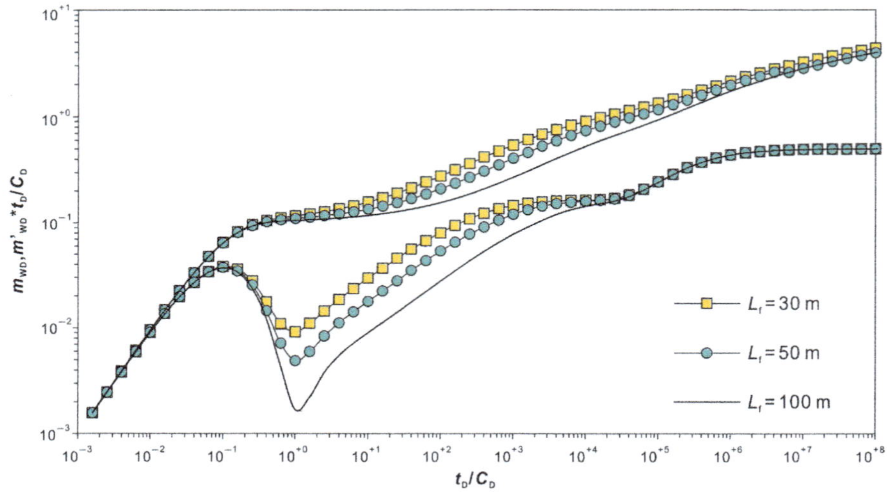

Fig. 4.7 Effect of fracture number on well production rate and cumulative production.

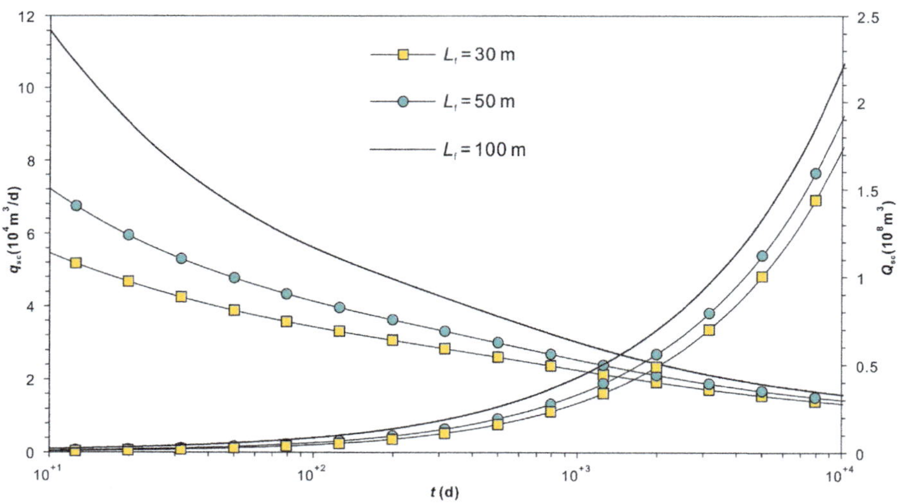

Fig. 4.8 Effect of fracture half length on well test type curves in a circular gas reservoir.

7000 m^3/d, which is a considerable amount for shale gas production. From the plot in Fig. 4.7, the production rates for different M become almost identical after 20 years' production. Similarly, the cumulative production is identical at well abandonment.

Figs. 4.8 and 4.9 show the effect of the fracture half-length on the well test type and production decline curves, respectively. Comparing Figs. 4.6 and 4.8, the effect of the fracture half-length has a similar effect to the fracture number on the well test type curves. The longer the L_f is, the lower the pseudo pressure and its derivative curves

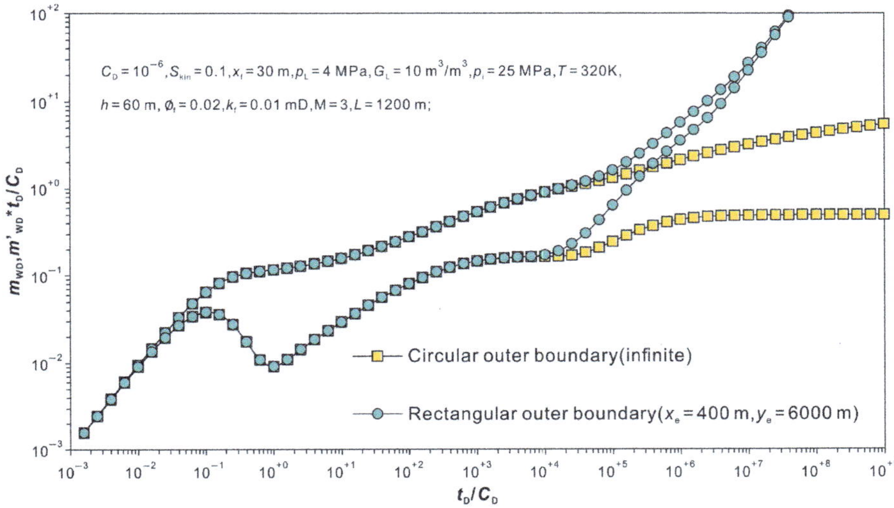

Fig. 4.9 Effect of fracture half length on well production rate and cumulative production curves in a circular gas reservoir.

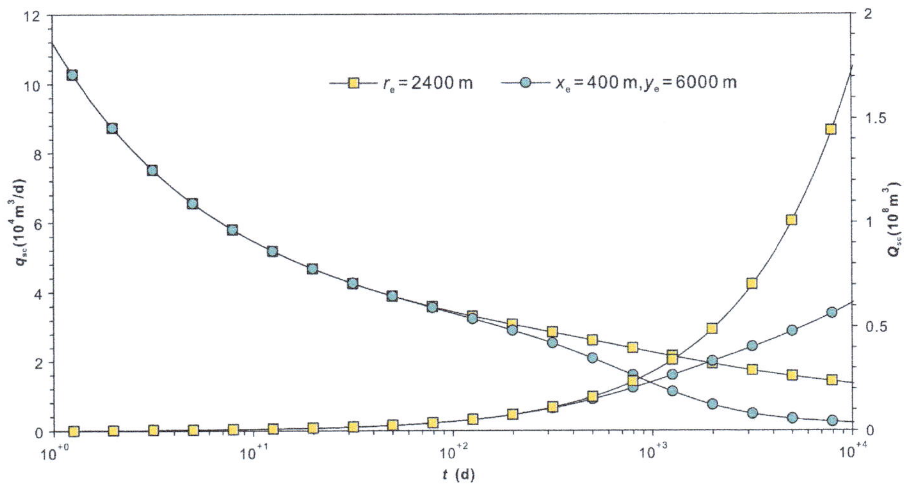

Fig. 4.10 Comparison of well test type curves for circular and rectangular boundary reservoirs.

show up in the plot in Fig. 4.8, which is also due to the bigger flow area from longer fractures. For the effect on well production, the production rate and cumulative production after 30 years' production are analyzed to better describe the difference. From the plot in Fig. 4.9, it can be seen that the longer the fracture half length is, the bigger difference the early time production is.

For a closed rectangular reservoir, the well test type and production decline curves can be analyzed through a similar method to that for a fractured vertical well. Fig. 4.10

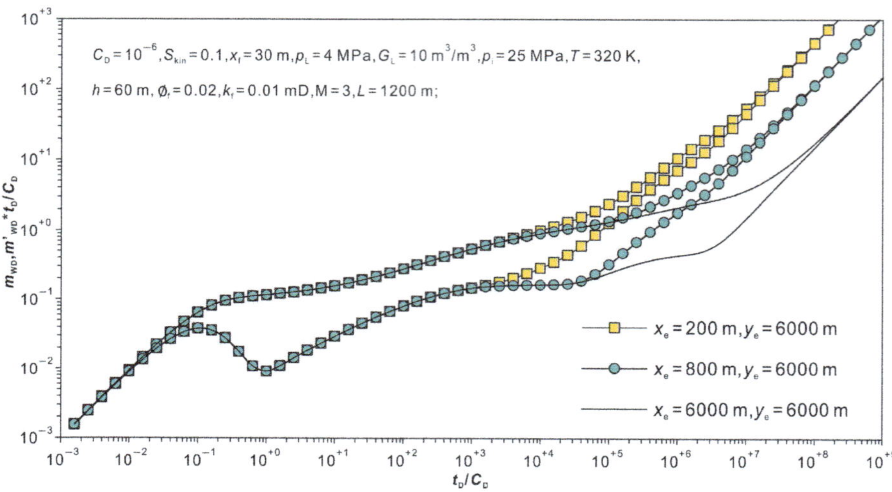

Fig. 4.11 Comparison of production and cumulative production curves for circular and rectangular boundary reservoirs.

illustrates the difference of the well test type curves between a closed rectangular reservoir and an infinite circular reservoir. In this plot, the pseudo pressure and its derivative curves overlay before pressure propagates to the nearest boundary of the rectangular reservoir. Thereafter, the pseudo pressure curves show up when the pressure reaches the boundary, which is the end of the early time radial flow. For such gas wells, it is very hard to observe the late time radial flow, and, therefore, it should be cautious to select a suitable well test analysis model. Fig. 4.11 shows the production curves. It can be obviously seen that the production rate from the circular reservoir is higher than that from the rectangular reservoir when the pressure propagation reaches the upper and lower boundaries. This is because the drainage area in the rectangular reservoir $2x_e \times h$ is smaller than that in the circular reservoir $2\pi r \times h$ $(r > x_e)$, when the pressure propagates to the nearest boundary.

Fig. 4.12 shows the effect of the rectangle dimensions on the well test type curves. For a fixed reservoir length 6000 m, which is identical to the well length, the smaller the reservoir width is, the earlier the boundary response happens. Also, the elliptical flow in the microfracture system and the late time radial flow are both concealed by the co-influence of linear flow and boundary dominated flow. As the reservoir width increases, the concealed flow stages gradually show up.

4.4.2 Microfracture + matrix macropores + steady state adsorption/desorption and diffusion

Table 4.2 lists the reservoir properties used for the mechanism Model 2 to analyze the well test type and production performance curves for a MFHW in circular and rectangular gas reservoirs. The corresponding pseudo pressure and pseudo pressure derivative curves are generated through computer programing.

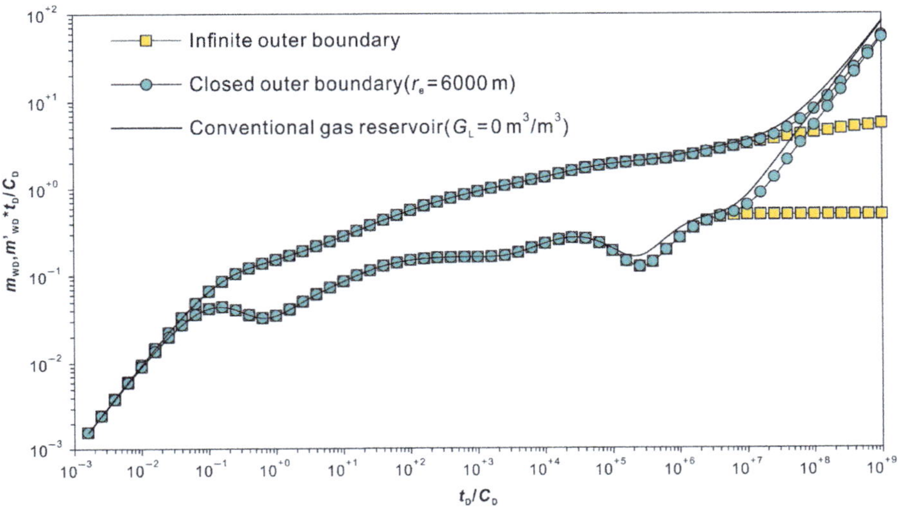

Fig. 4.12 Effect of rectangular reservoir dimensions on well test type curves.

Table 4.2 Shale reservoir properties

Reservoir properties	Value	Reservoir properties	Value
Microfracture system permeability, k_f (mD)	0.01	Microfracture system porosity, $\varnothing_f$	0.02
Macropore system permeability, k_m (mD)	0.0001	Macropore system porosity, $\varnothing_m$	0.12
Interporosity flow shape factor,α (1/m^2)	10^{-5}		

The well test type curves for a MFHW in a circular gas reservoir under the flow mechanism Model 2 are shown in Fig. 4.13. Compared to Model 1, a concave part on the pseudo pressure derivative curve caused by the interporosity flow from macropores to microfractures and the desorption and diffusion of the matrix adsorbed gas to micropores can be observed. The effect of an adsorbed gas volume on the type curves are illustrated in Fig. 4.14. Through the type curve comparison of shale gas and conventional gas reservoirs, it can be seen that the concave part for interporosity flow becomes wider and deeper with consideration of a shale gas adsorption and desorption effect. This is because the desorption of adsorbed gas can compensate for the pressure loss due to production, which slows down the pressure depletion with production, exhibiting as a deeper concave part on the pseudo pressure derivative curve. In addition, for the steady state ad-/desorption model, the equivalent adsorption and desorption compressibility weighs greater in the total reservoir compressibility, and the capability of pressure recharge of the adsorbed gas is stronger.

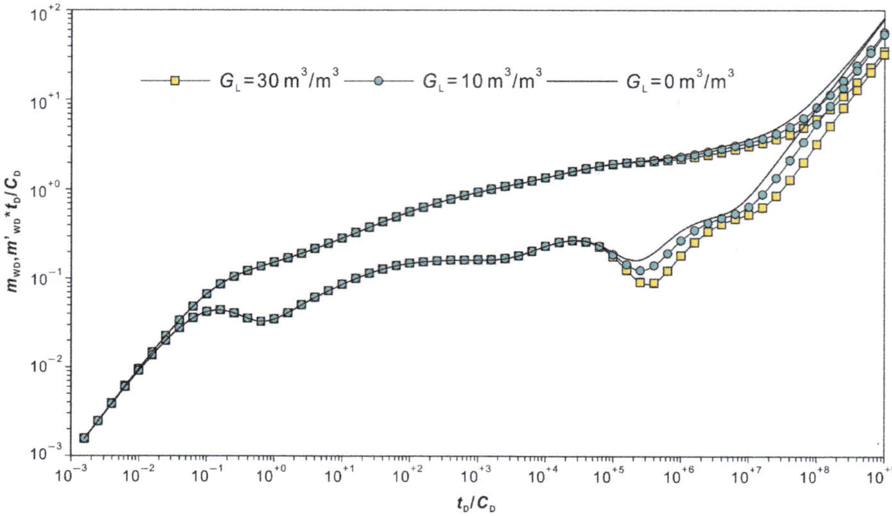

Fig. 4.13 Effect of outer boundary conditions on well test type curves of MFHW in a circular shale gas reservoir.

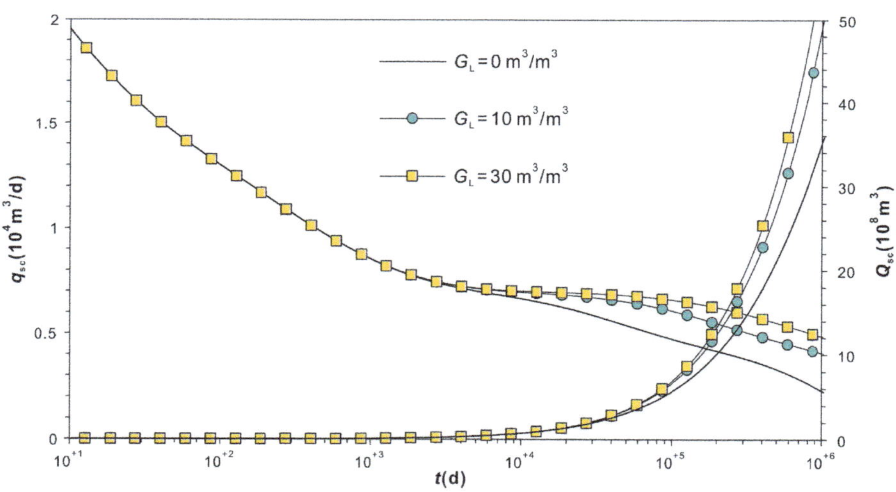

Fig. 4.14 Effect of adsorption gas volume on well test type curves for a circular gas reservoir.

As a result, the pseudo pressure derivative curve is lower in the plot in Fig. 4.14, and the time for the pressure to reach the boundary is longer.

Fig. 4.15 shows the effect of an adsorbed gas volume on the production rate and cumulative production. The microfacture and macropore systems, where free gas exists, are both assumed to exist in a shale reservoir. When the well is producing at a constant rate, free gas is produced first until the reservoir pressure drops to a

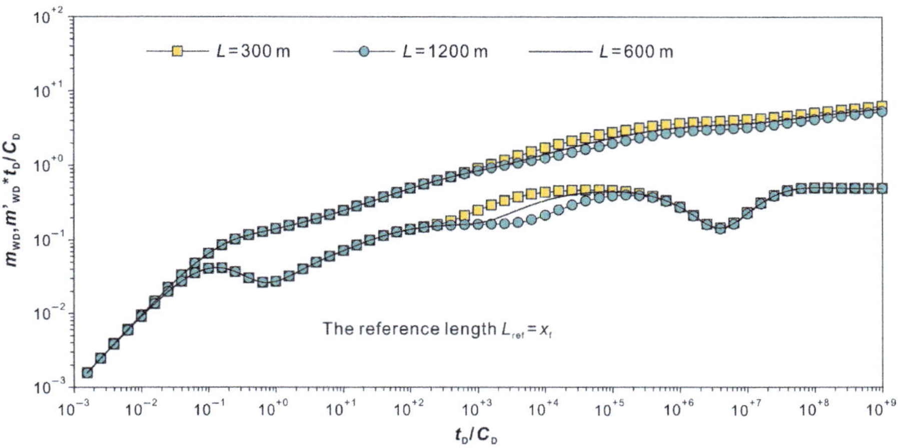

Fig. 4.15 Effect of adsorption gas volume on production of MFHW in a closed circular gas reservoir.

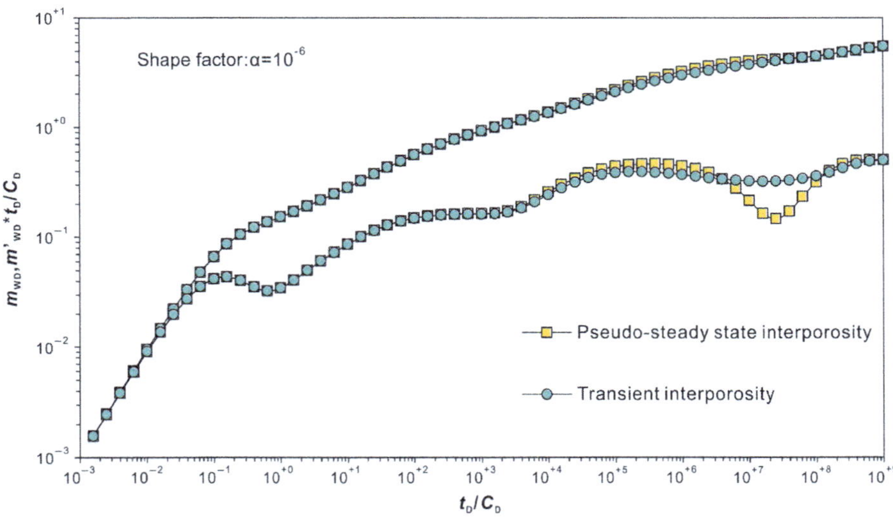

Fig. 4.16 Effect of fracture spacing on well test type curves.

certain level for adsorbed gas to desorb and diffuse. Therefore, the well production rates for different adsorption gas volumes are identical in the early production period until desorption and diffusion happen. With more and more desorbed gas in the produced gas, the higher volume of adsorbed gas there is, the higher rate the well produces at.

Fig. 4.16 shows the effect of fracture spacing on the well test type curves. According to the curves shown in this plot, fracture spacing mainly influences the early radial flow. For a constant fracture half length, the tighter the fractures distribute, the earlier the interference between the fractures happens and the more difficult the pseudo radial flow shows up in each individual fracture. Therefore, the early radial flow is less obvious on the pseudo pressure derivative curves.

Fig. 4.17 shows the well test type curves for a MFHW in an infinite circular gas reservoir under different diffusion models. From this plot, it can be seen that diffusion mainly affects the shape of curves during the interporosity flow period. For the pseudo steady state interporosity flow, an obvious concave part shows up on the curves; for transient flow, the pseudo pressure derivative curve is flat and slightly concave. In addition, the transient diffusion model is more sensitive to a reservoir pressure change, and, therefore, earlier interporosity flow can be observed from this plot.

Fig. 4.18 shows the comparison of the well test type curves for a MFHW in closed rectangular and infinite circular gas reservoirs. It can be seen that the fracture early radial flow is followed by linear flow for the closed rectangular reservoir because pressure propagates quickly to the rectangular boundary. There is a relatively long linear flow period on the type curve (slop = 0.5 for the pseudo pressure and pseudo pressure derivative curves). The linear flow is delayed with an increase in the reservoir size, and the pseudo pressure and pseudo pressure derivative curves become lower (as shown in Fig. 4.19) in the plot due to a less pressure drop.

Fig. 4.20 shows the effect of an interporosity flow coefficient on the well test type curves. Theoretically, this coefficient only influences the starting time of the interporosity flow concave part and does not influence the depth and width of this concave part. The higher the interporosity flow coefficient is, the earlier the concave part shows up on the curve. However, for a rectangular gas reservoir, the interporosity flow happens late and usually after the formation linear flow, and, therefore, the shape of the concave part is affected by multiple factors.

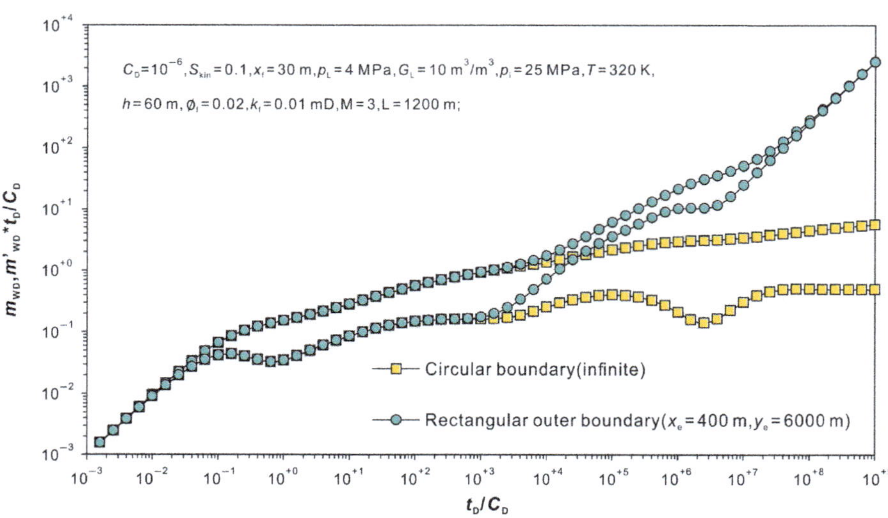

Fig. 4.17 Effect of diffusion models on well test type curves.

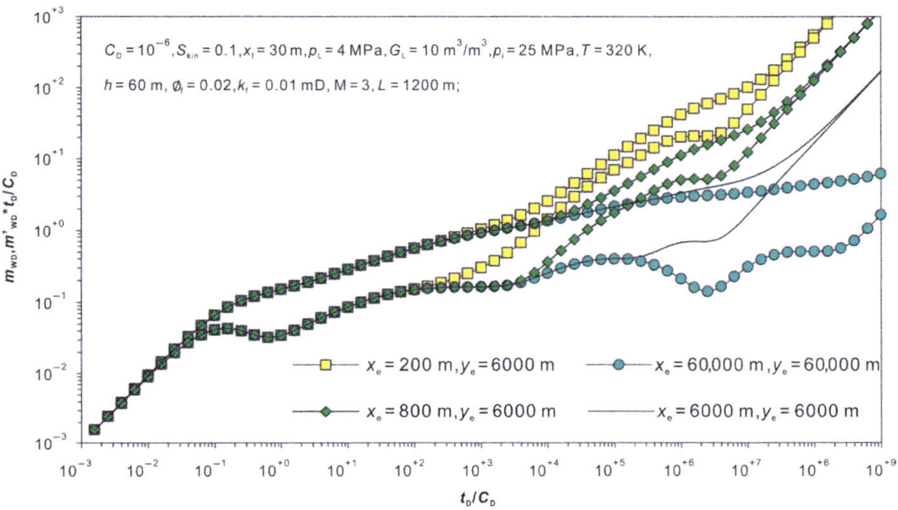

Fig. 4.18 Comparison of well test type curves of MFHW in closed rectangular and infinite circular gas reservoirs.

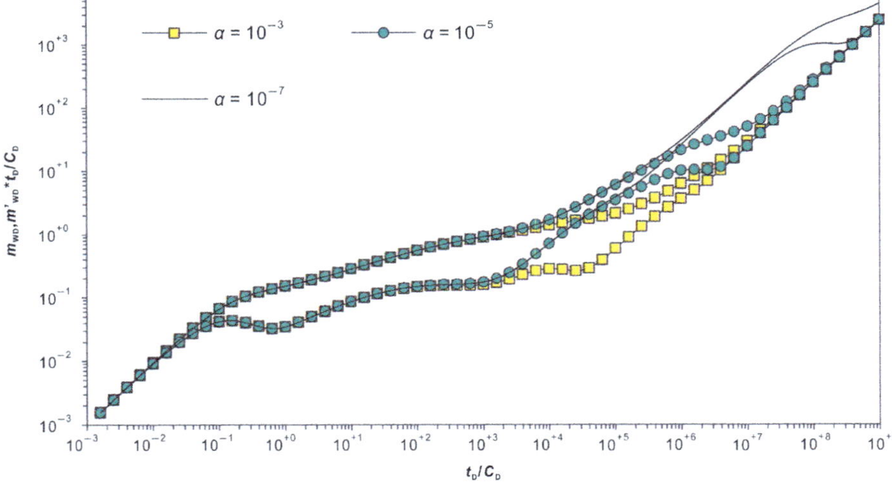

Fig. 4.19 Effect of reservoir size on well test type curves of MFHW in a rectangular gas reservoir.

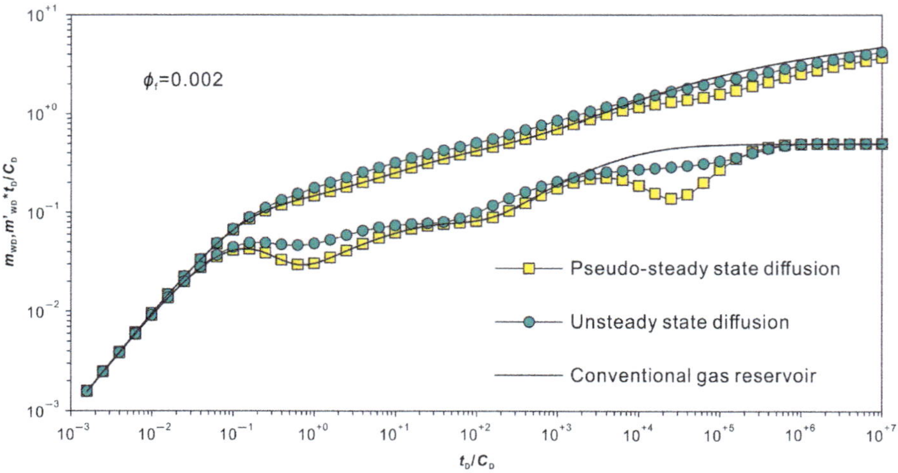

Fig. 4.20 Effect of the interporosity flow coefficient on the well-test type curves.

4.4.3 Microfractures + gas adsorption/desorption + Fick's diffusion

Table 4.3 lists the basic reservoir properties for the mechanism Model 3.

The well test type curves for a MFHW in an infinite circular gas reservoir are shown in Fig. 4.21. Compared to a conventional gas reservoir, there is an additional concave part on the curve for interporosity diffusion by using the mechanism Model 3. After a well produces for a certain period of time, there is a pressure difference between the microfracture system and the matrix system. The gas in the matrix diffuses into the natural fracture system in the pseudo steady state or transient mode, and then flows into the wellbore through hydraulic fractures. The concave part is caused by the diffusion process. For a conventional gas reservoir, there are no adsorbed gases in the matrix and no desorption and diffusion caused by a pressure difference; therefore, the concave part does not show up on the pseudo pressure derivative curve.

The well production rate and cumulative production with time for a well producing at constant BHP under different diffusion models are shown in Fig. 4.22. The early production for transient diffusion is higher than that for the pseudo steady state, which is because the unsteady state diffusion is more sensitive to the pressure in the fracture system. Once a pressure difference exists between the fracture system and the matrix, the adsorbed gas starts to charge the fracture system in the unsteady state mode, while this diffusion process has not happened in the pseudo steady state. After the pseudo steady state diffusion starts, the gas production becomes higher than that in the unsteady state. Finally, the production rates for the two diffusion models stabilize when the pressure in both the matrix and the fracture system decreases simultaneously. The extra gas supply from adsorption to the fracture system in a shale gas reservoir results in higher production than in a conventional gas reservoir no matter if it is the unsteady or pseudo steady state diffusion.

Table 4.3 Shale reservoir properties

Reservoir properties	Value	Reservoir properties	Value
Microfracture system permeability, k_f (mD)	0.01	Microfracture system porosity, $\varnothing_f$	0.02
Constant gas well production rate, $q_{sc}(m^3/d)$	1×10^4	$(R_m)^2/D_F$ Ratio	2×10^6

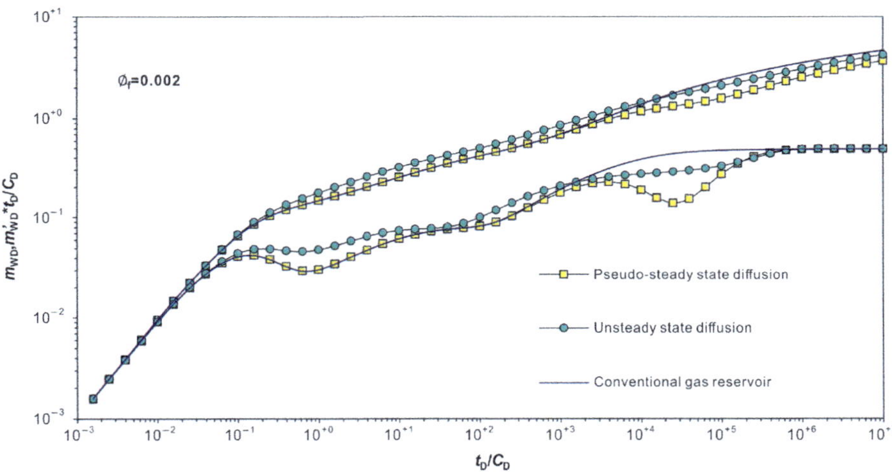

Fig. 4.21 Comparison of well test type curves for MFHW in a circular gas reservoir under different diffusion models.

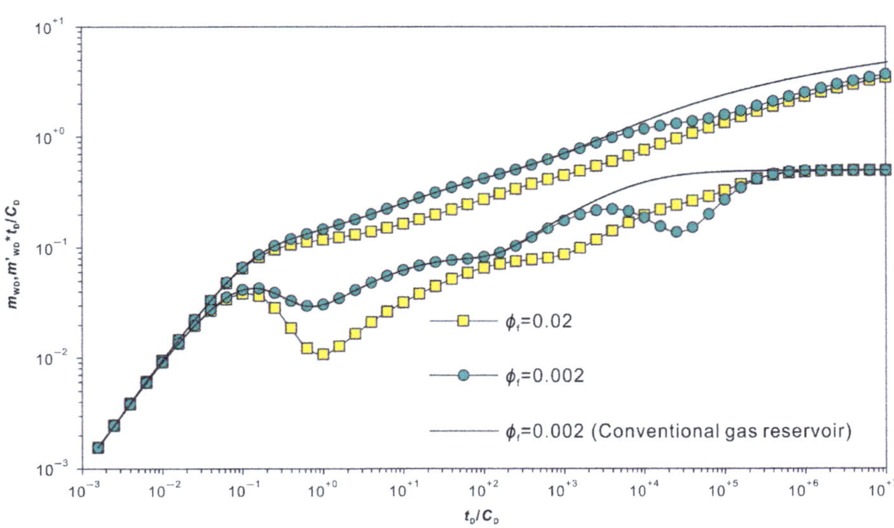

Fig. 4.22 Comparison of production and cumulative production curves for MFHW in a circular gas reservoir under different diffusion models.

Figs. 4.23 and 4.24 show, respectively, the effects of the fracture system porosity on the well test type and production curves. With an increase in the fracture system porosity, the amount of free gas in the fracture system increases and a smaller pressure drop is required when the well produces at a constant rate; therefore, the pseudo pressure and its derivative curves stay lower on the type curves. The higher the fracture system porosity is, the more easily the free gas supplies from a further reservoir region

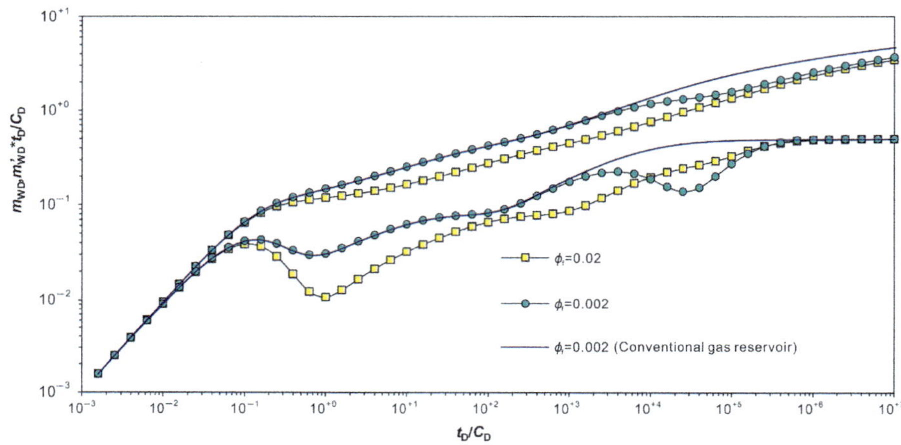

Fig. 4.23 Effect of microfracture system porosity on well test type curves of MFHW in a circular gas reservoir.

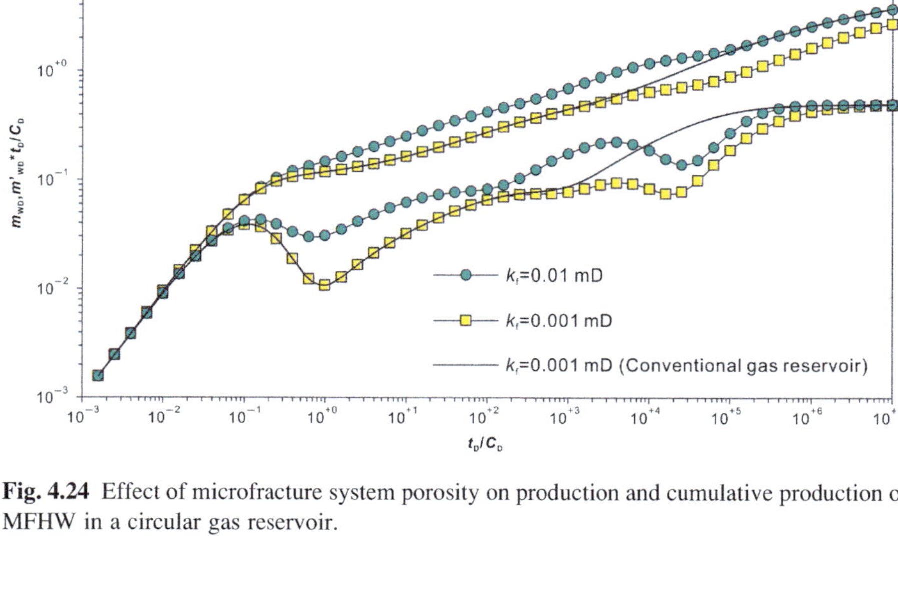

Fig. 4.24 Effect of microfracture system porosity on production and cumulative production of MFHW in a circular gas reservoir.

are used and less gas comes from desorption, which exhibits as a shallower interporosity flow concave part on the pseudo pressure derivative curves (According to the plot in Fig. 4.23, the concave part is merely seen for $\varnothing_f=0.02$). When the well produces at constant pressure, the initial rate is higher for higher fracture porosity, and the difference reduces with more and more free gas being produced. For the same fracture porosity, the initial rates from the shale reservoir and the conventional reservoir are almost equal, but after desorption and diffusion happen, the well production in the shale gas reservoir is obviously higher than that in the conventional reservoir.

Figs. 4.25 and 4.26 illustrate the effects of the fracture system permeability on the well test type and production curves. It can be seen from Fig. 4.25 that the higher the fracture system permeability is, the higher the pseudo pressure and its derivative curves on the plot (This does not mean that a higher pressure drop is required for a well producing at a constant rate in a gas reservoir with higher fracture permeability). According to Fig. 4.26, the fracture system permeability has a great effect on well production; the higher the permeability is, the higher the production is. For the shale gas reservoir and the conventional gas reservoir with the same fracture permeability, the early production rate is almost equal before adsorbed gas starts desorbing and diffusing, and then the production from the shale reservoir is obviously higher than that from the conventional reservoir and the higher production period lasts quite long.

Fig. 4.27 shows the well test type curves for a MFHW in a rectangular reservoir and an infinite circular gas reservoir. For the rectangular gas reservoir, the formation linear flow happens right after the fracture early time radial flow due to quick propagation of pressure to the closer boundary. In addition, the time period of linear flow is related to the L/W (length/width) ratio of the reservoir. After the linear flow lasts for a certain period of time, the concave part for shale gas desorption and diffusion shows up.

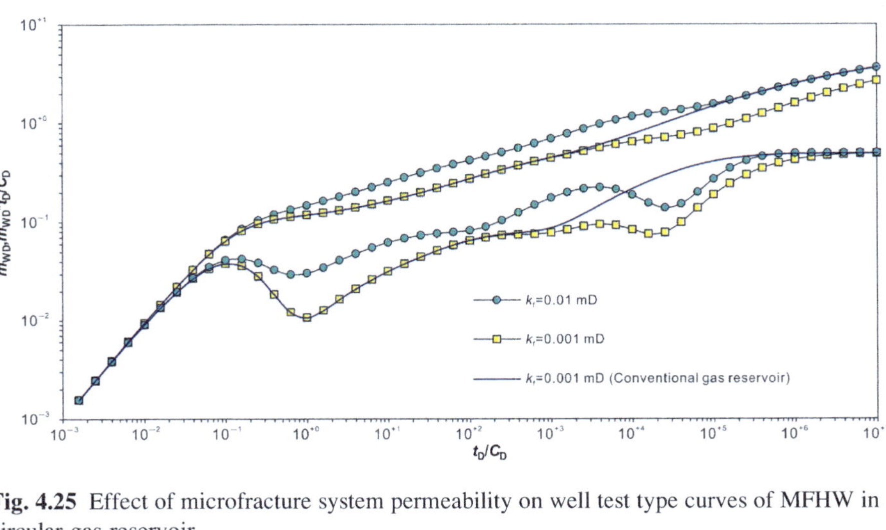

Fig. 4.25 Effect of microfracture system permeability on well test type curves of MFHW in a circular gas reservoir.

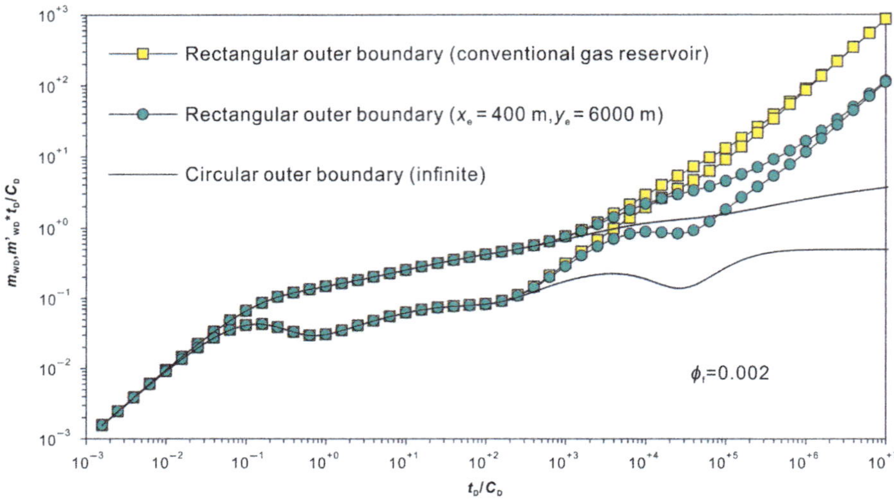

Fig. 4.26 Effect of microfracture system permeability on production and cumulative production of MFHW in a circular gas reservoir.

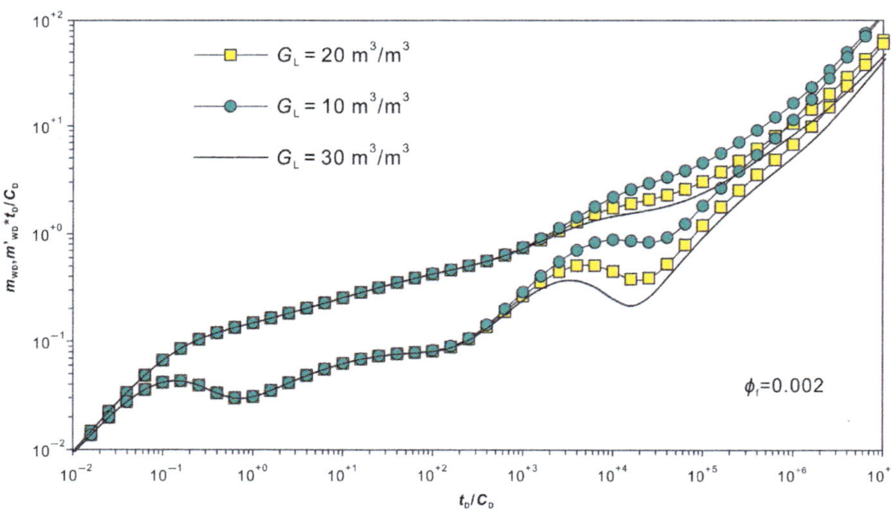

Fig. 4.27 Comparison of well test type curves for MFHW in closed rectangular and infinite circular reservoirs.

Compared to the conventional reservoir, the position of the pseudo pressure and its derivative on the plot is lower.

The effects of adsorption capacity (G_L) on the well test type and production curves are shown in Figs. 4.28 and 4.29, respectively. According to Fig. 4.28, G_L mainly affects the middle and late time flow stages. The bigger G_L is, the lower the pseudo

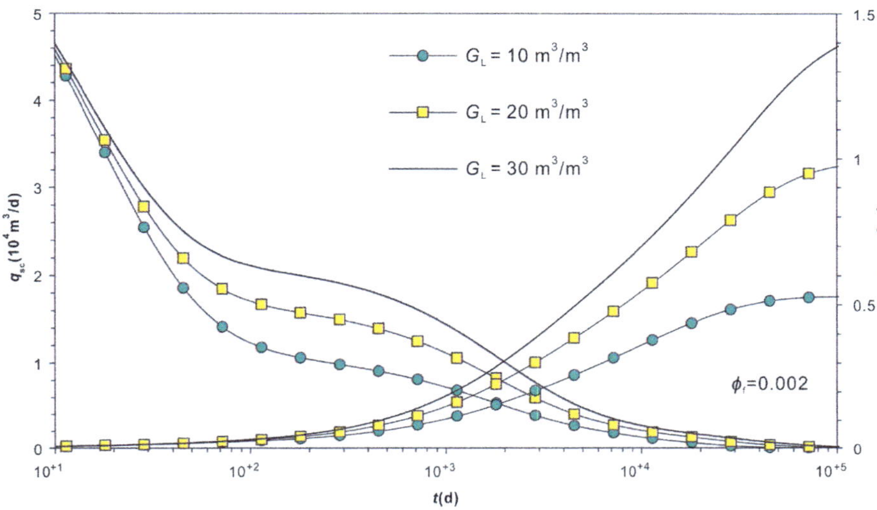

Fig. 4.28 Effect of isothermal adsorption capacity on well test type curves for MFHW in a closed rectangular gas reservoir.

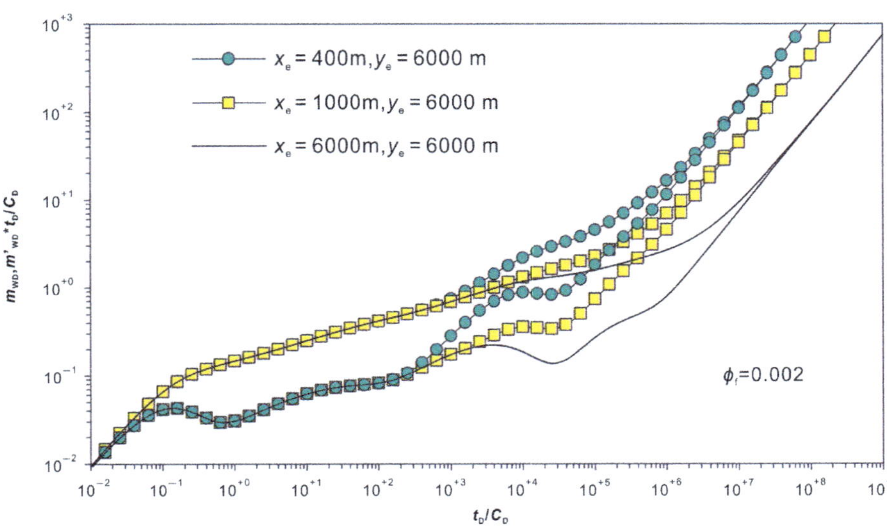

Fig. 4.29 Effect of isothermal adsorption capacity on production and cumulative production of MFHW in a closed rectangular gas reservoir.

pressure and its derivative curves are on the plot, indicating that a less of a pressure drop is required for well producing at a constant rate. When the well is producing at constant pressure, the bigger the G_L is, and the higher the production is after desorption, and diffusion happens.

Fig. 4.30 shows the effect of the reservoir width on the well test type curves. According to this plot, the wider the reservoir is, the higher position the pseudo pressure and its derivative curves are, indicating that a higher pressure drop is required for production. The narrower the reservoir is, the earlier the formation linear flow starts and the longer it lasts.

4.4.4 Microfractures + matrix macropores + gas adsorption/desorption + Fick's diffusion in nanopores

Table 4.4 lists the reservoir properties for the mechanism Model 4 input. The well test type and production curves are generated by these parameters.

Figs. 4.31 and 4.32 show the well test type and production curves for the pseudo steady state interporosity flow from the macropore system to the microfracture system and different types of adsorbed gas diffusion to macropores. According to these plots,

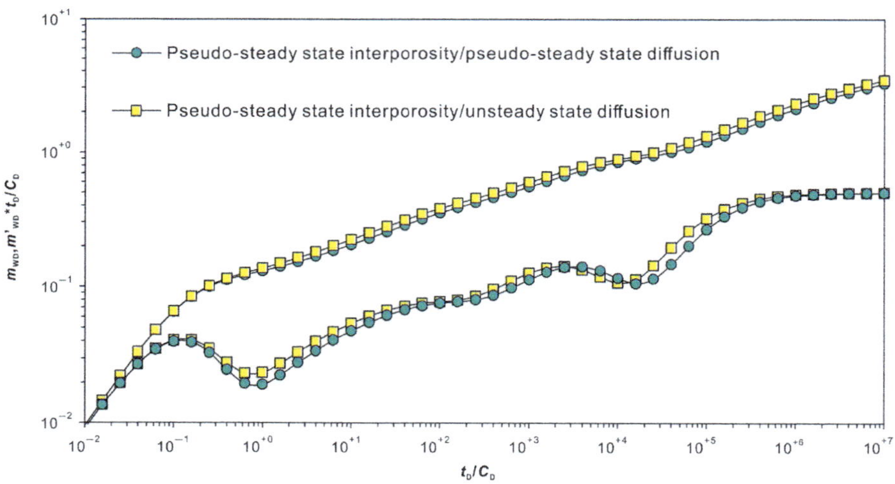

Fig. 4.30 Effect of reservoir size on well test type curves of MFHW in a closed rectangular gas reservoir.

Table 4.4 Shale reservoir properties

Reservoir properties	Value	Reservoir properties	Value
Microfracture system permeability, k_f (mD)	0.01	Microfracture system porosity, $\varnothing_f$	0.02
Macropore system permeability, k_m (mD)	0.0001	Macropore system porosity, $\varnothing_m$	0.12
Constant production rate, q_{sc}(m^3/d)	1×10^4	$(R_m)^2/D_F$ ratio	2×10^6
Interporosity coefficient, α (1/m^2)	10^{-3}		

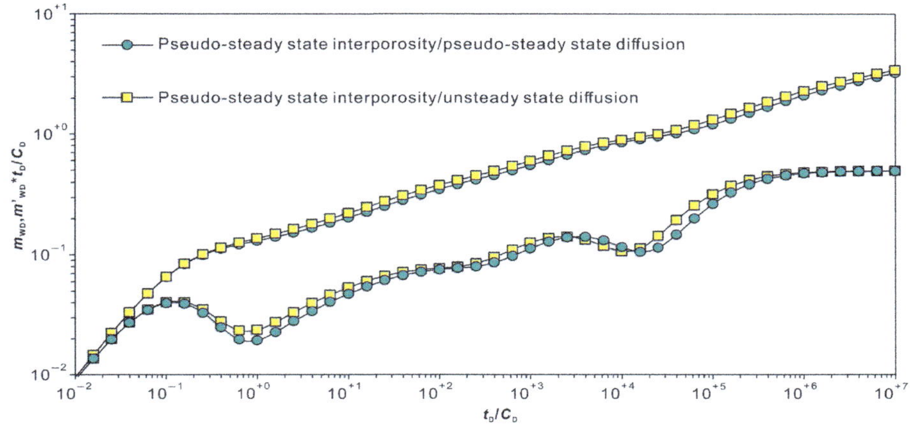

Fig. 4.31 Effect of diffusion model on well test type curves of MFHW in a circular gas reservoir under pseudo steady state interporosity flow.

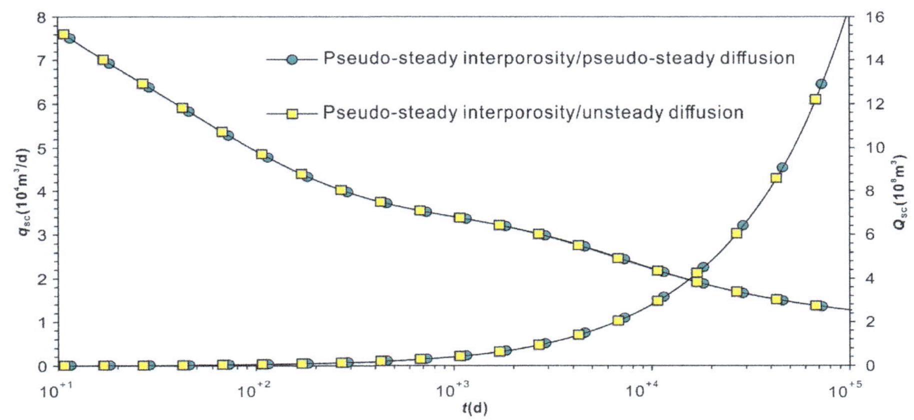

Fig. 4.32 Effect of diffusion model on production and cumulative production of MFHW in a circular gas reservoir under pseudo steady state interporosity flow.

once the interporosity flow from the macropore system to the microfracture system is in the pseudo steady state flow, the gas production is identical for both the pseudo steady state and unsteady state diffusion, while the pseudo pressure derivative is slightly higher during the early and middle time flow stages. This is caused by the different forms of the dimensionless parameters for the two models.

Fig. 4.33 shows the well test type curves for a MFHW without consideration of the wellbore storage and skin effect. Theoretically, there are two concave parts on the pseudo pressure derivative curve (one for interporosity flow from macropores to microfractures and the other for desorption and diffusion of adsorbed gas).

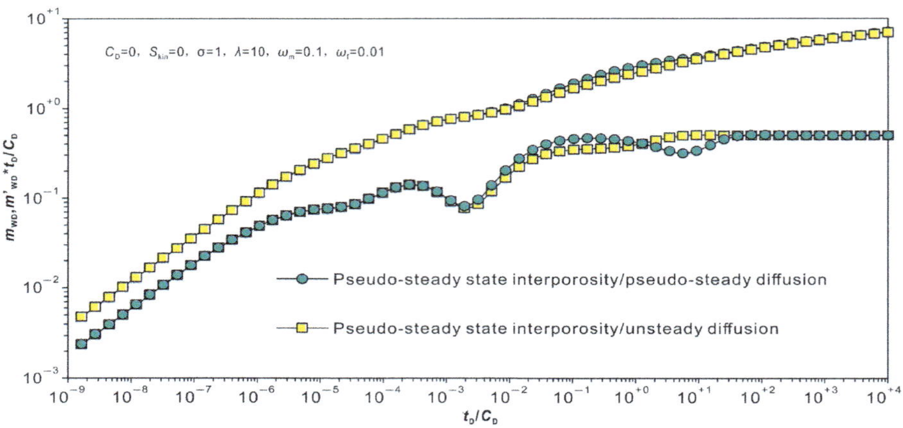

Fig. 4.33 Theoretical well test type curves for pseudo steady state interporosity model without consideration of wellbore storage and skin effect.

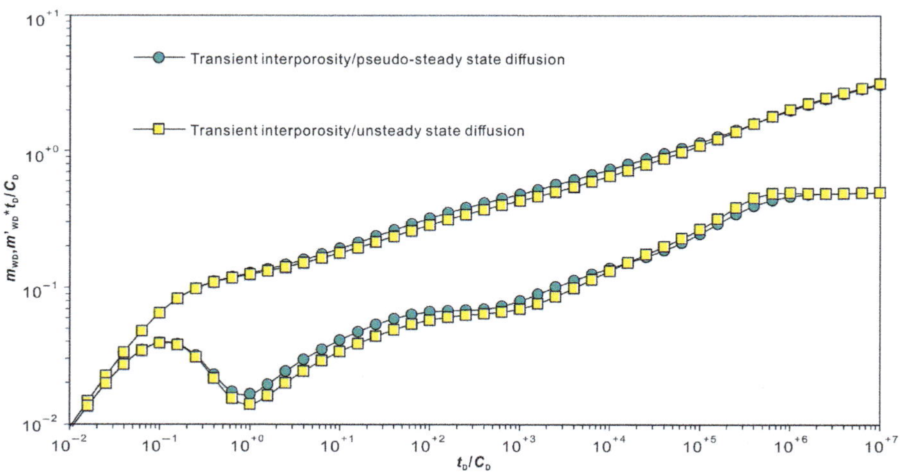

Fig. 4.34 Effect of diffusion model on well test type curves of MFHW under unsteady state interporosity flow.

Figs. 4.34 and 4.35 show the well test type and production curves for unsteady state interporosity flow and different types of diffusion. It can be seen that the unsteady state interporosity flow is sensitive to pressure in the microfracture system, and a gas diffusion volume in the same period of time is higher in the pseudo steady state than in the unsteady state. Therefore, the well rate is higher for the pseudo steady state diffusion during constant pressure production. The higher pseudo pressure difference on the well test type curves for the pseudo steady state diffusion is also caused by the definition of the dimensionless parameters.

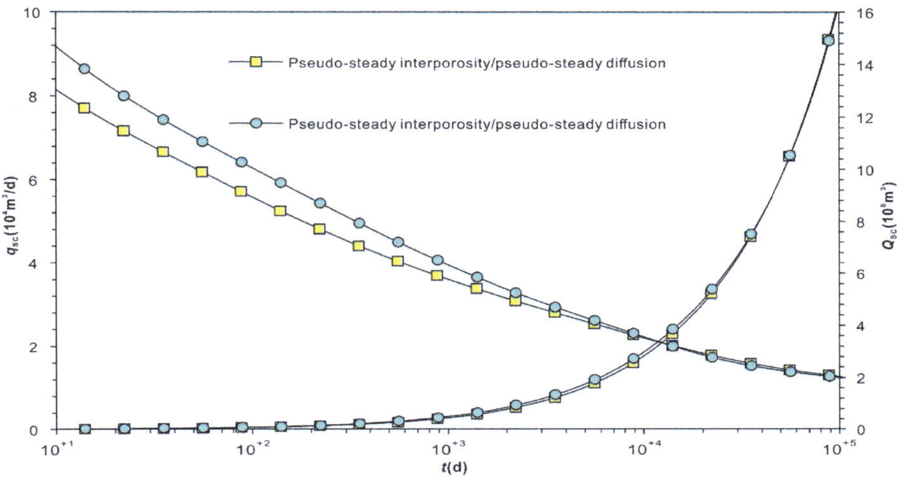

Fig. 4.35 Effect of diffusion model on production and cumulative production of MFHW under unsteady state interporosity flow.

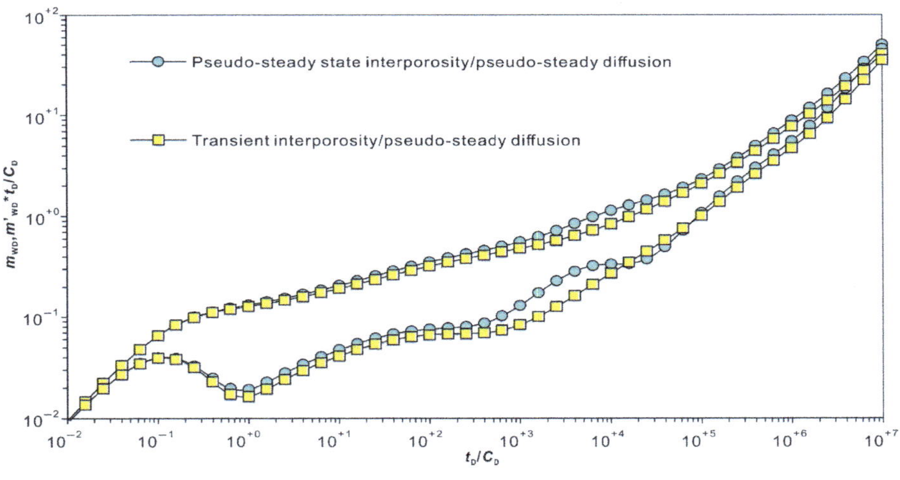

Fig. 4.36 Effect of interporosity flow models on well test type curves of MFHW in a rectangular gas reservoir.

Figs. 4.36 and 4.37 show the well test type and production curves for a MFHW under pseudo steady state diffusion and different types of interporosity flow. For a rectangular gas reservoir, similar to other mechanism models, only the early time fracture radial flow can be observed on these curves while the late time formation radial flow is missing, and the formation linear flow can be partially observed. This is caused by a high L/W ratio of the reservoir and quick pressure propagation to the closer boundary. For the pseudo steady state interporosity flow model, the interporosity flow

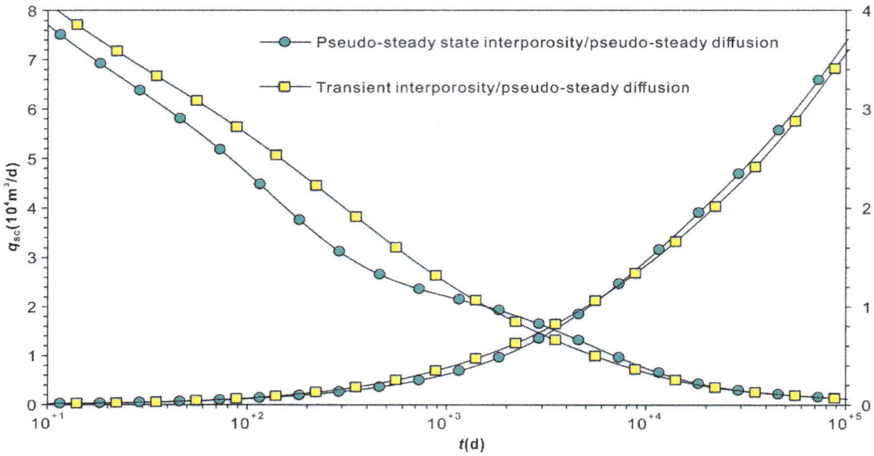

Fig. 4.37 Effect of interporosity flow models on production and cumulative production of MFHW in a rectangular gas reservoir.

concave part can be identified on the curves while the diffusion concave part is missing for the parameters used in this section. Similarly, because the unsteady state interporosity flow model is more sensitive to the microfracture system pressure, the flow rate and initial well production are higher in this state than in the pseudo steady state flow at the same fracture pressure condition. With a quicker pressure drop due to higher early time production, the well production becomes lower than in the pseudo steady state interporosity flow during the middle term until the pressure in the fracture and matrix systems drops simultaneously and formation flow enters into the pseudo steady state period.

4.4.5 Microfractures + gas adsorption/desorption + nanopore Knudsen diffusion

The mechanism model 5 can be used for gas flow under multiple flow mechanisms of Darcy's law, a slippage effect, Knudsen diffusion, and adsorbed gas desorption. Table 4.5 lists the parameters used for generating the well test and production type curves.

Fig. 4.38 illustrates the well test type curves for a well in a circular gas reservoir under different interporosity flow models. According to this plot, there is a concave part on the pseudo pressure derivative curves representing interporosity flow from the shale matrix to microfractures. For the pseudo steady state interporosity flow, the concave part is obvious; while, for the unsteady state diffusion in the matrix, the adsorbed and free gas in the matrix is more sensitive to a pressure change in the fracture system, and the interporosity flow between the matrix and fracture systems starts earlier than when the matrix diffusion is in the pseudo steady state, resulting in missing of the early fracture system radial flow on the curves. In addition, for the unsteady state diffusion

Table 4.5 Shale reservoir properties

Reservoir properties	Value	Reservoir properties	Value
Microfracture system permeability, k_f (mD)	0.01	Microfracture system porosity, $\varnothing_f$	0.02
Matrix permeability, k_m (mD)	0.0001	Matrix porosity, $\varnothing_m$	0.12
Interporosity flow coefficient, α (1/m^2)	10^{-5}	Knudsen diffusion, coefficient, D_k(1/m^2)	10^{-6}
Average radius of matrix micropores, r_n(m)	2×10^{-9}	α value for Eqs. (4.1)–(4.26), dimensionless	0.8

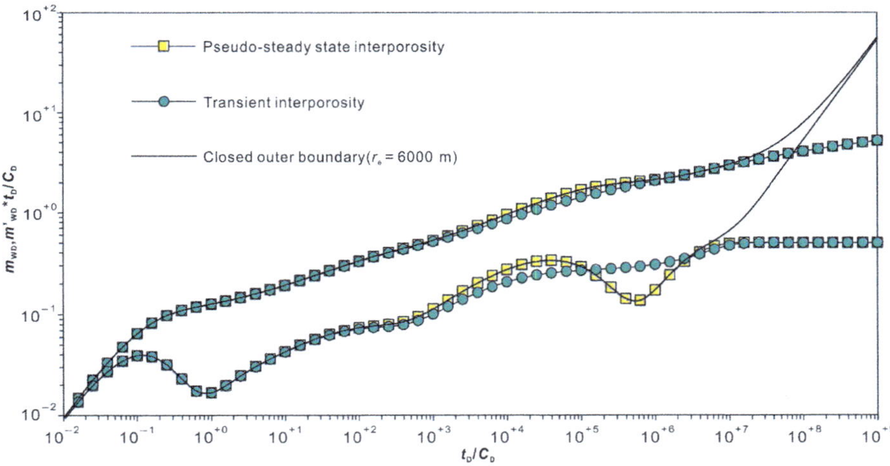

Fig. 4.38 Well test type curves of MFHW in a circular gas reservoir under different interporosity flow models.

in the matrix, the interval for interporosity flow on the derivative curve is a flat concave part other than a deep valley.

Figs. 4.39 and 4.40 show the effect of the slippage coefficient F_f on the well test type and production decline type curves. According to these plots, the higher F_f is, the earlier the concave part for interporosity flow shows up on the curves. This is because the apparent permeability of matrix is higher for a higher F_f, resulting in a bigger interporosity flow coefficient for the flow from matrix nanopores to the fracture system. Also, for well production, a higher F_f leads to a higher matrix permeability and, in consequence, a higher production rate.

Similar to the effect of a slippage coefficient, the effects of a Knudsen diffusion coefficient on the well test and production decline type curves are shown in Figs. 4.41 and 4.42, respectively. The bigger D_K is, the earlier the concave part happens and the higher the well production is.

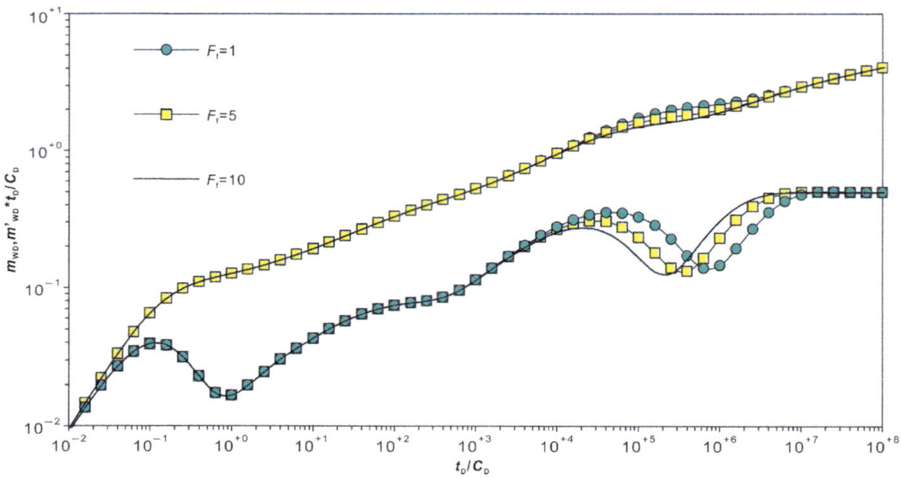

Fig. 4.39 Effect of slippage coefficient F_f on well test type curves of MFHW in a circular gas reservoir.

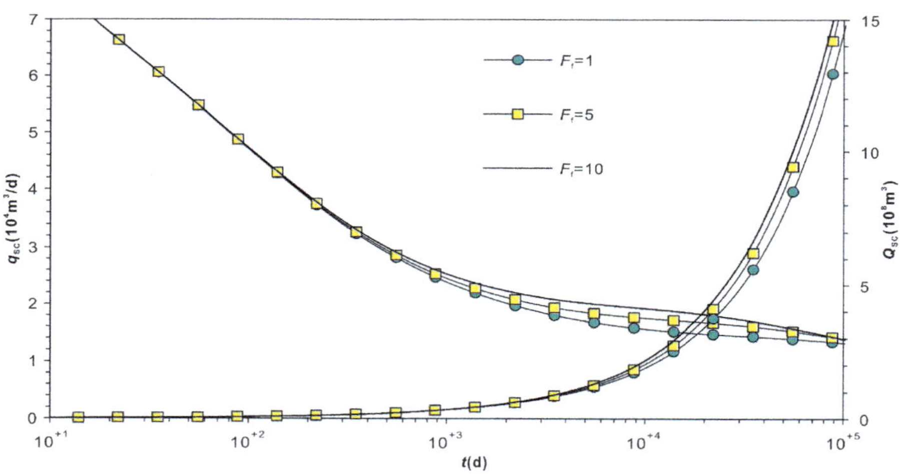

Fig. 4.40 Effect of slippage coefficient F_f on production and cumulative production of MFHW in a circular gas reservoir.

Fig. 4.43 shows the effect of the natural fracture system porosity ($\varnothing_f$) on the well test type curve. According to this plot, $\varnothing_f$ affects the curve shape of not only the interporosity flow but also the early time linear flow. The smaller $\varnothing_f$ is, the deeper the interporosity flow concave part is and the higher the pseudo pressure and its derivative curves for linear flow are on the plot. With an increase in $\varnothing_f$, the concave part becomes shallower and narrower, and the linear flow intervals on the pseudo pressure and its derivative curves move downward.

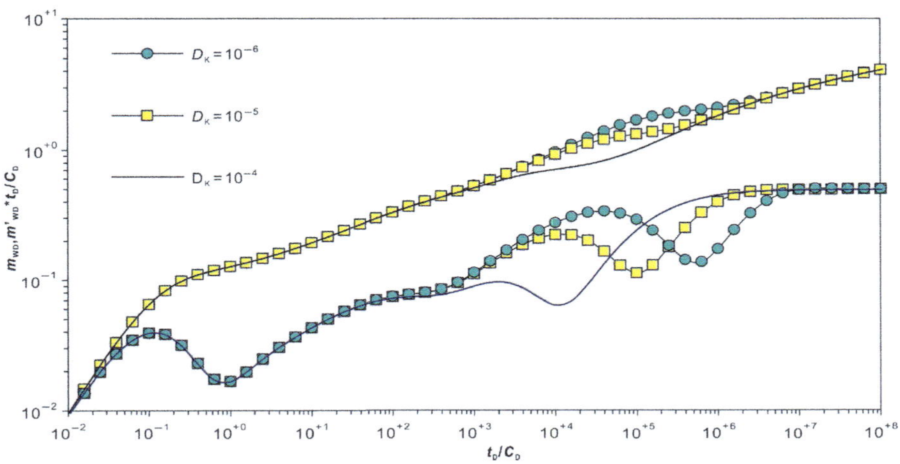

Fig. 4.41 Effect of Knudsen diffusion coefficient on well test type curves of MFHW in a circular gas reservoir.

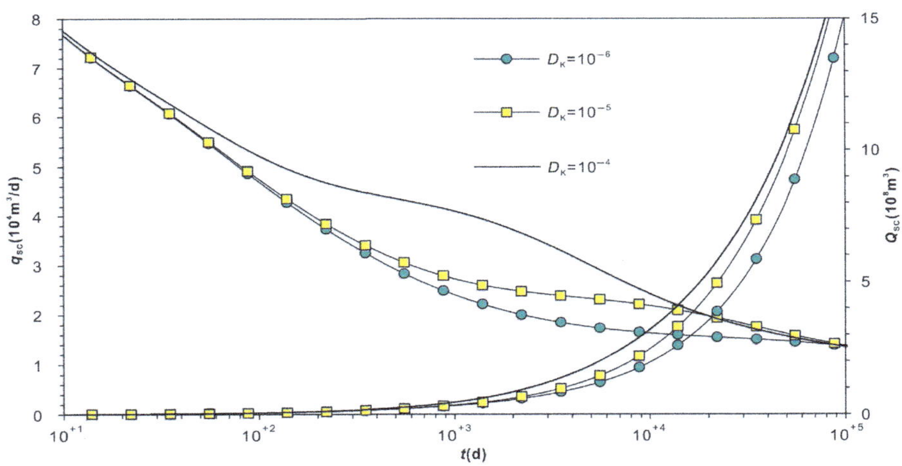

Fig. 4.42 Effect of Knudsen diffusion coefficient on production and cumulative production of MFHW in a circular gas reservoir.

Figs. 4.44 and 4.45 illustrate the effects of different factors on the well test and production decline type curves of a MFHW in a closed rectangular gas reservoir, respectively. As shown in these plots, the Knudsen diffusion in the matrix nanopores, the slippage effect and adsorbed gas desorption mainly influence the late time period of the formation linear flow and boundary dominated flow. This is because both the slippage effect and gas desorption are a gas supply to flow in the matrix nanopores. With more of these effects being considered, the matrix apparent permeability increases, the well production increases for a constant pressure drop, and the position of the pseudo pressure derivative curve on its plot becomes lower.

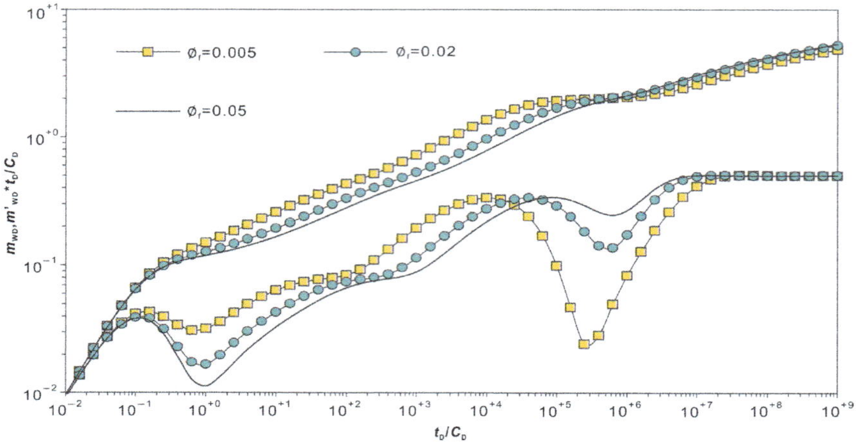

Fig. 4.43 Effect of natural fracture system porosity ($\varnothing_f$) on well test type curves of MFHW in a circular gas reservoir.

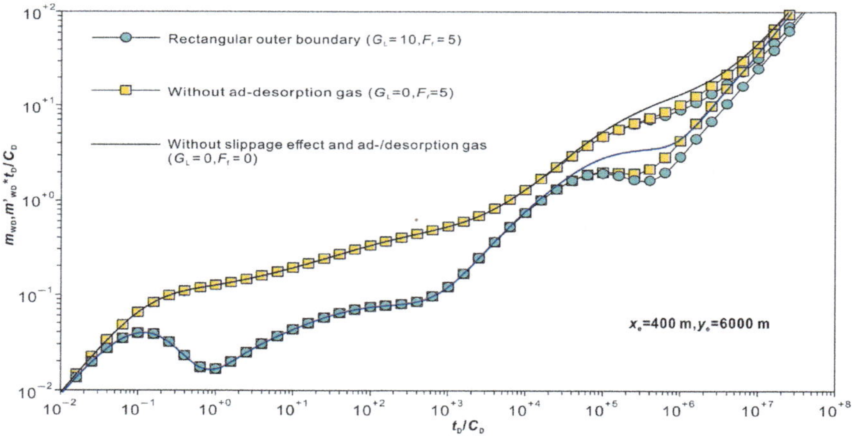

Fig. 4.44 Effect of different factors on well test type curves of MFHW in a closed rectangular gas reservoir.

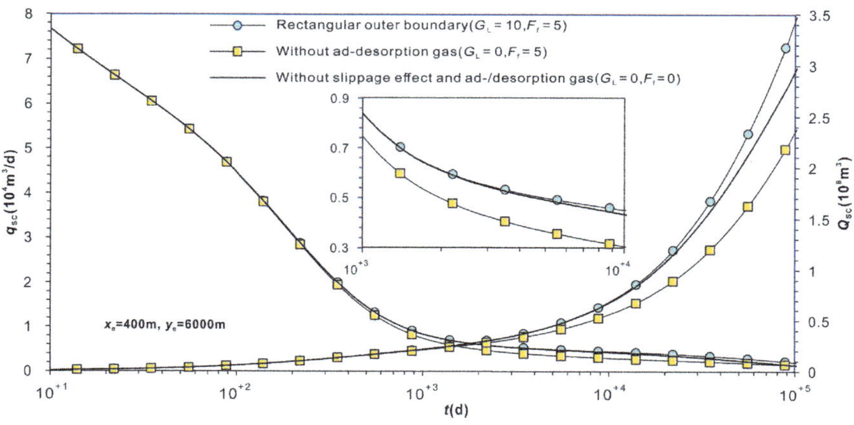

Fig. 4.45 Effect of different factors on production and cumulative production of MFHW.

Fractured vertical wells in circular gas reservoirs with circular SRV

5

Chapter Outline

5.1 Introduction 151
5.2 Continuous line source solutions in circular composite gas reservoirs 152
5.2.1 Derivation of instantaneous line source solutions 152
5.2.2 Continuous line source solutions 155
5.3 Fractured vertical wells in circular composite gas reservoirs 156
5.3.1 Physical model 156
5.3.2 Mathematical solutions 156
5.4 Analysis of pressure and production type curves 158
5.4.1 Microfractures + steady state adsorption/desorption and diffusion 158
5.4.2 Microfractures + matrix macro pores + steady state adsorption/desorption and diffusion 161
5.4.3 Microfractures + gas adsorption/desorption + matrix Fick's diffusion 164
5.4.4 Microfractures + matrix macro pores + gas adsorption/desorption + nanopore Fick's diffusion 168
5.4.5 Microfractures + gas adsorption/desorption + nanopore Knudsen diffusion 171

5.1 Introduction

In shale gas reservoirs, there are structural fractures, overpressured fractures, and diagenetic shrinking fractures developed with poor connectivity and distribution. But for shale gas reservoirs, the existence of these fractures makes economic development of shale gas possible. One of the hottest topics for shale reservoir development is to establish fracture networks throughout the reservoir, which is closely tied to not only fracturing technologies and equipment but also the microfractures distributed in the reservoir. Without these microfractures and hydraulic fracture networks, it is impossible to achieve economical development of a shale gas reservoir. After hydraulic fracturing, a certain number of secondary fractures are induced to connect natural fractures in a reservoir and establish fracture networks. The region developed with these fractures is called a SRV (stimulated reservoir volume) (Mayerhofer et al., 2008).

In the preceding two chapters, the transient flow for a fractured vertical well and a MFHW in circular and rectangular reservoirs under different complex flow mechanisms was introduced. However, the effect of a fracture network induced surrounding the wellbore and main fractures during hydraulic fracturing has not been considered.

Developments in Petroleum Science, Vol. 66. https://doi.org/10.1016/B978-0-444-64315-5.00005-X

Although a realistic fracture network is very complex, it can be represented by a dual- or multi-media model. Linear models (tri-linear or five-linear) whose regional permeability is increased to simulate a SRV can also be applied (Al-Ahmadi et al., 2010; Bello and Watenbargen, 2010; Brohi et al., 2011; Brown, 2009; Brown et al., 2011; Nobakht and Clarkson, 2012; Nobakht et al., 2012, 2013; Ozkan et al., 2011; Stalgorova and Mattar, 2012, 2013; Xu et al., 2012; Yao et al., 2013a,b). In this chapter, a composite model is proposed to overcome shortcomings of linear models in analyzing unsteady state flow for a fractured well in a shale gas reservoir.

First, a fully penetrated continuous line source in a circular composite gas reservoir under different complex flow mechanisms is considered, and the source function method is applied to analyze the unsteady state flow theory for a fractured vertical well in a circular gas reservoir with SRV. Then, for a rectangular gas reservoir with SRV, the Green function method is used to deal with each sub-region and the final model solution can be acquired by coupling all sub-region solutions.

5.2 Continuous line source solutions in circular composite gas reservoirs

Fig. 5.1 illustrates a continuous line source centered in a circular radial composite reservoir. We assume that the length of the line source equals the reservoir thickness; the dual-media concept is used to describe a SRV in an inner reservoir region; the upper and lower reservoir boundaries are closed; the outer reservoir boundary is infinite or closed; gas exists as free gas in the inner SRV and the outer microfracture system; adsorbed gas directly enters into the fracture system after desorption, or into macro pores and then the fracture system; the outer microfracture system is connected to the inner SRV region for gas flow from the outer system into the inner region while the flow from the outer matrix region to the inner matrix region or SRV is negligible (Zhao et al., 2015).

The different microscopic mechanism flow models introduced in the previous chapters are used to describe flow in the inner and outer reservoir regions in this composite gas reservoir. For given inner and outer boundary conditions and interface conditions, an instantaneous line source solution can be acquired and converted into a continuous line source solution by convolution integration.

5.2.1 Derivation of instantaneous line source solutions

To solve a composite model in a shale gas reservoir with consideration of a SRV, the different mechanism models introduced previously are used to establish flow models, which are normalized for the inner and outer fracture systems through the corresponding dimensionless variables (a detailed derivation process is given in Appendix B). Then a uniform format as below is acquired for all these mechanism models, which have different parameter groups $f_1(s)$ and $f_2(s)$ in practical applications.

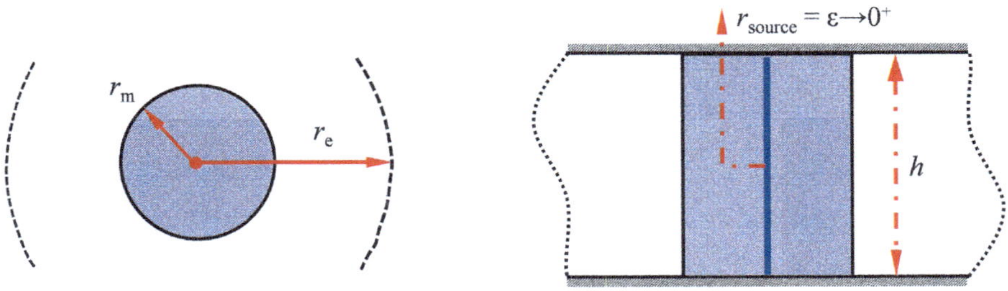

Fig. 5.1 A continuous line source in a composite gas reservoir with SRV.

For an inner SRV region, the simplified uniform format of a flow equation is:

$$\frac{1}{r_D}\frac{\partial}{\partial r_D}\left(r_D\frac{\partial \Delta \overline{m}_{f1}}{\partial r_D}\right)=f_1(s)\Delta \overline{m}_{f1} \tag{5.1}$$

For the outer natural fracture system, the flow equation is:

$$\frac{1}{r_D}\frac{\partial}{\partial r_D}\left(r_D\frac{\partial \Delta \overline{m}_{f2}}{\partial r_D}\right)=f_2(s)\Delta \overline{m}_{f2} \tag{5.2}$$

The outer boundary conditions for the infinite and closed boundaries are:

$$\Delta \overline{m}_{f2}(r_D, t_D)\big|_{r_D\to\infty}=0 \quad \text{(Infinite outer boundary)} \tag{5.3}$$

$$\left.\frac{\partial \Delta \overline{m}_{f2}(r_D, t_D)}{\partial r_D}\right|_{r_D=r_{eD}}=0 \quad \text{(Closed outer boundary)} \tag{5.4}$$

Moreover, the inner boundary condition is:

$$\lim_{\varepsilon\to 0^+}\left(r_D\frac{\partial \Delta \overline{m}_{f1}}{\partial r_D}\right)_{r_D=\varepsilon}=\frac{\overline{q}_s p_{sc} T}{\pi k_{f1} h T_{sc}}\frac{k_{2f}}{\Lambda L^2}\delta(t) \tag{5.5}$$

Since only the flow from the outer region microfracture system to the inner SRV is considered, the interface condition can be represented as:

$$\Delta \overline{m}_{f1}(r_D, t_D)\big|_{r_D=r_{mD}}=\Delta \overline{m}_{f2}(r_D, t_D)\big|_{r_D=r_{mD}} \tag{5.6}$$

$$\left.\frac{\partial \Delta \overline{m}_{f1}}{\partial r_D}\right|_{r_D=r_{mD}}=\frac{1}{M_{12}}\left.\frac{\partial \Delta \overline{m}_{f2}}{\partial r_D}\right|_{r_D=r_{mD}} \tag{5.7}$$

According to the general solution form of the Bessel equation, the general solutions of Eqs. (5.1) and (5.2) are:

$$\Delta \overline{m}_{f1}=A_1 \mathrm{I}_0\left(\sqrt{f_1(s)}r_D\right)+B_1 \mathrm{K}_0\left(\sqrt{f_1(s)}r_D\right) \tag{5.8}$$

$$\Delta\overline{m}_{f2} = A_2 I_0\left(\sqrt{f_2(s)}r_D\right) + B_2 K_0\left(\sqrt{f_2(s)}r_D\right) \tag{5.9}$$

The unknown variables in the general equations can be obtained by substituting the general equations into the inner and outer boundary conditions, respectively.

Infinite outer boundary. According to the outer boundary Eq. (5.3), there is

$$A_2 = 0 \tag{5.10}$$

Also, according to the inner boundary condition, there is:

$$\lim_{\varepsilon\to 0^+} r_D\left[A_1 I_1\left(\sqrt{f_1(s)}r_D\right) - B_1 K_1\left(\sqrt{f_1(s)}r_D\right)\right]_{r_D=\varepsilon} = -\frac{\overline{q}_s p_{sc} T}{\pi k_{f1} h\sqrt{f_1(s)}T_{sc}}\frac{k_{2f}}{\Lambda L^2} \tag{5.11}$$

For the Bessel function:

$$\lim_{x\to 0} x K_1(x) \to 1,\ \lim_{x\to 0} x I_1(x) \to 0 \tag{5.12}$$

$$B_1 = \frac{\overline{q}_s p_{sc} T}{\pi k_{f1} h T_{sc}}\frac{k_{2f}}{\Lambda L^2} \tag{5.13}$$

Introducing Eqs. (5.10) and (5.13) into Eqs. (5.6) and (5.7), respectively, then we see that:

$$A_1 I_0\left(\sqrt{f_1(s)}r_{mD}\right) + B_1 K_0\left(\sqrt{f_1(s)}r_{mD}\right) = B_2 K_0\left(\sqrt{f_2(s)}r_{mD}\right) \tag{5.14}$$

$$A_1 I_1\left(\sqrt{f_1(s)}r_{mD}\right) - B_1 K_1\left(\sqrt{f_1(s)}r_{mD}\right) = -B_2\frac{\sqrt{f_2(s)}}{M_{12}\sqrt{f_1(s)}}K_1\left(\sqrt{f_2(s)}r_{mD}\right) \tag{5.15}$$

Combining Eqs. (5.14) and (5.15) yields:

$$A_1 = B_1\frac{M_{12}\gamma_1 K_1(\gamma_1 r_{mD})K_0(\gamma_2 r_{mD}) - \gamma_2 K_0(\gamma_1 r_{mD})K_1(\gamma_2 r_{mD})}{M_{12}\gamma_1 I_1(\gamma_1 r_{mD})K_0(\gamma_2 r_{mD}) + \gamma_2 I_0(\gamma_1 r_{mD})K_1(\gamma_2 r_{mD})} \tag{5.16}$$

where $\gamma_1 = \sqrt{f_1(s)}$ and $\gamma_2 = \sqrt{f_2(s)}$. Hence we have:

$$\Delta\overline{m}_{f1} = \frac{\overline{q}_s p_{sc} T}{\pi k_{f1} h T_{sc}}\frac{k_{2f}}{\Lambda L^2}\left[A_c I_0\left(\sqrt{\gamma_1}r_D\right) + K_0\left(\sqrt{\gamma_1}r_D\right)\right] \tag{5.17}$$

where:

$$A_c = \frac{M_{12}\gamma_1 K_1(\gamma_1 r_{mD})K_0(\gamma_2 r_{mD}) - \gamma_2 K_0(\gamma_1 r_{mD})K_1(\gamma_2 r_{mD})}{M_{12}\gamma_1 I_1(\gamma_1 r_{mD})K_0(\gamma_2 r_{mD}) + \gamma_2 I_0(\gamma_1 r_{mD})K_1(\gamma_2 r_{mD})}$$

Closed outer boundary condition. Substituting Eq. (5.9) into the outer boundary condition Eq. (5.4), there is:

$$A_2 I_1(\gamma_2 r_{eD}) - B_2 K_1(\gamma_2 r_{eD}) = 0 \tag{5.18}$$

Also, introducing the general solutions of the inner and outer region Eqs. (5.8) and (5.9) into the interface Eqs. (5.6) and (5.7), we see that:

$$A_1\mathrm{I}_0(\gamma_1 r_{\mathrm{mD}})+B_1\mathrm{K}_0(\gamma_1 r_{\mathrm{mD}})=A_2\mathrm{I}_0(\gamma_2 r_{\mathrm{mD}})+B_2\mathrm{K}_0(\gamma_2 r_{\mathrm{mD}}) \tag{5.19}$$

$$A_1\mathrm{I}_1(\gamma_1 r_{\mathrm{mD}})-B_1\mathrm{K}_1(\gamma_1 r_{\mathrm{mD}})=\frac{\gamma_2}{M_{12}\gamma_1}[A_2\mathrm{I}_1(\gamma_2 r_{\mathrm{mD}})-B_2\mathrm{K}_1(\gamma_2 r_{\mathrm{mD}})] \tag{5.20}$$

According to the inner and outer boundary conditions, B_1 can be found as follows:

$$B_1=\frac{\overline{q}_{\mathrm{s}}p_{\mathrm{sc}}T}{\pi k_{\mathrm{f1}}hT_{\mathrm{sc}}}\frac{k_{\mathrm{2f}}}{\Lambda L^2} \tag{5.21}$$

From Eqs. (5.18) to (5.20), we have:

$$A_1=B_1\frac{\mathrm{K}_1(\gamma_1 r_{\mathrm{mD}})-D\mathrm{K}_1(\gamma_1 r_{\mathrm{mD}})}{\mathrm{I}_1(\gamma_1 r_{\mathrm{mD}})-D\mathrm{I}_1(\gamma_1 r_{\mathrm{mD}})} \tag{5.22}$$

where:

$$D=\frac{\gamma_2}{M_{12}\gamma_1}\frac{\mathrm{K}_1(\gamma_1 r_{\mathrm{eD}})\mathrm{I}_1(\gamma_2 r_{\mathrm{mD}})-\mathrm{I}_1(\gamma_2 r_{\mathrm{eD}})\mathrm{K}_1(\gamma_2 r_{\mathrm{mD}})}{\mathrm{K}_1(\gamma_2 r_{\mathrm{eD}})\mathrm{I}_0(\gamma_2 r_{\mathrm{mD}})+\mathrm{I}_1(\gamma_2 r_{\mathrm{eD}})\mathrm{K}_0(\gamma_2 r_{\mathrm{mD}})}$$

Substituting Eqs. (5.21) and (5.22) into Eq. (5.8) gives:

$$\Delta\overline{m}_{\mathrm{f1}}=\frac{\overline{q}_{\mathrm{s}}p_{\mathrm{sc}}T}{\pi k_{\mathrm{f1}}hT_{\mathrm{sc}}}\frac{k_{\mathrm{2f}}}{\Lambda L^2}\left[\mathrm{K}_0\left(\sqrt{\gamma_1}r_{\mathrm{D}}\right)+A_c\mathrm{I}_0\left(\sqrt{\gamma_1}r_{\mathrm{D}}\right)\right] \tag{5.23}$$

where:

$$A_c=\frac{\mathrm{K}_1(\gamma_1 r_{\mathrm{mD}})-D\mathrm{K}_1(\gamma_1 r_{\mathrm{mD}})}{\mathrm{I}_1(\gamma_1 r_{\mathrm{mD}})-D\mathrm{I}_1(\gamma_1 r_{\mathrm{mD}})}$$

5.2.2 Continuous line source solutions

Solution $\Delta\overline{m}_{\mathrm{f1}}$ is the instantaneous line source solution, and its continuous line source solution can be obtained through a convolution integral:

$$\Delta\overline{m}_{\mathrm{f1}}=\frac{p_{\mathrm{sc}}T}{\pi k_{\mathrm{f1}}hT_{\mathrm{sc}}}\frac{k_{\mathrm{2f}}}{\Lambda L^2}\int_0^t\overline{q}_{\mathrm{s}}(\tau)S_{\mathrm{c}}(t_{\mathrm{D}}-\tau)\mathrm{d}\tau \tag{5.24}$$

where $\overline{S}_{\mathrm{c}}=\mathrm{K}_0\left(\sqrt{\gamma_1}r_{\mathrm{D}}\right)+A_c\mathrm{I}_0\left(\sqrt{\gamma_1}r_{\mathrm{D}}\right)$. We recall the dimensionless time:

$$t_{\mathrm{D}}=\frac{k_{\mathrm{2f}}t}{\Lambda L^2} \tag{5.25}$$

Assume that the continuous line source produces at a constant rate; that is, the instantaneous line source intensity q_s is a constant. Using q_{scL} to represent the continuous line source intensity, then Eq. (5.24) becomes:

$$\Delta\overline{m}_{f1} = \frac{q_{scL}p_{sc}T}{\pi s k_{f1} h T_{sc}}\left[K_0\left(\sqrt{\gamma_1}r_D\right) + A_c I_0\left(\sqrt{\gamma_1}r_D\right)\right] \tag{5.26}$$

The above equation is the continuous line source function for a composite reservoir model in a shale gas reservoir considering a SRV.

5.3 Fractured vertical wells in circular composite gas reservoirs

In Section 5.2, the continuous line source function for a composite reservoir model in a shale gas reservoir considering a SRV induced around a fractured well is obtained. In this section, the continuous line source function is applied to analyze a fractured vertical well in a shale gas reservoir.

5.3.1 Physical model

Fig. 5.2 illustrates a fractured vertical well in a shale gas reservoir with SRV. Assume that the SRV is created by massively induced fractures around the main fractures to connect with the formation microfractures; the shape of the SRV is a cylinder with the well at the center and radius r_m; the main fractures are fully open and of infinite conductivity with negligible width and fracture half-length x_f; the well produces at a constant rate q_{sc}.

5.3.2 Mathematical solutions

A similar method to that with no SRV can be used to solve for bottom-hole pressure for a fractured vertical well in a composite reservoir with SRV. First, the bottom-hole pressure for a fractured well is obtained for a uniform flow rate distribution; then,

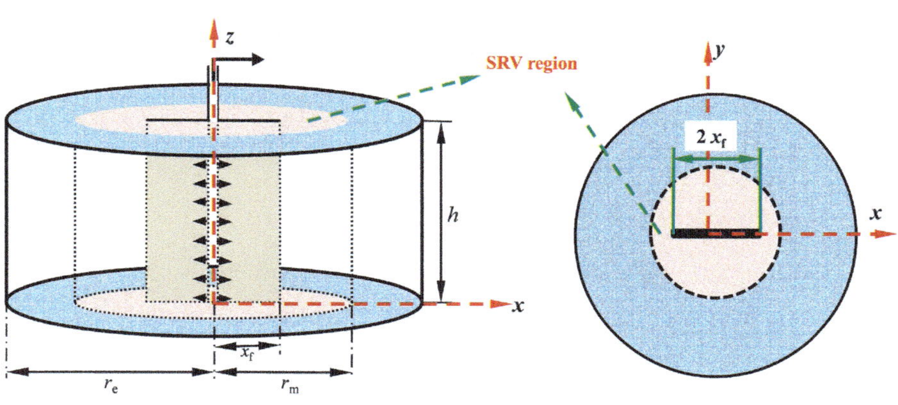

Fig. 5.2 A fractured vertical well in a shale gas reservoir with SRV.

following Gringarten et al. (1973), the solution at a special location ($x_D = 0.732$) along fractures can be used to approximate the bottom-hole pressure solution for an infinite conductivity fractured well. For a fractured well with a uniform flow rate distribution along a fracture, the bottom-hole pressure can be acquired through integration of the continuous line source along the fracture. That is:

$$\Delta \overline{m}_f = \frac{q_{scL} p_{sc} T}{\pi s k_{f1} h T_{sc}} \int_{-x_f}^{x_f} \left[K_0\left(\sqrt{\gamma_1} r_D\right) + A_c I_0\left(\sqrt{\gamma_1} r_D\right) \right] dx_w \tag{5.27}$$

For a fractured well with uniform flow distribution, there is a relationship between the well production rate q_{sc} and the continuous line source intensity q_{scL}:

$$q_{sc} = 2 x_f q_{scL} \tag{5.28}$$

The definition of the dimensionless pseudo pressure is:

$$m_{fD} = \frac{\pi k_{f2} h T_{sc}}{p_{sc} T q_{sc}} \Delta m_f \tag{5.29}$$

Substituting the dimensionless distance and Eqs. (5.28) and (5.29) into Eq. (5.27), there is:

$$\overline{m}_{fD} = \frac{1}{2 M_{12} s} \int_{-1}^{1} \left[K_0(\gamma_1 r_D) + A_C I_0(\gamma_1 r_D) \right] dx_{wD} \tag{5.30}$$

where:

$$r_D = \sqrt{(x_D - x_{wD})^2 + (y_D - y_{wD})^2} \quad \text{and} \quad M_{12} = \frac{k_{f1}/\mu_{gi}}{k_{f2}/\mu_{gi}}$$

Eq. (5.30) is the dimensionless pseudo pressure drop caused by a fractured vertical well with uniform flow distribution at any point of the reservoir with SRV. For an infinite conductivity fractured well, let $x_D = 0.732$ and $y_{wD} = y_D = 0$ due to the fracture surface locating along an x–z surface, and then substitute these parameters into Eq. (5.30) to have:

$$\overline{m}_{fD} = \frac{1}{2 M_{12} s} \int_{-1}^{1} \left[K_0(\gamma_1 |0.732 - \alpha|) + A_C I_0(\gamma_1 |0.732 - \alpha|) \right] d\alpha \tag{5.31}$$

Substituting an expression of A_C for an infinite or closed circular outer boundary into the above equation, then the model solution can be obtained by a similar method to that used for a fractured vertical well without consideration of a SRV. For the skin and wellbore storage effects, the method introduced in Section 3.4 can be also used to find a solution.

5.4 Analysis of pressure and production type curves

5.4.1 Microfractures + steady state adsorption/desorption and diffusion

For the microscopic mechanism flow model 1, the parameters in Table 5.1 are used to calculate and generate well test type and production curves.

Fig. 5.3 shows the well test type curves for a fractured vertical well in a shale gas reservoir with SRV assuming different boundary conditions. According to these curves, gas flow can be divided into the following stages:

Flow stage 1: the early time wellbore storage and its transition flow. The pressure curves overlap as a straight line with a slope of 1.
Flow stage 2: the early time linear flow from the formation to the fracture. After the well starts producing, free gas in the inner SRV flows toward the main fracture surface. On the plot, the pseudo pressure derivative exhibits as a straight line with a slope of 0.5.
Flow stage 3: the elliptical flow surrounding the fracture. In this stage, the derivative curve is a straight line with a slope of 0.36.
Flow stage 4: the radial flow in the SRV region. If the SRV region radius is far bigger than the main fracture half-length and pressure does not reach the SRV boundary after the elliptical flow stage ends, then the radial flow in the SRV region happens. In this stage, the derivative curve is a horizontal line with a value of $1/(2M_{12})$.
Flow stage 5: the transition flow between the inner region elliptical flow and the outer region radial flow. In this flow stage, gas in the outer region microfracture system has already started flowing toward the SRV region. Due to lower permeability of the outer reservoir region, more and more gas supply is required to charge the inner SRV, which has higher permeability under a constant well production rate. Therefore, a higher and higher pressure

Table 5.1 Shale reservoir properties

Reservoir properties	Value	Reservoir properties	Value
Initial gas reservoir pressure, p_i (MPa)	25	Gas reservoir temperature, T (K)	320
Reservoir thickness, h (m)	60	Fracture half-length, x_f (m)	50
Specific gas gravity, γ_g	0.65	Gas well production rate, q_{sc} (m^3/d)	1×10^4
Initial gas compressibility, C_{gi} (MPa^{-1})	0.02	Bottom-hole pressure, p_{wf} (MPa)	1
Permeability of SRV, k_{f1} (mD)	0.1	Porosity of SRV, $\varnothing_{f1}$ (fraction)	0.1
Permeability of microfracture system, k_{f2} (mD)	0.01	Porosity of microfracture system, $\varnothing_{f2}$ (fraction)	0.02
Skin factor, S_{kin} (dimensionless)	0.1	Dimensionless wellbore storage coefficient, C_D	10^{-3}
Langmuir pressure, P_L (MPa)	4	Langmuir volume, G_L (m^3/m^3)	10
Radius of SRV, r_m (m)	250	Outer boundary radius of closed circular gas reservoir, r_e (m)	1000

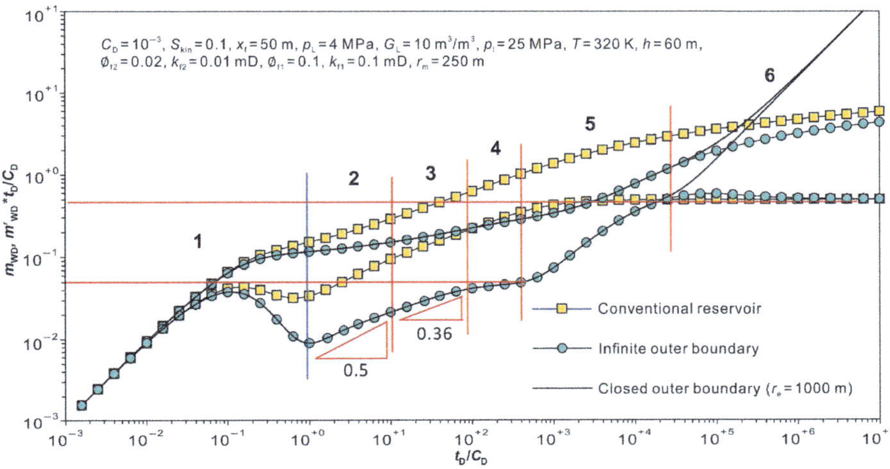

Fig. 5.3 Well test type curves of a fractured vertical well in a composite gas reservoir with SRV.

drop is created until the gas supply equals the well production rate, and then the transition flow stage ends.

Flow stage 6: the late-time radial flow or boundary dominated flow. After a complete ending of the transition flow, pressure propagates outside the SRV into the outer reservoir region, and flow in this period is called the formation radial flow, which shows up as a 0.5 horizontal line of the pseudo pressure derivative curve. When the pressure reaches the outer boundary, the pseudo pressure and its derivative curves overlap as a straight line with a slope of 1 if the boundary is closed. Note that if the outer boundary is far away enough, the formation radial flow can be observed before the boundary dominated flow starts.

Comparing the type curves of a fractured vertical well in a shale gas reservoir with SRV to those of a well in a conventional homogeneous gas reservoir, the pseudo pressure difference in the conventional gas reservoir is far higher under the same production rate. In addition, according to the comparison of production curves under constant production pressure, the plateau rate of the fractured well in the shale gas reservoir with SRV is higher than that of the well in the conventional homogeneous reservoir with the same basic reservoir properties by 5000 m^3/d (as shown in Fig. 5.4). Therefore, for a shale gas reservoir, the key for fracturing operation is to create a complex fracture network.

The effects of a SRV radius on the well test type and production curves are shown in Figs. 5.5 and 5.6, respectively. It can be seen that the main influence of the SRV radius is on the lasting time of the inner region radial flow and the start of the outer region radial flow. For the pseudo pressure curves, the bigger the SRV region is, the lower the pseudo pressure difference curve is on its plot, indicating that a smaller pressure drop is required for a constant production rate. As for the effect on a well production rate, the bigger the SRV radius is, the higher the mid- and late-time production rate is. The plateau production rate difference is about 5000 m^3/d.

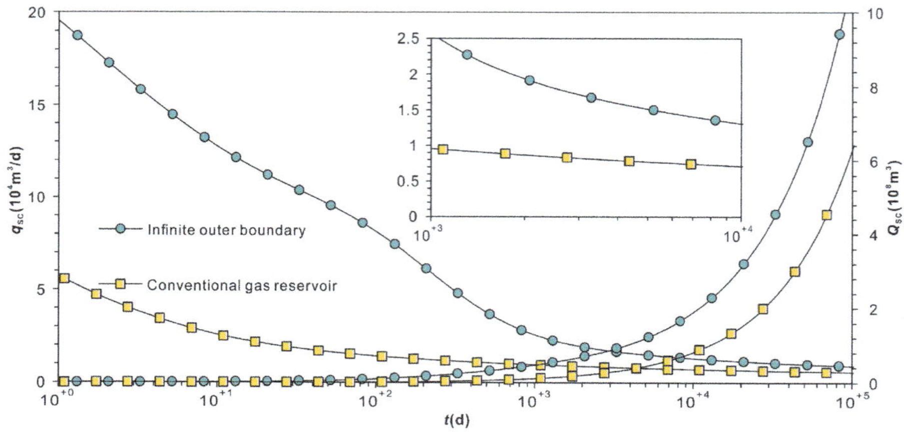

Fig. 5.4 Comparison of production curves of a fractured well in a shale gas reservoir with SRV and a well in a conventional homogeneous gas reservoir.

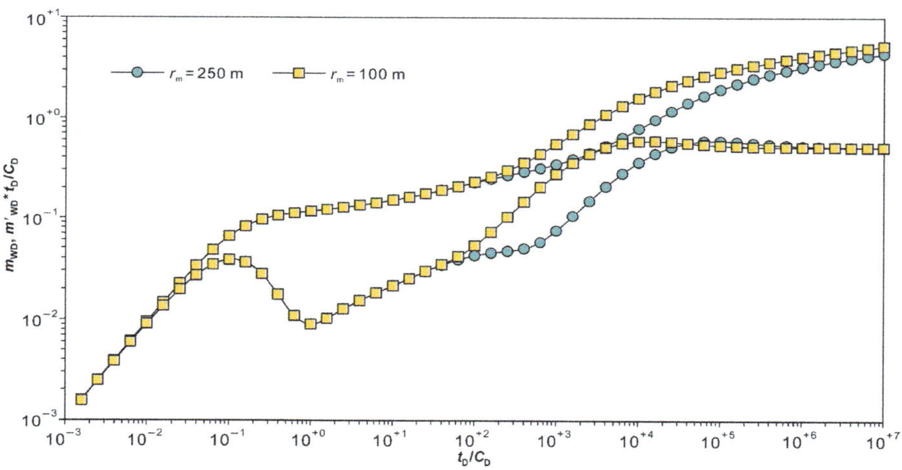

Fig. 5.5 Effect of SRV region radius on well test type curves.

For the effects of SRV permeability on the pressure and production curves, as shown in Figs. 5.7 and 5.8, the SRV permeability mainly affects the pressure curves' position during the mid-time formation linear flow and inner region radial flow stages. The higher the SRV permeability is, the lower the pseudo pressure and its derivative are on their plots. Also, the higher the SRV permeability is, the higher the well early time production rate is. After pressure propagates outside the inner region, due to the same outer region properties and SRV radius, there is only a small difference in the late-time well rate for different SRV permeability (the production rate difference is kept at about $1000\,m^3/d$ during the plateau production period for SRV permeability of 0.01 and 0.15 mD).

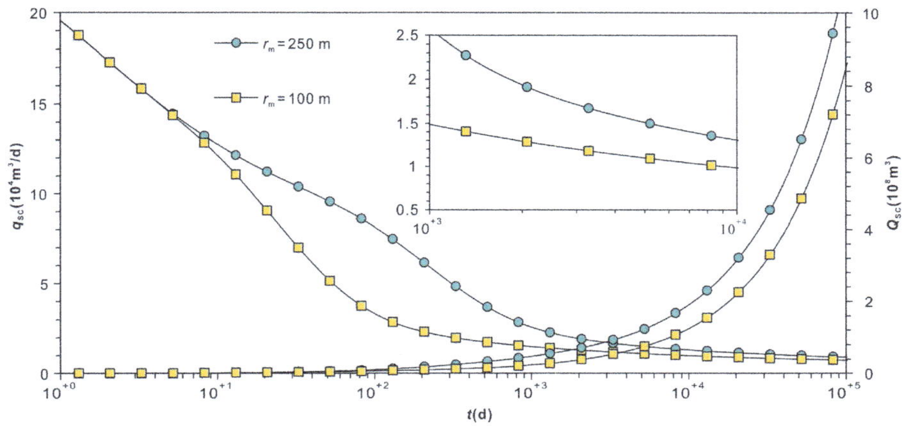

Fig. 5.6 Effect of SRV region radius on fractured well production.

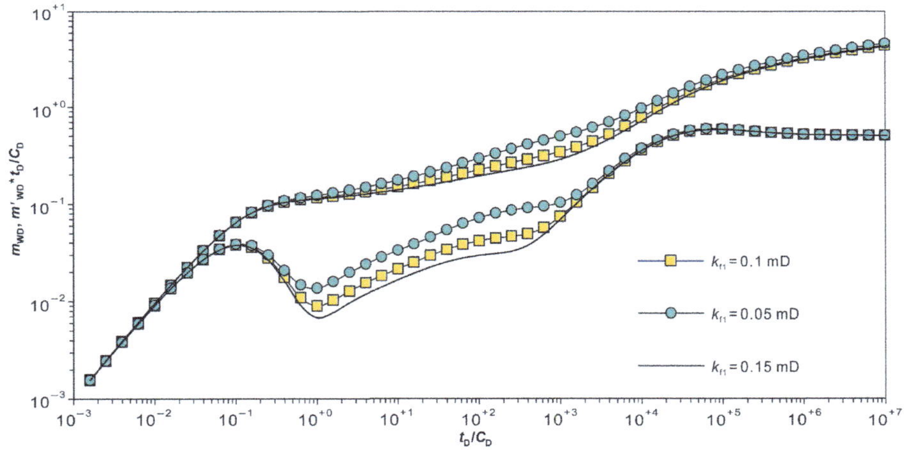

Fig. 5.7 Effect of SRV permeability on well test type curves.

5.4.2 Microfractures + matrix macro pores + steady state adsorption/desorption and diffusion

For the mechanism model 2, the parameters in Table 5.2 are used to generate the corresponding type curves.

Fig. 5.9 shows the effect of different interporosity flow models on the well test type curves. Compared to the mechanism model 1, an additional interporosity flow concave part is present on the pseudo pressure derivative curves. Similar to a conventional dual-porosity gas reservoir, the width and deepness of the concave part are related to the storage ratio, and the presence timing is related to the coefficient of interporosity flow from matrix to the microfracture system. In general, the smaller the fracture

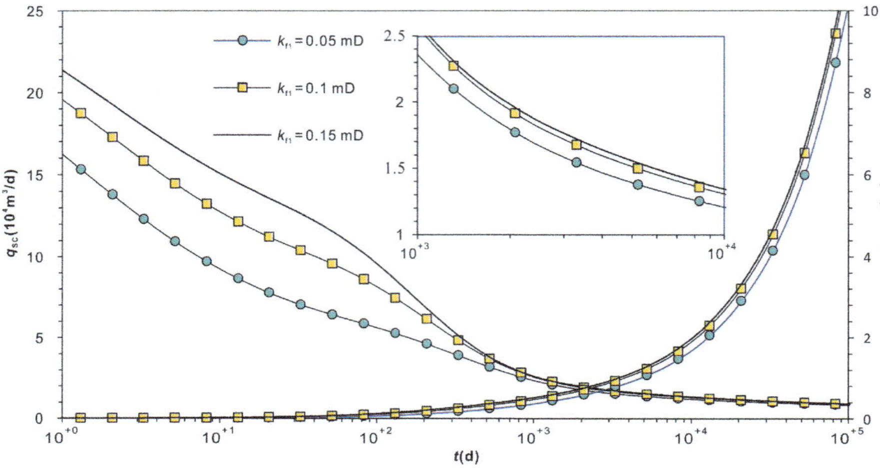

Fig. 5.8 Effect of SRV permeability on well production curves.

Table 5.2 Shale reservoir properties

Reservoir properties	Value	Reservoir properties	Value
Matrix permeability, k_m (mD)	0.0001	Matrix porosity, $\varnothing_m$	0.12
Permeability of microfracture system, k_{f2} (mD)	0.01	Porosity of microfracture system, $\varnothing_{f2}$ (fraction)	0.02
Permeability of SRV, k_{f1} (mD)	0.1	Porosity of SRV, $\varnothing_{f1}$ (fraction)	0.02
Geometry factor of interporosity flow, α (1/m^2)	10^{-4}	Radius of SRV, r_m (m)	250

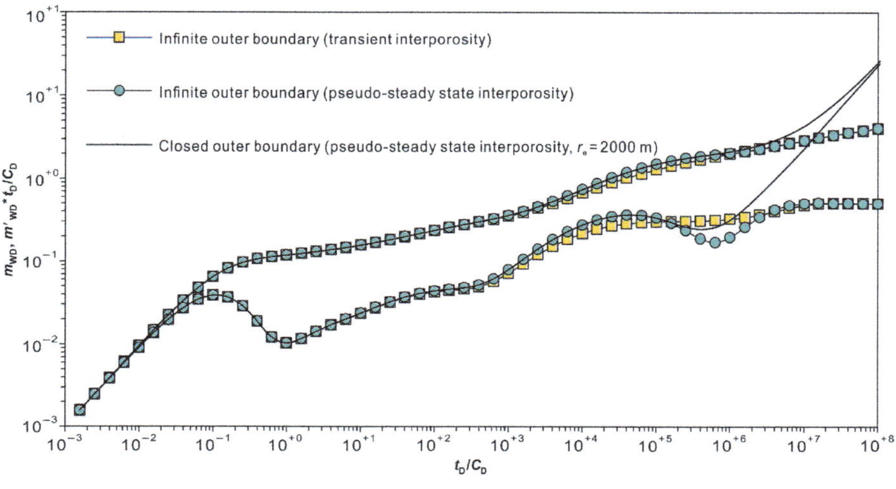

Fig. 5.9 Effect of interporosity flow models and boundary conditions on well test type curves.

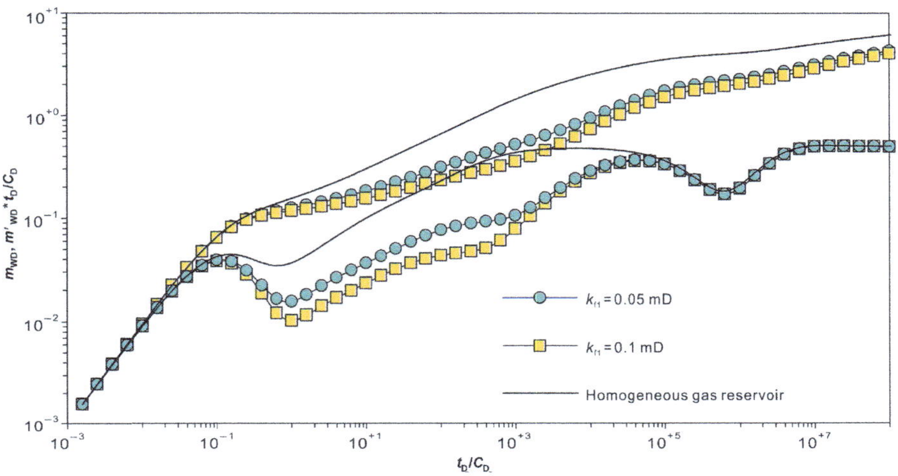

Fig. 5.10 Effect of SRV permeability on well test type curves.

system storage ratio is, the wider and deeper the concave part is; and the bigger the interporosity flow coefficient is, the earlier the concave part is present. According to the comparison of the pseudo pressure derivative curves of the pseudo steady state interporosity flow and unsteady state interporosity flow, the concave part shows up earlier for the unsteady state than for the pseudo steady state due to its higher sensitivity to the fracture system pressure.

Fig. 5.10 shows the effect of SRV permeability on the well test type curves. It can be seen that the higher the SRV permeability is, the lower the pressure difference curves are on the plot, indicating that a smaller pressure drop is required for the same constant production rate. The influence of SRV permeability on the pseudo pressure derivative curves shows up during the fracture linear flow and inner region radial flow periods. The more permeable the SRV is, the lower the derivative curves are on the plot. Compared to the well test type curves of a well in a homogeneous gas reservoir, the dimensionless pseudo pressure curves are obviously lower. Moreover, on the production curve plot, the plateau rate of the well with the SRV effect is higher than that of the well without SRV by 7000 m^3/d. For the well model with consideration of a SRV, the SRV permeability also affects a well production rate; when the SRV permeability increases from 0.05 to 0.1 mD, the plateau rate increases by 2000 m^3/d, which is especially considerable for production from a shale gas reservoir, as shown in Fig. 5.11.

The effects of a SRV radius on the well test type and production decline curves are shown in Figs. 5.12 and 5.13, respectively. It can be seen that a bigger SRV radius results in a longer time of the inner region radial flow and an earlier start of the transition flow from the inner to the outer region radial flow. For the production curves, similar to the mechanism model 1, a bigger SRV region corresponds to a higher well production rate at constant bottom-hole pressure.

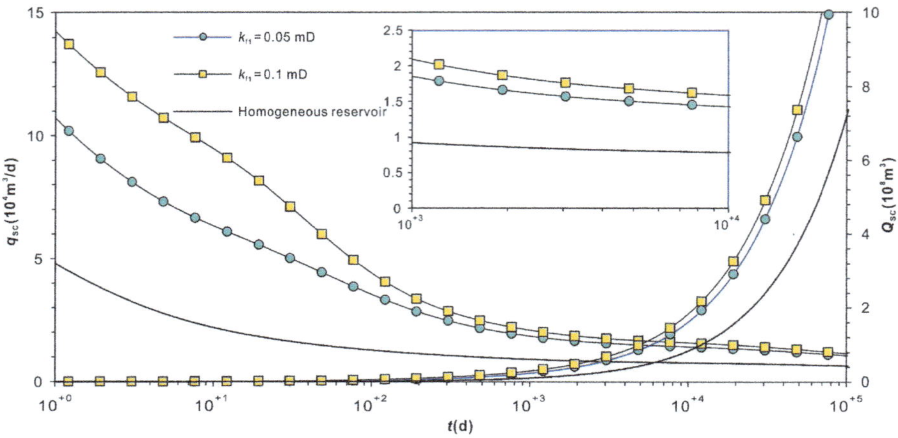

Fig. 5.11 Effect of SRV permeability on well production curves.

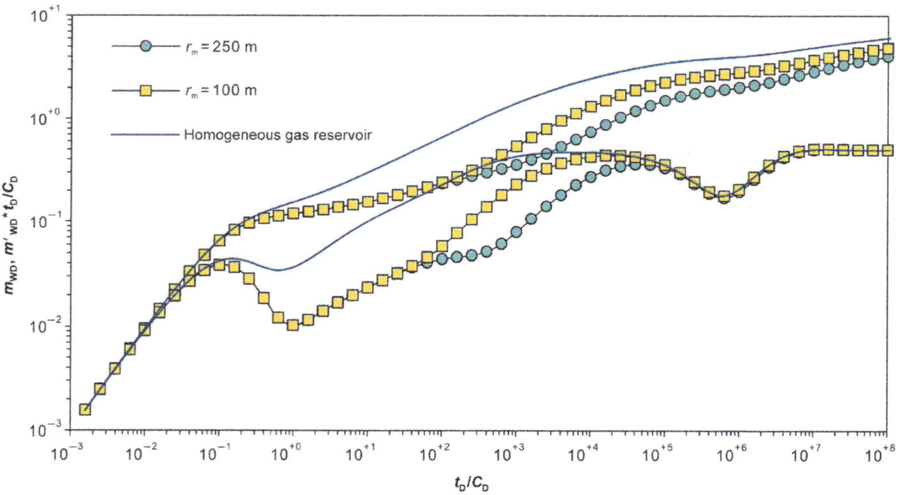

Fig. 5.12 Effect of SRV radius on well test type curves.

5.4.3 *Microfractures + gas adsorption/desorption + matrix Fick's diffusion*

The parameters in Table 5.3 are used to generate the pressure and production type curves for a fractured well in the center of a circular shale gas reservoir by the mechanism model 3.

For the pseudo steady state diffusion represented by the mechanism model 3, as shown in Fig. 5.14, there is a concave part of interporosity diffusion in the late-time period of the well test type curves. Different from the concave part of interporosity

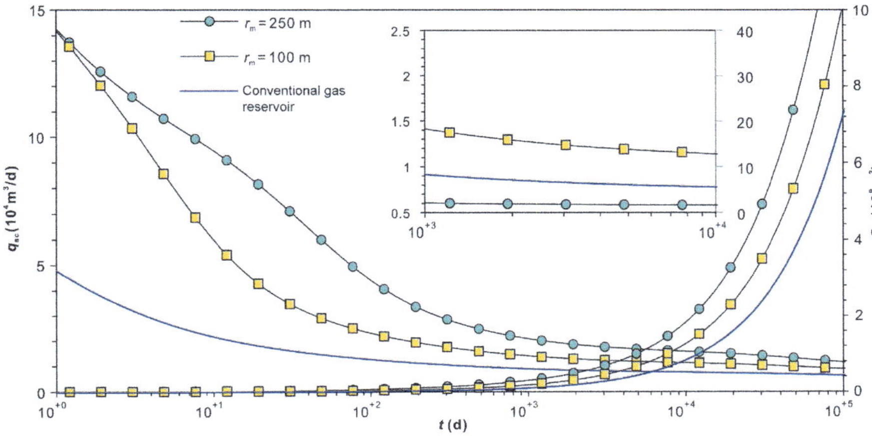

Fig. 5.13 Effect of SRV radius on well production curves.

Table 5.3 Shale reservoir properties

Reservoir properties	Value	Reservoir properties	Value
SRV region permeability, k_{f1} (mD)	0.1	SRV region porosity, $\varnothing_{f1}$ (fraction)	0.002
Microfracture permeability, k_{f2} (mD)	0.01	Microfracture system porosity, $\varnothing_{f2}$ (fraction)	0.002
Constant production rate, q_{sc} (m^3/d)	1×10^4	$(R_m)^2/D_F$	2×10^6

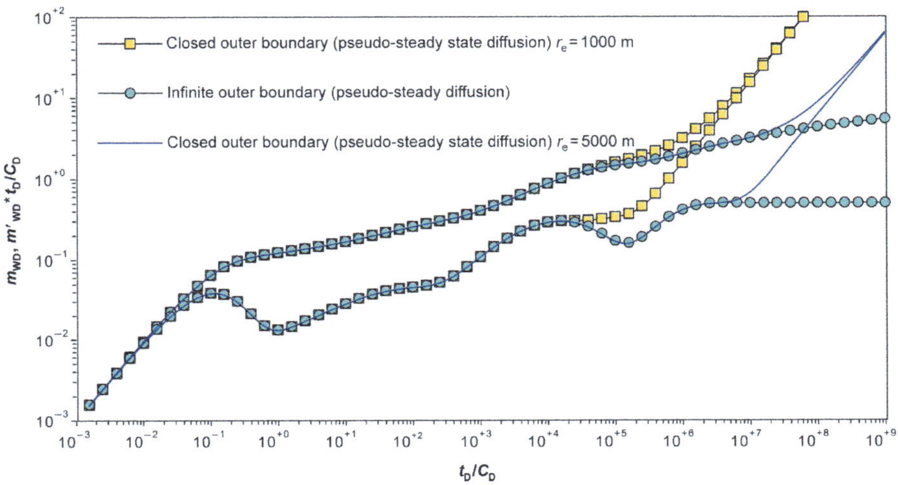

Fig. 5.14 Effect of boundary conditions on well test type curves.

flow from matrix pores to the microfracture system, this concave part is caused by the pseudo steady state Fick diffusion of desorbed gas from matrix to the microfracture system due to pressure depletion. If the radius of the outer boundary is relatively small, pressure propagation reaches the outer boundary at the same time of diffusive inter-porosity flow happening, and, therefore, the concave part trends upward under the co-effects of diffusion and boundary until the pseudo steady state flow happens.

Figs. 5.15 and 5.16 show the effects of the adsorption volume G_L on the well test type and production decline curves, respectively. G_L mainly affects the width and

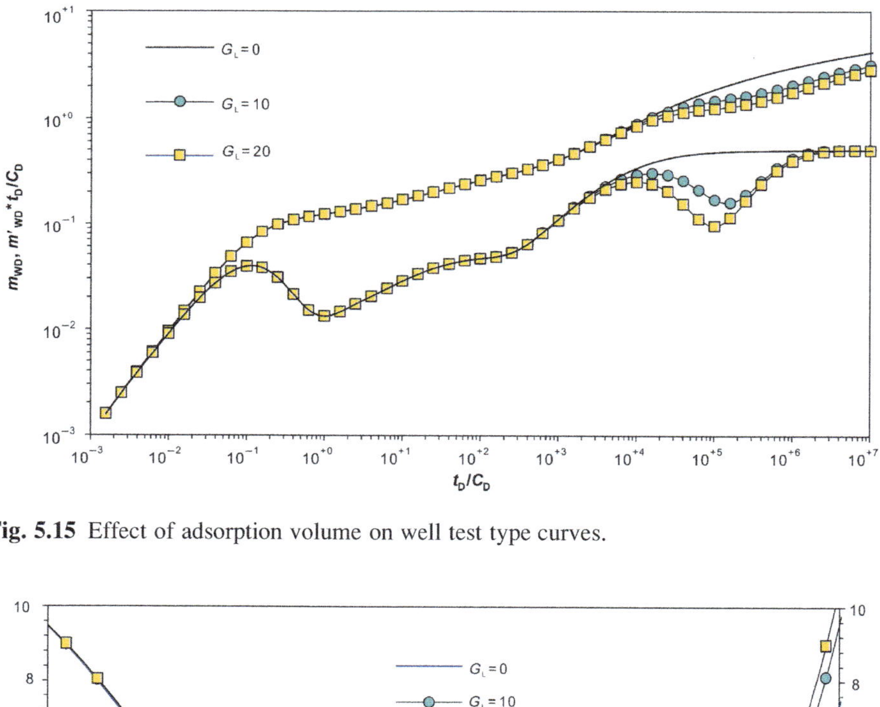

Fig. 5.15 Effect of adsorption volume on well test type curves.

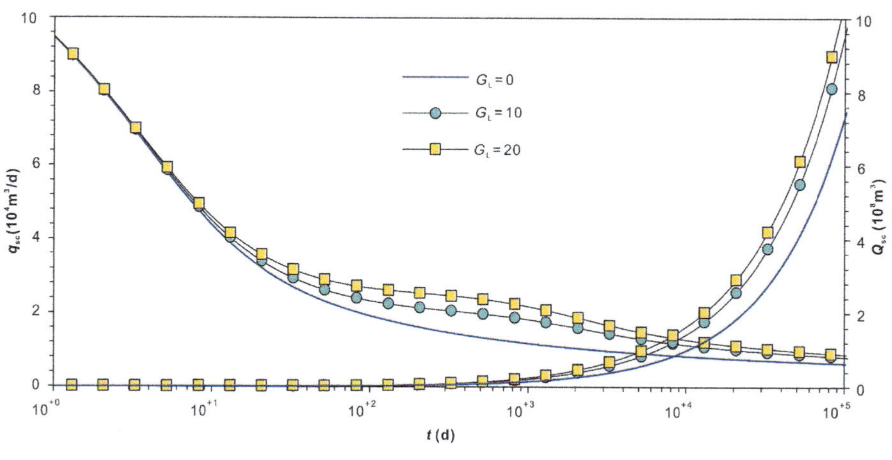

Fig. 5.16 Effect of adsorption volume on well production curves.

deepness of the diffusion concave portion. The bigger the G_L is, the wider and deeper the concave portion is and the earlier the diffusive interporosity flow happens. This is because G_L determines the adsorbed gas volume in shale matrix. The more adsorbed gas there is, the higher capacity of gas supply from desorption to the fracture system and the longer time the interporosity flow lasts. When a well produces at constant pressure, after the early production period, which is controlled by free gas content in SRV and natural fractures, the bigger the G_L is, the higher the plateau production is and the longer it lasts. Compared to a conventional composite gas reservoir without consideration of gas adsorption and desorption, a shale gas reservoir considering adsorbed gas has a higher production rate, which lasts quite a long time (after 10^4 d online, a well in the shale reservoir with $G_L = 10$ has higher production than a well in the conventional reservoir by $3000\,m^3/d$, and even at day 10^5, the production is still higher by $1500\,m^3/d$, as shown in Fig. 5.16).

For the effect of fracture half-length with the same SRV radius, as shown in Fig. 5.17, it mainly affects the inner region radial flow and the start of interporosity flow. The longer the fracture half-length is, the shorter time the inner region radial flow has. If the fracture half-length is almost identical to the SRV radius, the inner region radial flow is hard to observe. In addition, the longer the fracture half-length is, the earlier the interporosity flow starts. Meanwhile, as shown in Fig. 5.18, a bigger fracture half-length results in a higher production rate under constant production pressure. However, there is little difference of the late-time well production rate due to the same SRV radius. Therefore, compared to the SRV radius, the fracture half-length has a limited influence on well production improvement. For the development of a shale gas reservoir, creating massive complex fracture networks is the key target in hydraulic fracturing operations.

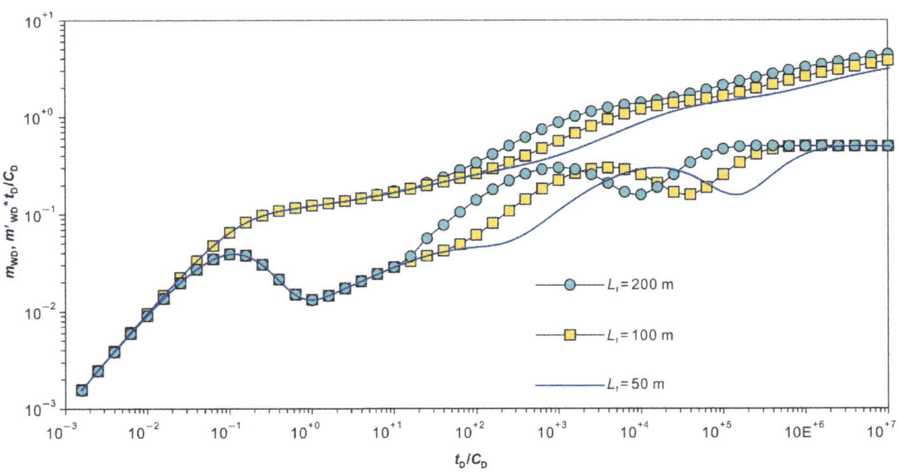

Fig. 5.17 Effect of fracture half-length on well test type curves.

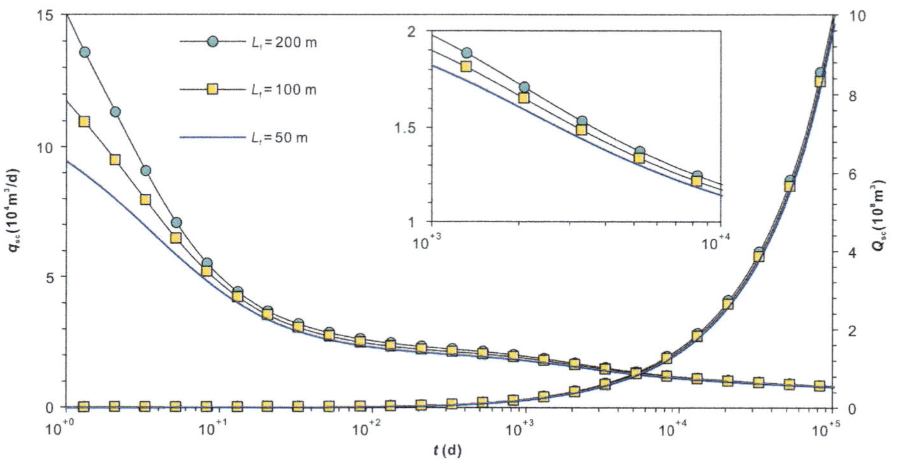

Fig. 5.18 Effect of fracture half-length on well production curves.

Table 5.4 Shale reservoir properties

Reservoir properties	Value	Reservoir properties	Value
Permeability of matrix macro pores, k_m (mD)	0.0001	Porosity of matrix macro pores, $\varnothing_m$	0.012
Permeability of fracture system, k_{f2} (mD)	0.01	Porosity of fracture system, $\varnothing_{f2}$ (fraction)	0.02
Permeability of SRV region, k_{f1} (mD)	0.1	Porosity of SRV region, $\varnothing_{f1}$ (fraction)	0.02
Geometry factor of interporosity flow, α (1/m²)	10^{-4}	$(R_m)^2/D_F$	2×10^6

5.4.4 Microfractures + matrix macro pores + gas adsorption/desorption + nanopore Fick's diffusion

For the mechanism model 4, the parameters in Table 5.4 are used to generate the well test type and production performance curves.

Shown in Fig. 5.19, there are the well test type curves of a fractured vertical well in a shale reservoir with SRV under the pseudo steady state interporosity flow of matrix macro pores to microfractures for different outer boundary conditions and diffusion models. The effect of a diffusion model is mainly on the concave portion of interporosity flow. Due to a higher sensitivity of unsteady state diffusion to a pressure change in the matrix macro pore system, the concave portion shows up a little bit earlier and is shallower than that of the pseudo steady state flow. If the reservoir outer boundary is sealed, the pseudo pressure derivative curves trend upward, leading to deformation of the interporosity flow concave portion.

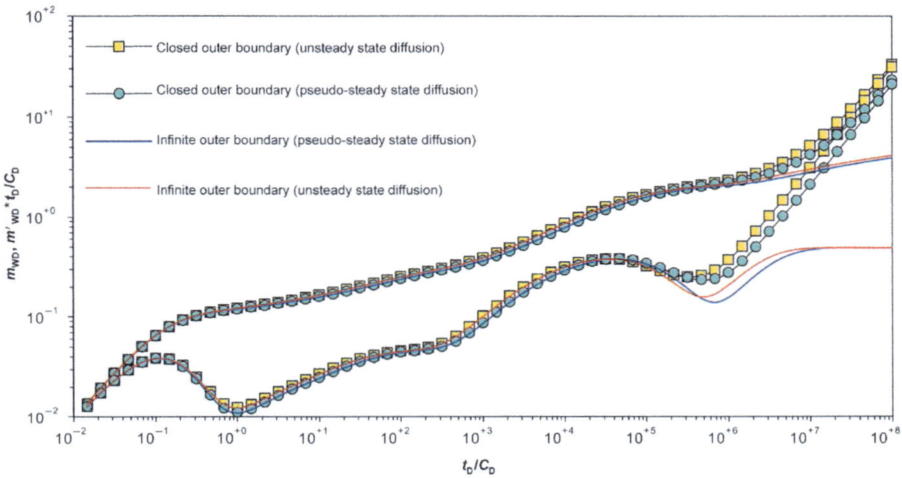

Fig. 5.19 Effect of outer boundary conditions and diffusion models on well test type curves.

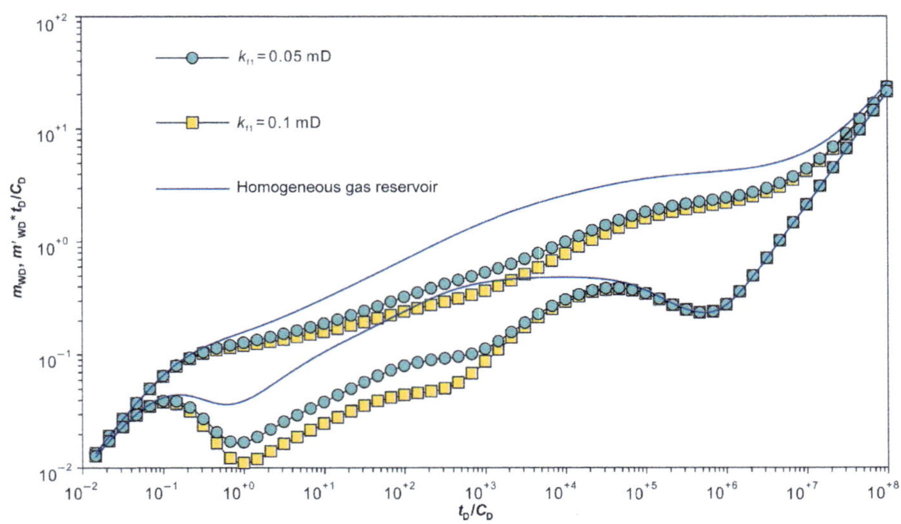

Fig. 5.20 Effect of SRV permeability on well test type curves.

Figs. 5.20 and 5.21 show the effects of SRV permeability on the well test type and production performance curves, respectively. It can be seen that the effect of SRV permeability is mainly on the early time flow period of the well test type curves. The higher the SRV permeability is, the less the pressure drop is required for the same production rate and the lower the pseudo pressure and its derivative curves are on their plots. When pressure propagates out to the outer reservoir region, the influence of SRV permeability becomes smaller. Comparing the curves in the homogeneous gas reservoir and the reservoir with SRV, although there is little difference for the

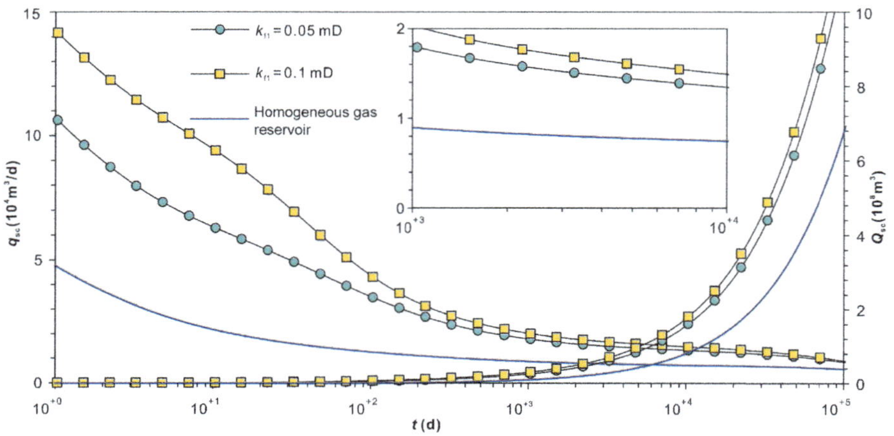

Fig. 5.21 Effect of SRV permeability on well production curves.

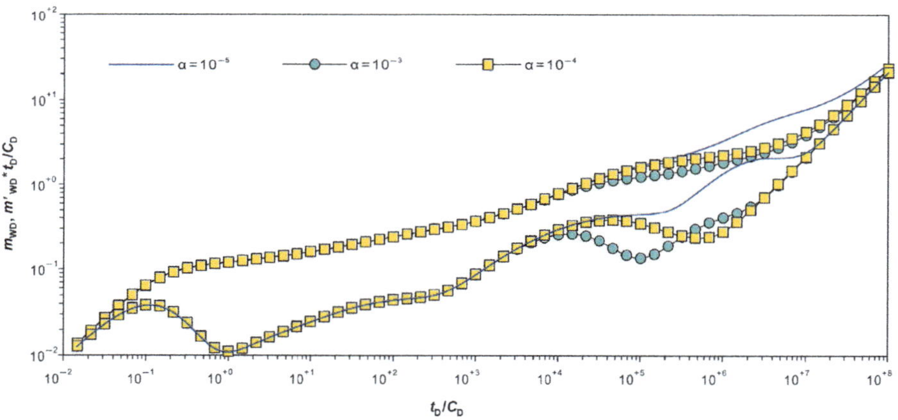

Fig. 5.22 Effect of matrix geometry factor on well test type curves.

late-time cures for a SRV of 5 and 10 times the matrix permeability (0.01 mD), the pseudo pressure curves in the homogeneous gas reservoir are obviously higher, indicating the effect of SRV on reducing flow resistance. Similarly, such an effect can be observed from the production performance curves.

The effects of a matrix geometry factor on the pressure and production curves are shown in Figs. 5.22 and 5.23, respectively. Since the matrix geometry factor mainly affects the interporosity flow from matrix macro pores to microfractures, the bigger the matrix geometry factor is, the bigger the interporosity flow coefficient is and the earlier the interporosity flow starts. For a well producing at constant pressure, the earlier start of the interporosity flow indicates earlier gas supply from nanometer pores and, consequently, a higher well production rate. With stabilization of the interporosity flow, the gas production rates become almost identical for different geometry factors.

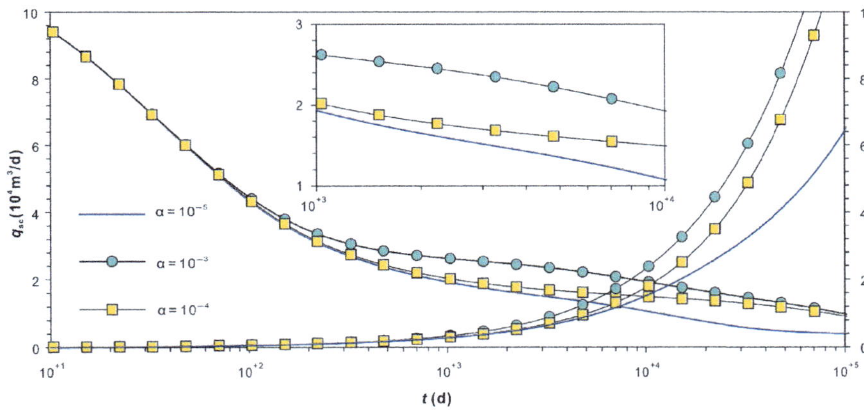

Fig. 5.23 Effect of matrix geometry factor on well production curves.

Table 5.5 Shale reservoir properties

Reservoir properties	Value	Reservoir properties	Value
Permeability of matrix macro pores, k_m (mD)	0.0001	Porosity of matrix macro pores, $\varnothing_m$	0.012
Permeability of microfracture system, k_{f2} (mD)	0.01	Porosity of microfracture system, $\varnothing_{f2}$ (fraction)	0.002
Permeability of SRV region, k_{f1} (mD)	0.1	Porosity of SRV region, $\varnothing_{f1}$ (fraction)	0.002
Geometry factor of interporosity flow, α (1/m²)	10^{-5}	Knudsen diffusion coefficient, D_k (1/m²)	10^{-6}
Average size of matrix macro pores, r_n (m)	2×10^{-9}	Value of α for Eq. (1.26) (dimensionless)	0.8

5.4.5 Microfractures + gas adsorption/desorption + nanopore Knudsen diffusion

The parameters in Table 5.5 are used to generate the well test type and production curves for the mechanism model 5.

The well test type curves of different interporosity models and outer boundary conditions are shown in Fig. 5.24. As shown, when the gas flow in a shale reservoir satisfies multiple flow mechanisms, such as Darcy's flow, Knudsen diffusion and a slippage effect, a concave portion of interporosity flow can be observed on the pseudo pressure derivative curves, whose shape is determined by all these mechanisms. If the outer boundary is sealed, the pseudo pressure and its derivative curve up in the late time. For the unsteady state interporosity flow, the concave part is also flat but happens earlier than for the pseudo steady state flow.

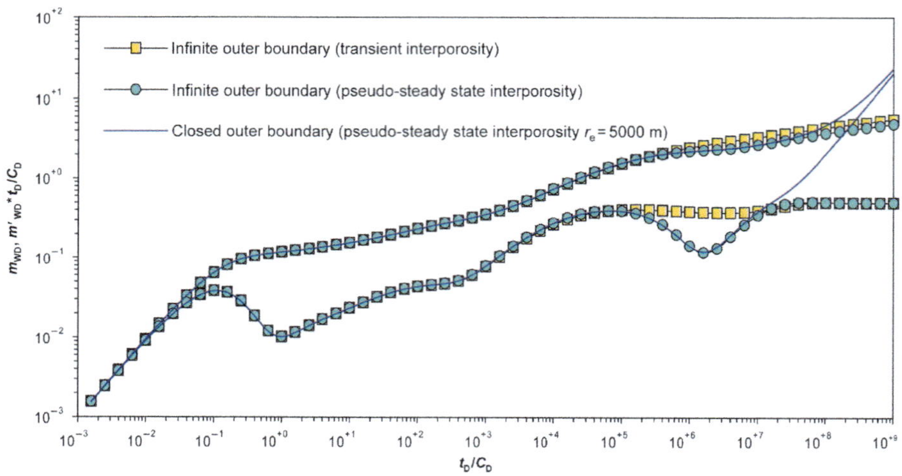

Fig. 5.24 Effect of interporosity flow models and outer boundary conditions on well test type curves.

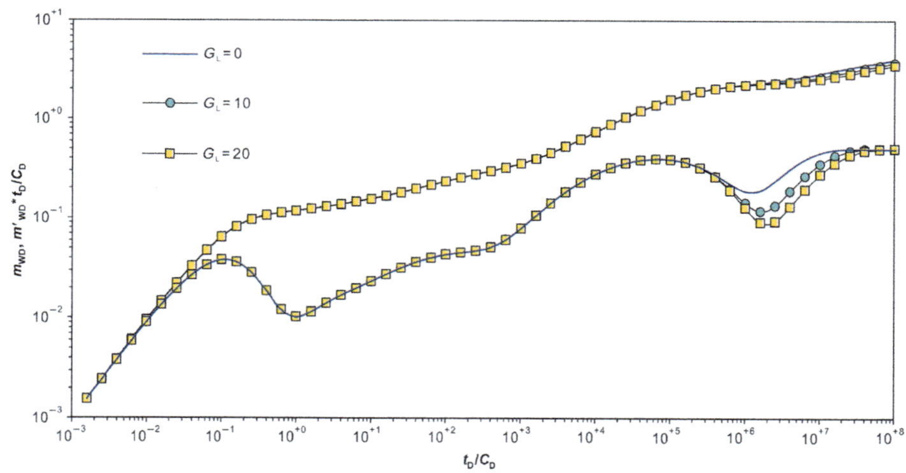

Fig. 5.25 Effect of gas adsorption volume on well test type curves.

The effects of the shale gas adsorption volume G_L on the well test type and production curves are shown in Figs. 5.25 and 5.26, respectively. As shown, the effect of G_L is visible through the concave part for the pseudo steady state interporosity flow. The bigger the G_L is, the deeper and wider the concave part is. This is because the supply from desorption of adsorbed gas slows down the pressure drop in the reservoir, leading to a deeper concave part, which also lasts a longer time. In the case where a well produces at constant pressure, only if the reservoir pressure drops below the Langmuir adsorption pressure, gas desorption happens. Before that, well production

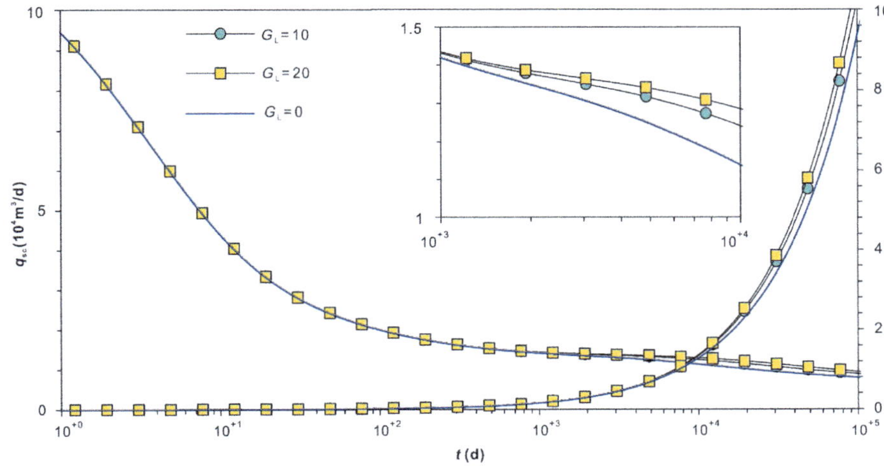

Fig. 5.26 Effect of gas adsorption volume on well production type curves.

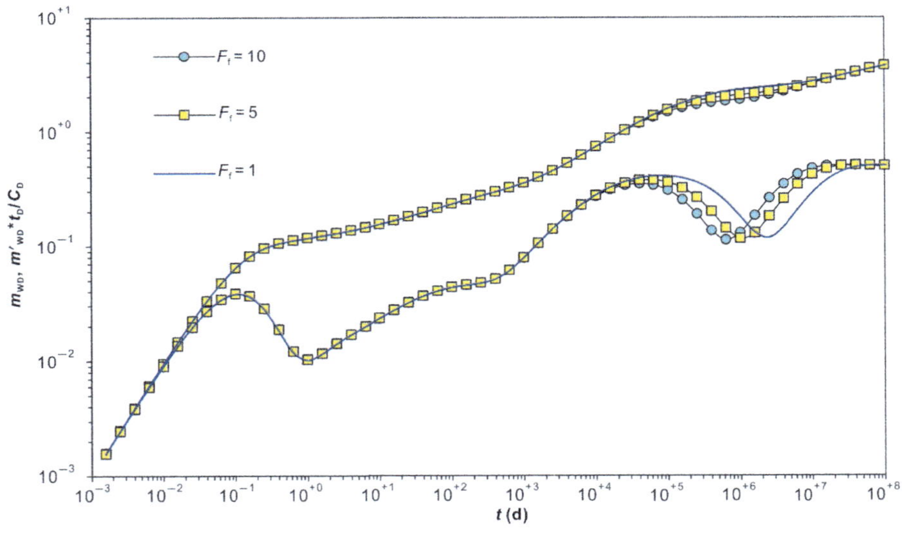

Fig. 5.27 Effect of slippage coefficient on well test type curves.

relies on free gas in reservoir microfractures and matrix pores. Under an infinite outer boundary condition, there is enough gas supply to well production, and, therefore, the well production rate only has a slight change in the late time.

The effects of a slippage coefficient on the well test type and production performance curves are shown in Figs. 5.27 and 5.28, respectively. Since a slippage effect increases the matrix apparent permeability, the bigger the coefficient is, the higher the matrix apparent permeability is, which consequently increases and accelerates

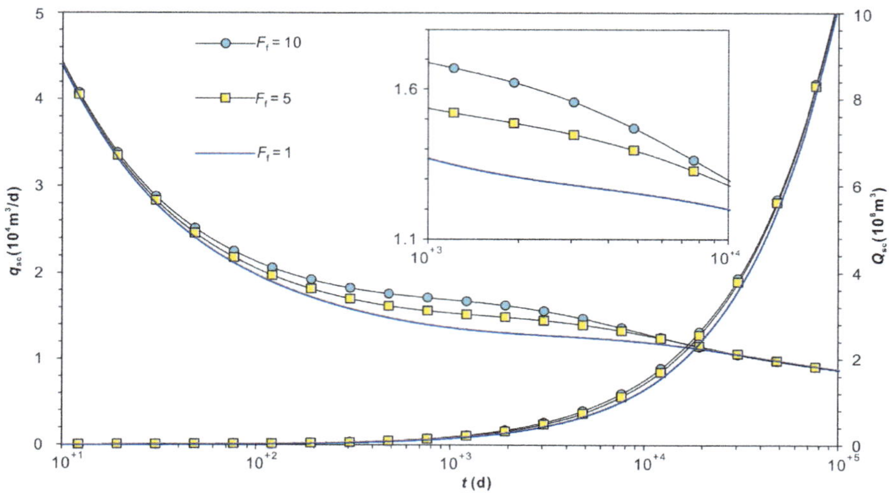

Fig. 5.28 Effect of slippage coefficient on well production curves.

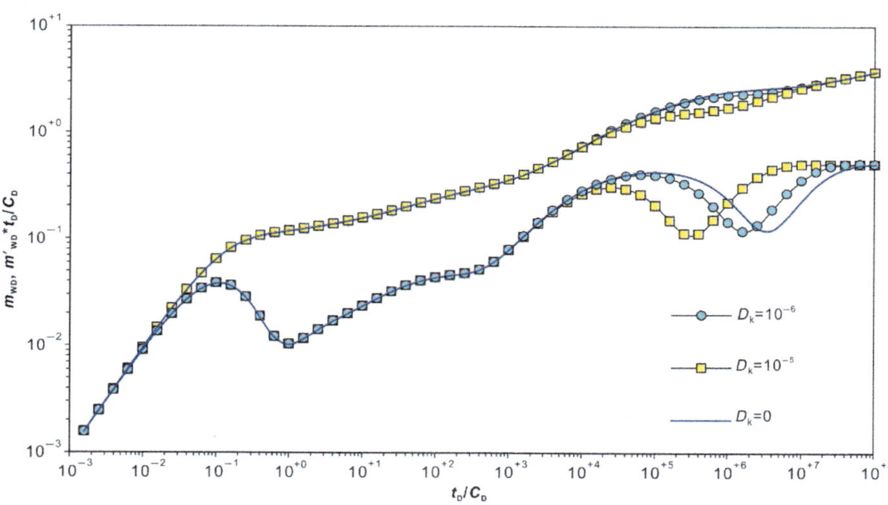

Fig. 5.29 Effect of Knudsen diffusion coefficient on well test type curves.

interporosity flow from matrix pores to microfractures. If a well produces at constant pressure, a bigger slippage coefficient leads to enhanced interporosity flow from matrix to microfractures and thus a higher well production rate.

As shown in Figs. 5.29 and 5.30, a Knudsen diffusion coefficient D_k also has an effect on the matrix apparent permeability. Similar to the effect of a slippage coefficient, the bigger the D_k is, the earlier the concave part of interporosity flow shows up and the higher production at constant pressure.

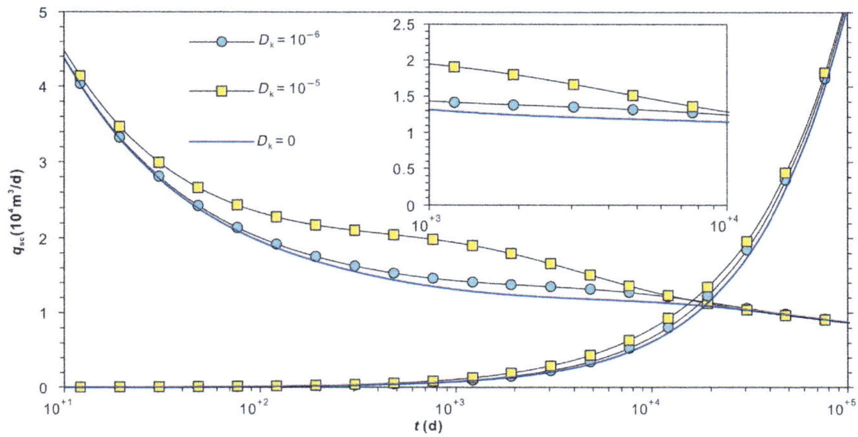

Fig. 5.30 Effect of Knudsen diffusion coefficient on well production curves.

Multi-stage fractured horizontal wells in circular reservoirs with SRV

6

Chapter Outline

6.1 Introduction 177
6.2 MFHWs in circular composite gas reservoirs 177
6.2.1 Physical model 178
6.2.2 Mathematical solution 179
6.3 Analysis of pressure and production type curves 180
6.3.1 Microfractures + steady state ad-/desorption and diffusion 180
6.3.2 Microfractures + gas adsorption/desorption + matrix Fick's diffusion 184
6.3.3 Microfractures + gas adsoption/desorption + nanopore Knudsen diffusion 188

6.1 Introduction

Multi-stage fractured horizontal wells (MFHWs) become a key to efficiently develop a shale gas reservoir. However, more and more practice indicates that a complex induced fracture network connecting a rock volume surrounding the main fractures is necessary to achieve economic and considerate production from such an ultra-low permeability reservoir. The reservoir region with increased permeability through these complex fracture networks is called a stimulated reservoir volume (SRV), as mentioned previously. Currently, flow models in MFHWs with consideration of a SRV are mainly linear flow models, which have different combinations of flow mechanisms based on a SRV size, a reservoir scale, and fracture spacing. A great amount of work has been done by researchers on such models, especially for a production decline analysis.

6.2 MFHWs in circular composite gas reservoirs

In this section, a composite flow model for a MFHW with consideration of a SRV is established through a dual-media model to represent flow in the SRV. Based on the continuous line source solution method, a discrete element method for hydraulic fractures similar to that used for a MFHW in a conventional reservoir is applied to solve this model. Moreover, the type curves of well tests and production are generated through numerical inversion and computer programming.

Developments in Petroleum Science, Vol. 66. https://doi.org/10.1016/B978-0-444-64315-5.00006-1

6.2.1 Physical model

The physical model for a MFHW in a circular reservoir with consideration of a SRV is shown in Fig. 6.1 (Zhao et al., 2014). A dual-media model is used to represent the SRV surrounding the MFHW. To facilitate the subsequent analysis, the following assumptions are made (Warren and Root, 1963):

1. M symmetrical or asymmetrical hydraulic fractures are distributed perpendicular to the horizontal wellbore, and their distribution can be even or uneven with the distance of L between the first and last fractures.
2. The SRV region is circular with radius r_m, and the MFHW locates at the center of the inner region.
3. The hydraulic fractures have full penetration with height being identical to the reservoir thickness h.
4. The fractures are of infinite conductivity, and the fracture width is negligible.
5. The reservoir fluid flows into the wellbore only through the main hydraulic fractures, and a pressure drop along the wellbore is zero.
6. The outer reservoir boundary is infinite or a closed circle with radius r_e.

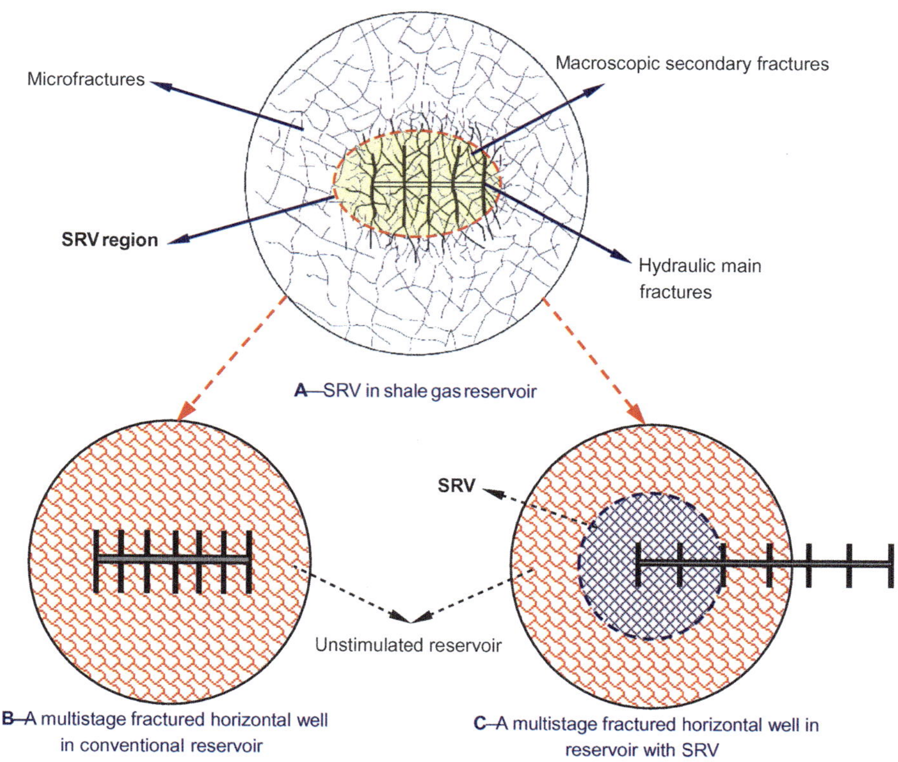

Fig. 6.1 A MFHW in a circular shale gas reservoir with consideration of SRV.

6.2.2 Mathematical solution

Similar to the solution of a conventional MFHW model, a discrete element method is used to solve the composite model considering a SRV. Each main fracture is evenly discretized into $2N$ elements, in which the flow rate is assumed uniform. Then a dimensionless pressure drop caused by each element at any point in the reservoir can be acquired through integration of a continuous line source solution along the element, and the drops in all elements are added up to acquire the total pressure drop based on the potential superposition principle. For M fractures, $M \times 2N$ linear equations are generated; together with a normalization equation, there are $M \times 2N+1$ equations for $M \times 2N+1$ unknowns.

According to the above solution method, the following dimensionless production rate and bottom-hole pressure are defined:

$$q_{\mathrm{D}i} = \frac{q_{\mathrm{scL}i} \Delta L_{\mathrm{f}i}}{q_{\mathrm{sc}}} \tag{6.1}$$

$$m_{\mathrm{fD}} = \frac{\pi k_{\mathrm{f2}} h T_{\mathrm{sc}}}{p_{\mathrm{sc}} T q_{\mathrm{sc}}} \Delta m_{\mathrm{f}} \tag{6.2}$$

where:

$q_{\mathrm{scL}i}$—continuous line source strength of the ith fracture discrete element (m^3/s);
$\Delta L_{\mathrm{f}i}$—length of the ith fracture discrete element (m);
q_{sc}—production rate of the MFHW (m^3/s).

By the superposition principle, the dimensionless pseudo pressure drop at the jth point is

$$\overline{m}_{\mathrm{D}}\left(x_{\mathrm{D}j}, y_{\mathrm{D}j}\right) = \sum_{i=1}^{M*2N} \overline{m}_{\mathrm{fD}i}\left(x_{\mathrm{D}j}, y_{\mathrm{D}j}\right) \tag{6.3}$$

where the dimensionless pseudo pressure drop caused by the ith discrete element at the jth point is

$$\Delta \overline{m}_{\mathrm{fD}i} = \frac{q_{\mathrm{D}i}}{s \Delta L_{\mathrm{fD}i}} \int_{-\frac{\Delta L_{\mathrm{fD}i}}{2}}^{\frac{\Delta L_{\mathrm{fD}i}}{2}} \begin{bmatrix} K_0\left(\xi_0 \sqrt{\left(x_{\mathrm{D}j} - x_{\mathrm{D}i} - \alpha\right)^2 + \left(y_{\mathrm{D}j} - y_{\mathrm{wD}i}\right)^2}\right) + \\ A_{\mathrm{C}} I_0\left(\xi_0 \sqrt{\left(x_{\mathrm{D}j} - x_{\mathrm{D}i} - \alpha\right)^2 + \left(y_{\mathrm{D}j} - y_{\mathrm{wD}i}\right)^2}\right) \end{bmatrix} \mathrm{d}\alpha \tag{6.4}$$

Since it is assumed that the fractures are of infinite conductivity and there is no pressure drop along the horizontal wellbore (Due to ultra-low permeability, well production in a shale gas reservoir is much lower than that in a conventional reservoir; then the pressure drop in the fractures and wellbore is negligible.), the pressure at all observation points is equal to the bottom-hole pressure of the gas well:

$$\overline{m}_{\mathrm{wD}} = m_{\mathrm{fD}j}, \quad (j = 1 \cdots M \times 2N) \tag{6.5}$$

For the MFHW, a sum of gas flow in all fracture elements equals the well production rate:

$$\sum_{i=1}^{M*2N} q_{Di} = 1 \tag{6.6}$$

Combining Eqs. (6.5) and (6.6), there is the matrix system of equations:

$$\mathbf{A}X = \mathbf{C} \tag{6.7}$$

where **A** and **C** have the same format as the matrix and vector coefficients in system (4.38).

6.3 Analysis of pressure and production type curves

6.3.1 Microfractures + steady state ad-/desorption and diffusion

By inserting the reservoir parameters in Table 6.1 into the matrix system derived above and applying Gaussian elimination, the well bottom-hole pseudo pressure and production rate with time can be acquired and used to generate the well test type and production decline curves.

For different boundary conditions, the well test type curves from the mechanism model 1 for a MFHW in a shale gas reservoir with SRV are shown in Fig. 6.2. Since the flow stage of the SRV region is merely observable based on the reservoir

Table 6.1 Shale reservoir properties

Reservoir properties	Value	Reservoir properties	Value
Initial reservoir pressure, p_i (MPa)	25	Reservoir temperature, T (K)	320
Reservoir thickness, h (m)	60	Specific gas gravity, γ_g (fraction)	0.65
Compressibility at initial reservoir conditions, C_{gi} (MPa^{-1})	0.02	Bottom-hole pressure, p_{wf} (MPa)	1
Number of fractures, M	4	Fracture half length, x_f (m)	30
Permeability of SRV region, k_{f1} (mD)	0.1	Porosity of SRV region, $\varnothing_{f1}$ (fraction)	0.1
Permeability of microfracture system, k_{f2} (mD)	0.01	Porosity of microfracture system, $\varnothing_{f2}$ (fraction)	0.002
Langmuir pressure, P_L (MPa)	4	Langmuir volume, G_L (m^3/m^3)	10
Radius of SRV region, r_m (m)	480	Outer radius of circular sealed gas reservoir, r_e (m)	6000
Effective length of horizontal well, L (m)	800	Dimensionless wellbore storage coefficient, C_D	10^{-5}
Skin factor, S_{kin} (dimensionless)	0.1		

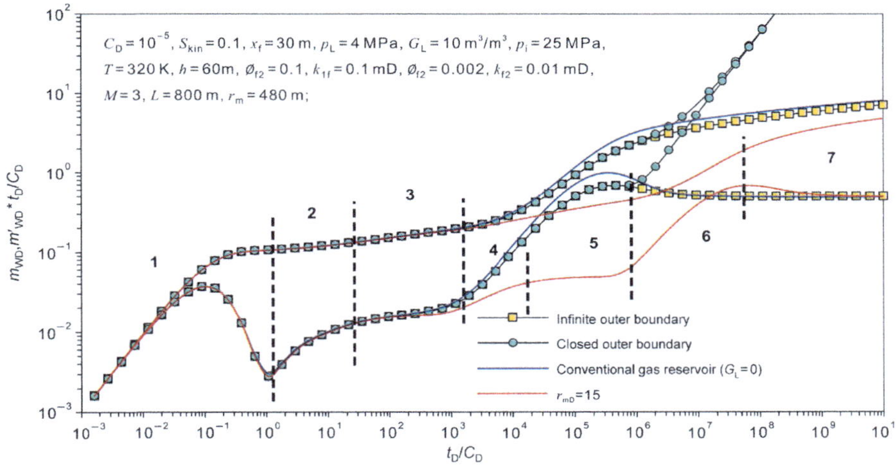

Fig. 6.2 Well test type curves of MFHW in a shale gas reservoir with SRV and different boundary conditions.

parameters in Table 6.1, the red dotted line on this plot is for a dimensionless SRV radius of 15. According to the plot, seven flow stages can be identified:

Stage 1: the early time wellbore storage and its transition flow. For pure wellbore storage, the pseudo pressure and its derivative overlap as a straight line with a slope of 1, while, during the transition flow, the derivative becomes a hump whose value and width are determined by both the wellbore storage and skin effects.

Stage 2: the fracture system early time linear flow. Affected by the wellbore storage and skin effects, the linear flow is not obvious due to a short time of duration. To clearly illustrate this stage, Fig. 6.3 shows the type curves without the wellbore storage and skin effects. It can be seen that the pseudo pressure and its derivative are straight lines with a slope of 0.5.

Stage 3: the fracture early time radial flow. If the fracture spacing is big enough, after the ending of the early linear flow and before interference happening between fractures, this early time radial flow can be observed, exhibiting as a horizontal line at a value of $1/(2M_{12}*M)$ on the pseudo pressure derivative curves.

Stage 4: the fracture interference and transition flow. Following the fracture early time radial flow, if pressure has not propagated outside the area between fractures, this stage happens.

Stage 5: the inner region (SRV region) radial flow. During this flow stage, the pseudo pressure derivative is a horizontal line at a value of $1/(2*M_{12})$.

Stage 6: the transition flow from the inner region radial flow to the outer region radial flow. Due to much lower permeability of the outer region, pressure exhibits behavior similar to hitting a sealed boundary, which is curving up on the pseudo pressure and its derivative curves. If the storativity ratio of the outer region is greater than that of the inner region, there is a hump showing up on the late time derivative curves, and its height and width are determined by the inner and outer region storativity ratio.

Stage 7: the outer region radial flow. After pressure completely propagates out into the outer reservoir region, this flow stage happens. On the type curves, the pseudo pressure derivative is a horizontal line at a value of 0.5.

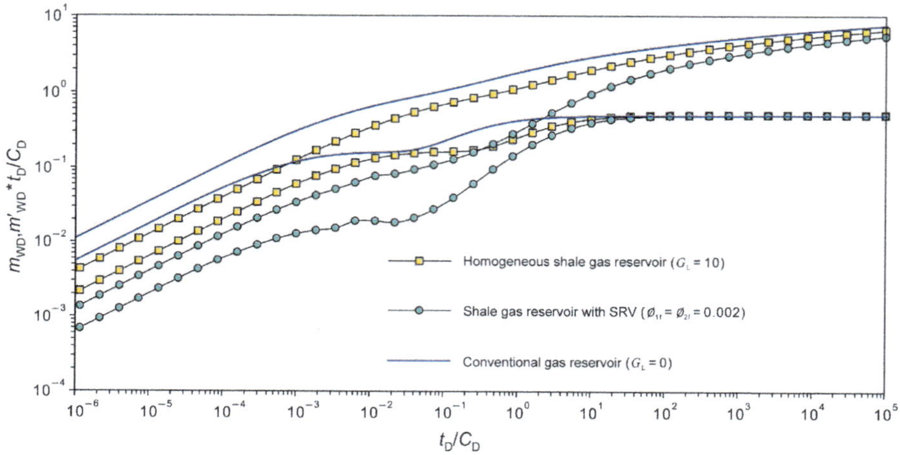

Fig. 6.3 Well test type curves for conventional and shale gas reservoirs without consideration of wellbore storage and skin effects.

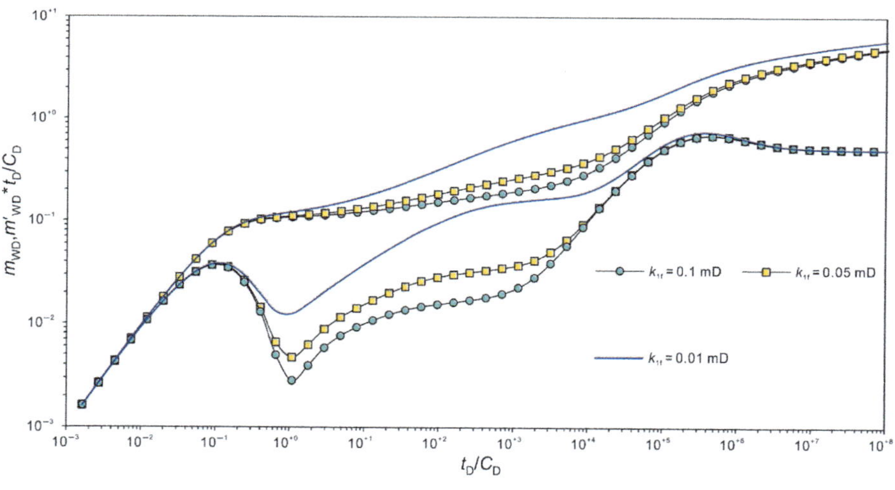

Fig. 6.4 Effect of SRV permeability on well test type curves.

Note that if the outer reservoir boundary is sealed, the pseudo steady state flow will be observed during the late time. Thus, the pseudo pressure and its derivative overlap as a straight line with a slope of 1.

As shown in Fig. 6.2, if the SRV size is comparable to the horizontal well half length, the flow stages 4 and 5 cannot be observed. In addition, if the outer reservoir boundary is sealed and pressure quickly reaches the boundary after propagating into the outer region, the flow stage 7 is barely observable, either.

Figs. 6.4 and 6.5 show the effects of the SRV permeability on the well test type and production performance curves, respectively. As shown, the SRV permeability mainly affects the early time fracture linear flow and radial flow. The higher the SRV permeability is, the lower pressure drawdown is required for a constant production rate and the lower the pseudo pressure and its derivative curves are on their plots. Since the MFHW in this model is assumed to be at the center of the SRV, the gas flow is primarily controlled by the SRV reservoir properties. Once pressure reaches the SRV boundary, the late time well test type curves overlap with each other due to the exactly same reservoir properties in the outer region. Similarly, if the well produces at constant pressure, the early production rate increases with an increase in SRV permeability; while, during the middle to late time production period, although pressure already propagates outside into the outer reservoir region, the well production rate is still higher for higher SRV permeability due to lower flow resistance in the inner region.

For the effects of a SRV size on the well test type and production curves, as shown in Figs. 6.6 and 6.7, it mainly reflects the time of duration of the early time fracture

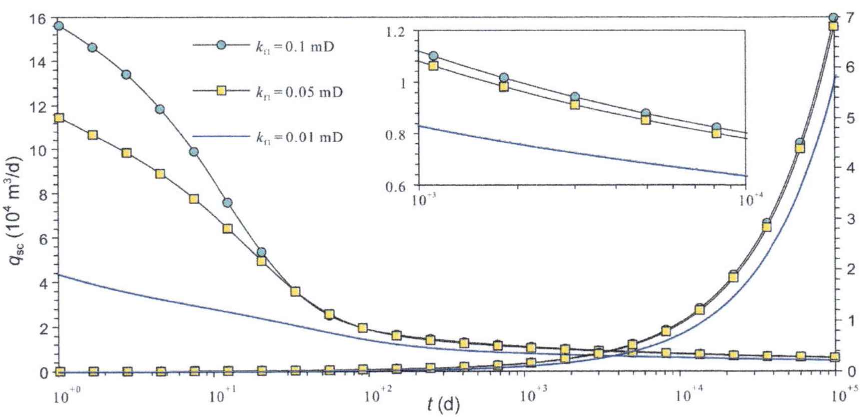

Fig. 6.5 Effect of SRV permeability on well production curves.

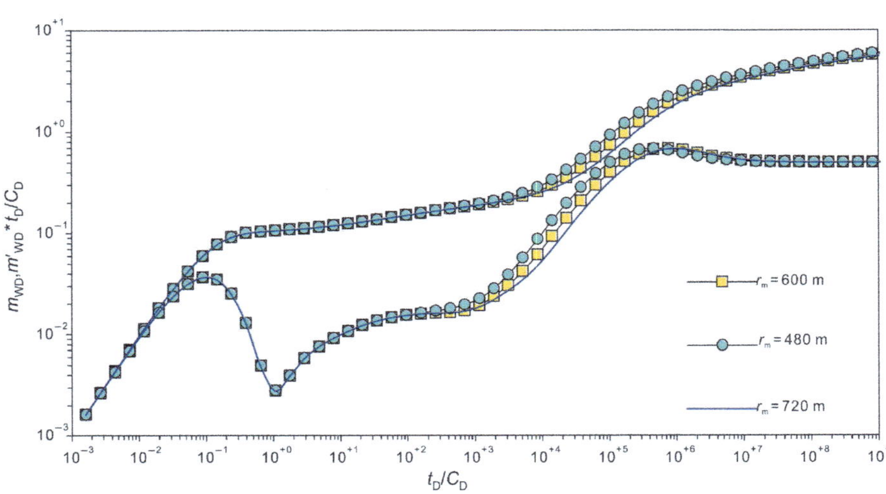

Fig. 6.6 Effect of SRV size on well test type curves.

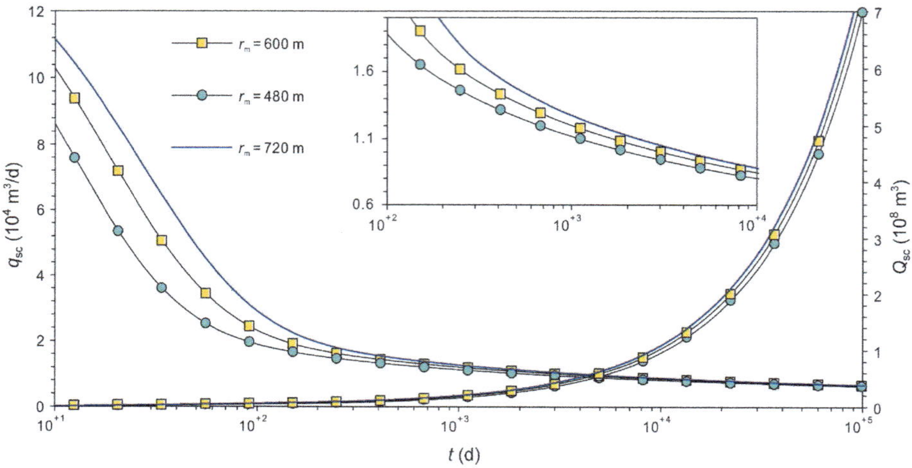

Fig. 6.7 Effect of SRV size on well production curves.

radial flow. If the gas well produces at constant bottom-hole pressure, the size of a SRV directly affects the flow area of gas flowing from the outer region into the more permeable inner region. In addition, the flow area is positively proportional to the square of the SRV radius. Therefore, the well middle to late time production rate is higher for a bigger SRV size. Under the infinite outer boundary conditions, after pressure completely propagates outside into the outer reservoir region, due to the low-flow capacity caused by ultra-low permeability in the outer region, well production becomes identical for different SRV sizes.

6.3.2 Microfractures + gas adsorption/desorption + matrix Fick's diffusion

For the mechanism model 3, the parameters in Table 6.2 are used to calculate and generate the well test type and production decline curves.

Figs. 6.8 and 6.9 show the well test type curves of a MFHW for different boundary conditions based on the mechanism model 3. Through a comparison of the conventional and unconventional well test type curves, it can be seen that a concave part of diffusive interporosity flow shows up on the pseudo pressure derivative curves for the shale gas reservoir considering adsorbed gas desorption and Fick's diffusion. However, this concave part is not observable in many cases due to properties and strong heterogeneity of actual gas reservoirs. For a well producing at constant pressure, the production rate from a shale gas reservoir considering adsorbed gas desorption is obviously higher and declines more slowly than that from a conventional gas reservoir. Therefore, there may be only a slight difference in the well test type curves for conventional and unconventional gas reservoirs, but there is an obvious difference in the production curves.

Table 6.2 Shale reservoir properties

Reservoir properties	Value	Reservoir properties	Value
Numbers of fracture, M	4	Fracture half length, x_f (m)	30
SRV permeability, k_{f1} (mD)	0.1	SRV porosity, $\varnothing_{f1}$ (fraction)	0.01
Permeability of micro fracture system, k_{f2} (mD)	0.01	Porosity of microfracture system, $\varnothing_{f2}$ (fraction)	0.002
Gas production at constant pressure, q_{sc}(m^3/d)	1×10^4	$(R_m)^2/D_F$	2×10^6

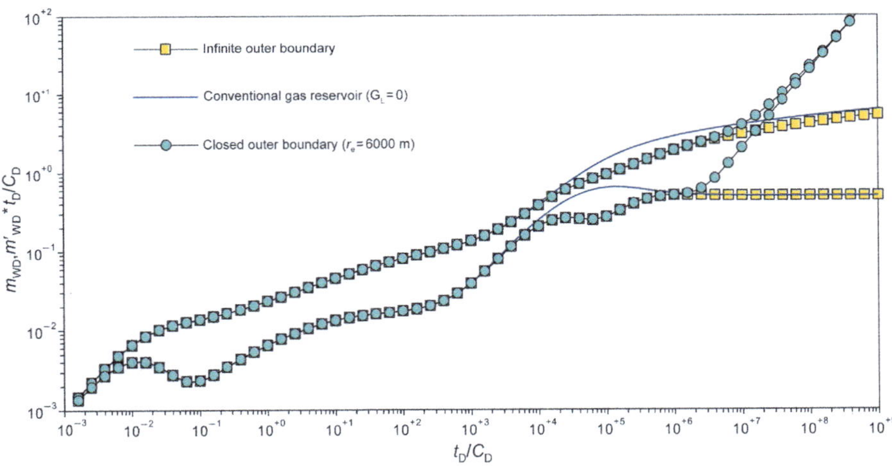

Fig. 6.8 Well test type curves of shale and conventional gas reservoirs with different boundary conditions.

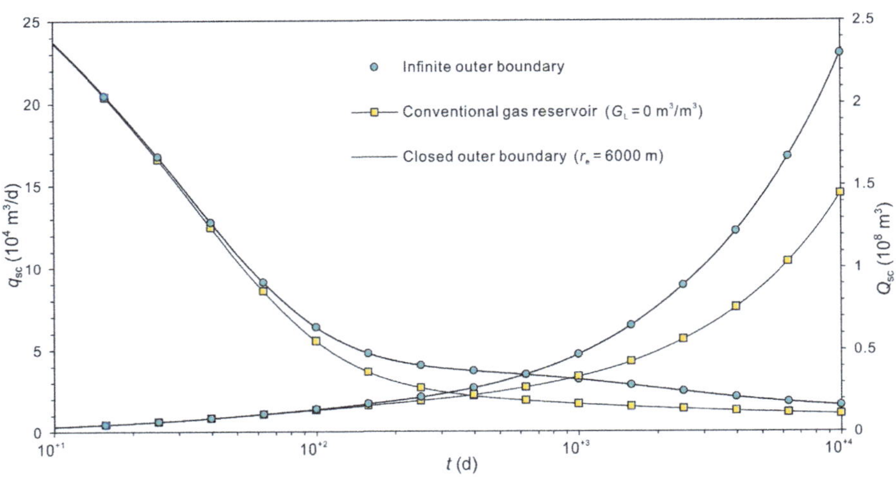

Fig. 6.9 Well production curves of shale and conventional gas reservoirs with different boundary conditions.

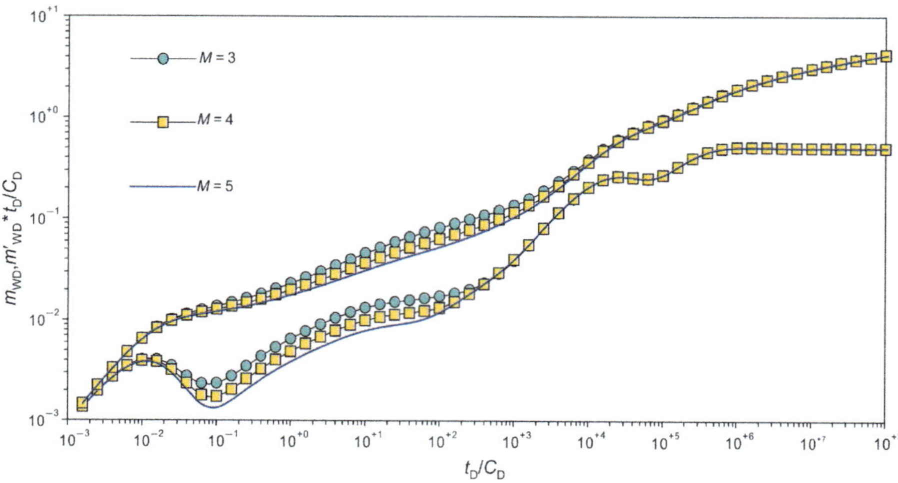

Fig. 6.10 Effect of fracture numbers on well test type curves.

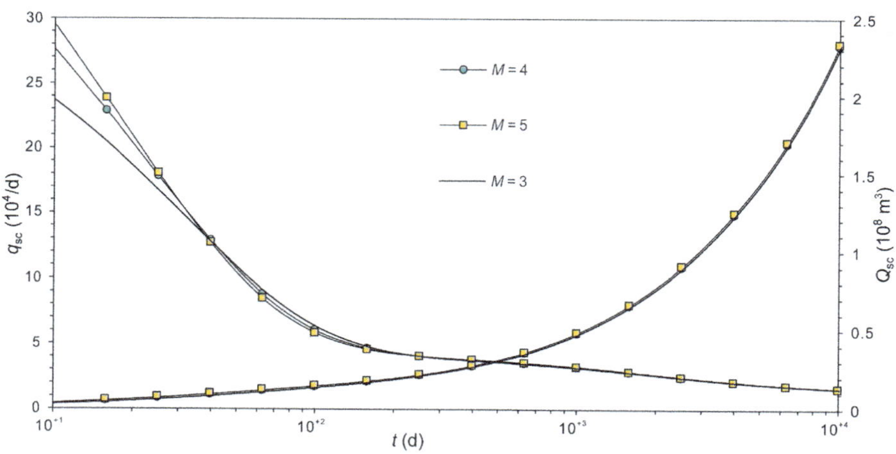

Fig. 6.11 Effect of fracture numbers on well production curves.

It can be seen from Figs. 6.10 and 6.11 that the effect of the fracture numbers is the same as that from the mechanism model 1, which is reflected only in the early time flow period. Similarly, for a well producing at constant pressure, the number of fractures mainly affects the gas well early production; once pressure reaches the outer reservoir region, the well production is dominated by the outer reservoir properties and tends to become identical for different numbers of fractures. Therefore, optimizing fracture numbers based on well conditions and reservoir properties is important for efficient and economic development.

The effects of desorption time on the well test type curves and production performance are shown in Figs. 6.12 and 6.13, respectively. Since desorption time directly

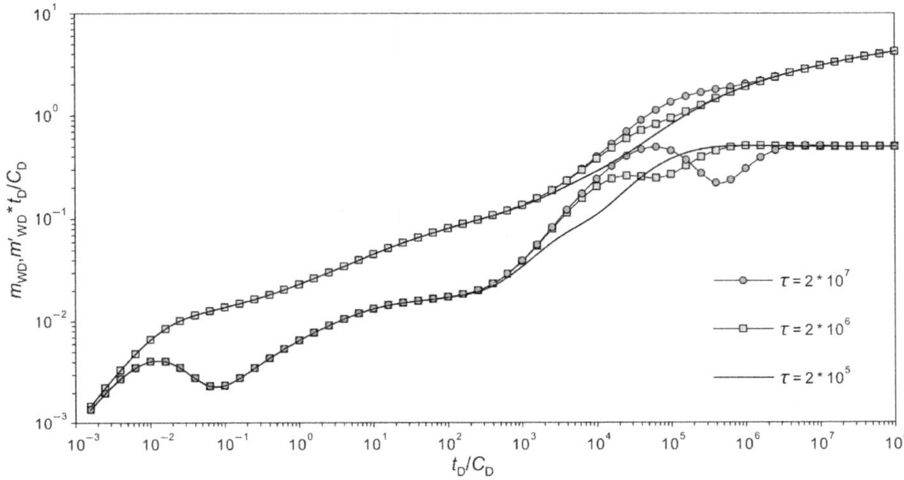

Fig. 6.12 Effect of gas desorption time on well-test type curves.

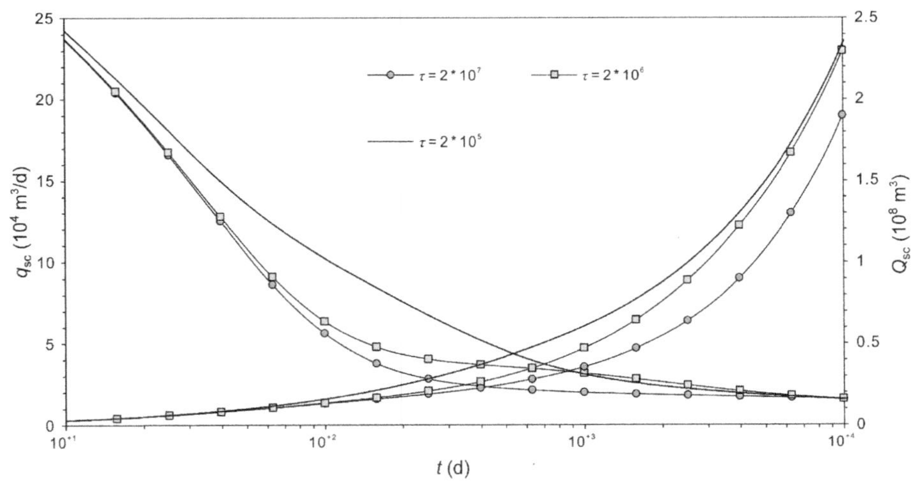

Fig. 6.13 Effect of gas desorption time on well production type curves.

affects the time of gas entering into micro fractures from desorption, the shorter the desorption time is, the bigger the volume of gas desorption is in unit time and the smaller a pressure drawdown is required for a constant production rate or the higher production is at constant pressure. As shown on the well test type curves, for a short desorption time, desorption happens quickly after pressure propagates outside into the outer reservoir region, and, therefore, the pseudo pressure and it derivative curves become lower in position during the transition flow stage. For well production, a shorter desorption time corresponds to a higher production rate and a lower decline.

6.3.3 Microfractures + gas adsoption/desorption + nanopore Knudsen diffusion

For using the mechanism model 5 to calculate and generate the well test type and production curves, the parameters are listed in Table 6.3.

For the mechanism model 5, the corresponding well test and production type curves for different boundary conditions are shown in Figs. 6.14 and 6.15, respectively. Although multiple flow mechanisms, including Darcy's flow, a slippage effect and Knudsen diffusion, are considered in the mechanism model 5, gas in matrix micro pores flows into micro fractures under a certain pressure difference. Therefore, a concave part of interporosity flow from matrix pores to micro fractures can be identified on the well test type curves. The size of the supply boundary directly affects the start of boundary dominated flow; the smaller the supply boundary is, the earlier the pseudo

Table 6.3 Shale reservoir properties

Reservoir properties	Value	Reservoir properties	Value
Matrix macropore system permeability, k_m (mD)	0.0001	Matrix macropore system porosity, $\varnothing_m$	0.012
Micro fracture system permeability, k_{f2} (mD)	0.01	Microfracture system porosity, $\varnothing_{f2}$ (fraction)	0.002
SRV region permeability, k_{f1} (mD)	0.1	SRV region porosity, $\varnothing_{f1}$ (fraction)	0.02
Geometry factor of interporosity flow, α (1/m^2)	10^{-6}	Knudsen diffusion coefficient, D_k (1/m^2)	10^{-6}
Average matrix pore size, r_n (m)	2×10^{-9}	α in Eq. (1.26) (dimensionless)	0.8

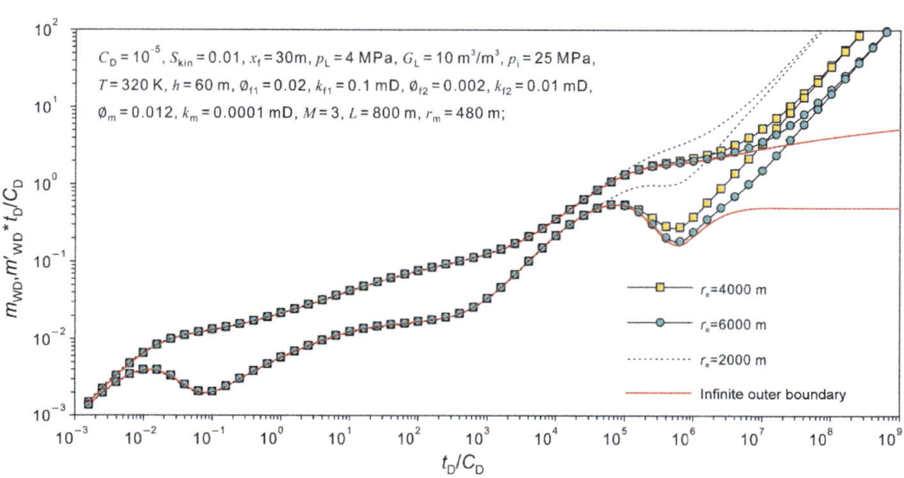

Fig. 6.14 Well test type curves of MFHW in a shale gas reservoir with different boundary sizes.

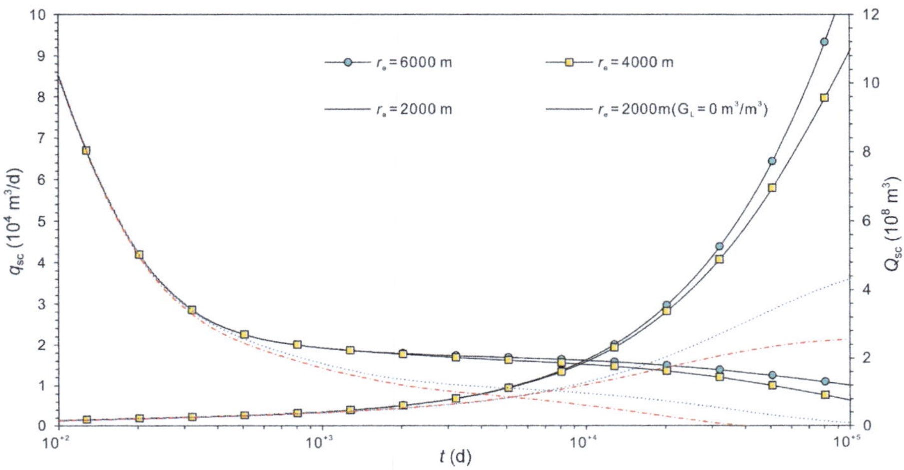

Fig. 6.15 Well production curves of MFHW in a shale gas reservoir with different boundary sizes.

pressure derivative rises. Moreover, under the co-effects of interporosity flow and boundary dominated flow, the position and shape of the interporosity concave portion change to be higher, shallower and narrower with the boundary becoming smaller. If a well produces under constant pressure, the bigger the boundary is, the higher the middle to late time production is, which is because of a bigger gas supply volume from a bigger area. By comparing well production from the conventional and shale gas reservoirs with the same radius of 2000 m, it can be seen that the late time production from the shale gas reservoir is much higher than that from the conventional reservoir, which is caused by the late time adsorbed gas desorption.

Figs. 6.16 and 6.17 illustrate the effects of a slippage coefficient on the well test and production type curves. The slippage coefficient mainly affects the matrix apparent permeability; a bigger slippage coefficient corresponds to higher matrix apparent permeability and thus a higher capacity of interporosity flow from matrix to the micro fracture system. This exhibits as an earlier start of the interporosity concave part on the well test type curves. Combining with the boundary effect, the bigger the slippage coefficient is, the earlier the interporosity concave part shows up and the lower position it has. For a well producing at constant pressure, a bigger slippage coefficient results in a higher mid-time production rate, and the production for different slippage coefficients trend to become identical after the pseudo steady state flow starts.

Similar to the slippage coefficient, the effect of a Knudsen diffusion coefficient on the well test type curves is shown in Fig. 6.18. The bigger the Knudsen diffusion coefficient is, the higher the matrix apparent permeability is and the higher the capacity of interporosity flow from matrix pores to micro fractures. This reflects an earlier start of the interporosity concave part on the type curves. Combining with the boundary effect, not only does the shape of the concave part change with the diffusion coefficient, but also the position of the concave part becomes lower on the curves with an increase in the diffusion coefficient.

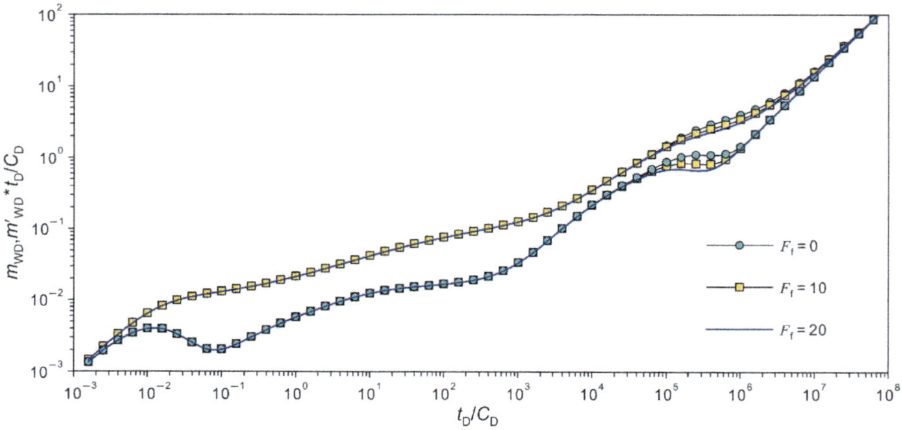

Fig. 6.16 Effect of slippage coefficient on well test type curves.

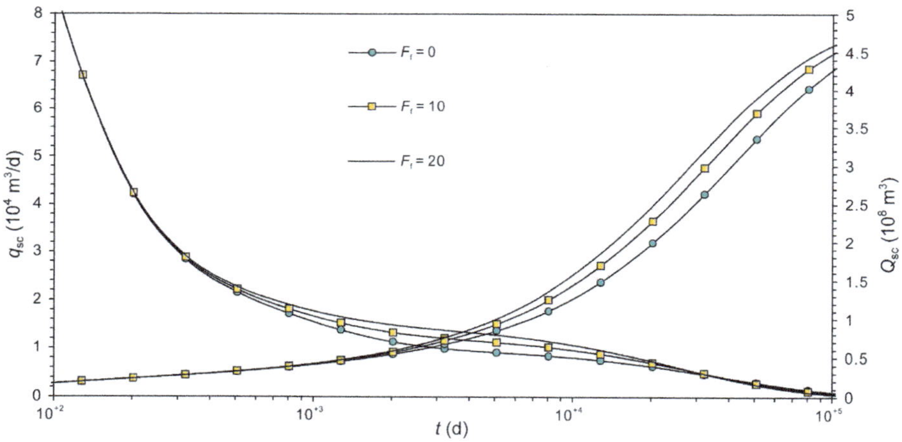

Fig. 6.17 Effect of slippage coefficient on well production curves.

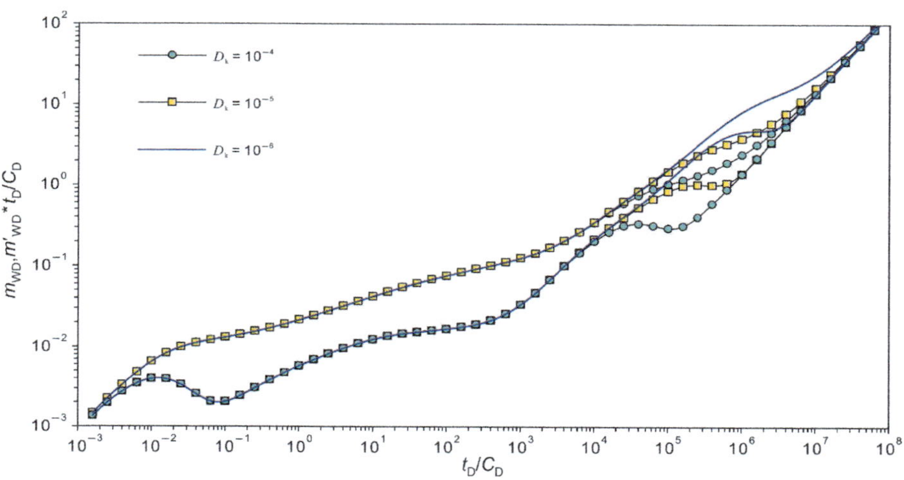

Fig. 6.18 Effect of Knudsen diffusion coefficient on well test type curves.

Fractured wells in rectangular gas reservoirs with SRV

7

Chapter Outline

7.1 Introduction 191
7.2 Boundary element models in composite gas reservoirs with SRV 194
7.3 Fractured vertical wells in rectangular composite shale gas reservoirs with SRV 197
7.3.1 Physical model 198
7.3.2 Pressure and production type curve analysis 198
7.4 MFHW in rectangular composite shale gas reservoirs with global SRV 210
7.4.1 Physic model 210
7.4.2 Pressure and production type curve analysis 212
7.5 MFHW in shale gas reservoirs with local SRV 222
7.5.1 Physical model 223
7.5.2 Pressure and production curve analysis 224

7.1 Introduction

In the previous two chapters, the transient flow models for fractured vertical wells and multi-stage fractured horizontal wells (MFHWs) in composite shale gas reservoirs with consideration of a SRV (Stimulated reservoir volume) have been analyzed. However, these theoretical models may not be practical, especially for MFHWs, because a hydraulically induced fracture network usually cannot extend to a horizontal well half-length in reality. Therefore, an application of such composite models could result in relatively significant errors. Although the models are comparatively acceptable for fractured vertical wells, certain errors still exist. To better represent induced fracture networks, fractures, wellbore and reservoirs, most researchers propose rectangular models to investigate a SRV and a reservoir boundary (A physical model is shown in Fig. 7.1).

Since a source function and the Green formula were introduced in the research of unsteady state flow in oil and gas reservoirs by Gringarten et al. (1973), their method has been widely used in the petroleum industry and is still one of the important methods for an oil and gas transient flow analysis and a well test analysis. However, reservoir boundary conditions are required for this method to acquire source function solutions in different reservoir dimensions, and it is difficult to

Developments in Petroleum Science, Vol. 66. https://doi.org/10.1016/B978-0-444-64315-5.00007-3

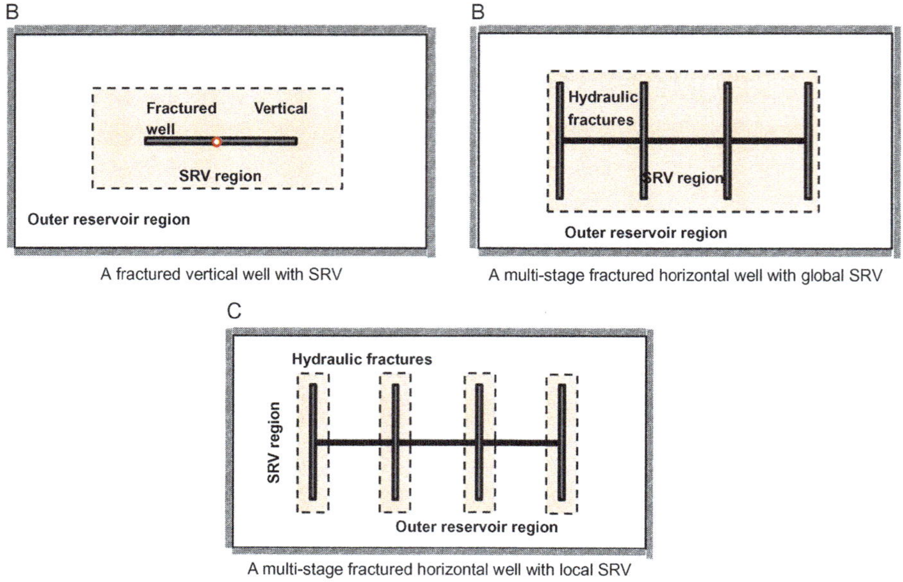

Fig. 7.1 Fractured vertical and horizontal wells with SRV.

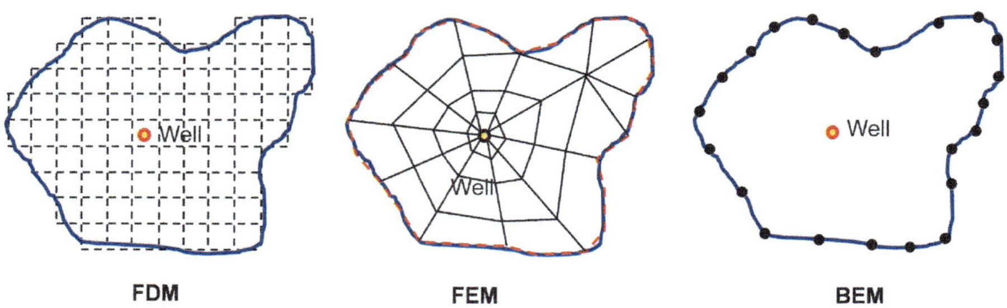

Fig. 7.2 Schemes of discretization for various numerical methods.

derive continuous point source solutions for reservoirs with complex boundary conditions. Even if a solution can be obtained, it can have a complex expression; for wells with a complex wellbore structure, integration of a continuous point source solution along a sink surface is usually required. These issues lead to a difficult and time-consuming solution process due to massive series sums and trigonometric functions being involved.

A solution of the physical model in Fig. 7.2 is very complex. It is almost impossible to solve it though analytical or semi-analytical methods. Therefore, some researchers brought up various linear flow models for an analysis of fractured wells in shale gas reservoirs with SRV. It is known that linear flow models not only neglect fracture interference and simplify flow regimes but also impose some constrains on the dimension of fractures, SRVs, and reservoirs. Thus, an application of linear flow models is limited due to their big differences from real geological models. For fractured wells in rectangular reservoirs with consideration of a SRV, the previous method to acquire bottom-hole pressure through integration of a continuous point source or line source solution along a fracture surface is not applicable any more. Therefore, based on the derivation of a continuous point source solution in an infinite shale gas reservoir, the boundary element method (BEM) is applied to solve the two models mentioned above. The BEM is used mostly for vertical well flow models by researchers; in this chapter, it is applied to solve models for fractured vertical and horizontal wells in composite gas reservoirs based on its previous applications.

Although analytical and semi-analytical methods can solve many well and reservoir models to analyze a transient pressure response of MFHWs in various reservoirs, most of the existing papers have concentrated on wells located in reservoirs with a regular shape boundary. For a horizontal well with multiple fractures, a drainage area not only depends on a well location and a well pattern type but also has an intimate connection with the fracture length and distribution, which can result in an irregular drainage boundary. For the pressure and production performance of a well in such a reservoir, most analytical and semi-analytical methods are helpless. A few of them can treat these problems, but their solution process is very complicated.

Due to the limitations existing in analytical and semi-analytical solution processes, numerical simulation methods, such as the finite difference method (FDM), the finite element method (FEM), and the boundary element method (BEM), are introduced into solutions of petroleum problems. Although the FDM and FEM can be used to analyze the pressure of a MFHW in an arbitrarily shaped gas reservoir, both of them discretize a full reservoir into a grid of small blocks (a "domain type" method). On the contrary, the BEM solution based on Green's function in a free space, called the fundamental solution, satisfies a governing partial differential equation at a point source without satisfying any of the prescribed boundary conditions. Meanwhile, because the total grid number needed for the BEM is much smaller than that in the FDM and FEM, the BEM can efficiently and accurately solve various boundary value problems. Compared to the FDM and FEM, the BEM has a great advantage in handling problems with complex boundary geometries since there is no interior of the domain under consideration. The BEM is also superior to Green's function method in that it features the flexibility of being applicable to any reservoir boundary condition and to any arbitrary shape of a reservoir (Chen et al., 2003).

In the review of the petroleum literature, many papers have been concentrated on vertical wells or fractured vertical wells in 2D domains (Gao et al., 2007; Jongkittinarukorn and Tiab, 1998; Kikani and Horne, 1989; Kryuchkov and Sanger, 2004; Pecher and Stanislav, 1997; Sato and Horne, 1993; Wang and Zhang, 2009; Yin et al., 2005). With the BEM, arbitrarily shaped reservoir boundaries can be easily handled. For a complex well, such as a MFHW, the related reports are rare. Therefore, in this chapter, the transient pressure behavior of MFHWs in arbitrarily shaped gas reservoirs is analyzed by the BEM.

7.2 Boundary element models in composite gas reservoirs with SRV

According to the previously derived unsteady-state flow models considering different flow mechanisms, the flow equation in an inner fracture region with the term of a source can be expressed in a rectangular coordinate system as:

$$\frac{\partial \Delta \overline{m}_{f1}}{\partial x_D^2}+\frac{\partial \Delta \overline{m}_{f1}}{\partial y_D^2}+\frac{\partial \Delta \overline{m}_{f1}}{\partial z_D^2}+\frac{2Tp_{sc}}{k_{f1}T_{sc}h}\frac{q_{v1}}{s}\delta\left(x_D-x'_{s1D},y_D-y'_{s1D},z_D-z'_{s1D}\right)=f_1(s)\Delta \overline{m}_{f1} \quad (7.1)$$

The flow equation in an outer natural fracture system is:

$$\frac{\partial \Delta \overline{m}_{f2}}{\partial x_D^2}+\frac{\partial \Delta \overline{m}_{f2}}{\partial y_D^2}+\frac{\partial \Delta \overline{m}_{f2}}{\partial z_D^2}+\frac{2Tp_{sc}}{k_{f2}T_{sc}h}\frac{q_{v2}}{s}\delta\left(x_D-x'_{s1D},y_D-y'_{s1D},z_D-z'_{s1D}\right)=f_2(s)\Delta \overline{m}_{f2} \quad (7.2)$$

The joint interface conditions of the inner and outer regions are:

$$\Delta \overline{m}_{f1}=\Delta \overline{m}_{f2},(x_D,y_D,z_D)\in S_{interface} \quad (7.3)$$

$$\frac{\partial \Delta \overline{m}_{f1}}{\partial n_D}=-\frac{1}{M_{12}}\frac{\partial \Delta \overline{m}_{f2}}{\partial n_D},(x_D,y_D,z_D)\in S_{interface} \quad (7.4)$$

The outer boundary conditions are:

$$\frac{\partial \Delta \overline{m}_{f2}}{\partial n}=0,(x_D,y_D,z_D)\in S_{interface} \quad \text{(Closed)} \quad (7.5)$$

$$\Delta \overline{m}_{f2}=0,(x_D,y_D,z_D)\in S_{interface} \quad \text{(Constant pressure)} \quad (7.6)$$

$$\beta_1\Delta \overline{m}_{f2}+\beta_2\frac{\partial \Delta \overline{m}_{f2}}{\partial n}=0,(x_D,y_D,z_D)\in S_{interface} \quad \text{(mixed boundary)} \quad (7.7)$$

By using the same dimensionless variable definitions, Eqs. (7.1)–(7.7) can be reformed as:

$$\frac{\partial m_{f1D}}{\partial x_D^2}+\frac{\partial m_{f1D}}{\partial y_D^2}+\frac{\partial m_{f1D}}{\partial z_D^2}=f_1(s)m_{f1D}-\frac{2\pi q_{v1D}}{s\,M_{12}}\delta\left(x_D-x'_{s1D},y_D-y'_{s1D},z_D-z'_{s1D}\right) \quad (7.8)$$

$$\frac{\partial m_{f2D}}{\partial x_D^2}+\frac{\partial m_{f2D}}{\partial y_D^2}+\frac{\partial m_{f2D}}{\partial z_D^2}=f_2(s)m_{f2D}-\frac{2\pi q_{v2D}}{s}\delta\left(x_D-x'_{s1D},y_D-y'_{s1D},z_D-z'_{s1D}\right) \quad (7.9)$$

$$m_{f1D}=m_{f2D},(x_D,y_D,z_D)\in S_{interface} \quad (7.10)$$

$$\frac{\partial m_{f1D}}{\partial n_D}=-\frac{1}{M_{12}}\frac{\partial m_{f2D}}{\partial n_D},(x_D,y_D,z_D)\in S_{interface} \quad (7.11)$$

The above equations are the mathematical model in the Laplace space for composite shale gas reservoirs with any boundary conditions.

To solve the above model by the BEM, the fundamental solutions of its differential equations are used. Assume that the fundamental solutions of the outer and inner reservoir regions are $E_1(R_D, R_D', s)$ and $E_2(R_D, R_D', s)$, respectively, and the following equations are satisfied:

$$\nabla^2 E_1(R_D, R_D', s) - f_1(s) E_1(R_D, R_D', s) + 2\pi\delta(R_D, R_D') = 0 \tag{7.12}$$

$$\nabla^2 E_2(R_D, R_D', s) - f_2(s) E_1(R_D, R_D', s) + 2\pi\delta(R_D, R_D') = 0 \tag{7.13}$$

The hydraulic fractures investigated in this paper are fully penetrated, and thus the fundamental solutions E_1 and E_2 can be expressed as:

$$E_1 = \mathrm{K}_0\left[\sqrt{f_1(s)}(R_D - R_D')\right] \tag{7.14}$$

$$E_2 = \mathrm{K}_0\left[\sqrt{f_2(s)}(R_D - R_D')\right] \tag{7.15}$$

Multiply m_{f1D} and m_{f2D} to both sides of Eqs. (7.12) and (7.13), respectively, to obtain:

$$m_{f1D}\nabla^2 E_1 - f_1(s) m_{f1D} E_1 + 2\pi m_{f1D}\delta(R_D, R_D') = 0 \tag{7.16}$$

$$m_{f2D}\nabla^2 E_2 - f_2(s) m_{f2D} E_1 + 2\pi m_{f2D}\delta(R_D, R_D') = 0 \tag{7.17}$$

Multiply E_1 and E_2 to both sides of Eqs. (7.8) and (7.9), respectively, to result in:

$$E_1\nabla^2 m_{f1D} = f_1(s) E_1 m_{f1D} - E_1 \frac{2\pi q_{v1D}}{s\, M_{12}}\delta(R_D, R_{wD}') \tag{7.18}$$

$$E_2\nabla^2 m_{f2D} = f_2(s) E_2 m_{f2D} - \frac{2\pi q_{v2D}}{s} E_2\delta(R_D, R_{wD}') \tag{7.19}$$

Subtract Eq. (7.16) from Eq. (7.18) and Eq. (7.19) from Eq. (7.17), respectively, and then manipulate to have:

$$m_{f1D}\nabla^2 E_1 - E_1\nabla^2 m_{f1D} + 2\pi m_{f1D}\delta(R_D, R_D') = \frac{2\pi q_{v1D}}{s\, M_{12}} E_1\delta(R_D, R_{s1D}') \tag{7.20}$$

$$m_{f2D}\nabla^2 E_2 - E_2\nabla^2 m_{f2D} + 2\pi m_{f2D}\delta(R_D, R_D') = \frac{2\pi q_{v2D}}{s} E_2\delta(R_D, R_{s2D}') \tag{7.21}$$

For Eqs. (7.20) and (7.21), integrate on R_D' in the inner region Ω_1 and the outer region Ω_2, respectively:

$$\int_{\Omega_1}\left[m_{f1D}\nabla^2 E_1 - E_1\nabla^2 m_{f1D} + 2\pi m_{f1D}\delta(R_D, R_D') - \frac{2\pi q_{v1D}}{s\, M_{12}} E_1\delta(R_D, R_{s1D}')\right] d\Omega_1 = 0 \tag{7.22}$$

$$\int_{\Omega_2} \left[m_{\text{f2D}} \nabla^2 E_2 - E_2 \nabla^2 m_{\text{f2D}} + 2\pi m_{\text{f2D}} \delta(R_{\text{D}}, R'_{\text{D}}) - \frac{2\pi q_{v2\text{D}}}{s} E_2 \delta(R_{\text{D}}, R'_{\text{s2D}}) \right] \text{d}\Omega_2 = 0 \quad (7.23)$$

Note the following Green's second formula:

$$\int_{\Omega} \left(u\nabla^2 v - v\nabla^2 u \right) \text{d}\Omega' = \int_{s} (u\nabla v - v\nabla u) \text{d}S' \quad (7.24)$$

where $S\left(S = \sum_i S_i\right)$ is all the boundaries in the Ω area.

Using Eq. (7.24) and the Delta function properties, the differential equations in the inner and outer reservoir regions can be transformed into integral equations on the boundaries, and hence the boundary integral equations of Eqs. (7.22) and (7.23) are:

$$m_{\text{f1D}}(R_{\text{D}}) = \frac{1}{2\pi} \int_{S_1} [E_1 \nabla m_{\text{f1D}} - m_{\text{f1D}} \nabla E_1] \text{d}S'_1 + \frac{1}{M_{12} s} \int_{\Omega_1} q_{v1\text{D}} \left(R'_{\text{s1D}} \right) E_1 \text{d}\Omega'_1 \quad (7.25)$$

$$m_{\text{f2D}}(R_{\text{D}}) = \frac{1}{2\pi} \int_{S_2} [E_2 \nabla m_{\text{f2D}} - m_{\text{f2D}} \nabla E_2] \text{d}S'_2 + \frac{1}{s} \int_{\Omega_2} q_{v2\text{D}} \left(R'_{\text{s2D}} \right) E_2 \text{d}\Omega'_2 \quad (7.26)$$

For a composite gas reservoir with two boundaries, the outer boundary is discretized into NO elements and the inner boundary into NI elements, and then the total number of elements is NO + NI. In addition, the numbering of these elements should follow the pattern: clockwise for the inner boundary elements and counter clockwise for the outer boundary elements (the boundary discretization and element numbering are shown in Fig. 7.3).

The boundary integral equation R_{D} represents any point on the boundaries; R'_{D} involves any point within the reservoir region, and $R_{1\text{SD}}$ and $R_{2\text{SD}}$ are the

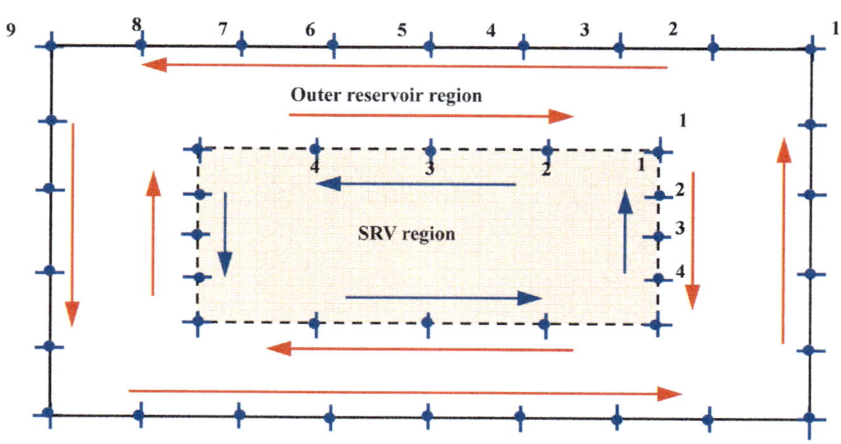

Fig. 7.3 Inner and outer boundary discretization and numbering of rectangular composite gas reservoir.

dimensionless coordinates of the source points in the inner and outer regions, respectively. If $R_D^{'}$ is expected to represent any point not only within the reservoir region but also on the boundaries, Eqs. (7.25) and (7.26) need to be transformed as:

$$\theta m_{f1D}(R_D) = \frac{1}{2\pi}\int_{S_1}[E_1\nabla m_{f1D} - m_{f1D}\nabla E_1]dS_1' + \frac{1}{M_{12}s}\int_{\Omega_1} q_{v1D}\left(R_{s1D}'\right)E_1 d\Omega_1' \quad (7.27)$$

$$\theta m_{f2D}(R_D) = \frac{1}{2\pi}\int_{S_2}[E_2\nabla m_{f2D} - m_{f2D}\nabla E_2]dS_2' + \frac{1}{s}\int_{\Omega_2} q_{v2D}\left(R_{s2D}'\right)E_2 d\Omega_2' \quad (7.28)$$

where θ is a constant depending on the geometric shape of $R_D^{'}$ point and its expression is:

$$\theta = \frac{\alpha}{2\pi} \quad (7.29)$$

where α is the intersection angle of the boundary tangent line at $R_D^{'}$ point.

$$\theta = \begin{cases} 1/2 & \text{for smooth boundary}\ \alpha = \pi \\ \alpha/2\pi & \text{for nonsmooth boundary} \\ 1 & \text{for point inside the domain}\ \alpha = 2\pi \end{cases} \quad (7.30)$$

Through taking the fundamental solutions into Eqs. (7.27) and (7.28) and solving for $R_D^{'}$ along the boundary points and fracture points one by one, a series of linear equation systems can be generated and combined to obtain the dimensionless pseudo-pressure type curves in fractured wells.

To discretize the reservoir outer boundary Γ into N_b elements, the common methods are the constant element method, the linear element method, and the curved element method. In the constant element method, each element central point is taken as a node, and a function value is constant in the same element; in the linear element method, each element end point is taken as a node, and any value within an element is determined by a linear interpolation. The curved element method is to better represent irregular geometry. Here, the linear element method is used for boundary discretization. A detailed derivation is not stated, and the reader who wants to study it can refer to the following papers: Gao et al. (2007), Jongkittinarukorn and Tiab (1998), Kikani and Horne (1989), Kryuchkov and Sanger (2004), Pecher and Stanislav (1997), Sato and Horne (1993), Wang and Zhang (2009), Yin et al. (2005).

7.3 Fractured vertical wells in rectangular composite shale gas reservoirs with SRV

According to the previous introduction, during a reservoir evaluation phase, a lot of vertical wells are drilled to save operation costs, and hydraulic fracturing is performed to acquire economic gas production from these vertical wells. In practice, main fractures are usually not created during these fracturing operations; instead, massive secondary fractures around natural fractures are induced. A region containing such secondary fractures is called a SRV or a fracture network region. In the following sections, well-test type curves and production curves are introduced and analyzed for a fractured vertical well in a rectangular gas reservoir with a SRV under different flow mechanism models.

7.3.1 Physical model

Fig. 7.4 shows a physical model of a fractured vertical well in a rectangular shale gas reservoir with a SRV. It is assumed that the fractured vertical well is located in the center of the closed rectangular reservoir; a hydraulic fracture is a symmetric or asymmetric bi-wing fracture with infinite conductivity. The fracture half-lengths are x_{fL} and x_{fR}, respectively. The reservoir length and width are x_{e} and y_{e}. The fracture height and reservoir thickness are identical to h; the SRV region is around the fracture with length x_{m} and width y_{m} and is represented by a dual-media model.

From the previous analysis, the mechanisms in mechanism models 2 and 5 are almost the same. The only difference is that both the gas slippage effect and Knudsen diffusion caused by a concentration difference are considered for flow in nanopores in mechanism model 5. Therefore, in this chapter, the unsteady-state flow models 1, 3, and 5 are used for analyses of fractured vertical and horizontal wells in gas reservoirs with SRV.

7.3.2 Pressure and production type curve analysis

7.3.2.1 Microfractures + steady state adsorption/desorption and diffusion

For mechanism model 1, the BEM can be used to acquire bottom-hole pseudo-pressure response curves and production and cumulative production curves at constant pressure for a fractured vertical well in a rectangular composite gas reservoir based on the parameters in Table 7.1.

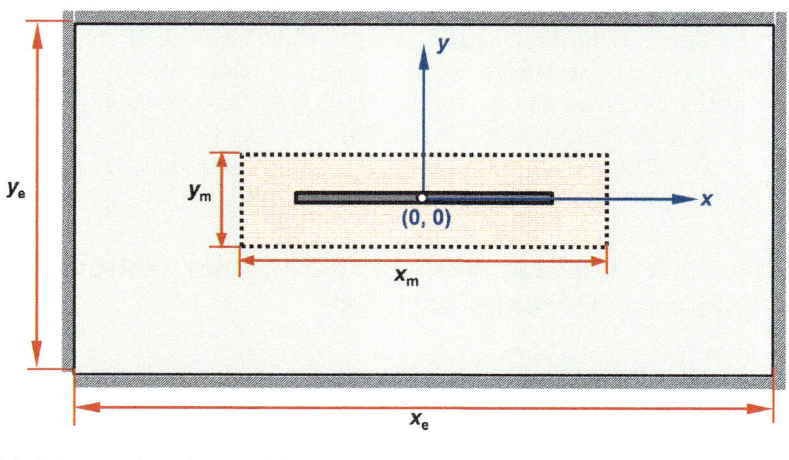

Fig. 7.4 A fractured vertical well in composite rectangular gas reservoir with SRV.

Table 7.1 Shale reservoir properties

Reservoir properties	Value	Reservoir properties	Value
Original gas reservoir pressure, p_i (MPa)	25	Reservoir temperature, T (K)	320
Reservoir thickness, h (m)	60	Specific gas gravity, γ_g (fraction)	0.65
Original reservoir gas compressibility, C_{gi} (MPa^{-1})	0.02	Bottom hole pressure, p_{wf} (MPa)	1
Langmuir pressure, P_L (MPa)	4	Langmuir volume, G_L (m^3/m^3)	10
Reservoir outer boundary length, x_e (m)	1500	Reservoir outer boundary width, y_e (m)	500
Permeability of reservoir microfractures, k_{f1} (mD)	0.01	Porosity of reservoir microfractures, $\varnothing_{f1}$ (fraction)	0.02
Length of SRV region, x_m (m)	300	Width of SRV region, y_m (m)	100
Permeability of SRV, k_{f2} (mD)	0.1	Porosity of SRV, $\varnothing_{f2}$ (fraction)	0.1
Fracture half length, x_f (m)	100	Dimensionless wellbore storage coefficient, C_D	10^{-3}
Skin factor, S_{kin} (dimensionless)	0.01		

Fig. 7.5 shows the calculated well-test type curves. To theoretically identify all flow stages from these curves, it is assumed that the reservoir outer boundary length and width are both 10,000 m, and the SRV length and width are both 1000 m for the dotted curves. Therefore, the following stages can be identified:

Flow stage 1: An early time wellbore storage and skin effect. In this stage, the pseudo pressure and its derivative curves have the same characteristics as a conventional gas reservoir and the previously introduced models.

Flow stage 2: Early time linear flow from SRV to main fractures. In this stage, a derivative curve is a straight line with a slop of 0.5 on a log–log plot.

Flow stage 3: An elliptical flow stage. After ending of the early time linear flow, this stage is visible if pressure has not propagated outside the SRV region. Its derivative curve has a slope of 0.36 in this stage.

Flow stage 4: Early time radial flow in the SRV region. Since the given SRV size is big enough, this flow stage can be observed. In this section, the reservoir permeability is the reference for defining dimensionless variables; therefore, the pseudo-pressure derivative curve in this stage has a value of $1/(2M_{12})$.

Flow stage 5: Transition flow from the early time radial flow to the late time formation radial flow.

Flow stage 6: Late time formation radial flow. When pressure reaches the outer reservoir region and the gas flow rate from the outside to the SRV equals the well production rate, this stage happens. In this stage, the pseudo-pressure derivative curve is a horizontal straight line with a value of 0.5.

Flow stage 7: A boundary response. After pressure reaches the boundary, a boundary response or boundary dominated flow happens. For the equivalent reservoir length and width given in this case, the pressure reaches the four boundaries at the same time, and its derivative curve has a slope of 1.

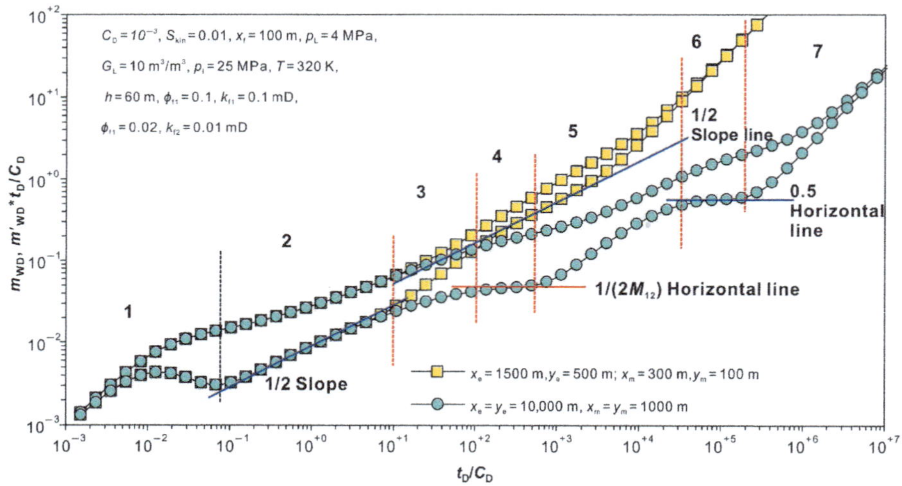

Fig. 7.5 Well test type curves of a fractured vertical well in a rectangular composite gas reservoir with SRV.

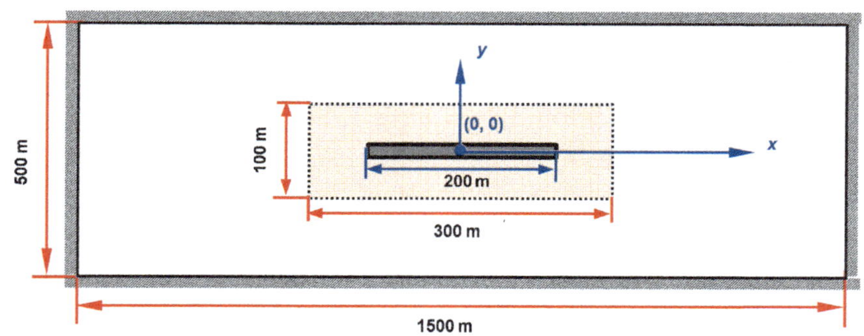

Fig. 7.6 Physical model based on parameters in Table 7.1.

These type curves are based on theoretical reservoir and SRV sizes, which are bigger than realistic cases. If the parameters in Table 7.1 are used, the generated curves are shown as lines with squares in Fig. 7.5. It can be seen that the radial flow stage in the SRV region is hard to observe due to the small SRV size. Moreover, pressure also quickly reaches the horizontal reservoir boundaries due to a small reservoir width, resulting in missing of the formation radial flow. However, since the reservoir length is bigger than its width, linear flow along the horizontal direction can be observed, which exhibits as a straight line section with a slope of 1/2 on the pseudo pressure and its derivative curves (shown as a blue dotted line in the plot). In the later stage, after pressure propagates to the further boundaries, pseudo radial flow happens.

For a further analysis of well test type curves and production curves based on the parameters in Table 7.1, the physical model is illustrated in Fig. 7.6.

Figs. 7.7 and 7.8 show the effects of SRV permeability on well test type and production curves, respectively. According to these plots, the higher the SRV permeability, the lower the pseudo pressure and its derivative curves on the plots. Theoretically, the SRV permeability has influence on the early time SRV linear flow and radial flow, but the radial flow is unobservable due to a small size of the SRV region. When the well is producing at constant pressure, the SRV permeability has a big effect on the early time production rate; after pressure reaches the outer reservoir region, the gas supply to the inner region is limited due to the ultra-low reservoir permeability, and thus the SRV permeability has little effect on the late time production rate.

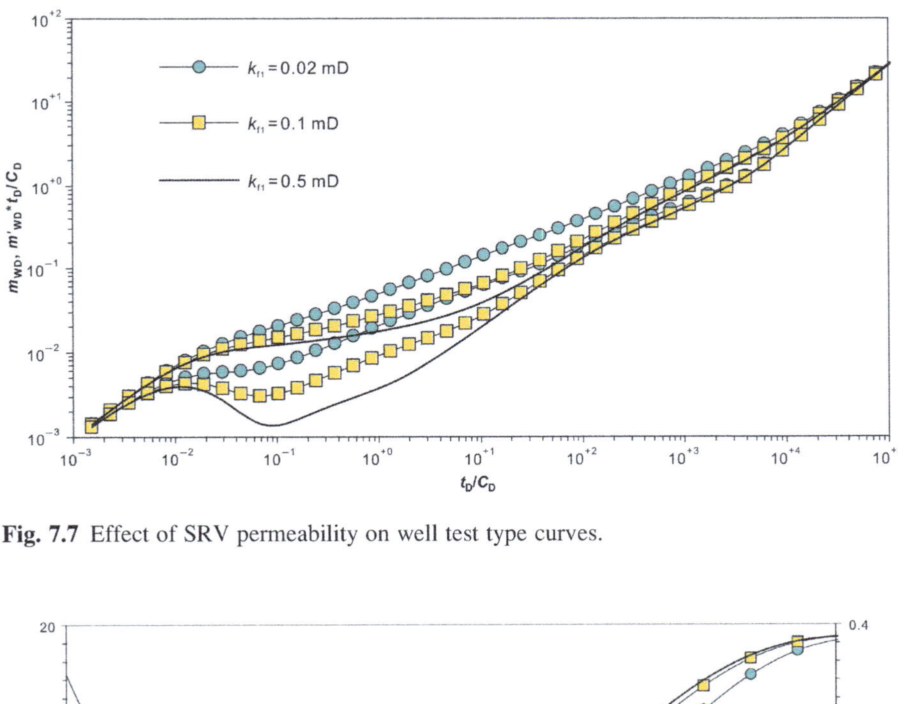

Fig. 7.7 Effect of SRV permeability on well test type curves.

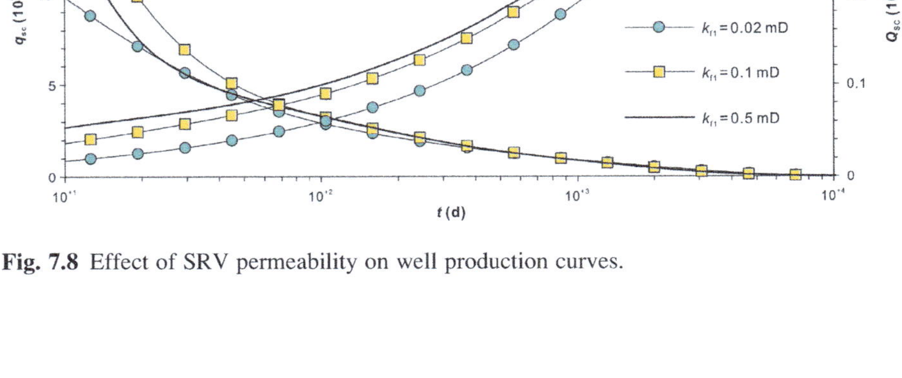

Fig. 7.8 Effect of SRV permeability on well production curves.

Figs. 7.9 and 7.10 exhibit the effect of an adsorbed gas volume on the well test type and production curves. It can be seen from these plots that the effect of an adsorbed gas volume on well test type curves is mainly at the middle and late time stages. The bigger the adsorbed gas volume, the higher the pseudo-pressure curves on the plots, and the earlier pressure reaches the outer boundary. This is because adsorbed gas supplies the microfracture system to delay the pressure propagation. Additionally, an adsorbed gas volume is positively proportional to well production, and this effect lasts a quite long time.

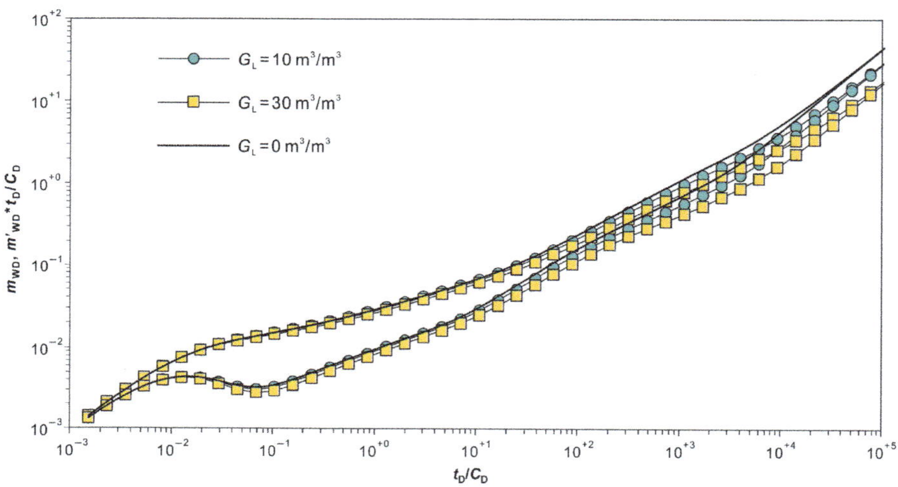

Fig. 7.9 Effect of adsorbed gas volume on well test type curves.

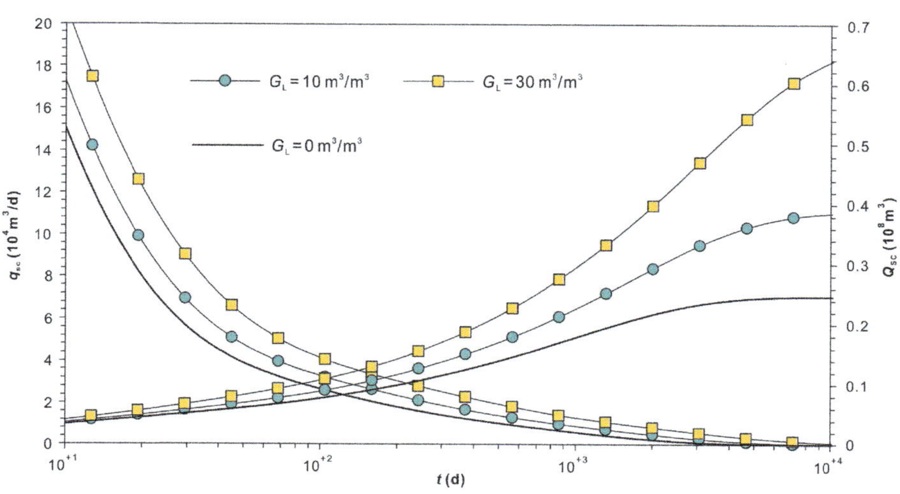

Fig. 7.10 Effect of adsorbed gas volume on well production curves.

The effect of a SRV size on well test and production curves is presented in Figs. 7.11 and 7.12, respectively. According to these plots, a SRV size only affects the middle and late time stages of the well test curves. If the length of the SRV region is big enough, a certain period of linear flow in the SRV region will present after ending of the SRV early time linear flow. During this period, the longer the SRV region, the lower the pseudo pressure and its derivative curves on the plots. An increase in the SRV region size only increases a free gas volume within the inner region. Thus, the SRV region size is positively proportional to a well early time production rate and has little effect on the well rate after pressure reaches the outer reservoir region.

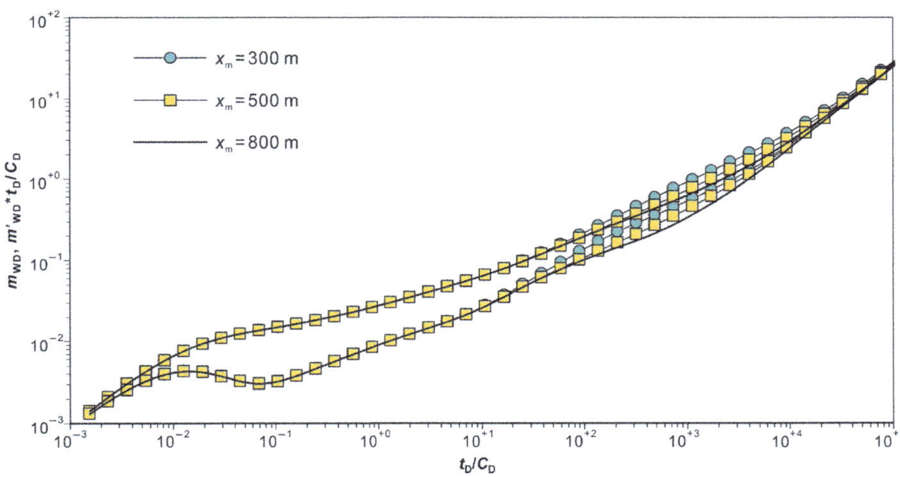

Fig. 7.11 Effect of SRV region length on well test type curves.

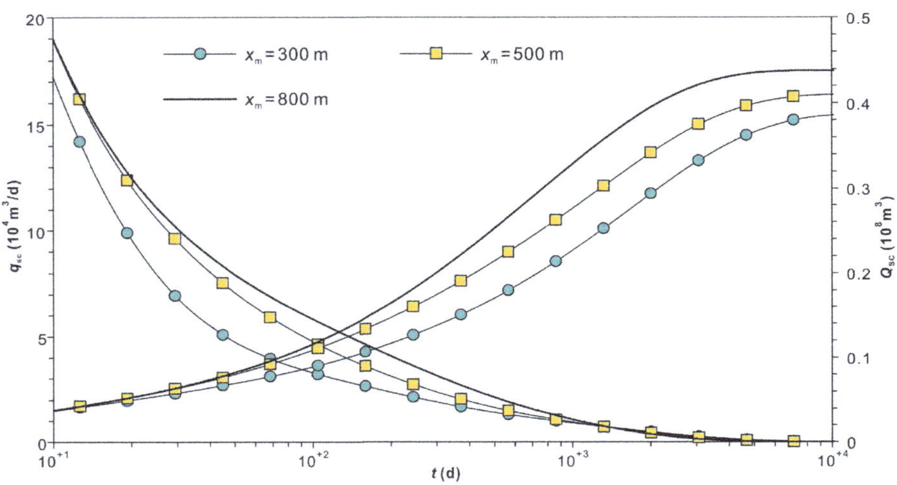

Fig. 7.12 Effect of SRV region length on well production curves.

7.3.2.2 *Microfractures + gas adsorption/desorption + matrix Fick's diffusion*

Similarly, for mechanism model 3, the BEM is used to acquire bottom-hole pseudo pressure response curves and production and cumulative production curves at constant pressure for a fractured vertical well in a rectangular composite gas reservoir with SRV based on the parameters in Table 7.2.

Fig. 7.13 shows the well test curves of a fractured vertical well centered in a closed rectangular gas reservoir with consideration of the SRV effect. It can be seen that if the sizes of the reservoir and SRV region are assumed big enough, the flow stages introduced earlier for model 1 can be observed. Theoretically, a concave portion for gas diffusion should be observed on the pseudo-pressure derivative curves, which is conceived due to the boundary effect and reservoir parameters. For a realistic model, due to the small sizes of the reservoir and SRV region, pressure quickly reaches the closer

Table 7.2 Shale reservoir properties

Reservoir properties	Value	Reservoir properties	Value
Permeability of microfracture system, k_{f1} (mD)	0.01	Porosity of microfracture system, $\varnothing_{f1}$ (fraction)	0.02
Length of SRV region, x_m (m)	300	Width of SRV region, y_m (m)	100
Permeability of SRV region, k_{f2} (mD)	0.1	Porosity of SRV region, $\varnothing_{f2}$ (fraction)	0.1
$(R_m)^2/D_F$ ratio	2×10^6	Production rate under constant pressure, q_{sc} (m^3/d)	2×10^4

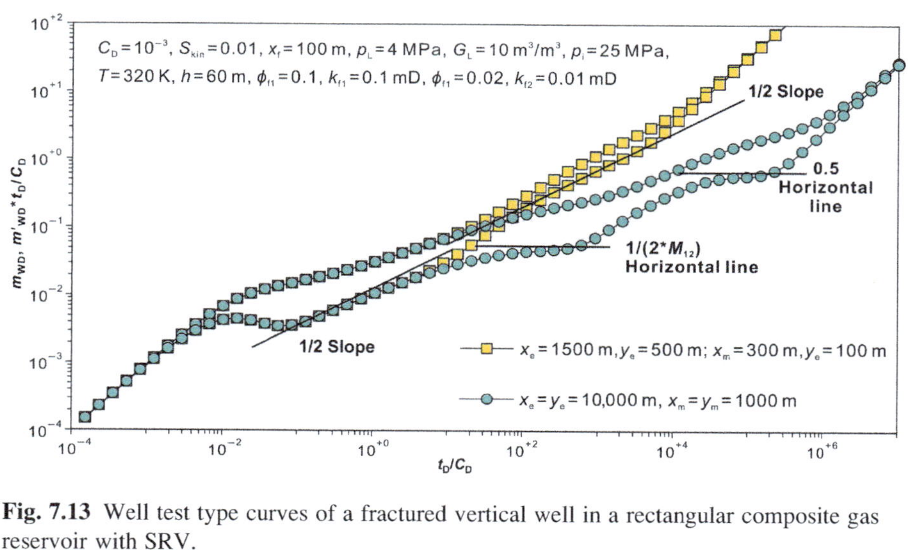

Fig. 7.13 Well test type curves of a fractured vertical well in a rectangular composite gas reservoir with SRV.

boundaries after propagating outside the SRV region, leading to a relatively long linear flow period observable on the pseudo pressure and its derivative curves. During the late period of the linear flow, slight downward moving of the curves can be observed before the pseudo radial flow, which is due to the desorption and interporosity flow of adsorbed gas from the matrix surface.

An effect of the SRV permeability on well test and production curves is shown in Figs. 7.14 and 7.15. Permeability of the SRV region mainly affects the middle and late time response of the well test curves, especially in the period of pressure propagation in the SRV region. The higher the SRV permeability, the lower the pseudo pressure

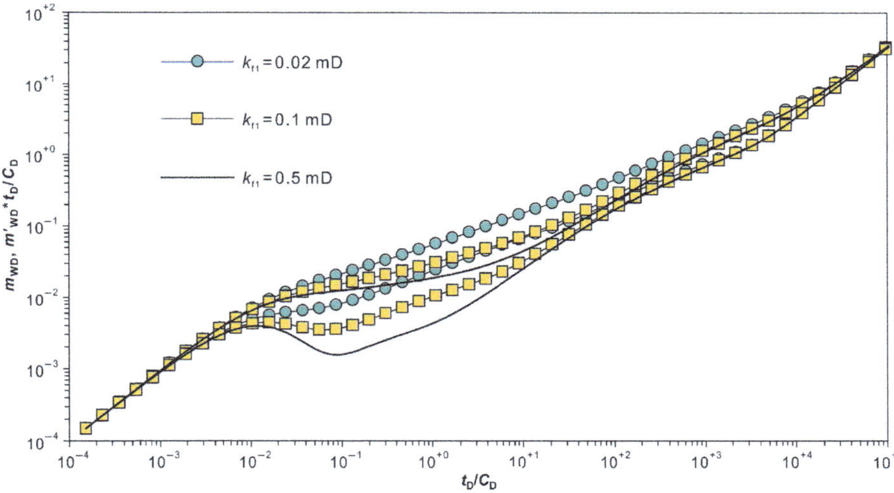

Fig. 7.14 Effect of SRV region permeability on well test type curves.

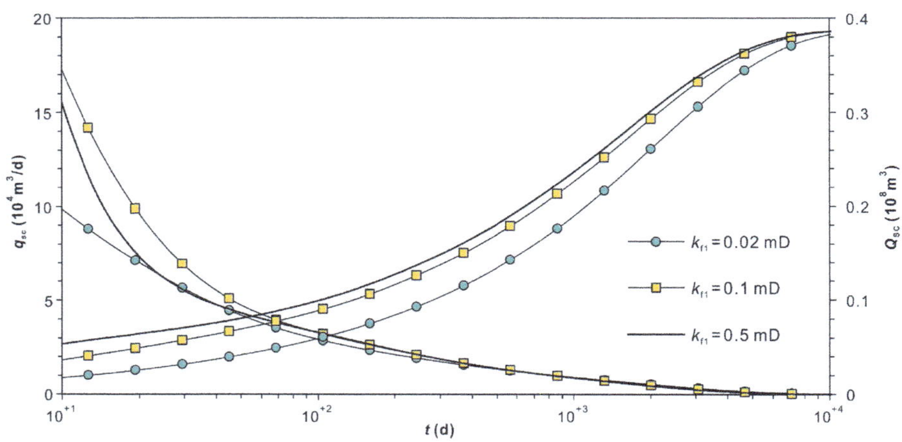

Fig. 7.15 Effect of SRV region permeability on well production curves.

and its derivative on these plots indicating lower production drawdown. After the pressure response gets into the linear flow period, the higher the SRV permeability. The lower the pseudo-pressure curve on the plot, the less change is in the derivative curve. When the producer is producing at constant pressure, the higher the SRV permeability; the higher well early production rate leads to higher cumulative production within the same time period. This also causes higher pressure depletion, and, thereafter, a lower rate after a certain period than that with lower SRV permeability. For cumulative production, the higher the SRV permeability, the higher the cumulative production.

7.3.2.3 Microfractures + gas adsorption/desorption + nanopore Knudsen diffusion

Mechanism model 5 is used to represent gas flow in a reservoir under multiple mechanisms of Darcy's flow, a slippage effect, Knudsen diffusion, and adsorbed gas desorption. Again, the well test type and production curves can be generated for a fractured vertical well in a rectangular shale gas reservoir by the BEM based on the parameters given in Table 7.3.

According to the well test type curves shown in Fig. 7.16 for mechanism model 5, the pseudo pressure and its derivative are similar to those in the previous two models if the reservoir size and SRV region size are big enough. Theoretically, the concave portion for interporosity flow from matrix micropores to microfractures can be observed. But due to an interporosity flow coefficient and the difference in reservoir permeability, it is conceived directly by the boundary effect. While, for realistic reservoir sizes, flow behavior in the SRV and outer reservoir regions is hard to observe from the type curves. Instead, a relatively long linear flow period presents on the pseudo pressure and its derivative curves after the early time linear flow stage because pressure quickly propagates to the closer boundaries.

An effect of the SRV region permeability on well test and production curves for mechanism model 5 is illustrated in Figs. 7.17 and 7.18. Similar to the previous

Table 7.3 Shale reservoir properties

Reservoir properties	Value	Reservoir properties	Value
Permeability of microfracture system, k_{f1} (mD)	0.01	Porosity of microfracture system, $\varnothing_{f1}$ (fraction)	0.02
Length of SRV region, x_m (m)	300	Width of SRV region, y_m (m)	100
Permeability of SRV region, k_{f2} (mD)	0.1	Porosity of SRV region, $\varnothing_{f2}$ (fraction)	0.1
Shale matrix permeability, k_m (mD)	0.0001	Shale matrix porosity, $\varnothing_m$	0.012
Shape factor of interporosity flow, α (1/m^2)	10^{-5}	Knudsen diffusion coefficient, D_k (1/m^2)	10^{-6}
Average size of matrix micro-pores, r_n (m)	2×10^{-9}	α value of Eq. (1.26) (dimensionless)	0.8

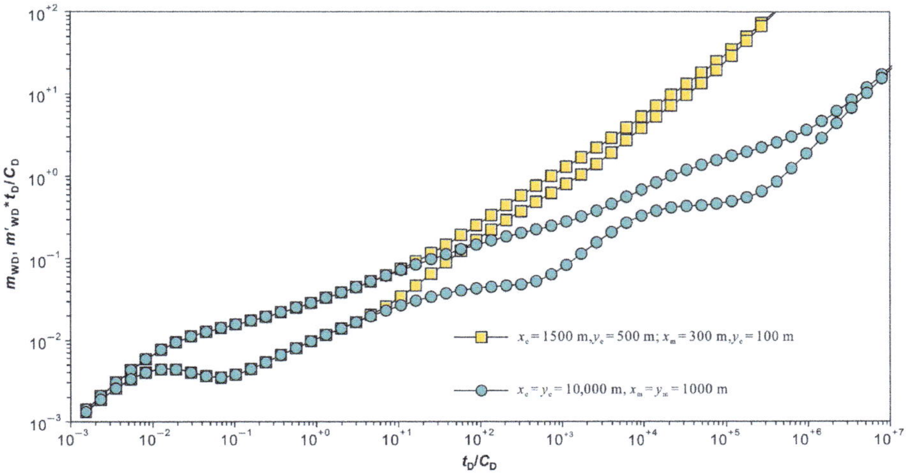

Fig. 7.16 Well test type curves for different sizes of SRV region and reservoir boundary.

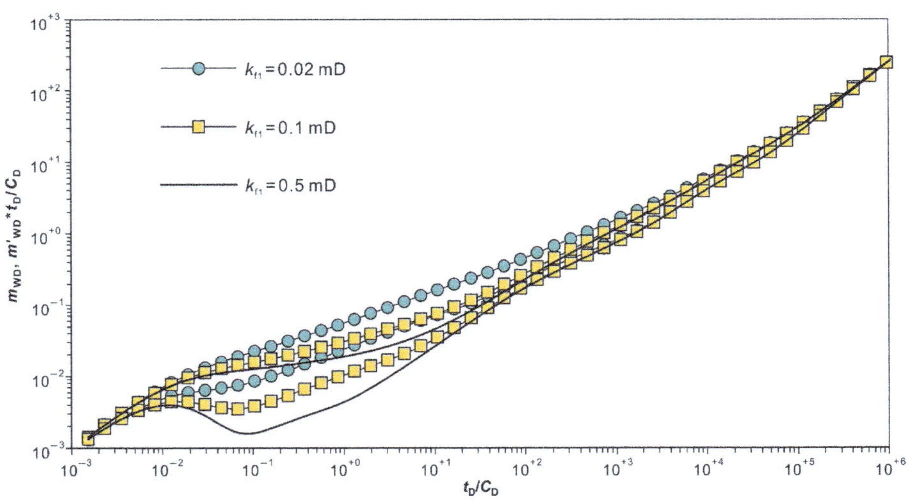

Fig. 7.17 Effect of SRV permeability on well test type curves.

models, the SRV permeability mainly affects the middle to late time flow curves. The higher the SRV permeability, the lower the curves of the early time linear flow, and the lower the pseudo-pressure curve of the late time formation linear flow on the plot. For well production performance, a higher SRV permeability results in a higher well early production rate, which becomes reverse during the middle production period due to a reservoir pressure drop. A higher SRV permeability finally results in higher cumulative production.

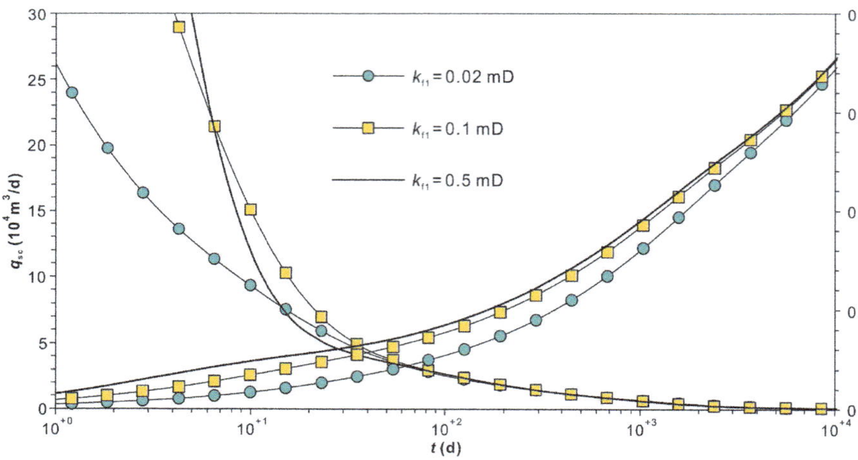

Fig. 7.18 Effect of SRV permeability on well production curves.

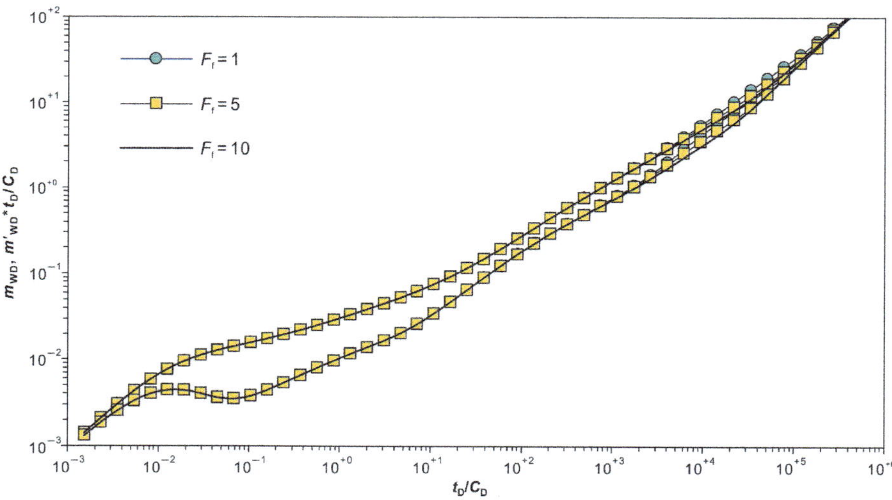

Fig. 7.19 Effect of slippage coefficient on well test type curves.

The effect of a slippage coefficient on well test type and production curves is shown in Figs. 7.19 and 7.20. In mechanism model 5, the microfracture system is considered. The permeability of this system is 2 orders of magnitude higher than that of the micropore system, and, therefore, gas supply from the micropore system to microfractures is far below that from the outer microfracture system, leading to an unobservable effect of the slippage coefficient on either the well test or production type curves. Only after the pressure propagation reaches the reservoir boundaries, the higher the slippage coefficient, the lower the pseudo pressure and its derivative on the plots and the higher the well production rate and cumulative production.

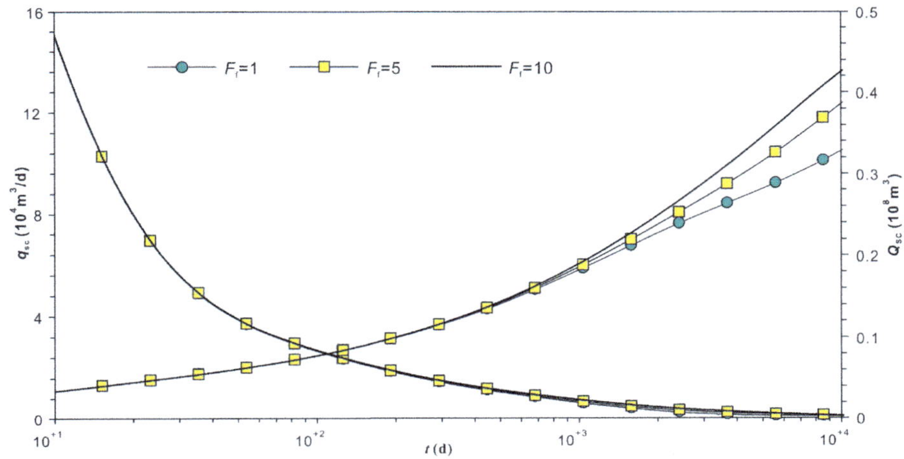

Fig. 7.20 Effect of slippage coefficient on well production curves.

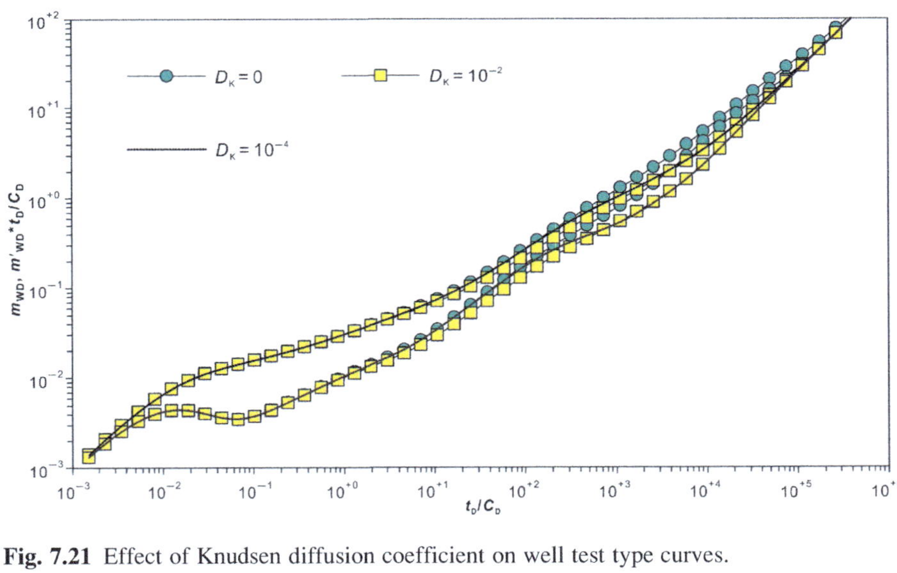

Fig. 7.21 Effect of Knudsen diffusion coefficient on well test type curves.

Figs. 7.21 and 7.22 show the effect of a diffusion coefficient on well test type and production curves, respectively. The bigger the Knudsen constant, the lower the middle and late time pseudo-pressure and its derivative curves on their plots. The downward moving of the middle and late time pseudo-pressure derivative curve is caused by the co-effect of interporosity flow from micropores to microfractures and the boundary. For a production rate, $D_K = 0$ and $D_K = 10^{-4}$ almost have the same outcome, which indicates that a relatively small Knudsen constant has little effect on

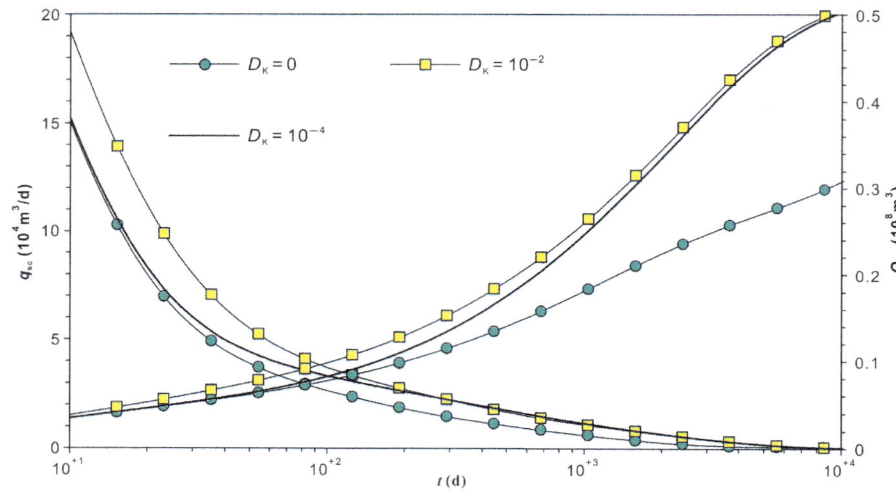

Fig. 7.22 Effect of Knudsen diffusion coefficient on well production curves.

early time production. The production rate, $D_K = 10^{-2}$, has an obvious improvement because the interporosity flow from matrix to the SRV region happens earlier due to the obviously higher apparent permeability of the micropore system.

7.4 MFHW in rectangular composite shale gas reservoirs with global SRV

In the previous sections, the radial composite model of a MFHW in a circular gas reservoir with consideration of a circular SRV has been established. A semi-analytical solution was obtained from a continuous line source solution by source function derivation. However, according to microseismic interpretation results of MFHWs in shale gas reservoirs, there are rare scenarios where the assumption of a circular composite reservoir is valid. For most MFHWs, a SRV region is far smaller than the assumed inner circular region due to a long horizontal well length. Based on previous studies, it was thought that the assumption of a rectangular region is more reasonable to represent a SRV, and a lot of linear flow models were established by assuming rectangular composite models. In this section, a rectangular composite reservoir model is used for unsteady state flow of a MFHW in a shale gas reservoir with SRV; the previously introduced BEM is applied to acquire well test and production type curves for this model under different flow mechanisms.

7.4.1 Physic model

According to the previous analysis, the brittleness of reservoir rock is one of the basic requirements to realize volumetric fracturing. Therefore, the establishment of a SRV relies on the content of brittle minerals such as quartz and carbonate; the higher content the brittle minerals have, the more easily complex fracture networks are established. Since the mineral components of reservoir rock vary a lot in different formations and in different areas, the fracturing techniques need to be screened and optimized according to rock geo-mechanics of a target reservoir. For example, the Barnett shale in North America is composed of 37.38% quartz, 19.13% carbonate and 41.13% clay minerals which are mainly illite/smectite mixtures. Hence secondary fractures are easy to be created to form networks with natural fractures through hydraulic fracturing operations in such a formation which is rich in brittle minerals. Fig. 7.23 shows a SRV distribution in the Barnett shale after a fracturing job.

In this section, a global SRV means that hydraulic fracture networks connect between the main fractures in a formation, as shown in Fig. 7.24, where a multistage MFHW is surrounded by the SRV. This model is physically similar to a trilinear flow model, but the trilinear flow model only represents linear flow in the main fractures, SRV and outer reservoir region, and does not stand for other more complex flow regimes. Meanwhile, a trilinear flow model assumes an even distribution of identical and symmetrical main fractures along the horizontal wellbore. In a word, a trilinear flow model is simple but has a lot of constraints or limitations regarding representing a MFHW in a shale gas reservoir with SRV. Therefore, to overcome these shortages, a rectangular composite gas reservoir model is established and solved by the BEM in this section.

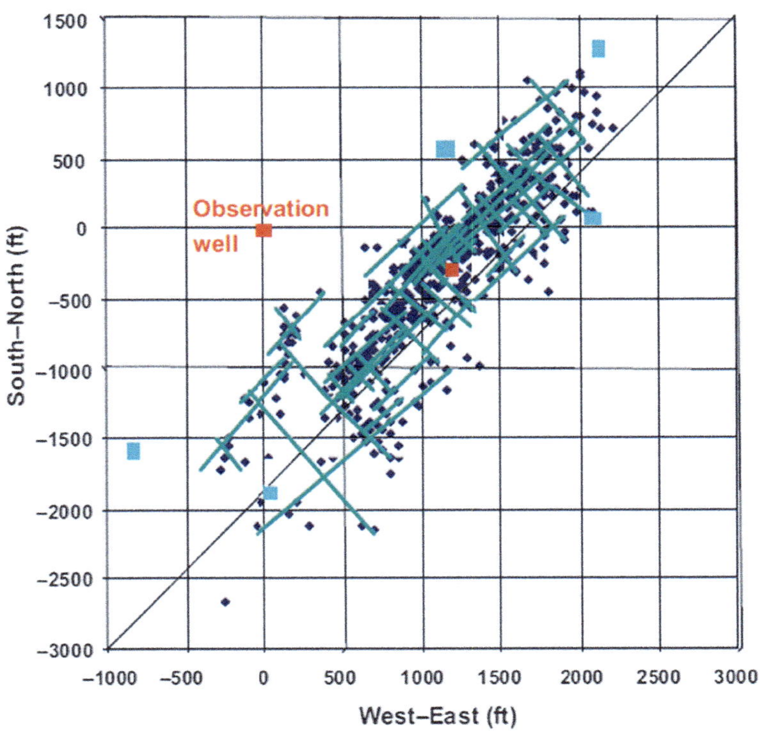

Fig. 7.23 Microseismic interpretation of hydraulically fractured Barnett shale.

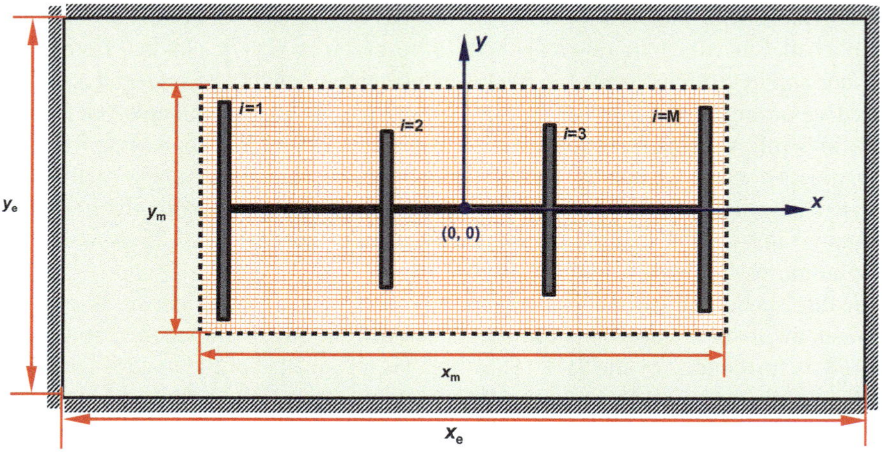

Fig. 7.24 A MFHW in a rectangular composite gas reservoir with global SRV.

To simplify the analysis, the following assumptions are applied: A multi-stage horizontal well is centered in the closed rectangular reservoir and the SRV region. The reservoir length, width, and height are x_e, y_e, and h, respectively; hydraulic fractures are perpendicular and randomly distributed along the wellbore. Each fracture can be symmetrical or asymmetrical, and the fracture half-lengths are L_{fui} and L_{fli}. The effective well length is L (distance between the most left and right fractures); the length and width of the SRV region are x_m and y_m. There are M infinite conductivity fractures; no flow occurs from the reservoir to the horizontal wellbore, and the pressure drop of gas flow along the wellbore is ignored.

7.4.2 Pressure and production type curve analysis

7.4.2.1 Microfractures + steady state adsorption/desorption and diffusion

For mechanism model 1, the BEM is used to acquire type curves of dimensionless bottom-hole pressure, a production rate and cumulative production of the well producing at constant pressure based on the parameters in Table 7.4.

From the well test type curves shown in Fig. 7.25, the following seven flow stages can be identified if the sizes of the reservoir and SRV region are assumed big enough.

Flow stage 1: Early time wellbore storage and its transition flow. The initial section in this stage is a straight line from the coordinate origin, and a hump follows for the transition flow. The height and width of the hump depend on the co-effect of wellbore storage and skin.

Flow stage 2: Early time linear flow in the fracture system. In this flow stage, the pseudo-pressure derivative curve is a straight line with a slop of 1/2. The position of this section on the plot is determined by SRV permeability, and the lasting time is determined by the inner region permeability, fracture length, and fracture spacing.

Table 7.4 Shale reservoir properties

Reservoir properties	Value	Reservoir properties	Value
Permeability of microfracture system, k_{f1} (mD)	0.01	Porosity of microfracture system, $\varnothing_{f1}$ (fraction)	0.02
Length of SRV, x_m (m)	1600	Width of SRV, y_m (m)	400
Permeability of SRV, k_{f2} (mD)	0.1	Porosity of SRV, $\varnothing_{f2}$ (fraction)	0.1

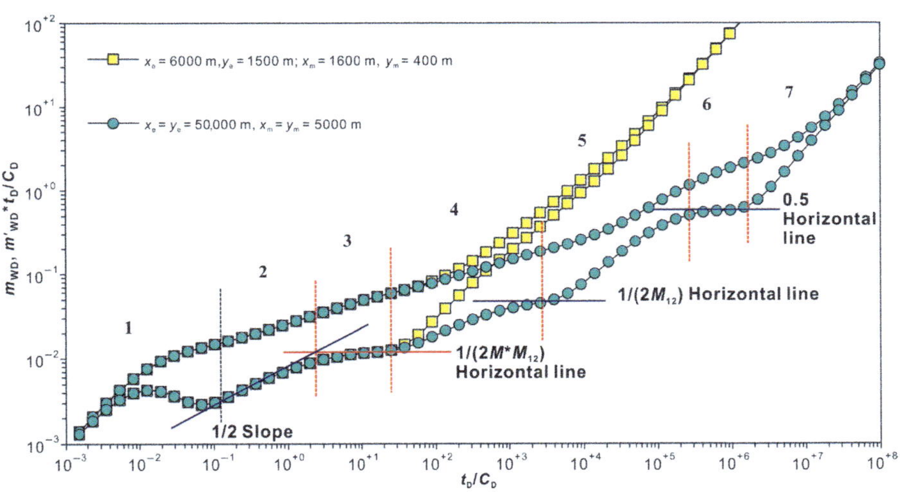

Fig. 7.25 Effect of reservoir size on well test type curves of a MFHW in gas reservoir with global SRV.

Flow stage 3: Early time radial flow in the fracture system. On the plot, the pseudo-pressure derivative in this stage is a section of the horizontal straight line with a value of $1/(2M^*M_{12})$. To be noted, if the SRV height is close to a fracture length, the line section in this stage is conceived or deviates due to an effect of outer reservoir region properties.

Flow stage 4: SRV transition flow and radial flow. In this flow stage, the pseudo-pressure derivative is a section of the horizontal straight line with a value of $1/(2M_{12})$. The higher the SRV permeability, the lower the position of the line.

Flow stage 5: Transition flow between inner and outer region radial flow.

Flow stage 6: Outer reservoir radial flow. Since all the dimensionless parameters refer to the permeability in the outer region micro-fracture system, the derivative curve of the radial flow is a line with a value of 0.5.

Flow stage 7: boundary dominated flow. In this flow stage, the pseudo pressure and its derivative curves overlap and become a straight line with a slope of 1. Due to the identical reservoir length and width given in this case, formation linear flow is not observed.

If realistic sizes are used for the gas reservoir and SRV region, the pressure curves are shown in the next plot as lines with yellow squares. Right after the ending of early time fracture linear flow, pressure reaches the closer reservoir boundaries, which exhibits as parallel upward moving of the pseudo pressure and its derivative curves, but the slope of these curves is not ½. When pressure fills all the boundaries, the curves overlap.

An effect of the main fracture number on well test and production curves is shown in Figs. 7.26 and 7.27, respectively. According to these plots, for the given reservoir and SRV sizes, the number of fractures only affects the middle time behavior of the well test type curves. The more main fractures there are, the lower position the pseudo pressure and its derivative curves on the plots, indicating smaller drawdown for early production. When the producer is at constant bottom-hole pressure, the more the main

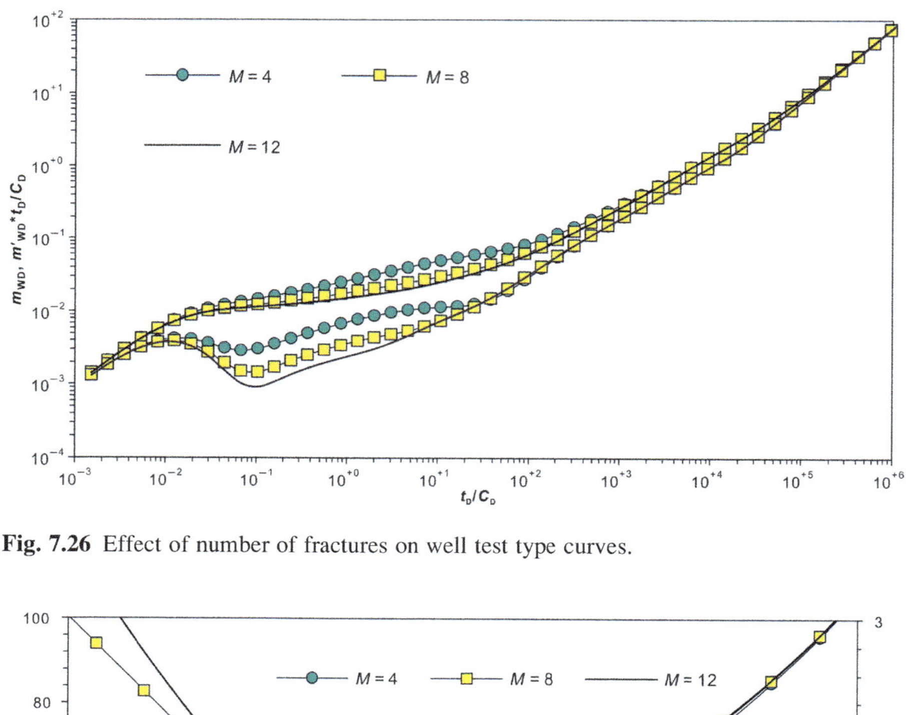

Fig. 7.26 Effect of number of fractures on well test type curves.

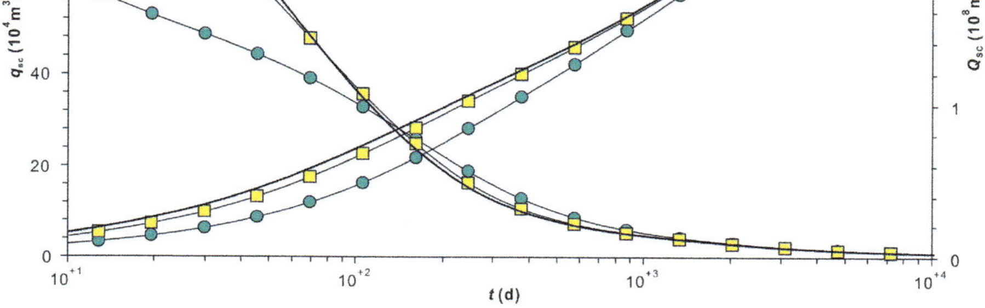

Fig. 7.27 Effect of number of fractures on well production curves.

fractures there are and the higher the early production. As shown in the next plot, there is a small difference between the curves of $M=8$ and $M=12$, but these curves are obviously higher than those of $M=4$. Therefore, there is an optimal number of fractures for given reservoir and SRV sizes.

Figs. 7.28 and 7.29 show the effect of fracture spacing on the well test type and production curves, respectively. For a given SRV size, a bigger fracture spacing means a greater SRV region of each individual fracture. Theoretically, a fracture spacing has no effect on the early time fracture linear flow. But for a given number of fractures, the bigger the fracture spacing and the longer the well horizontal section, which

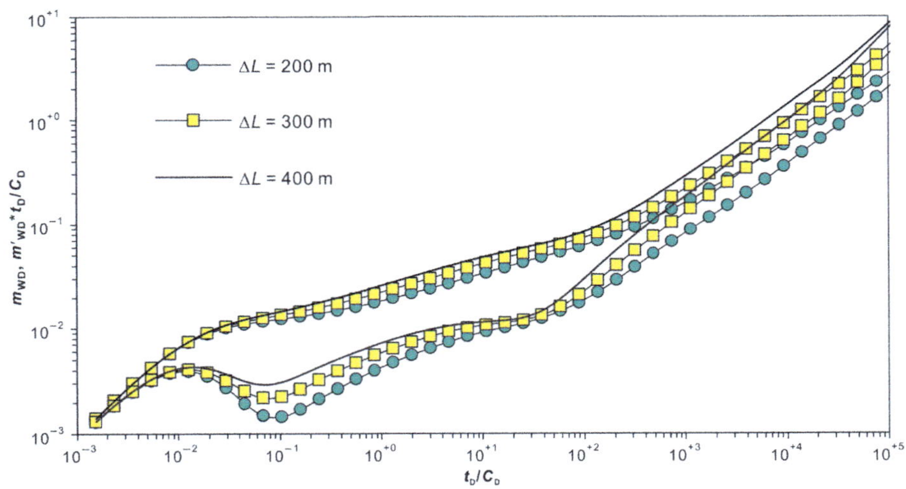

Fig. 7.28 Effect of fracture spacing on well test type curves.

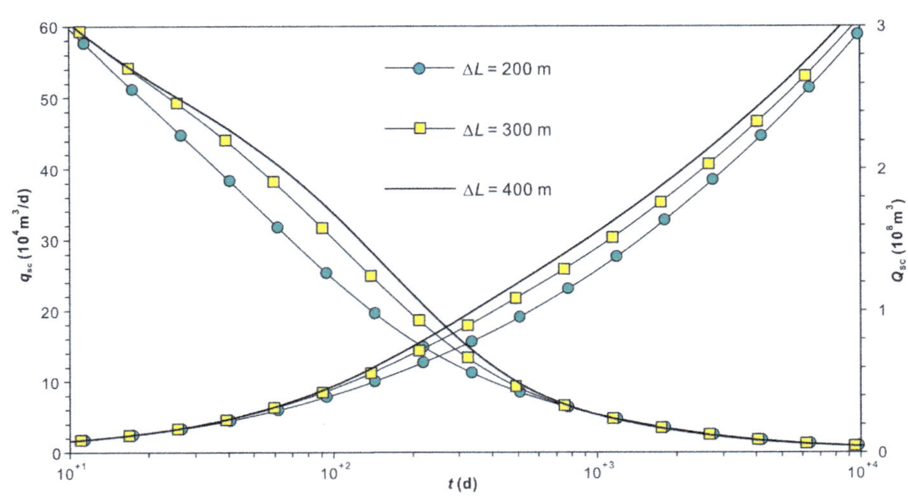

Fig. 7.29 Effect of fracture spacing on well production curves.

has pseudo pressure and its derivative curves behaving as shown in the plots. From the production plot, it is obvious that a bigger fracture spacing results in higher early production.

7.4.2.2 Microfractures + gas adsorption/desorption + matrix Fick's diffusion

For mechanism model 3, the well test type and production curves are acquired for a multistage MFHW in a rectangular composite shale gas reservoir with a global SRV based on the parameters in Table 7.5.

Fig. 7.30 shows the effect of the reservoir and SRV sizes on the well test type curves. For a gas reservoir with a big enough size, the flow stages observable from the curves are similar to those in model 1. For smaller reservoir sizes, the curves quickly enter into the co-effect of linear flow and boundary dominated flow after ending of SRV flow.

Table 7.5 Shale reservoir properties

Reservoir properties	Value	Reservoir properties	Value
Permeability of microfracture system, k_{f1} (mD)	0.01	Porosity of microfracture system, $\varnothing_{f1}$ (fraction)	0.02
Length of SRV, x_m (m)	1600	Width of SRV, y_m (m)	400
Permeability of SRV, k_{f2} (mD)	0.1	Porosity of SRV, $\varnothing_{f2}$ (fraction)	0.1
$(R_m)^2/D_F$ ratio	2×10^{-6}	Gas production rate at constant pressure, q_{sc} (m^3/d)	1×10^4

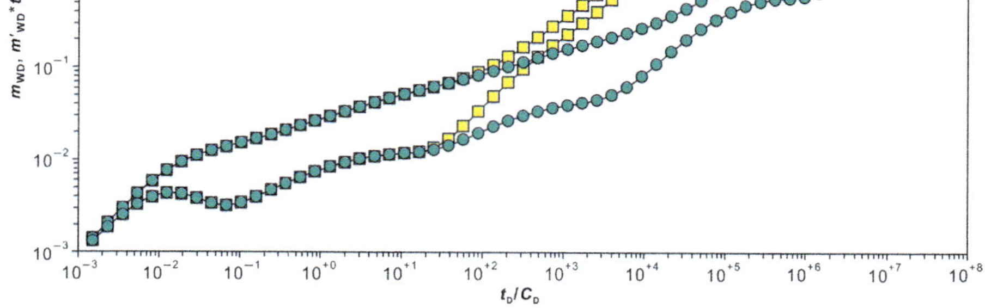

Fig. 7.30 Effect of reservoir and SRV sizes on well test type curves.

The effect of the number of fractures on the well test type and production curves is shown in Figs. 7.31 and 7.32. For a given SRV size, the number of fractures mainly affects early time radial flow and the radial flow period. The effect gets less after pressure propagates outside the SRV region. As shown in the production plot, for the same SRV permeability, the more fractures there are, the bigger the flow area is, and the higher the early production is. After pressure reaches the outer reservoir region, the well production depends on reservoir properties and becomes identical.

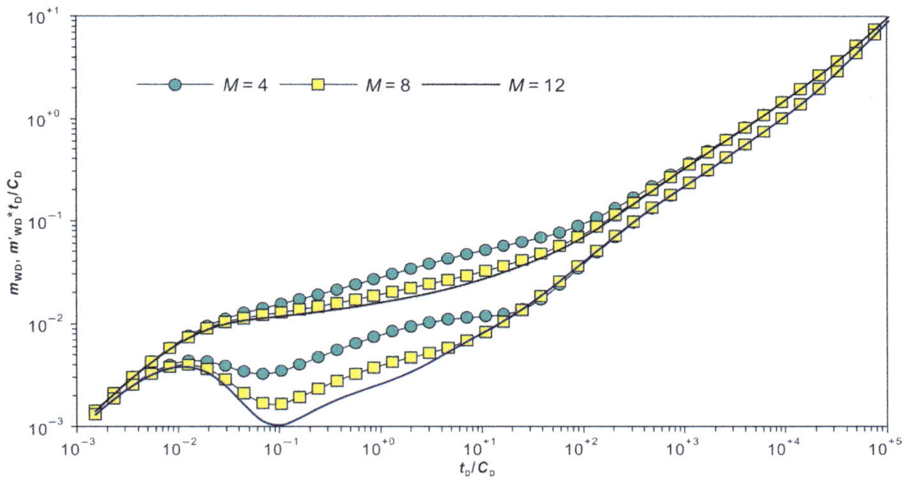

Fig. 7.31 Effect of number of fractures on well test type curves.

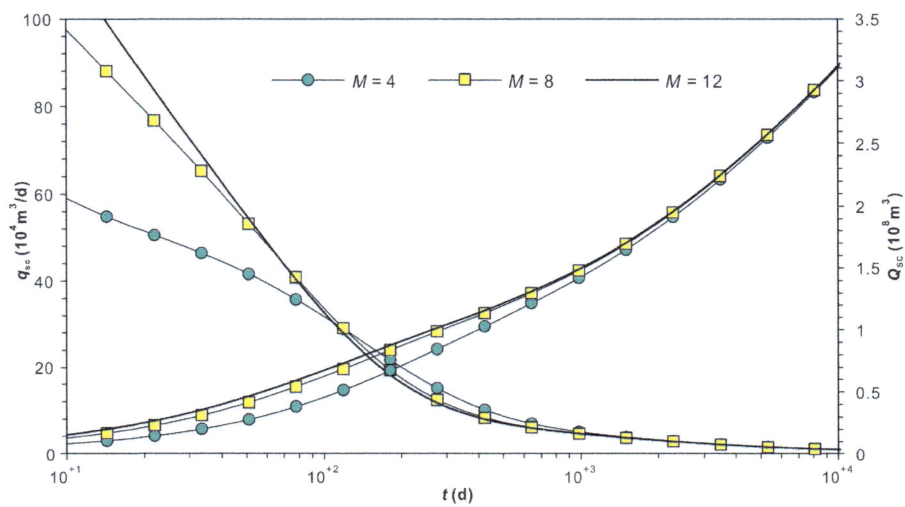

Fig. 7.32 Effect of number of fractures on well production curves.

The effect of the SRV permeability is shown in Figs. 7.33 and 7.34 for the well test type and production curves. The SRV permeability mainly affects the middle and late time flow stages. The more permeable the SRV region is, the lower the curves are on the plots. But after pressure reaches outside the SRV, the difference becomes smaller and smaller until it disappears. For the well production rate, the higher the SRV permeability is, the higher the early production rate is, but with less influence on the late time production rate. For cumulative production, a higher SRV permeability results in greater cumulative production.

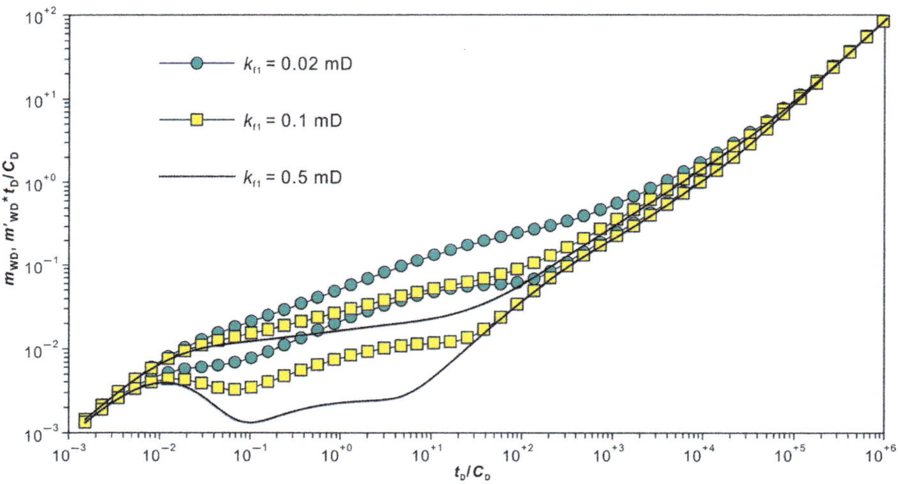

Fig. 7.33 Effect of SRV permeability on well test type curves.

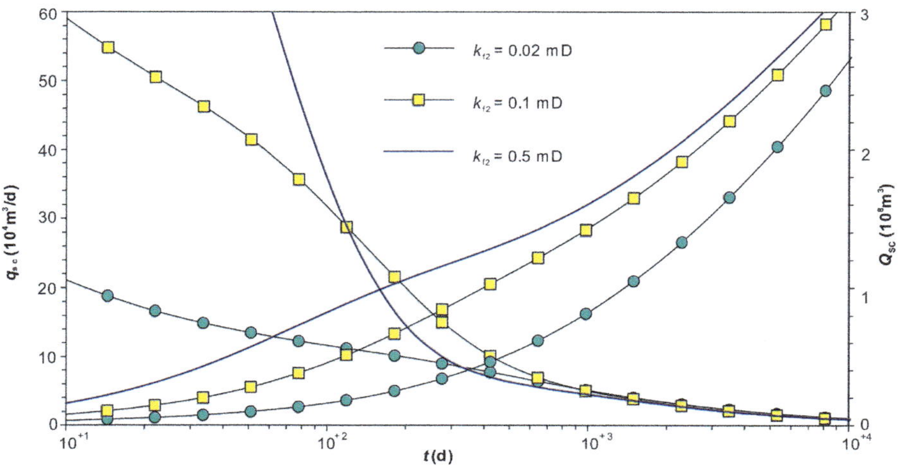

Fig. 7.34 Effect of SRV permeability on well production curves.

Figs. 7.35 and 7.36 illustrate an effect of the permeability of the microfracture system in the outer reservoir region. Since the dimensionless variables are defined based on the microfracture system permeability, a higher curve position for a higher microfracture system permeability does not mean higher drawdown required for early production. In theory, before pressure propagates outside the SRV region, the pseudo pressure difference is identical for different microfracture properties. When the well is producing at constant pressure, a higher permeability results in higher production due to sufficient supply from the outside microfracture system.

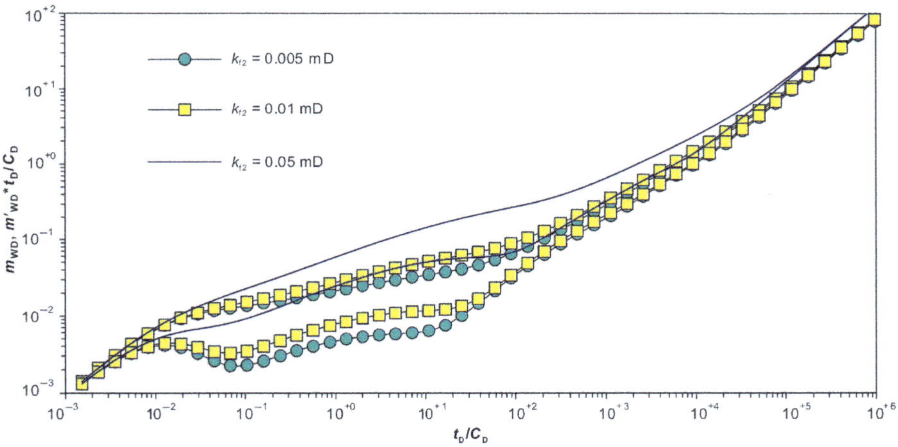

Fig. 7.35 Effect of microfracture system permeability on well test type curves.

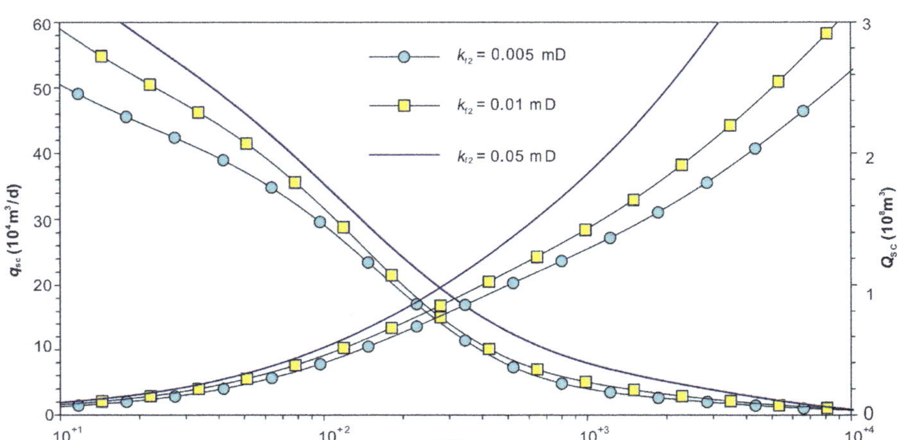

Fig. 7.36 Effect of microfracture system permeability on well production curves.

7.4.2.3 Microfractures + gas adsorption/desorption + nanopore Knudsen diffusion

Mechanism model 5 is used to consider the presence of Darcy's flow, a slippage effect, Knudsen diffusion, and adsorbed gas desorption in a shale gas reservoir. Based on the parameters in Table 7.6, the BEM is used to generate the well test type and production curves for a MFHW in a rectangular composite gas reservoir with a global SRV.

The well test type curves are shown in Fig. 7.37. It can be seen that there is a platform-like shape on the middle to late time pseudo-pressure derivative curves due to the interporosity flow from the micropore system to the microfracture system. It is the co-effect of the interporosity flow and boundary response instead of the outer reservoir radial flow.

Table 7.6 Shale reservoir properties

Reservoir properties	Value	Reservoir properties	Value
Permeability of microfracture system, k_{f1} (mD)	0.01	Porosity of microfracture system, $\varnothing_{f1}$ (fraction)	0.02
Length of SRV, x_m (m)	1600	Width of SRV, y_m (m)	400
Permeability of SRV, k_{f2} (mD)	0.1	Porosity of SRV, $\varnothing_{f2}$ (fraction)	0.1
Matrix permeability, k_m (mD)	0.0001	Matrix porosity, $\varnothing_m$	0.12
Geometry factor of interporosity flow, α (1/m²)	10^{-5}	Knudsen diffusion coefficient, D_k (1/m²)	10^{-6}
Average radius of matrix micropores, r_n (m)	2×10^{-9}	Reference parameter α (dimensionless) of Eq. (1.26)	0.8

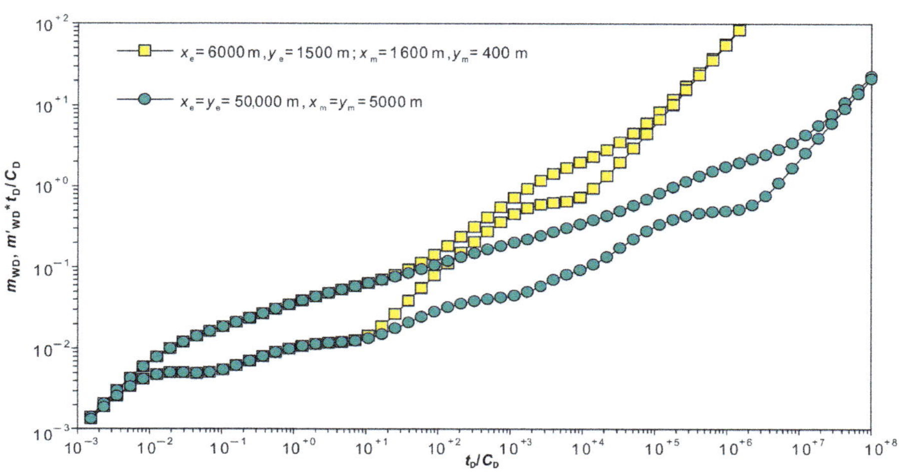

Fig. 7.37 Effect of reservoir and SRV sizes on well test type curves.

The effect of a slippage coefficient is shown in Figs. 7.38 and 7.39 for the well test type and production curves. $F_f = 1$ means no slippage effect, and a bigger F_f indicates a more severe slippage effect for gas flow in matrix micropores. In general, the slippage coefficient only affects the apparent permeability of matrix, and, in consequence, affects the interporosity flow from matrix micropores to the microfracture system. The bigger the slippage coefficient is, the earlier the interporosity flow happens; a higher well production rate is due to gas supply from matrix to microfractures.

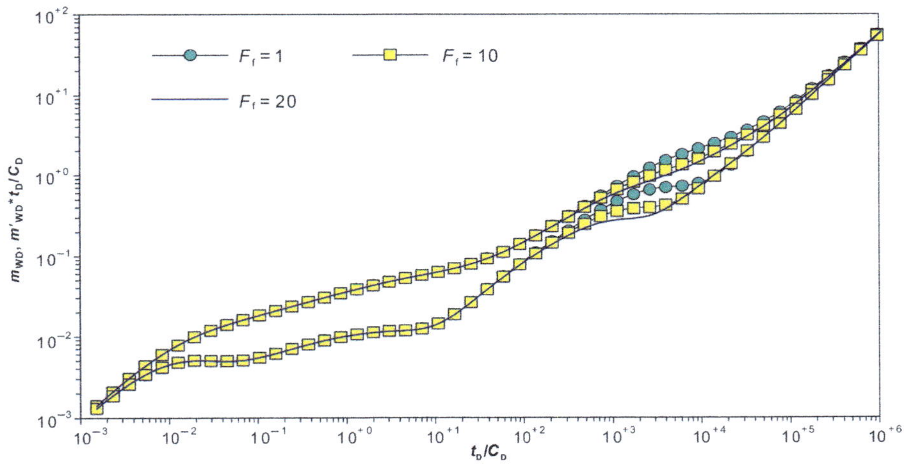

Fig. 7.38 Effect of slippage coefficient on well test type curves.

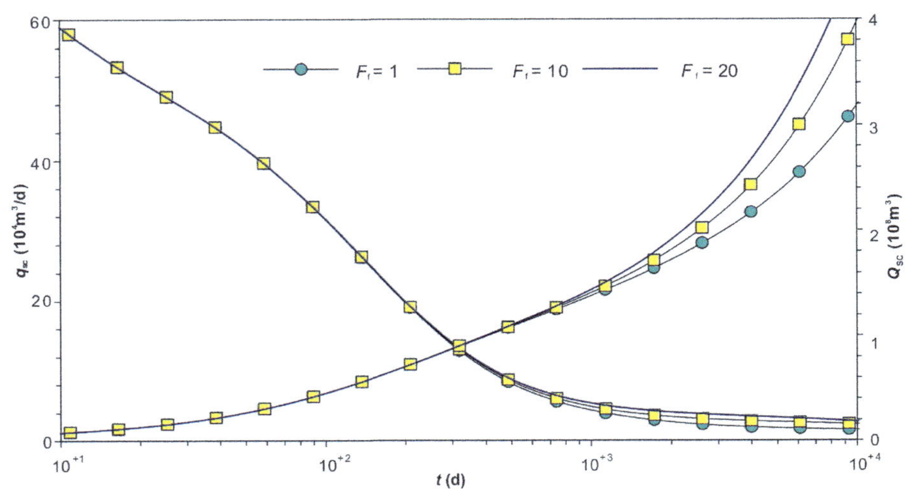

Fig. 7.39 Effect of slippage coefficient on well production curves.

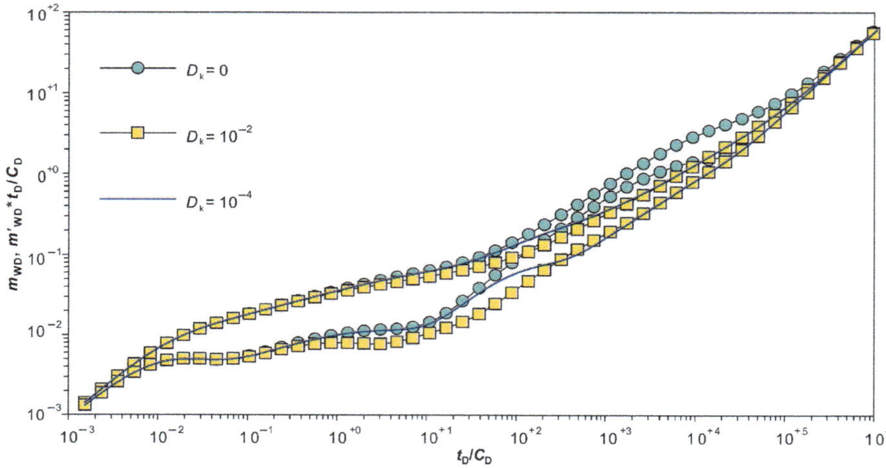

Fig. 7.40 Effect of Knudsen diffusion coefficient on well test type curves.

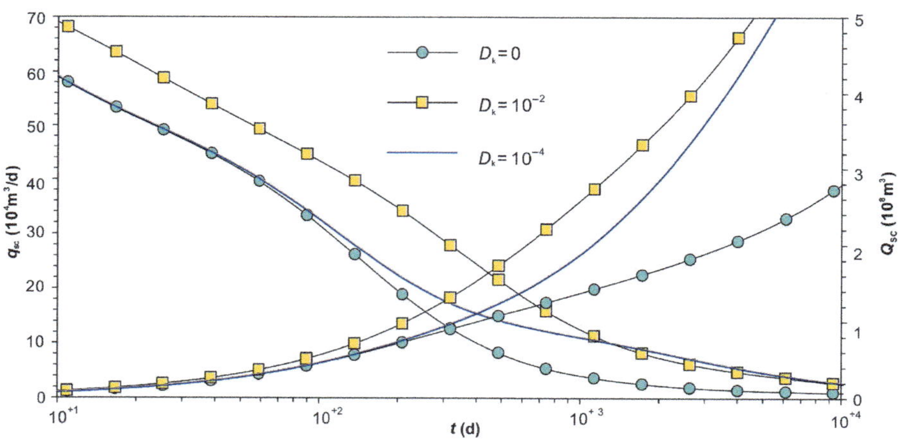

Fig. 7.41 Effect of Knudsen diffusion coefficient on well production curves.

For an effect of the Knudsen diffusion coefficient, as shown in Figs. 7.40 and 7.41, the bigger the D_K is, the lower the middle to late time pseudo-pressure and its derivative curves are on the plots and the higher the production rate is.

7.5 MFHW in shale gas reservoirs with local SRV

In the previous section, the unsteady state flow models of a multistage MFHW under different flow mechanisms with consideration of a global SRV are solved and analyzed by the BEM. Compared to the trilinear flow model, these models are more representative. However, if fracture spacing is big or a SRV is constrained, the reservoir

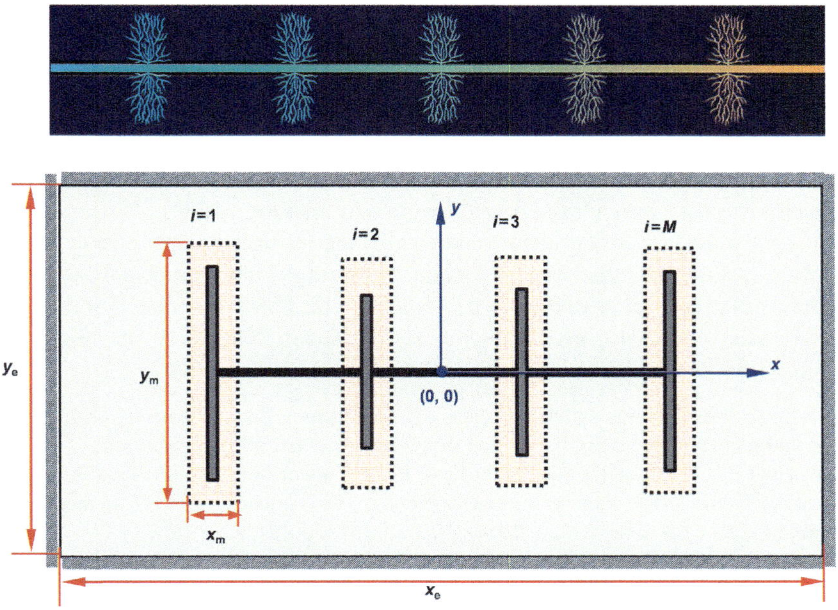

Fig. 7.42 A MFHW in a shale gas reservoir with localized SRV.

region between the main fractures may not be connected thoroughly; instead, a localized SRV surrounding the main fractures develops. This situation will be analyzed using a model of a multistage MFHW in a shale reservoir with a localized SRV, which is shown in Fig. 7.42. Physically, this model is similar to a five-linear flow model, which can be analyzed using the same method as for the trilinear flow model with an additional analysis for gas linear flow in the reservoir and SRV. Similar to trilinear flow models, five-linear flow models have some restriction conditions regarding fracture spacing, half-length, and supply boundaries, and are only valid under special circumstances. Therefore, in order to more accurately represent such models, an unsteady state flow mathematical model of a MFHW in a closed rectangular gas reservoir considering a localized SRV is investigated in this section, and the corresponding well test type and production curves are analyzed through the BEM.

7.5.1 Physical model

The physical model of a MFHW in a closed rectangular reservoir with a localized SRV is shown in Fig. 7.42. To analyze this model, the following assumptions are applied: The well is centered in the reservoir; the reservoir length, width, and height are x_e, y_e, and h, respectively; M infinite conductivity fractures are perpendicular to and randomly distributed along the wellbore with the upper and lower fracture half-length L_{fui} and L_{fli}; the effective horizontal well length is L (the distance between the most left and right fractures); flow from the reservoir to the horizontal wellbore and a pressure drop for flow along the wellbore are ignored.

7.5.2 Pressure and production curve analysis

7.5.2.1 Microfractures + steady state adsorption/desorption and diffusion

For mechanism model 1, the parameters in Table 7.7 are used to generate the dimensionless BHP response curves and the curves of production and cumulative production at constant pressure through the boundary element method (BEM).

Fig. 7.43 shows an effect of the number of fractures on the well test type curves. According to this plot, there are two radial flow stages observable on the pseudo-pressure derivative curves for the model of the localized SRV, which is different from the three radial flow platforms of the global SRV model. Six flow stages can be identified from the plot as follows:

Flow stage 1: Early time wellbore storage and its transition flow. The behavior of the pressure and its derivative is similar to that of the model in the previous section;

Flow stage 2: Early time fracture linear flow. It is the linear flow from the localized SRV to the main fractures. The slope of the pseudo-pressure derivative curve is 1/2. In theory, there is an elliptical flow stage with a slope of 0.36 following this linear flow;

Table 7.7 Shale reservoir properties

Reservoir properties	Value	Reservoir properties	Value
Permeability of microfracture system, k_{f1} (mD)	0.01	Porosity of microfracture system, $\varnothing_{f1}$ (fraction)	0.02
Permeability of SRV, k_{f2} (mD)	0.1	Porosity of SRV, $\varnothing_{f2}$ (fraction)	0.1
Number of fractures, M	4	Fracture half-length, x_f (m)	50
Length of SRV, x_m (m)	100	Width of SRV, y_m(m)	400

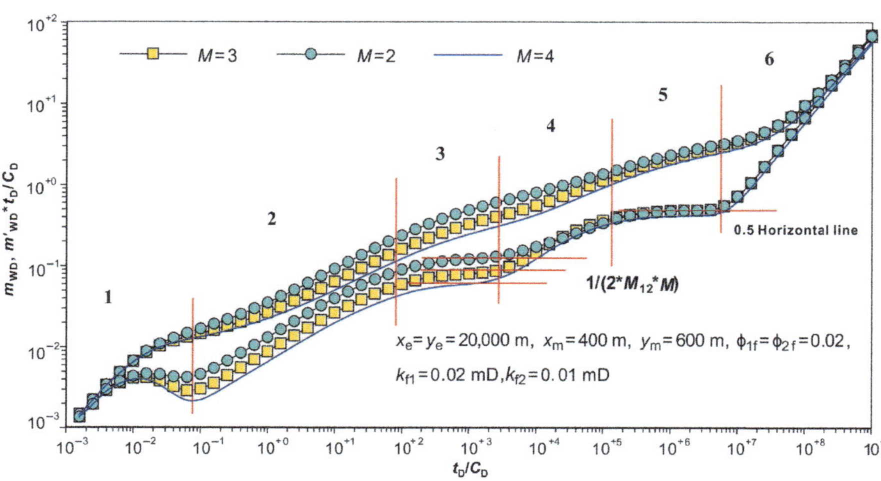

Fig. 7.43 Effect of main fracture number on well test type curves.

Flow stage 3: Radial flow within the SRV. In this flow stage, the pseudo-pressure derivative curve is a horizontal line with a value of $1/2{*}M_{12}{*}M$;
Flow stage 4: Transition flow;
Flow stage 5: Radial flow in the outer reservoir microfracture system. Gas flow enters into this stage when pressure propagation completely reaches outside the SRV. During this stage, the pseudo pressure derivative is a horizontal line with a value of 0.5;
Flow stage 6: boundary dominated flow. In this stage, the pseudo pressure and its derivative curve overlap as a straight line with a slop of 1.

As shown in Fig. 7.44C, for the given SRV size, the model becomes a global SRV model when $M=4$. However, the theoretical inner region radial flow is not present on the correspondent curves in Fig. 7.43, which is caused by a small SRV size.

An effect of the SRV permeability on the well test type and production curves is shown, respectively, in Figs. 7.45 and 7.46. The SRV permeability mainly affects the early time flow period and has little effect on the middle to late time flow, because gas flow is dominated by the outer reservoir region properties after pressure reaches

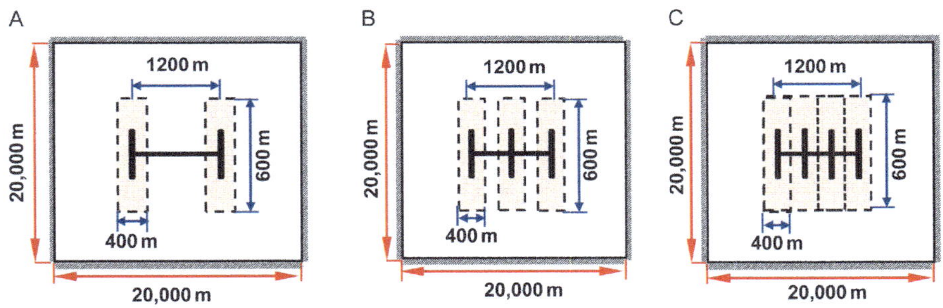

Fig. 7.44 Diagram of horizontal well with different numbers of fractures.

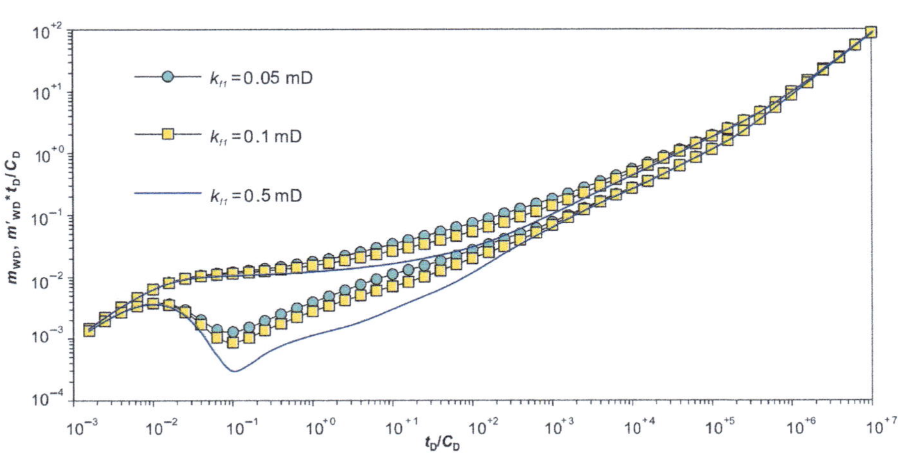

Fig. 7.45 Effect of SRV permeability on well test type curves of a MFHW in shale reservoir with localized SRV.

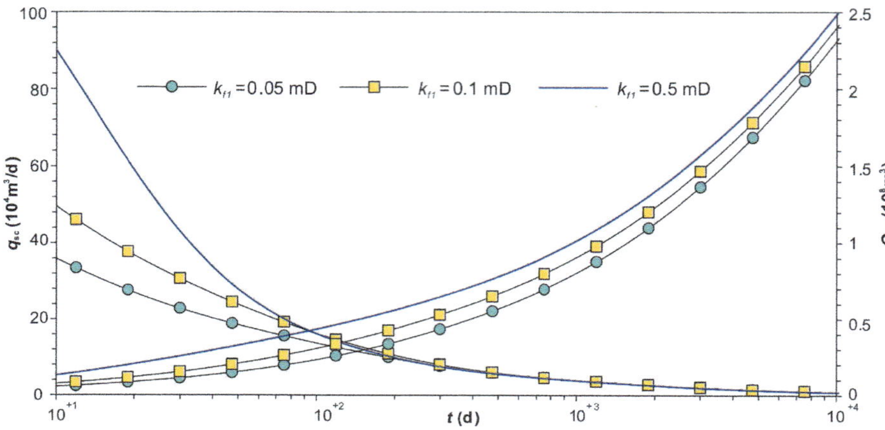

Fig. 7.46 Effect of SRV permeability on well production curves of a MFHW in shale reservoir with localized SRV.

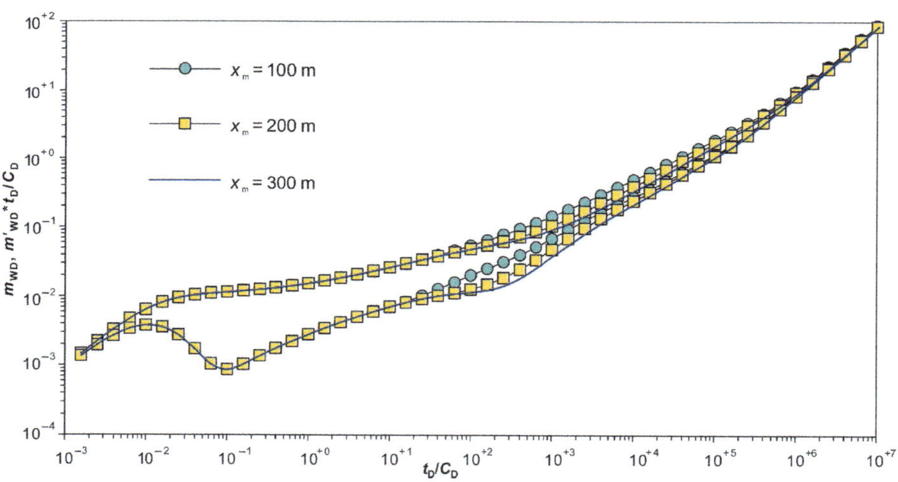

Fig. 7.47 Effect of SRV width on well test type curves.

outside the SRV. The higher the SRV permeability is, the lower the pseudo-pressure derivative curves are on the plot. For a well producing at constant pressure, the SRV permeability affects the early production rate, and cumulative production increases with an increase in the SRV permeability.

An effect of the SRV width on the well test type and production curves is shown in Figs. 7.47 and 7.48, respectively. According to the physical model, the wider the SRV region, the longer pressure propagates in this region, which leads to a section of inner SRV radial flow on the pseudo-pressure derivative curves before entering into linear

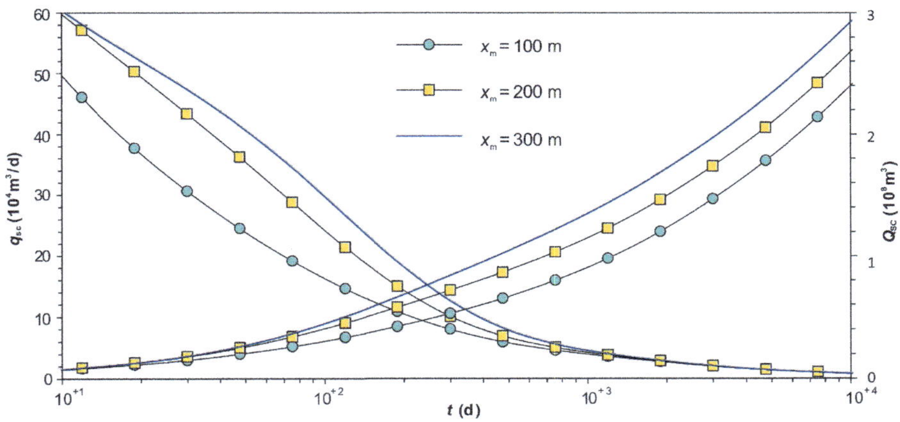

Fig. 7.48 Effect of SRV width on well production curves.

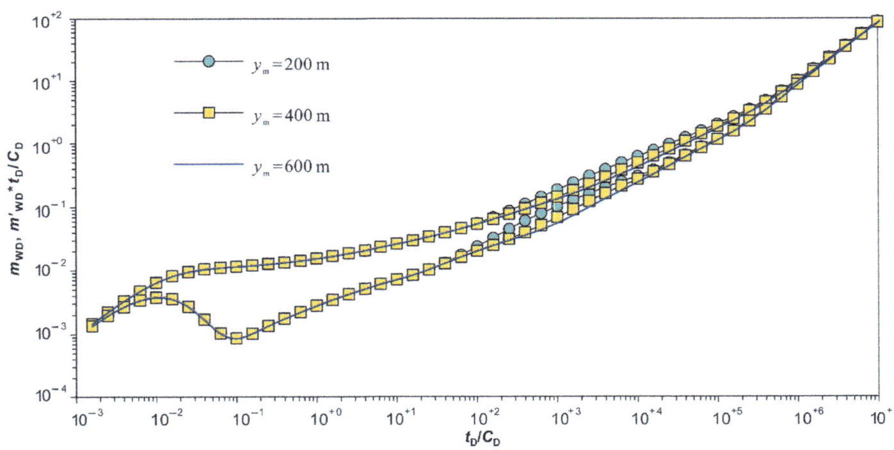

Fig. 7.49 Effect of SRV length on well test type curves of a MFHW.

flow. For a narrow SRV region, pressure quickly reaches the outer reservoir region, whose poorer reservoir properties, contrasted with the inner reservoir properties, result in upward moving of the pseudo-pressure derivative curves until pressure reaches the closer boundaries.

Figs. 7.49 and 7.50 illustrate the effect of the SRV length on the above curves. As it is shown in these plots, for a given SRV width, the SRV length mainly affects the middle time flow period. In general, the longer the SRV length, the bigger the horizontal flow area and the smaller a pressure drawdown is required for a constant production rate, exhibiting as a lower position of the pressure curves. For a well producing at constant pressure, due to a bigger flow area from a bigger SRV length, the production rate during the middle to late time is higher.

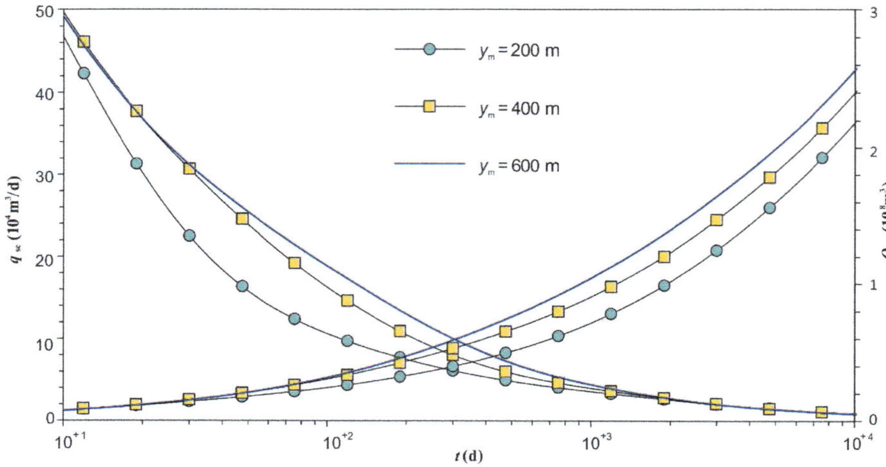

Fig. 7.50 Effect of SRV length on well production curves of a MFHW

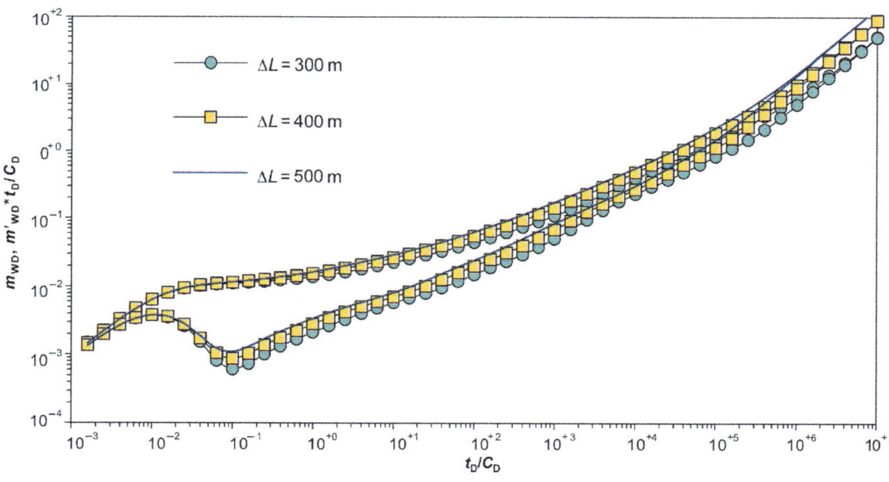

Fig. 7.51 Effect of fracture spacing on well test type curves of a MFHW.

The effect of fracture spacing on the well test and production curves is shown, respectively, in Figs. 7.51 and 7.52. For a given number of fractures, different fracture spacing corresponds to a different effective well length; therefore, the tighter the fractures are, the lower the pressure curves are present on the plots. For well production, bigger fracture spacing results in a bigger reservoir volume affected by the SRV of a single fracture, and, consequently, a higher mid-term production rate. After pressure completely reaches the outer boundaries, gas flow enters into a pseudo-steady state period, and the production rate becomes identical.

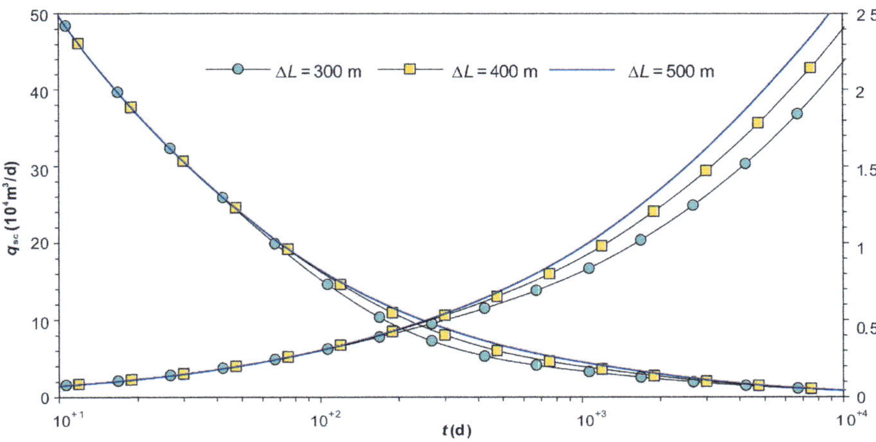

Fig. 7.52 Effect of fracture spacing on well production curves of a MFHW.

Table 7.8 Shale reservoir properties

Reservoir properties	Value	Reservoir properties	Value
Permeability of microfracture system, k_{f1} (mD)	0.01	Porosity of microfracture system, $\varnothing_{f1}$ (fraction)	0.02
Permeability of SRV, k_{f2} (mD)	0.1	Porosity of SRV, $\varnothing_{f2}$ (fraction)	0.1
$(R_m)^2/D_F$ ratio	2×10^6	Gas production at constant pressure, q_{sc} (m^3/d)	1×10^4

7.5.2.2 Microfractures + gas adsorption/desorption + matrix Fick's diffusion

For mechanism model 3, the parameters in Table 7.8 are applied to generate the well test type and production curves of a MFHW in a shale gas reservoir with a localized SRV.

Fig. 7.53 shows the well test type curves of a MFHW in a reservoir of different sizes and a reservoir with a localized SRV of different sizes. Compared with the curves of mechanism model 1, the behavior of the pseudo pressure and its derivative is similar except a slight concave portion on the derivative during the late period of the late time radial flow. This is because the co-effect of adsorbed gas desorption, diffusion, and a boundary response deforms the interporosity flow concave part into a slight downward turning. For a smaller reservoir and SRV, due to the early happening of the outer region flow and boundary dominated flow, the curves start moving upward right after the early time fracture linear flow.

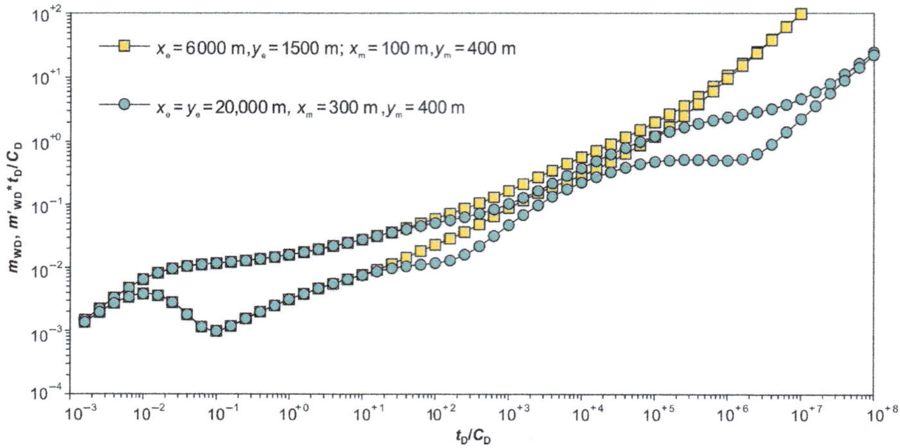

Fig. 7.53 Effect of reservoir and SRV sizes on well test type curves of a MFHW.

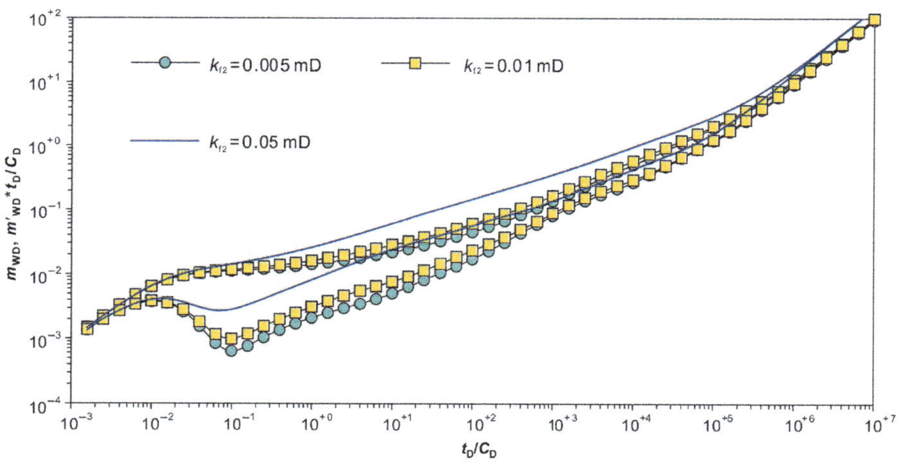

Fig. 7.54 Effect of SRV permeability on well test type curves of a MFHW in a shale gas reservoir with localized SRV.

For the effect of permeability of the outer region micro-fracture system, as shown in Figs. 7.54 and 7.55, the higher the permeability is, the higher the curves are on the pressure plot, and the higher the well early production is. The production tends to become equal after entering into pseudo-steady state flow.

Figs. 7.56 and 7.57 are the well test type and production curves illustrating the effect of desorption time, respectively. As shown in Fig. 7.56, the desorption time has little effect on the well test type curves for the parameters given here. For a gas well producing at constant pressure, the effect of desorption time on production diminishes when the desorption time is greater than a certain value.

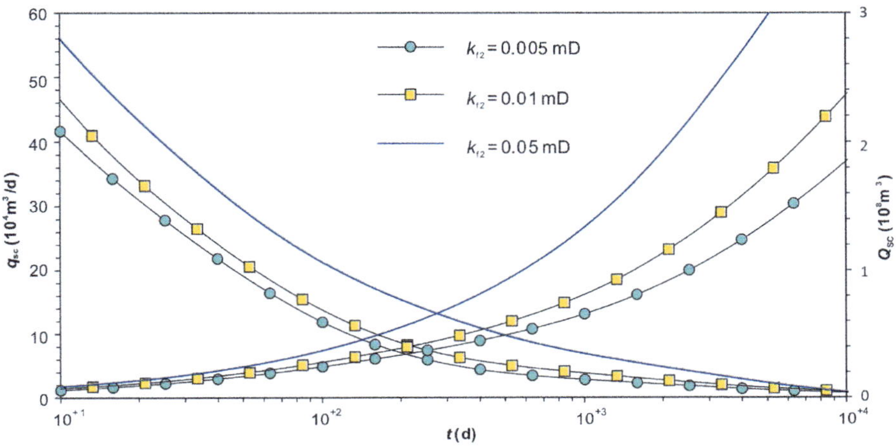

Fig. 7.55 Effect of SRV permeability on well production curves of a MFHW in a shale gas reservoir with localized SRV.

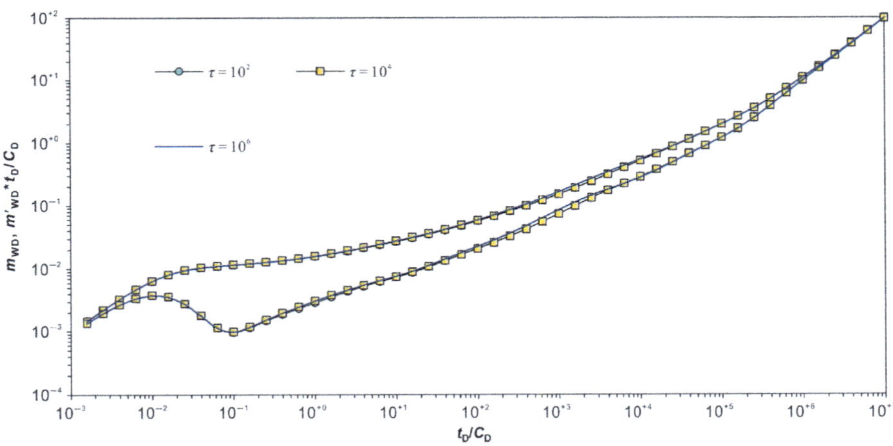

Fig. 7.56 Effect of desorption time on well test type curves of a MFHW in a shale gas reservoir with localized SRV.

7.5.2.3 Microfractures + gas adsorption/desorption + nanopore Knudsen diffusion

Mechanism model 5 is used to represent co-existence of Darcy's flow, a slippage effect, Knudsen diffusion, and adsorbed gas desorption in nanometer pores of a shale reservoir. The parameters used to generate pressure and production curves are listed in Table 7.9.

The effect of a reservoir and SRV size on the well test type curves is shown in Fig. 7.58. For relatively big sizes, the difference from the previous two models is

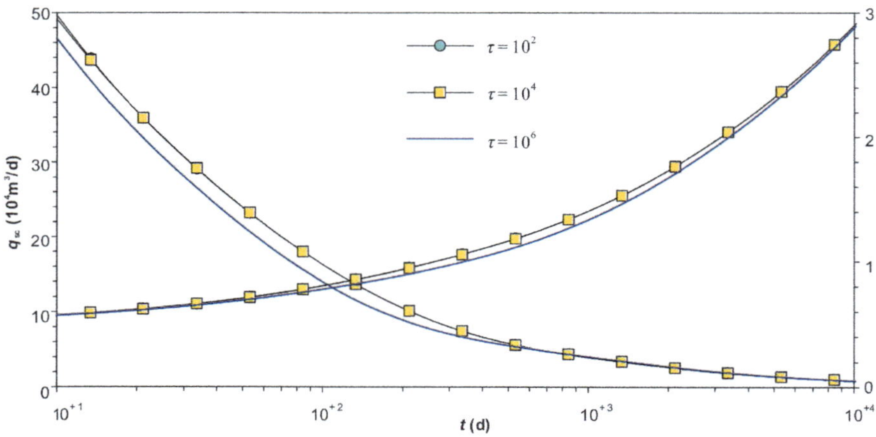

Fig. 7.57 Effect of desorption time on well production curves of a MFHW in shale gas reservoir with localized SRV.

Table 7.9 Shale reservoir properties

Reservoir properties	Value	Reservoir properties	Value
Permeability of microfracture system, k_{f1} (mD)	0.01	Porosity of microfracture system, $\varnothing_{f1}$ (fraction)	0.02
Permeability of SRV, k_{f2} (mD)	0.1	Porosity of SRV, $\varnothing_{f2}$ (fraction)	0.1
Permeability of matrix pores, k_m (mD)	0.0001	Porosity of matrix pores, $\varnothing_m$	0.012
Geometry factor of interporosity flow, α (1/m^2)	10^{-5}	Knudsen diffusion coefficient, D_k (1/m^2)	10^{-6}
Average size of matrix micropores, r_n (m)	2×10^{-9}	Reference value α of Eq. (1.26) (dimensionless)	0.8

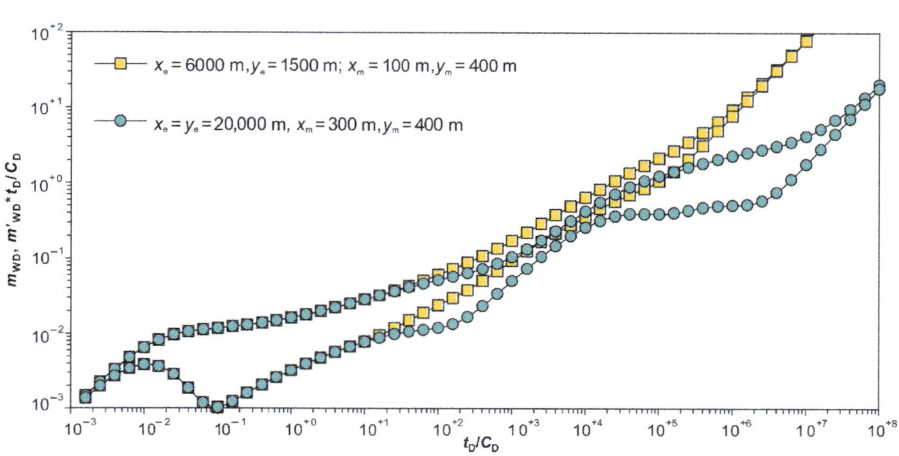

Fig. 7.58 Effect of reservoir and SRV sizes on well test type curves of a MFHW in a shale gas reservoir with localized SRV.

an earlier start of radial flow in the outer reservoir region presenting a concave portion of interporosity flow. For small sizes, the pseudo pressure and its derivative curves are almost the same as those of the previous models.

For the effect of fracture numbers, as shown in Figs. 7.59 and 7.60, the more fractures there are along the same effective horizontal well length, a smaller production drawdown is required for a well producing at a constant rate. However, it can be seen that the curves for 8 and 10 fractures almost overlap, which indicates an optimal fracture number for the given effective well length.

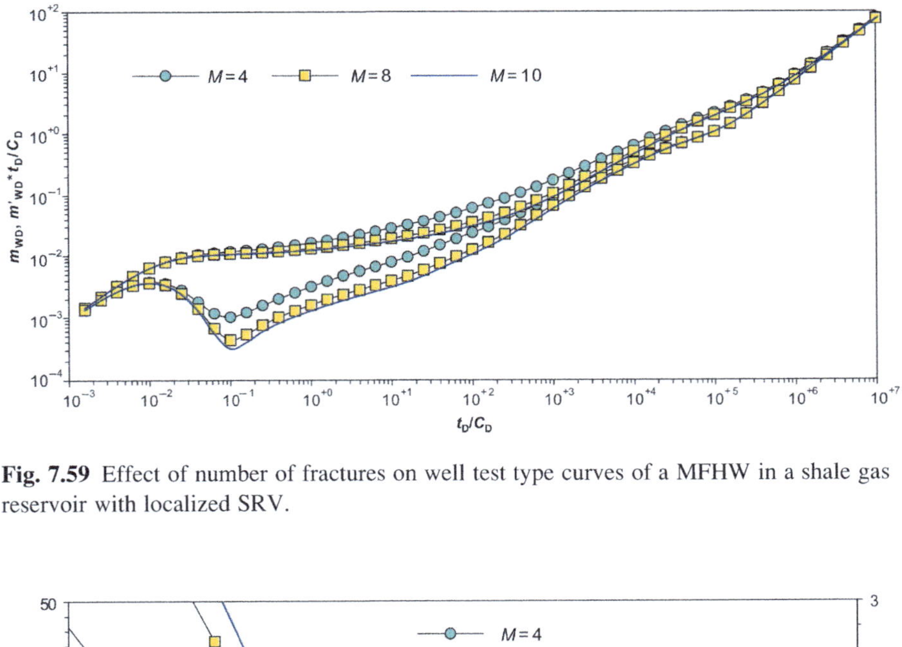

Fig. 7.59 Effect of number of fractures on well test type curves of a MFHW in a shale gas reservoir with localized SRV.

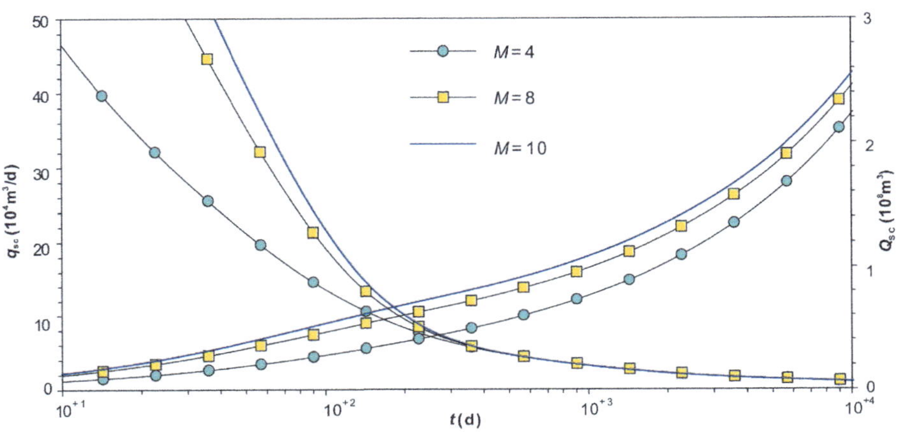

Fig. 7.60 Effect of number of fractures on well production curves of a MFHW in a shale gas reservoir with localized SRV.

The effect of a slippage coefficient on the well test type and production curves are shown in Figs. 7.61 and 7.62, respectively. Similar to other mechanism models, a bigger slippage coefficient results in a lower position of the middle and late time pseudo-pressure derivative curves on the plot. This is because bigger slippage coefficients correspond to higher matrix apparent permeability and thus a better supply capacity of the matrix to the natural microfracture system. Similarly, the effect of a

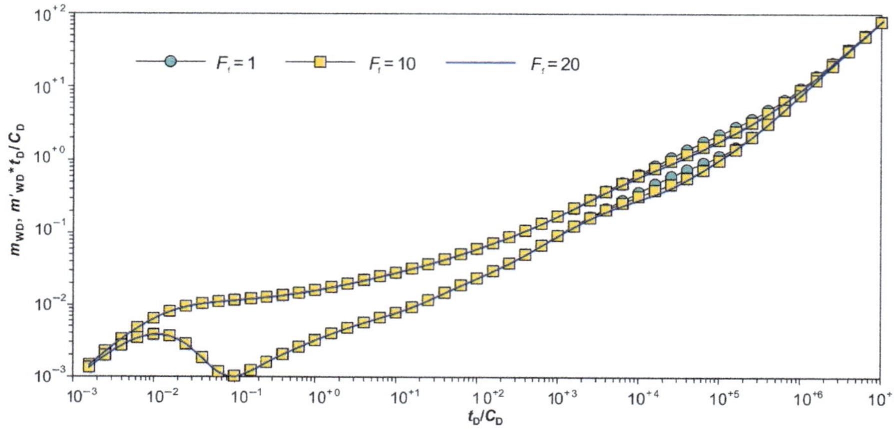

Fig. 7.61 Effect of a slippage coefficient on well test type curves of a MFHW in a shale gas reservoir with localized SRV.

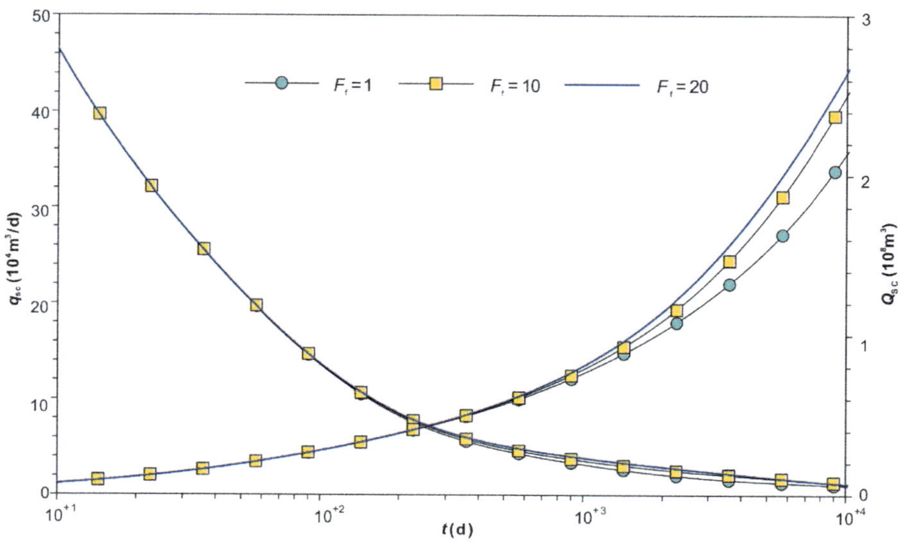

Fig. 7.62 Effect of a slippage coefficient on well production type curves of a MFHW in a shale gas reservoir with localized SRV.

slippage coefficient on well production is also in the middle and late time flow periods; the bigger the coefficient, the higher the production and cumulative production.

Figs. 7.63 and 7.64 show the effect of a Knudsen diffusion coefficient on the well test type and production curves. It can be seen that, similar to the effect of a slippage coefficient, the effect of a Knudsen diffusion coefficient is mainly in the middle and

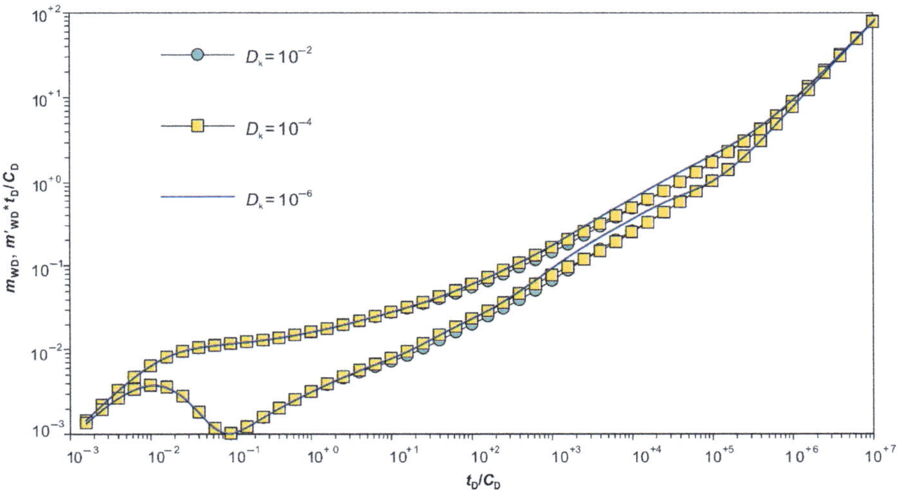

Fig. 7.63 Effect of Knudsen diffusion coefficient on well test type curves of a MFHW in a shale gas reservoir with localized SRV.

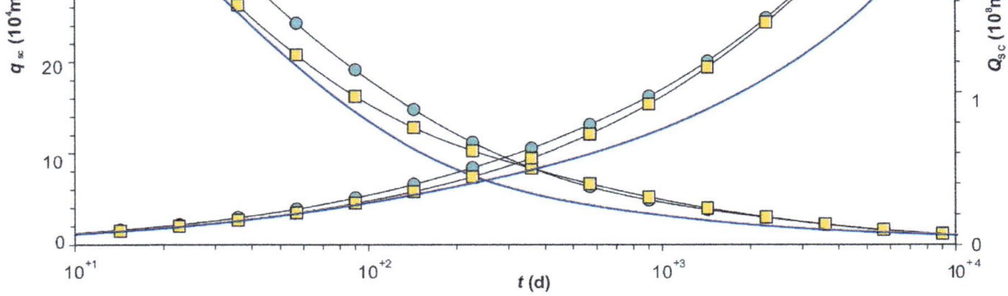

Fig. 7.64 Effect of Knudsen diffusion coefficient on well production curves of a MFHW in a shale gas reservoir with localized SRV.

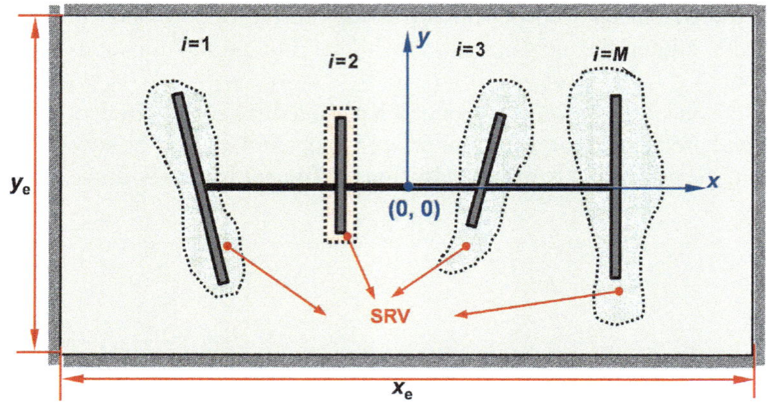

Fig. 7.65 Model of a MFHW with irregular fractures and SRV.

late time flow periods. The bigger the coefficient, the lower the pressure curves on the plot and the higher the production and cumulative production.

It is emphasized that although the MFHW and SRV models analyzed in this section are all regular and symmetrical, the analysis method can also be applied to solving well models with a nonuniform fracture distribution, asymmetrical fractures, and irregular SRVs. As a physic model shown in Fig. 7.65, after discretization of the boundaries and a treatment of the irregular boundaries as linear elements, the method introduced in this chapter can be used to solve this model.

Numerical simulation of MFHWs in shale gas reservoirs based on CVFEM

8

Chapter Outline

8.1 Introduction 237
8.2 A MFHW in a single porosity medium in a shale gas reservoir 238
8.2.1 Model assumptions and mathematical model 238
8.2.2 Numerical solutions 240
8.2.3 Production performance analysis 244
8.3 A MFHW in a dual-continuum medium in a shale gas reservoir 249
8.3.1 Model assumptions and mathematical model 249
8.3.2 Numerical solution 252
8.3.3 Production performance analysis 257
8.3.4 Effects of a fracture network on production 261
8.4 Tri-porosity models in shale gas reservoirs 271
8.4.1 Model assumptions and mathematical model 272
8.4.2 Numerical solutions 273
8.4.3 Production performance analysis 281

8.1 Introduction

In the past 80 years, scientists have investigated the fluid transport properties of various porous rocks, including sandstones and carbonates, which are the matrix of most conventional reservoirs. The main purpose of the investigation was to describe and understand the migration of hydrocarbons from a source rock to a reservoir rock. (During the investigation, shales were considered to be only a source rock or cap rock.) The biggest difference between the rocks of a conventional reservoir and a shale reservoir is the magnitude of their properties, which are permeability, porosity, and micro- to macro-pore size distribution.

The unique reservoir properties of shale make the gas flow mechanisms in shale different from those in conventional reservoirs. To accurately describe the movement of gas through shale (Askarieh et al., 2000; Ekeroth et al., 2006), it is necessary first to define a correct conceptual model that best represents empirical data. For clay-rich

Developments in Petroleum Science, Vol. 66. https://doi.org/10.1016/B978-0-444-64315-5.00008-5

rocks, four primary phenomenological models can be defined to describe gas flow as proposed by Marschall et al. (2005):

(1) Gas movement by solution and/or diffusion is governed by Henry's and Fick's laws, respectively, within interstitial fluids along prevailing hydraulic gradients.
(2) Gas flow in the original porosity of a fabric is governed by a generalized form of Darcy's law, commonly referred to as viscous capillary (or two-phase) flow.
(3) Gas flow is along localized dilatant pathways (microfissuring), which may or may not interact with the continuum stress field and whose permeability is dependent on an interplay between local gas pressure and the effective stress state.
(4) Gas flow is along macrofractures similarly in form to those observed in hydraulic fracturing activities during reservoir stimulation, where fracture initiation occurs when gas pressure exceeds the sum of the minor principle stress and tensile strength.

Due to the difference of gas flow behavior between conventional and shale gas reservoirs, the numerical models used to represent flow in conventional reservoirs are inefficient to represent flow in shale reservoirs. A successful numerical simulation model for shale gas production should appropriately incorporate the above various mechanisms and be able to model complex hydraulic fracture networks. An important area of research is the inclusion of non-Darcy's effects in numerical simulators, such as gas slippage and diffusion (Swami and Settari, 2012). Some authors (Clarkson et al., 2012) choose to model these effects by incorporating apparent permeability (or transmissibility) changes as a function of pressure in conventional simulators. Javadpour (2009) noted that non-Darcy's effects manifested themselves as apparent permeability changes that are functions of gas composition, temperature, and pressure. Additional areas of research include coupled geomechanical and flow models to enable modeling of a hydraulic fracture network creation during stimulation, multi-component adsorption and diffusion of gases, and phase behavior of fluids in nanoporous materials.

In view of all these features, this chapter proposes comprehensive models to describe flow in a MFHW in a shale gas reservoir. These models take into account the multiple flow mechanisms in a shale gas reservoir such as water-gas two-phase flow; adsorption, diffusion, and viscous flow in matrix; stress sensitivity of a natural fracture network and SRV. The control volume finite element method (CVFEM) based on unstructured 3D tetrahedral meshes is used to obtain numerical solutions. We briefly summarize the current research of numerical simulation for shale gas production from four aspects: ① governing equations; ② flow models; ③ methods representing multiscale fractures; and ④ application of numerical simulation for shale gas production.

8.2 A MFHW in a single porosity medium in a shale gas reservoir

8.2.1 Model assumptions and mathematical model

In this section, the production performance of a MFHW in a shale gas reservoir is analyzed. It is assumed that the outer boundary of a reservoir is a rectangle and is of no flow as shown in Fig. 8.1.

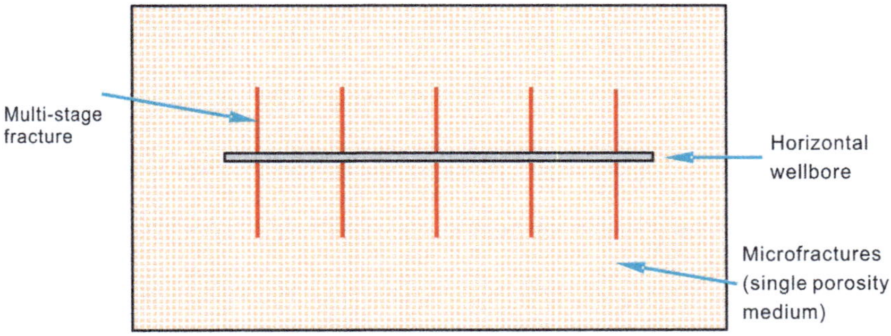

Fig. 8.1 A MFHW in a closed rectangular gas reservoir with single porosity medium.

A single porosity fracture system is composed of a natural shale gas reservoir and a fractured region. The following assumptions and descriptions are applied: ① gas flow is isothermal and of single phase, considering the nonlinearity of properties with pressure changes; ② pressure loss along a horizontal well is negligible; ③ a microfracture system is composed of natural fracture and hydraulic fracture networks, and this system is the main reservoir space and flow paths for free gas; ④ the adsorbed gas mainly exists on the surfaces of the matrix and obeys the isothermal Langmuir law, and gas goes directly into the flow paths after desorbing form the matrix; ⑤ the flow mechanisms in this system include Darcy's flow and Knudsen diffusion (Table 8.1).

The flow equation in this system based on the above model assumptions can be written as:

$$\nabla \cdot \left[\frac{k_{\mathrm{app}}}{\mu_{\mathrm{g}} B_{\mathrm{g}}} \nabla p \right] + q + q^* = \frac{\partial}{\partial t} \left[\frac{\phi}{B_{\mathrm{g}}} \right] \tag{8.1}$$

where:

q—the term of source and sink, m^3/s;
q^*—desorption flux, m^3/s;
k_{app}—apparent permeability, m^2.

According to the isothermal Langmuir law (Langmuir, 1918), with the definition of adsorption volume V_{E} as the adsorption volume per unit volume of the matrix under the standard conditions, there is:

$$q^* = -(1-\phi)\frac{\partial V_{\mathrm{E}}}{\partial t} = -(1-\phi)\frac{\partial}{\partial t}\left(V_{\mathrm{L}} \frac{p}{p_{\mathrm{L}} + p} \right) \tag{8.2}$$

Using an equivalent apparent permeability to represent the co-effect of Darcy's flow and Knudsen diffusion (Ozkan et al., 2010a,b), there is:

$$k_{\mathrm{app}} = k_{\infty}\left(1 + \mu_{\mathrm{g}} D_{\mathrm{g}} C_{\mathrm{gm}}\right) = k_{\infty}\left(1 + \frac{33.93 u_{\mathrm{g}} c_{\mathrm{g}} k_{\infty}^{0.67}}{M_{\mathrm{g}}^{0.5} k_{\infty}}\right) \tag{8.3}$$

Table 8.1 Comparison of reservoir properties between conventional and unconventional reservoir rocks (Gensterblum et al., 2015)

Property	Conventional (sandstones/carbonates)	Unconventional (considering only shales with TOC up to 20%, clay up to 70%)	Sources
Porosity	10%–40%	0.1%–20%	
Permeability (unfractured)	10^{13} m^2 up to 10^{11} m^2	10^{23} m^2 up to 10^{19} m^2	
Sorption capacity (mole of CH_4/kg of rock)	Negligible $\ll$0.01	0.01–0.4	Gasparik et al. (2013, 2014)
Compressive strength (MPa)	650	200–300	Jizba (1991)
Shear strength (MPa)	85	60	Jizba (1991)
Static bulk modulus (GPa)	30	15	Jizba (1991)
Poison ratio	Carbonates: ~0.3 Sandstones: ~0.2	>0.3	
Effective stress coefficient χ	1.00, Pottsville sandstone, Bernabe (1987)	0.15–0.8	Heller et al. (2014)
	2.2, Berea sandstone with clay content: up to 4, Zoback and Byerlee (1975)	0.99	Kwon et al. (2001)

8.2.2 Numerical solutions

In this section, as Fig. 8.2 shows, the area of a single MFHW in a shale gas reservoir can be discretized by pseudo 3D unstructured grids. Monteagudo and Firoozabadi (2004) and Chen (2005) pointed out that the traditional Finite Element Method (FEM) cannot ensure local conservation; there are numerical dispersions in the solution of a strongly nonlinear flow equation. Thus, in this section, the CVFEM is used to analyze the flux exchanges between elements based on a virtual finite volume grid, which can ensure local mass conservation and ultimately result in numerical solutions for strongly nonlinear multiscale flow in a shale gas reservoir.

Flow potential variables are approximated inside each element (a Delaunay triangle) by linear approximations:

$$\tilde{p}(x,y) = \sum_{l=i}^{k} N_l(x,y)p_l \tag{8.4}$$

$$N_l = \frac{1}{2A}(a_l + b_l x + c_l y) \quad l = i,j,k \tag{8.5}$$

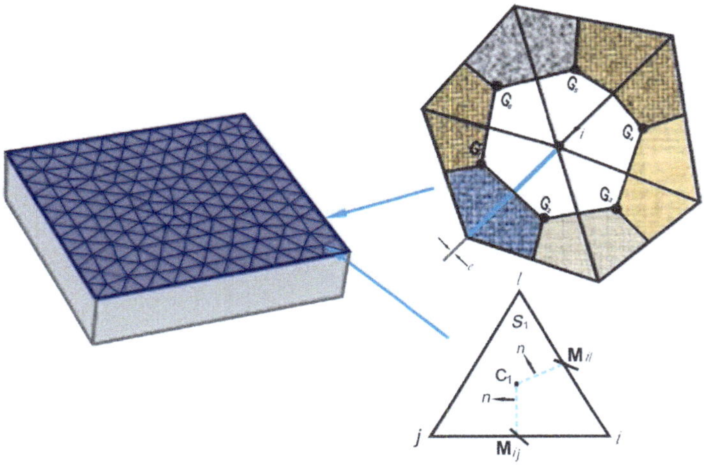

Fig. 8.2 Pseudo 3D unstructured grid cartogram.

where A is the area of each triangle. Based on the Galerkin method, the weak form in pressure can be transformed by the Gauss theorem:

$$\iint_{V_i} -\nabla\cdot(\lambda\cdot\nabla p)\mathrm{d}V = -\int_{ac+cd} \lambda\nabla p\boldsymbol{n}d\Gamma = F_i \tag{8.6}$$

where:

Γ—the boundary of V_i,
$\boldsymbol{n}$—the outer normal vector of Γ.

Chen (2006) pointed out that the flux F_i is continuous across the surface of the control volume V_i. Furthermore, Eq. (8.6) indicates that the CVFEM is locally conservative. The permeability tensor can be expressed as:

$$\bar{\bar{k}} = \begin{bmatrix} k_x & & \\ & k_y & \\ & & k_z \end{bmatrix} \tag{8.7}$$

By taking the barycenter and the outer normal vector formula into Eq. (8.6), the result is:

$$\begin{aligned} f_i &= -\int_{ac+cd} \lambda\nabla\tilde{p}\boldsymbol{n}d\Gamma = (y_c - y_a, x_a - x_c)\left(\vec{\vec{\lambda}}\nabla\tilde{p}\right) + (y_d - y_c, x_c - x_d)\left(\vec{\vec{\lambda}}\nabla\tilde{p}\right) \\ &= (y_d - y_a, x_a - x_d)\left(\vec{\vec{\lambda}}\nabla\tilde{p}\right) = \left(\frac{y_k - y_j}{2}, \frac{x_j - x_k}{2}\right)\left(\vec{\vec{\lambda}}\nabla\tilde{p}\right) \\ &= |A|\,\vec{\vec{\lambda}}\,\nabla N_i\cdot\sum_{l=i}^{k}\nabla N_l p_l \end{aligned} \tag{8.8}$$

By substituting Eqs. (8.4) and (8.5) into Eq. (8.8), the expansion of the flow equation similar to the form of a finite difference can be obtained as:

$$f_{\mathrm{i}} = \sum_{l=j,k} T_{il}(p_{\mathrm{wm}l} - p_{\mathrm{wm}i}) \tag{8.9}$$

where T_{il} is a transmissibility coefficient:

$$T_{il} = \frac{1}{4A}\left(\frac{k_x k_r}{\mu B} b_i b_l + \frac{k_y k_r}{\mu B} c_i c_l\right) \tag{8.10}$$

For nodes j and k of a triangular element (**e**), the convection term can be expanded based on the above logic, resulting in the flux expression over control volume element V_i in element (**e**):

$$\mathbf{T}_{3\times 3}\mathbf{P}_{3\times 1} = \begin{pmatrix} T_{ij}+T_{ik} & -T_{ij} & -T_{ik} \\ -T_{ji} & T_{ji}+T_{jk} & -T_{jk} \\ -T_{ki} & -T_{kj} & T_{ki}+T_{kj} \end{pmatrix} \begin{pmatrix} p_i \\ p_j \\ p_k \end{pmatrix} \tag{8.11}$$

Compared to the mass conservation equation in a single porosity conventional gas reservoir, the form of this equation in a shale gas reservoir is the same except the addition of a desorption gas term. Solving for the partial derivatives of the equivalent apparent permeability by an implicit iteration can be performed during the solution of a transmissibility Jacobi matrix without changing the matrix structure.

Based on the CVFEM and the unstructured triangular grids, the element discrete matrix of the desorption gas term can be expressed as:

$$\frac{A(1-\phi)}{3\Delta t}\begin{pmatrix} (V_{\mathrm{E}})_1^{n+1} - (V_{\mathrm{E}})_1^{n} \\ (V_{\mathrm{E}})_2^{n+1} - (V_{\mathrm{E}})_2^{n} \\ (V_{\mathrm{E}})_3^{n+1} - (V_{\mathrm{E}})_3^{n} \end{pmatrix} \tag{8.12}$$

An iterative approximation is given by:

$$\begin{aligned} \frac{A(1-\phi)}{3\Delta t}\begin{pmatrix} (V_{\mathrm{E}})_1^{n+1} - (V_{\mathrm{E}})_1^{n} \\ (V_{\mathrm{E}})_2^{n+1} - (V_{\mathrm{E}})_2^{n} \\ (V_{\mathrm{E}})_3^{n+1} - (V_{\mathrm{E}})_3^{n} \end{pmatrix} &= \frac{A(1-\phi)}{3\Delta t}\begin{pmatrix} (V_{\mathrm{E}})_1^{k} - (V_{\mathrm{E}})_1^{n} \\ (V_{\mathrm{E}})_2^{k} - (V_{\mathrm{E}})_2^{n} \\ (V_{\mathrm{E}})_3^{k} - (V_{\mathrm{E}})_3^{n} \end{pmatrix} \\ &+ \frac{A(1-\phi)}{3\Delta t}\begin{pmatrix} \partial V_{\mathrm{E}}/\partial p|p_1^k & 0 & 0 \\ 0 & \partial V_{\mathrm{E}}/\partial p|p_2^k & 0 \\ 0 & 0 & \partial V_{\mathrm{E}}/\partial p|p_3^k \end{pmatrix}\begin{pmatrix} \delta p_1 \\ \delta p_2 \\ \delta p_3 \end{pmatrix}^k \end{aligned} \tag{8.13}$$

Substituting Eq. (8.13) into the discrete mass equation of gas in a conventional reservoir, the element matrix in a single porosity shale gas reservoir can be expressed as:

$$\left[\boldsymbol{T}_{(e)}+\left(\frac{\partial \boldsymbol{T}_{(e)}}{\partial \boldsymbol{p}_{(e)}}\right)-\left(\frac{\partial \boldsymbol{N}_{(e)}}{\partial \boldsymbol{p}_{(e)}}\right)-\left(\frac{\partial \boldsymbol{V}_{(e)}}{\partial \boldsymbol{p}_{(e)}}\right)\right]^{k}\boldsymbol{\delta p}_{(e)}^{k}=-\boldsymbol{R}_{(e)}^{k}$$
$$\boldsymbol{R}_{(e)}^{k}=\boldsymbol{T}_{(e)}^{k}\boldsymbol{p}_{(e)}^{k}-\boldsymbol{N}_{(e)}^{k}-\boldsymbol{V}_{(e)}^{k} \tag{8.14}$$

where:

$$\boldsymbol{\delta p}_{(e)}=(\delta p_1\ \delta p_2\ \delta p_3)^{\mathrm{T}} \tag{8.15}$$

$$\boldsymbol{p}_{(e)}=(p_1\ \ p_2\ \ p_3)^{\mathrm{T}} \tag{8.16}$$

$$\boldsymbol{R}_{(e)}=(R_1\ \ R_2\ \ R_3)^{\mathrm{T}} \tag{8.17}$$

$\boldsymbol{T}_{(e)}$ is the transmissibility matrix:

$$T_{(e)}=\begin{pmatrix} T_{12}^{k}+T_{13}^{k} & -T_{12}^{k} & -T_{13}^{k} \\ -T_{21}^{k} & T_{21}^{k}+T_{23}^{k} & -T_{23}^{k} \\ -T_{31}^{k} & -T_{32}^{k} & T_{31}^{k}+T_{32}^{k} \end{pmatrix} \tag{8.18}$$

$\boldsymbol{N}_{(e)}$ is the matrix of an element accumulation term:

$$\boldsymbol{N}_{(e)}=\frac{A}{3\Delta t}\begin{pmatrix} (\phi/B_{\mathrm{g}})_1-(\phi/B_{\mathrm{g}})_1^{n} \\ (\phi/B_{\mathrm{g}})_2-(\phi/B_{\mathrm{g}})_2^{n} \\ (\phi/B_{\mathrm{g}})_3-(\phi/B_{\mathrm{g}})_3^{n} \end{pmatrix} \tag{8.19}$$

$\boldsymbol{V}_{(e)}$ is the matrix of the gas adsorption term:

$$\boldsymbol{V}_{(e)}=\frac{A}{3\Delta t}\begin{pmatrix} [(1-\phi)V_{\mathrm{E}}]_1-[(1-\phi)V_{\mathrm{E}}]_1^{n} \\ [(1-\phi)V_{\mathrm{E}}]_2-[(1-\phi)V_{\mathrm{E}}]_2^{n} \\ [(1-\phi)V_{\mathrm{E}}]_3-[(1-\phi)V_{\mathrm{E}}]_3^{n} \end{pmatrix} \tag{8.20}$$

$\partial \boldsymbol{T}_{(e)}/\partial \boldsymbol{p}_{(e)}$ is the partial derivative matrix of an element transmissibility term:

$$\frac{\partial \boldsymbol{T}_{(e)}}{\partial \boldsymbol{p}_{(e)}}=\begin{pmatrix} \begin{bmatrix} \dfrac{\partial T_{12}}{\partial p_1}(p_1-p_2) \\ +\dfrac{\partial T_{13}}{\partial p_1}(p_1-p_3) \end{bmatrix} & \dfrac{\partial T_{12}}{\partial p_2}(p_1-p_2) & \dfrac{\partial T_{13}}{\partial p_3}(p_1-p_3) \\ \dfrac{\partial T_{12}}{\partial p_1}(p_2-p_1) & \begin{bmatrix} \dfrac{\partial T_{12}}{\partial p_2}(p_2-p_1) \\ +\dfrac{\partial T_{23}}{\partial p_2}(p_2-p_3) \end{bmatrix} & \dfrac{\partial T_{23}}{\partial p_3}(p_2-p_3) \\ \dfrac{\partial T_{13}}{\partial p_1}(p_3-p_1) & \dfrac{\partial T_{23}}{\partial p_2}(p_3-p_2) & \begin{bmatrix} \dfrac{\partial T_{13}}{\partial p_3}(p_3-p_1) \\ +\dfrac{\partial T_{23}}{\partial p_3}(p_3-p_2) \end{bmatrix} \end{pmatrix} \tag{8.21}$$

$\partial \boldsymbol{N}_{(e)}/\partial \boldsymbol{p}_{(e)}$ is the partial derivative matrix of an element accumulation term:

$$\frac{\partial \boldsymbol{N}_{(e)}}{\partial \boldsymbol{p}_{(e)}}=\frac{A}{3\Delta t}\begin{pmatrix} \left.\frac{\partial(\phi/B_{\mathrm{g}})}{\partial p}\right|_{p_1} & 0 & 0 \\ 0 & \left.\frac{\partial(\phi/B_{\mathrm{g}})}{\partial p}\right|_{p_2} & 0 \\ 0 & 0 & \left.\frac{\partial(\phi/B_{\mathrm{g}})}{\partial p}\right|_{p_3} \end{pmatrix} \tag{8.22}$$

$\partial \boldsymbol{V}_{(e)}/\partial \boldsymbol{p}_{(e)}$ is the partial derivative matrix of an element gas adsorption term:

$$\frac{\partial \boldsymbol{V}_{(e)}}{\partial \boldsymbol{p}_{(e)}}=\frac{A}{3\Delta t}\begin{pmatrix} \left.\frac{\partial[(1-\phi)V_{\mathrm{E}}]}{\partial p}\right|_{p_1} & 0 & 0 \\ 0 & \left.\frac{\partial[(1-\phi)V_{\mathrm{E}}]}{\partial p}\right|_{p_2} & 0 \\ 0 & 0 & \left.\frac{\partial[(1-\phi)V_{\mathrm{E}}]}{\partial p}\right|_{p_3} \end{pmatrix} \tag{8.23}$$

By assembling the above matrices, the general discretization equation of the differential flow equation in a single porosity shale gas reservoir can be expressed as:

$$\left[\boldsymbol{T}+\left(\frac{\partial \boldsymbol{T}}{\partial \boldsymbol{p}}\right)-\left(\frac{\partial \boldsymbol{N}}{\partial \boldsymbol{p}}\right)-\left(\frac{\partial \boldsymbol{V}}{\partial \boldsymbol{p}}\right)\right]^k \delta \boldsymbol{p}^k=-\boldsymbol{R}^k \tag{8.24}$$

$$\boldsymbol{R}^k=\boldsymbol{T}^k\boldsymbol{p}^k-\boldsymbol{N}^k-\boldsymbol{V}^k \tag{8.25}$$

In this section, it is assumed that the reservoir region is discretized into *NN* elements and *MM* nodes. Thus the residue matrix is:

$$\boldsymbol{R}=(R_1R_2\ldots R_{MM})$$

$$\boldsymbol{\delta p}=(\delta p_1\ \ \delta p_2\ \ \cdots\ \ \delta p_{MM})^{\mathrm{T}};\ \boldsymbol{p}=(p_1\ \ p_2\ \ \cdots\ \ p_{MM})^{\mathrm{T}}$$

$$\zeta=\sum_1^{NN}\zeta_{(e)};\ \frac{\partial \zeta}{\partial \boldsymbol{p}_{(e)}}=\sum_1^{NN}\frac{\partial \zeta_{(e)}}{\partial \boldsymbol{p}_{(e)}},\quad \zeta=\boldsymbol{T},\boldsymbol{N},\boldsymbol{V}$$

For the inner boundary conditions, the fractures are treated as line sources and substituted into the general equilibrium equation for solution.

8.2.3 Production performance analysis

The basic physical parameters of a gas reservoir are shown in Table 8.2. Here it is assumed that the Original Gas In Place (OGIP) controlled by the single well is $4.28\times10^8\,\mathrm{m}^3$. The total volume of gas adsorption is $2.23\times10^8\,\mathrm{m}^3$, which is about

Table 8.2 Parameters used for the single porosity mode

Reservoir parameter	Value	Reservoir parameter	Value
Formation thickness, h, m	50	Reservoir boundary, $X \times Y$, m	1200×400
Initial reservoir pressure, p_i, MPa	20	Reservoir temperature, T, °C	100
Gas specific gravity, r_g, fraction	0.6	Gas production rate, q_{sc}, m^3/d	10×10^4
Horizontal wellbore length, L, m	1000	Hydraulic fracture numbers, N	11
Half fracture length, x_f, m	100	Microfracture system permeability, k, mD	0.005
Microfracture system porosity, Φ_f, %	5	Langmuir pressure, P_L, MPa	4
Langmuir volume, V_L, m^3/m^3	10		

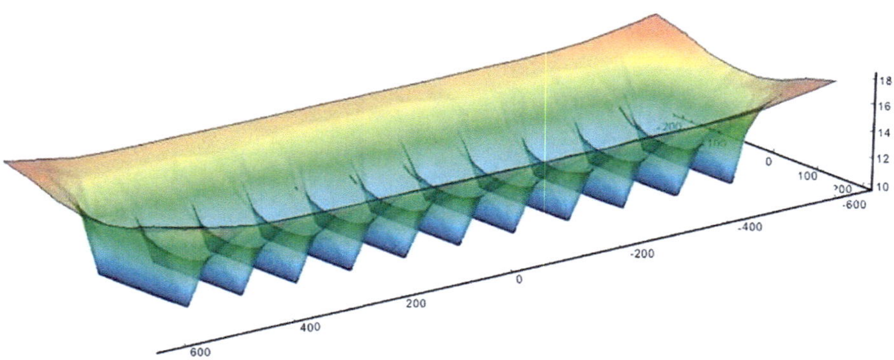

Fig. 8.3 Pressure distribution and 3D projection graph after 600 days.

52% of the total volume. The constant production under standard conditions is $10 \times 10^4 m^3/d$. Fig. 8.3 shows that a pressure contour in the MFHW is close to a rectangle at most flow periods. There is a low pressure zone around the hydraulic fractures, and the bottom of the low pressure zone is similar to a gully shape. As shown in Fig. 8.4, the pressure in the hydraulic fracture zone decreased rapidly; the characteristic of ultra-low permeability leads to slow pressure propagation across the reservoir.

In addition, production performance at constant bottom-hole pressure of 4 MPa is analyzed. Fig. 8.5 shows the typical production decline curve and the cumulative gas production curve. After the early high production period, which is about 30 days, the production rate rapidly decreases to $20 \times 10^4 m^3/d$; later, due to the supplement of gas desorption from the rock surface into the fracture system, the production rate is stable between $1–8 \times 10^4 m^3/d$. After 3000 days, the production rate of the well is $0.61 \times 10^4 m^3/d$; the cumulative gas production is $1.77 \times 10^8 m^3$, among which the adsorption gas volume is $0.61 \times 10^8 m^3$, and the recovery is up to 41.3%.

Fig. 8.6 shows the influence of a hydraulic fracture number on production performance of the horizontal well. With an increase in the hydraulic fracture number, gas

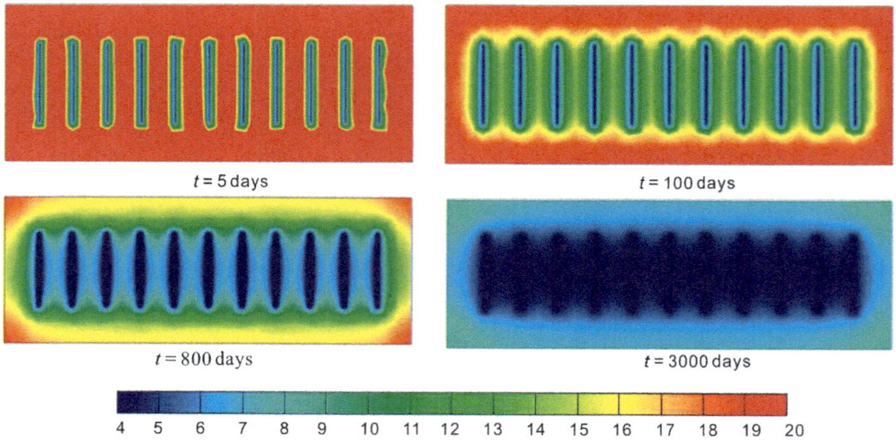

Fig. 8.4 Pressure distribution at a different time.

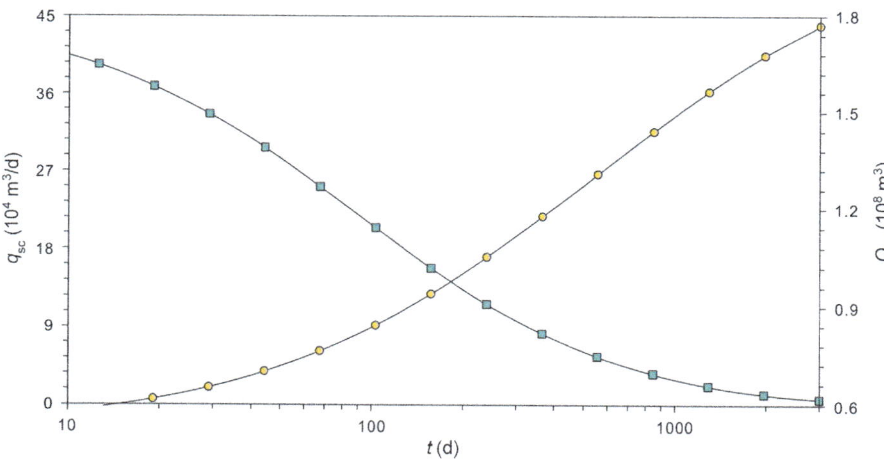

Fig. 8.5 Production rate and cumulative production curves of a MFHW in a shale gas reservoir.

production increases; such an effect becomes smaller with a further increase in the fracture number. For the same fracture and horizontal well length, there is an optimal number of hydraulic fractures. At the late production period, flow forms an approximately rectangular area of low pressure, which can be considered as a potential body. Thus the effect of the number of fractures on production becomes very small. After 500 days, the production rates of the horizontal well with different numbers of fractures are almost equal, and according to the material balance principle, the cumulative production of the well with different numbers is the same after a long time (Fig. 8.7).

Fig. 8.8 shows the effect of the fracture length on the production of a MFHW in a shale gas reservoir. As the fracture length increases, the production rate and

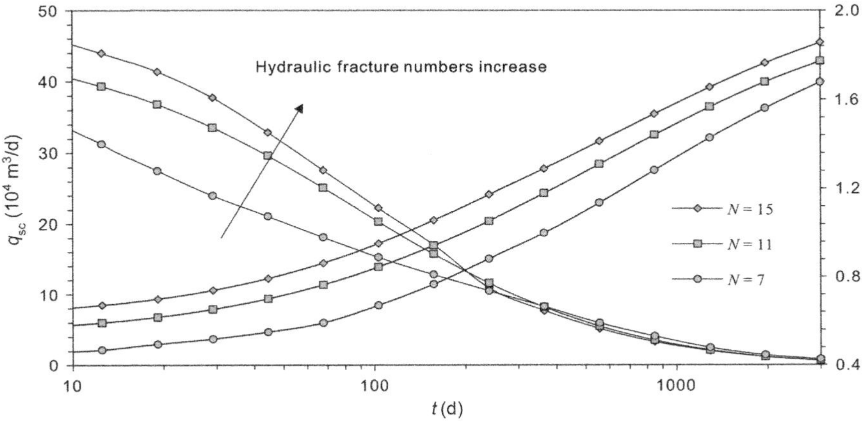

Fig. 8.6 Effect of hydraulic fracture number increase on production of a MFHW in a shale gas reservoir.

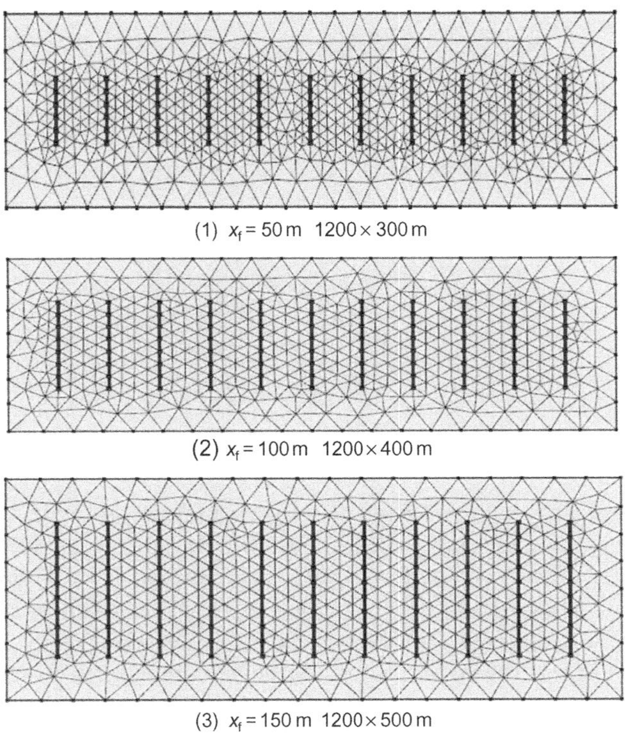

Fig. 8.7 Unstructured grid cartogram of domain areas with different fracture half-length.

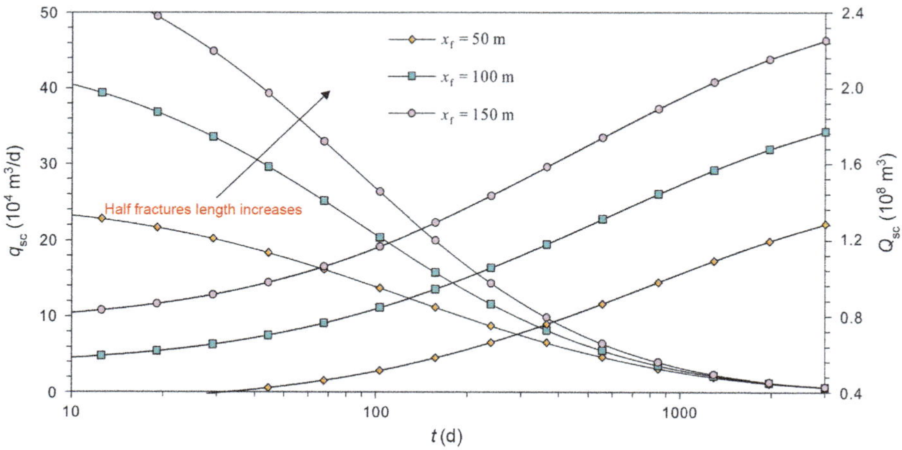

Fig. 8.8 Effect of fracture length on production of a shale gas reservoir.

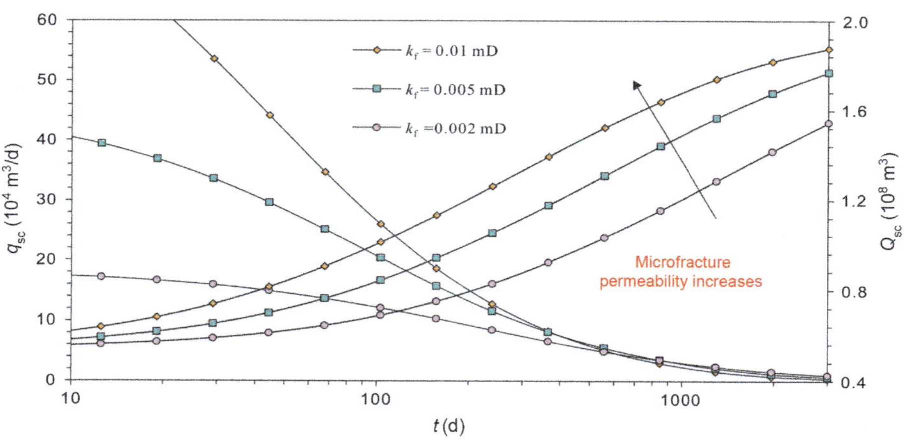

Fig. 8.9 Effect of permeability of microfracture system on production of a shale gas reservoir.

cumulative production increase. This is because a fracture half-length increase will significantly improve the single well controlling area with the same horizontal well length. Fig. 8.8 shows the production performance of a MFHW with fracture half-length of 50, 100, and 150, corresponding, respectively, to the model regions of 1200×300, 1200×400, and 1200×500 m (Fig. 8.7). Therefore, in the development of a shale gas reservoir, an extension of the fracture length can yield higher production.

Fig. 8.9 shows the effect of permeability of the microfracture system on production. The production rate and cumulative production of the MFHW increase with an increase in the microfracture system permeability. The permeability of the microfracture system mainly affects the early period due to the improved reservoir

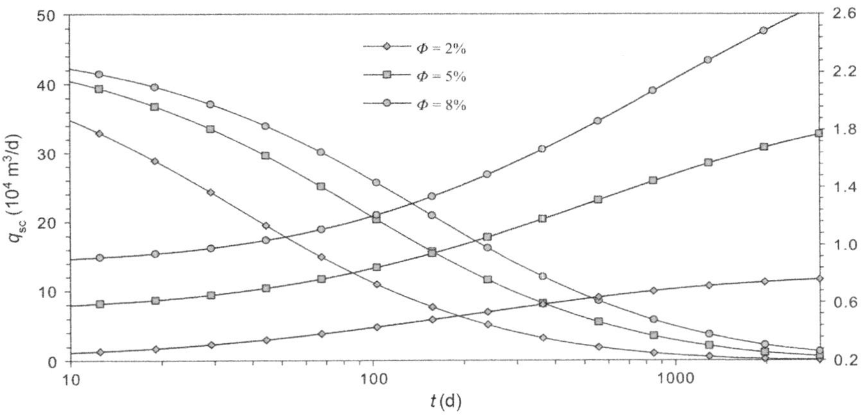

Fig. 8.10 Effect of microfracture system porosity on production of a MFHW in a shale gas reservoir.

properties around fractures. When pressure reaches the boundary, the productivity of the high permeability reservoir region begins to decrease, and the cumulative production increase slows down. Therefore, in practical development, an increase in the shale reservoir permeability can improve the recovery of a defined reservoir region.

The effect of the microfracture system porosity on production of a MFHW in a shale gas reservoir is shown in Fig. 8.10. As the porosity increases, the gas production rate and cumulative production increase significantly. On one hand, a porosity increase can increase gas storage in matrix pores; on the other hand, it can also increase a gas adsorption volume. However, for an ultra-low porosity of a shale reservoir, such an increase in the gas adsorption volume is ignorable for gas production. When $\Phi = 2\%$, the OGIP controlled is calculated as $3.18 \times 10^8\,m^3$, while adsorbed gas is $2.35 \times 10^8\,m^3$; when $\Phi = 8\%$, the controlled OGIP is $5.42 \times 10^8\,m^3$, while adsorbed gas is $2.21 \times 10^8\,m^3$. Field practice shows that large amounts of gas adsorbed in a low porosity rock should be desorbed with depressurization; moreover, according to the relevant principles of porosity and permeability, low porosity will inevitably lead to poor flow conditions and the adsorbed gas is more difficult to produce than free gas.

8.3 A MFHW in a dual-continuum medium in a shale gas reservoir

8.3.1 Model assumptions and mathematical model

With developed natural fractures and high content of brittle minerals, a shale gas reservoir can be hydraulically fractured to acquire favorable flow capacity for gas production by connecting with the natural fractures. However, due to the high contrast of permeability and porosity between shale matrix and microfractures, free gas in the microfracture system is produced first, and then gas is recharged in the matrix with an increase in the pressure difference between the matrix and microfracture system.

In this section, the structure of a microfracture system and a matrix system is represented by a dual-continuum medium model. The CVFEM and implicit NR (Newton–Raphson) iterative method are combined to obtain numerical solution for a MFHW in a shale gas reservoir. The outer rectangular boundary condition of the reservoir is assumed to be no flow, as shown in Fig. 8.11.

A dual-medium model is composed of matrix and a fracture system. Large amounts of free gas is stored in the fracture system; the matrix system is tight and has a large specific surface area, in which there are small portions of the free gas and adsorbed gas, as shown in Fig. 8.12. The fracture system is the main flow channel because of

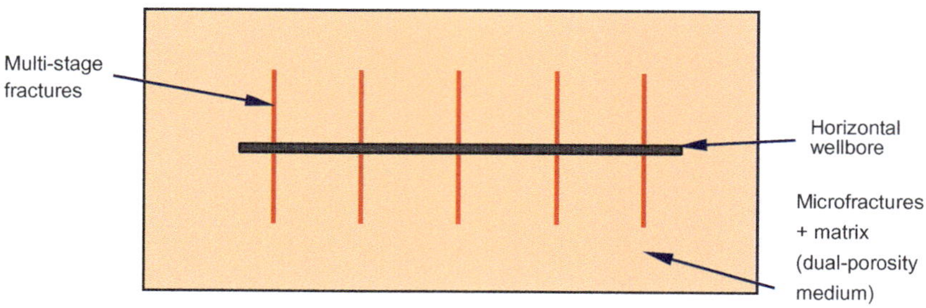

Fig. 8.11 Physical model of a MFHW in a sealed rectangular dual-porosity gas reservoir.

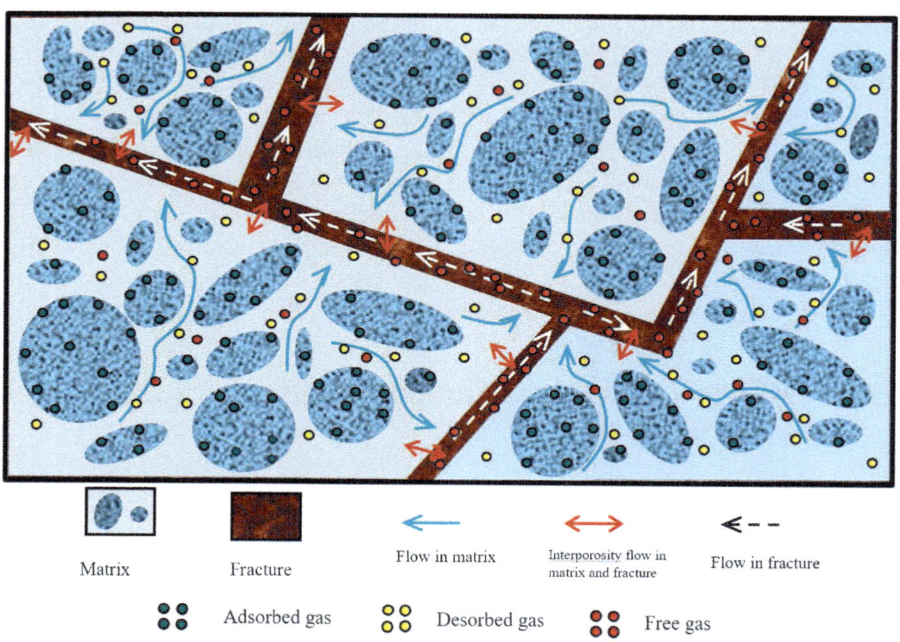

Fig. 8.12 Gas storage and flow mechanisms in a dual-porosity shale gas reservoir.

high permeability and large diameter pore throats. However, with a pressure drop, the conductivity of natural fractures without proppants decreases a lot, which can be seen as a stress sensitivity of the natural fracture system. Adsorbed gas desorption happens in the matrix system, and at the same time, desorbed gas flows from the matrix into the fracture system under a pressure difference, which includes multiple flow mechanisms, viscous flow, Knudsen diffusion, and slippage.

In this section, for the flow in a multistage fracturing horizontal well in a dual-continuum shale gas reservoir, the basic assumptions different from those for a single porosity model are as follows: (1) the fracture system is composed of natural fracture and hydraulic fracture networks, which is the main storage space and flow paths for free gas, and the flow mechanism in this system is viscous flow with stress sensitivity (Fig. 8.13); (2) gas is mainly adsorbed on the surface of shale matrix, and the desorbed gas cannot directly flow into the flow paths, but flows within the matrix under multiple flow mechanisms (viscous flow, Knudsen diffusion, and slippage) (Fig. 8.14); (3) the flow between the matrix and fracture system follows Fick's first law.

Based on the above physical model and assumptions, for single phase flow, considering the mass conservation being equivalent to the volume conservation under the same conditions, the flow differential equations can be described as:

$$\nabla \cdot \left(\frac{k_{\mathrm{fe}}}{\mu B_{\mathrm{g}}} \nabla p_{\mathrm{f}}\right) + \alpha \frac{k_{\mathrm{me}}}{\mu B_{\mathrm{g}}}(p_{\mathrm{m}} - p_{\mathrm{f}}) + q = \frac{\partial(\phi_{\mathrm{f}}/B_{\mathrm{g}})}{\partial t} \tag{8.26}$$

$$\nabla \cdot \left(\frac{k_{\mathrm{me}}}{\mu B_{\mathrm{g}}} \nabla p_{\mathrm{m}}\right) - \alpha \frac{k_{\mathrm{me}}}{\mu B_{\mathrm{g}}}(p_{\mathrm{m}} - p_{\mathrm{f}}) + q^{*} = \frac{\partial(\phi_{\mathrm{m}}/B_{\mathrm{g}})}{\partial t} \tag{8.27}$$

where q is the source sink term, q^* is the desorption rate, and k_{me} and k_{fe} are the permeability for the matrix and fracture systems.

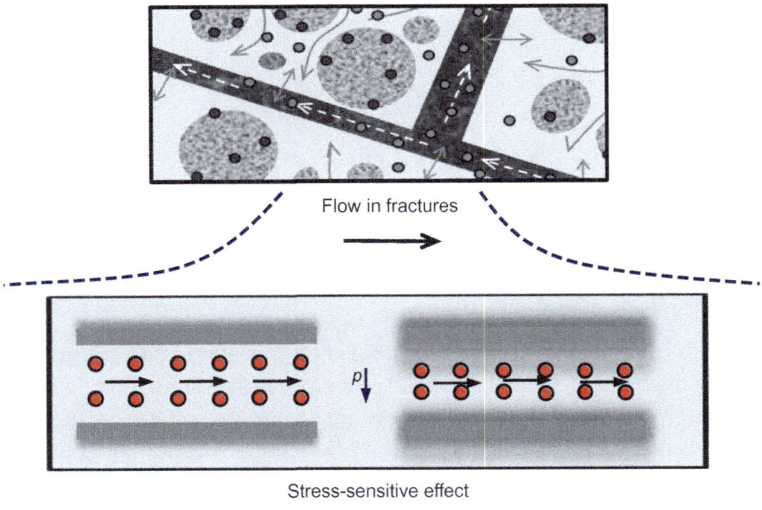

Fig. 8.13 Flow mechanisms in microfractures in a dual-porosity shale gas reservoir.

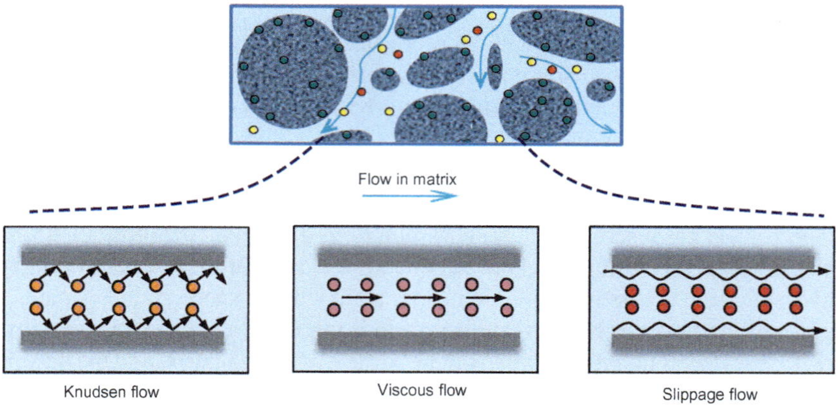

Fig. 8.14 Flow mechanism in the matrix in a dual-porosity shale gas reservoir.

Eqs. (8.26) and (8.27) are, respectively, for the fracture and matrix systems. According to the Langmuir isothermal adsorption and desorption under equilibrium, there is:

$$q^* = -(1-\phi_{\mathrm{m}}-\phi_{\mathrm{f}})\frac{\partial V_{\mathrm{E}}}{\partial t} \tag{8.28}$$

For the effect of stress sensitivity (Chin et al., 2000; Cho et al., 2013), there is:

$$k_{\mathrm{fe}} = k_{\mathrm{f}}\mathrm{e}^{-\theta(p_{\mathrm{fi}}-p_{\mathrm{f}})} \tag{8.29}$$

Taking the multiscale flow in the matrix system in a shale gas reservoir into account, Darcy's flow, Knudsen diffusion flow, and a slippage effect (Civan et al., 2011; Jiang and Rami, 2015; Li et al., 2014) can be integrated to represent the apparent permeability of matrix as:

$$k_{\mathrm{me}} = k_{\mathrm{m}}(1+\beta(K_n)K_n)\left(1+\frac{4K_n}{1-bK_n}\right)+\mu_{\mathrm{g}}D_{\mathrm{g}}C_{\mathrm{g}} \tag{8.30}$$

$$\beta(K_n) = \frac{128}{15\pi^2}\tan^{-1}(4K_n)^{0.4} \tag{8.31}$$

8.3.2 Numerical solution

Compared with the gas mass conservation equation in a single porosity medium, the mass equation of shale gas in a dual-porosity medium is the same except with an addition of the interporosity flow term. According to the discretization theory of the CVFEM, the discretization of the interporosity flow item over triangular unstructured grids can be expressed as:

$$\frac{\alpha A}{3}\begin{pmatrix} (k_{\mathrm{m}}/\mu B_{\mathrm{g}})_1 & 0 & 0 \\ 0 & (k_{\mathrm{m}}/\mu B_{\mathrm{g}})_2 & 0 \\ 0 & 0 & (k_{\mathrm{m}}/\mu B_{\mathrm{g}})_3 \end{pmatrix}\begin{pmatrix} p_{\mathrm{m}1}-p_{\mathrm{f}1} \\ p_{\mathrm{m}2}-p_{\mathrm{f}2} \\ p_{\mathrm{m}3}-p_{\mathrm{f}3} \end{pmatrix} \tag{8.32}$$

Combining the element discretization form of the single porosity shale gas reservoir with Eq. (8.19), the discretization forms of the dual-medium differential flow equations over triangular elements can be acquired.

For the fracture system:

$$\boldsymbol{T}_{\mathrm{f}}^{(e)}\boldsymbol{p}_{\mathrm{f}}^{(e)}+\boldsymbol{W}^{(e)}\left(\boldsymbol{p}_{\mathrm{m}}^{(e)}-\boldsymbol{p}_{\mathrm{f}}^{(e)}\right)=\left[\left(\boldsymbol{N}_{\mathrm{f}}^{(e)}\right)^{n+1}-\left(\boldsymbol{N}_{\mathrm{f}}^{(e)}\right)^{n}\right] \tag{8.33}$$

For the matrix system:

$$\boldsymbol{T}_{\mathrm{m}}^{(e)}\boldsymbol{p}_{\mathrm{m}}^{(e)}-\boldsymbol{W}^{(e)}\left(\boldsymbol{p}_{\mathrm{m}}^{(e)}-\boldsymbol{p}_{\mathrm{f}}^{(e)}\right)-\left[\left(\boldsymbol{V}_{\mathrm{E}}^{(e)}\right)^{n+1}-\left(\boldsymbol{V}_{\mathrm{E}}^{(e)}\right)^{n}\right]=\left[\left(\boldsymbol{N}_{\mathrm{m}}^{(e)}\right)^{n+1}-\left(\boldsymbol{N}_{\mathrm{m}}^{(e)}\right)^{n}\right] \tag{8.34}$$

where:

$$\eta=\mathrm{m},\mathrm{f};\ \boldsymbol{p}_{\eta}^{(e)}=(p_1\ \ p_2\ \ p_3)_{\eta}^{\mathrm{T}}$$

$\boldsymbol{T}_{\eta}^{(e)}$ is the element transmissibility matrix:

$$\boldsymbol{T}_{\eta}^{(e)}=\begin{pmatrix} T_{12}^{k}+T_{13}^{k} & -T_{12}^{k} & -T_{13}^{k} \\ -T_{21}^{k} & T_{21}^{k}+T_{23}^{k} & -T_{23}^{k} \\ -T_{31}^{k} & -T_{32}^{k} & T_{31}^{k}+T_{32}^{k} \end{pmatrix}_{\eta}$$

$\boldsymbol{N}_{\eta}^{(e)}$ is the element cumulative matrix:

$$\boldsymbol{N}_{\eta}^{(e)}=\frac{A}{3\Delta t}\begin{pmatrix} \left(\frac{\phi}{B_{\mathrm{g}}}\right)_1 \\ \left(\frac{\phi}{B_{\mathrm{g}}}\right)_2 \\ \left(\frac{\phi}{B_{\mathrm{g}}}\right)_3 \end{pmatrix}_{\eta}$$

$\boldsymbol{W}^{(e)}$ is the element interporosity flow matrix:

$$\boldsymbol{W}^{(e)}=\frac{\alpha A}{3}\begin{pmatrix} (k_{\mathrm{m}}/\mu B_{\mathrm{g}})_1 & 0 & 0 \\ 0 & (k_{\mathrm{m}}/\mu B_{\mathrm{g}})_2 & 0 \\ 0 & 0 & (k_{\mathrm{m}}/\mu B_{\mathrm{g}})_3 \end{pmatrix}$$

$\boldsymbol{V}_{\mathrm{E}}^{(e)}$ is the element adsorption matrix:

$$\boldsymbol{V}_{\mathrm{E}}^{(e)}=\frac{A}{3\Delta t}\begin{pmatrix}[(1-\phi_{\mathrm{t}})V_{\mathrm{E}}]_1\\ [(1-\phi_{\mathrm{t}})V_{\mathrm{E}}]_2\\ [(1-\phi_{\mathrm{t}})V_{\mathrm{E}}]_3\end{pmatrix}_{\mathrm{m}}$$

The element discrete equations are composed of Eqs. (8.33) and (8.34). Due to gas compressibility, there is strong nonlinearity of the diffusion, adsorption, and interporosity terms in these equations, causing a difficulty in solving them through a sequential decoupling method for compressible gas. Therefore, the implicit NR iterations are used to solve for unknown variables simultaneously.

The implicit iteration format is as follows: Let the approximate solution at the $(k+1)$ th iteration be $p^{n+1}\approx p^{k+1}=p^{k}+\delta p^{k}$, and for all nonlinear variables at this iteration there is $\zeta^{n+1}\approx\zeta^{k+1}=\zeta^{k}+\delta\zeta^{k}$.

For the fracture system:

$$\begin{aligned}&\left[\boldsymbol{T}_{\mathrm{f}}^{(e)}+\boldsymbol{\delta T}_{\mathrm{f}}^{(e)}-\left(\boldsymbol{W}^{(e)}+\boldsymbol{\delta W}^{(e)}\right)\right]^{k}\left(\boldsymbol{p}_{\mathrm{f}}^{(e)}+\boldsymbol{\delta p}_{\mathrm{f}}^{(e)}\right)^{k}+\left(\boldsymbol{W}^{(e)}+\boldsymbol{\delta W}^{(e)}\right)^{k}\left(\boldsymbol{p}_{\mathrm{m}}^{(e)}+\boldsymbol{\delta p}_{\mathrm{m}}^{(e)}\right)^{k}\\&=\left(\boldsymbol{N}_{\mathrm{f}}^{(e)}+\boldsymbol{\delta N}_{\mathrm{f}}^{(e)}\right)^{k}-\left(\boldsymbol{N}_{\mathrm{f}}^{(e)}\right)^{n}\end{aligned}\tag{8.35}$$

For the matrix system:

$$\begin{aligned}&\left[\boldsymbol{T}_{\mathrm{m}}^{(e)}+\boldsymbol{\delta T}_{\mathrm{m}}^{(e)}-\left(\boldsymbol{W}^{(e)}+\boldsymbol{\delta W}^{(e)}\right)^{k}\right]^{k}\left(\boldsymbol{p}_{\mathrm{m}}^{(e)}+\boldsymbol{\delta p}_{\mathrm{m}}^{(e)}\right)^{k}+\left(\boldsymbol{W}^{(e)}+\boldsymbol{\delta W}^{(e)}\right)^{k}\left(\boldsymbol{p}_{\mathrm{f}}^{(e)}+\boldsymbol{\delta p}_{\mathrm{f}}^{(e)}\right)^{k}\\&=\left(\boldsymbol{V}_{\mathrm{E}}^{(e)}+\boldsymbol{\delta V}_{\mathrm{E}}^{(e)}\right)^{k}-\left(\boldsymbol{V}_{\mathrm{E}}^{(e)}\right)^{n}+\left(\boldsymbol{N}_{\mathrm{m}}^{(e)}+\boldsymbol{\delta N}_{\mathrm{m}}^{(e)}\right)^{k}-\left(\boldsymbol{N}_{\mathrm{m}}^{(e)}\right)^{n}\end{aligned}\tag{8.36}$$

Expanding the above equations, ignoring higher order terms, and expressing them in residual form, there are:

$$\begin{aligned}&\left[\boldsymbol{T}_{\mathrm{f}}^{(e)}\boldsymbol{\delta p}_{\mathrm{f}}^{(e)}+\boldsymbol{\delta T}_{\mathrm{f}}^{(e)}\boldsymbol{p}_{\mathrm{f}}^{(e)}-\boldsymbol{\delta W}^{(e)}\boldsymbol{p}_{\mathrm{f}}^{(e)}-\boldsymbol{W}^{(e)}\boldsymbol{\delta p}_{\mathrm{f}}^{(e)}\right]^{k}+\left(\boldsymbol{W}^{(e)}\boldsymbol{\delta p}_{\mathrm{m}}^{(e)}+\boldsymbol{\delta W}^{(e)}\boldsymbol{p}_{\mathrm{m}}^{(e)}\right)^{k}-\left(\boldsymbol{\delta N}_{\mathrm{f}}^{(e)}\right)^{k}\\&=-\left\{\left[\boldsymbol{T}_{\mathrm{f}}^{(e)}\boldsymbol{p}_{\mathrm{f}}^{(e)}+\boldsymbol{W}^{(e)}\left(\boldsymbol{p}_{\mathrm{m}}^{(e)}-\boldsymbol{p}_{\mathrm{f}}^{(e)}\right)\right]^{k}-\left(\boldsymbol{N}_{\mathrm{f}}^{(e)}\right)^{k}+\left(\boldsymbol{N}_{\mathrm{f}}^{(e)}\right)^{n}\right\}=-\left(\boldsymbol{R}_{\mathrm{f}}^{(e)}\right)^{k}\end{aligned}\tag{8.37}$$

$$\begin{aligned}&\left[\boldsymbol{T}_{\mathrm{m}}^{(e)}\boldsymbol{\delta p}_{\mathrm{m}}^{(e)}+\boldsymbol{\delta T}_{\mathrm{m}}^{(e)}\boldsymbol{p}_{\mathrm{m}}^{(e)}-\boldsymbol{\delta W}^{(e)}\boldsymbol{p}_{\mathrm{m}}^{(e)}-\boldsymbol{W}^{(e)}\boldsymbol{\delta p}_{\mathrm{m}}^{(e)}\right]^{k}+\left(\boldsymbol{W}^{(e)}\boldsymbol{\delta p}_{\mathrm{f}}^{(e)}+\boldsymbol{\delta W}^{(e)}\boldsymbol{p}_{\mathrm{f}}^{(e)}\right)^{k}-\left(\boldsymbol{\delta N}_{\mathrm{m}}^{(e)}\right)^{k}-\left(\boldsymbol{\delta V}_{\mathrm{E}}^{(e)}\right)^{k}\\&=-\left\{\left[\boldsymbol{T}_{\mathrm{m}}^{(e)}\boldsymbol{p}_{\mathrm{m}}^{(e)}+\boldsymbol{W}^{(e)}\left(\boldsymbol{p}_{\mathrm{m}}^{(e)}-\boldsymbol{p}_{\mathrm{f}}^{(e)}\right)\right]^{k}-\left(\boldsymbol{N}_{\mathrm{m}}^{(e)}\right)^{k}+\left(\boldsymbol{N}_{\mathrm{m}}^{(e)}\right)^{n}-\left(\boldsymbol{V}_{\mathrm{E}}^{(e)}\right)^{k}+\left(\boldsymbol{V}_{\mathrm{E}}^{(e)}\right)^{n}\right\}\\&=-\left(\boldsymbol{R}_{\mathrm{m}}^{(e)}\right)^{k}\end{aligned}\tag{8.38}$$

where:

$$\boldsymbol{\delta p}_{\eta}^{(e)} = (\delta p_1 \quad \delta p_2 \quad \delta p_3)_{\eta}^{\mathrm{T}}; \quad \boldsymbol{R}_{\eta}^{(e)} = (R_1 \quad R_2 \quad R_3)_{\eta}^{\mathrm{T}}$$

$$\boldsymbol{\delta T}_{\eta}^{(e)} \boldsymbol{p}_{\eta}^{(e)} = \boldsymbol{\delta} \tilde{\boldsymbol{T}}_{\eta}^{(e)} \delta \boldsymbol{p}_{\eta}^{(e)}; \quad \boldsymbol{\delta N}_{\eta}^{(e)} = \boldsymbol{\delta} \tilde{\boldsymbol{N}}_{\eta}^{(e)} \delta \boldsymbol{p}_{\eta}^{(e)}$$

$$\boldsymbol{\delta W}^{(e)} \boldsymbol{p}_{\eta}^{(e)} = \boldsymbol{\delta} \tilde{\boldsymbol{W}}^{(e)} \delta \boldsymbol{p}_{\eta}^{(e)}; \quad \boldsymbol{\delta V}_{\mathrm{E}}^{(e)} = \boldsymbol{\delta} \tilde{\boldsymbol{V}}_{\mathrm{E}}^{(e)} \delta \boldsymbol{p}_{\mathrm{m}}^{(e)}; \quad \eta = \mathrm{m, f}$$

$\boldsymbol{\delta}\tilde{\boldsymbol{T}}_{\eta}^{(e)}$ is the element partial derivative matrix of transmissibility:

$$\boldsymbol{\delta} \tilde{\boldsymbol{T}}_{\eta}^{(e)} = \begin{pmatrix} \begin{bmatrix} \frac{\partial T_{12}}{\partial p_1}(p_1 - p_2) \\ + \frac{\partial T_{13}}{\partial p_1}(p_1 - p_3) \end{bmatrix} & \frac{\partial T_{12}}{\partial p_2}(p_1 - p_2) & \frac{\partial T_{13}}{\partial p_3}(p_1 - p_3) \\ \frac{\partial T_{12}}{\partial p_1}(p_2 - p_1) & \begin{bmatrix} \frac{\partial T_{12}}{\partial p_2}(p_2 - p_1) \\ + \frac{\partial T_{23}}{\partial p_2}(p_2 - p_3) \end{bmatrix} & \frac{\partial T_{23}}{\partial p_3}(p_2 - p_3) \\ \frac{\partial T_{13}}{\partial p_2}(p_3 - p_1) & \frac{\partial T_{23}}{\partial p_2}(p_3 - p_2) & \begin{bmatrix} \frac{\partial T_{13}}{\partial p_3}(p_3 - p_1) \\ + \frac{\partial T_{23}}{\partial p_3}(p_3 - p_2) \end{bmatrix} \end{pmatrix}_{\eta}$$

$\boldsymbol{\delta}\tilde{\boldsymbol{N}}_{\eta}^{(e)}$ is the element partial derivative matrix of the cumulative item:

$$\boldsymbol{\delta} \tilde{\boldsymbol{N}}_{\eta}^{(e)} = \frac{A}{3\Delta t} \begin{pmatrix} \frac{\partial}{\partial p}\left(\frac{\phi_\eta}{B_{\mathrm{g}}}\right)\Big|_{p_{\eta i}^k} & 0 & 0 \\ 0 & \frac{\partial}{\partial p}\left(\frac{\phi_\eta}{B_{\mathrm{g}}}\right)\Big|_{p_{\eta j}^k} & 0 \\ 0 & 0 & \frac{\partial}{\partial p}\left(\frac{\phi_\eta}{B_{\mathrm{g}}}\right)\Big|_{p_{\eta k}^k} \end{pmatrix}$$

$\boldsymbol{\delta}\tilde{\boldsymbol{W}}^{(e)}$ is the element partial derivative matrix of the interporosity flow:

$$\boldsymbol{\delta} \tilde{\boldsymbol{W}}^{(e)} = \frac{\alpha A}{3} \begin{pmatrix} (k/\mu B_{\mathrm{g}})_1 & 0 & 0 \\ 0 & (k/\mu B_{\mathrm{g}})_2 & 0 \\ 0 & 0 & (k/\mu B_{\mathrm{g}})_3 \end{pmatrix}_{\mathrm{m}}$$

$\delta\widetilde{V}_{\rm E}^{(e)}$ is the element partial derivative matrix of the gas adsorption item:

$$\delta\widetilde{V}_{\rm E}^{(e)} = \frac{A}{3\Delta t}\begin{pmatrix} \partial[(1-\phi_{\rm t})V_{\rm E}]/\partial p|_{p_{\rm mi}^k} & 0 & 0 \\ 0 & \partial[(1-\phi_{\rm t})V_{\rm E}]/\partial p|_{p_{\rm mj}^k} & 0 \\ 0 & 0 & \partial[(1-\phi_{\rm t})V_{\rm E}]/\partial p|_{p_{\rm mk}^k} \end{pmatrix}$$

Then results are:

$$\left(\boldsymbol{T}_{\rm f}^{(e)}+\boldsymbol{\delta}\tilde{\boldsymbol{T}}_{\rm f}^{(e)}-\boldsymbol{W}^{(e)}-\boldsymbol{\delta}\tilde{\boldsymbol{W}}^{(e)}-\boldsymbol{\delta}\tilde{\boldsymbol{N}}_{\rm f}^{(e)}\right)^k\left(\boldsymbol{\delta p}_{\rm f}^{(e)}\right)^k+\left(\boldsymbol{W}^{(e)}+\boldsymbol{\delta}\tilde{\boldsymbol{W}}^{(e)}\right)^k\left(\boldsymbol{\delta p}_{\rm m}^{(e)}\right)^k=-\left(\boldsymbol{R}_{\rm f}^{(e)}\right)^k \tag{8.39}$$

$$\left[\boldsymbol{T}_{\rm m}^{(e)}+\boldsymbol{\delta}\tilde{\boldsymbol{T}}_{\rm m}^{(e)}-\boldsymbol{W}^{(e)}-\boldsymbol{\delta}\tilde{\boldsymbol{W}}^{(e)}-\boldsymbol{\delta}\tilde{\boldsymbol{N}}_{\rm m}^{(e)}-\boldsymbol{\delta}\widetilde{\boldsymbol{V}}_{\rm E}^{(e)}\right]^k\left(\boldsymbol{\delta p}_{\rm m}^{(e)}\right)^k+\left(\boldsymbol{W}^{(e)}+\boldsymbol{\delta}\tilde{\boldsymbol{W}}^{(e)}\right)^k\left(\boldsymbol{\delta p}_{\rm f}^{(e)}\right)^k=-\left(\boldsymbol{R}_{\rm m}^{(e)}\right)^k \tag{8.40}$$

Eqs. (8.39) and (8.40) are the iterative format of the element discrete matrix for the dual-medium system, which can be assembled into the general matrix in the solution domain through the assembly method for a finite element matrix:

$$\left(\boldsymbol{T}_{\rm f}+\boldsymbol{\delta}\tilde{\boldsymbol{T}}_{\rm f}-\boldsymbol{W}-\boldsymbol{\delta}\tilde{\boldsymbol{W}}-\boldsymbol{\delta}\tilde{\boldsymbol{N}}_{\rm f}\right)^k(\boldsymbol{\delta p}_{\rm f})^k+\left(\boldsymbol{W}+\boldsymbol{\delta}\tilde{\boldsymbol{W}}\right)^k(\boldsymbol{\delta p}_{\rm m})^k=-(\boldsymbol{R}_{\rm f})^k \tag{8.41}$$

$$\left(\boldsymbol{T}_{\rm m}+\boldsymbol{\delta}\tilde{\boldsymbol{T}}_{\rm m}-\boldsymbol{W}-\boldsymbol{\delta}\tilde{\boldsymbol{W}}-\boldsymbol{\delta}\tilde{\boldsymbol{N}}_{\rm m}-\boldsymbol{\delta}\widetilde{\boldsymbol{V}}_{\rm E}\right)^k(\boldsymbol{\delta p}_{\rm m})^k+\left(\boldsymbol{W}+\boldsymbol{\delta}\tilde{\boldsymbol{W}}\right)^k(\boldsymbol{\delta p}_{\rm f})^k=-(\boldsymbol{R}_{\rm m})^k \tag{8.42}$$

In this section, it is assumed that the whole reservoir region is discretized into NN elements and MM nodes. Thus the residue matrix is:

$$\mathbf{R}_\eta = (R_1 \;\; R_2 \;\; \cdots \;\; R_{MM})_\eta^{\rm T}$$

$$\boldsymbol{\delta p}_\eta = (\delta p_1 \;\; \delta p_2 \;\; \cdots \;\; \delta p_{MM})_\eta^{\rm T}$$

$$\boldsymbol{p}_\eta = (p_1 \;\; p_2 \;\; \cdots \;\; p_{MM})_\eta^{\rm T}$$

$$\zeta = \sum_1^{NN}\zeta^{(e)};\;\; \boldsymbol{\delta}\zeta = \sum_1^{NN}\boldsymbol{\delta}\zeta^{(e)},\;\; \zeta = \boldsymbol{T}_\eta, \boldsymbol{N}_\eta, \boldsymbol{W}, \boldsymbol{V}_E$$

The solution of the simultaneous equations is:

$$\begin{Bmatrix} A_{11} & A_{12} \\ A_{21} & A_{22} \end{Bmatrix}\begin{pmatrix} \delta\tilde{\boldsymbol{P}}_{\rm f}^k \\ \delta\tilde{\boldsymbol{p}}_{\rm m}^k \end{pmatrix} = \begin{pmatrix} -\tilde{\boldsymbol{R}}_{\rm f}^k \\ -\tilde{\boldsymbol{R}}_{\rm m}^k \end{pmatrix} \tag{8.43}$$

where:

$$\boldsymbol{A}_{11} = \boldsymbol{T}_{\rm f}+\boldsymbol{\delta}\tilde{\boldsymbol{T}}_{\rm f}-\boldsymbol{W}-\boldsymbol{\delta}\tilde{\boldsymbol{W}}-\boldsymbol{\delta}\tilde{\boldsymbol{N}}_{\rm f};\;\; \boldsymbol{A}_{12}=\boldsymbol{A}_{21}=\boldsymbol{W}+\boldsymbol{\delta}\tilde{\boldsymbol{W}}$$

$$\boldsymbol{A}_{22} = \boldsymbol{T}_{\rm m}+\boldsymbol{\delta}\tilde{\boldsymbol{T}}_{\rm m}-\boldsymbol{W}-\boldsymbol{\delta}\tilde{\boldsymbol{W}}-\boldsymbol{\delta}\tilde{\boldsymbol{N}}_{\rm m}-\boldsymbol{\delta}\widetilde{V}_{\rm E}$$

Table 8.3 Shale reservoir properties

Reservoir property	Value	Reservoir property	Value
Microfracture system porosity, Φ_f, %	3	Microfracture system permeability, k_f, mD	0.01
Matrix porosity, Φ_m, %	1	Matrix permeability, k_m, mD	0.0005

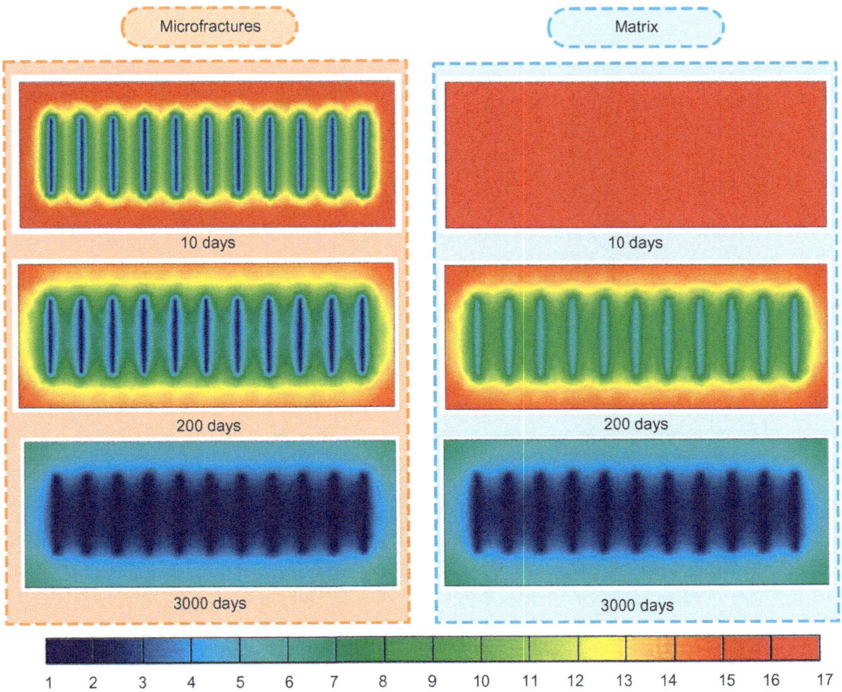

Fig. 8.15 Pressure distribution in each system in a dual-porosity shale gas reservoir.

8.3.3 Production performance analysis

The basic properties of a shale gas reservoir are the same as in the previous section. In this section, we add the data for a dual-continuum medium as shown in Table 8.3.

Fig. 8.15 shows the pressure distribution changes of a MFHW in the dual-continuum medium. After production, free gas around the hydraulic fractures and microfracture system flows along the fracture direction into the bottom of the well. The pressure field in this early linear flow period is similar to a rectangle. Pressure in the matrix system remains in the initial status during this period, and the pressure difference between the matrix system and the fracture system is not obvious. After that, interference among these hydraulic fractures happens, and interporosity flow occurs. However, the matrix pressure is still higher than the fracture pressure. Until a late flow period, the matrix and fracture systems form a unified pressure system, within which the flow characteristics are equivalent to those in a single porosity medium.

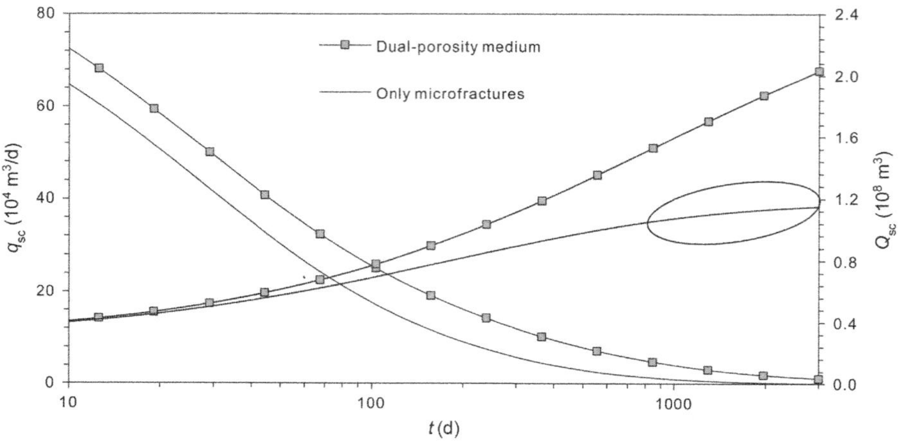

Fig. 8.16 Production decline curve of a MFHW in a dual-porosity shale gas reservoir.

Fig. 8.16 shows the production decline curve of a MFHW in a dual-porosity reservoir. It can be seen from this figure that in the early period, free gas in the fracture system improves the production rate. As the free gas is produced, the pressure difference between the microfracture and matrix systems becomes large; gas stored in the matrix begins flowing into the fracture system, and the absorbed gas starts desorbing. After a period of production, free gas has been exhausted; desorbed gas becomes the main source of gas production. After 3000 days, the cumulative production is $2.28 \times 10^8 \, m^3$; the recovery factor is up to 58.3%. By assuming the matrix permeability equal to 0, this model is simplified to a single porosity model. The plot comparison shows that due to the lack of interporosity flow supplement, the production rate and cumulative production in a single porosity medium are both lower than those in a dual-porosity medium. After 3000 days, the cumulative production for the former is only $1.15 \times 10^8 \, m^3$.

The influence of a Langmuir volume on production performance is shown in Fig. 8.17. It is clearly observed that V_L mainly influences the curves in the middle and later flow periods, which is the interporosity flow between the matrix and natural fracture systems. The larger the V_L, the greater the production rate with a constant BHP; after a quick decline period, the production curve becomes flat. This is because a bigger value of the Langmuir volume represents a larger amount of adsorbed gas in shale matrix, which decreases the formation pressure drop and supplements the gas capacity.

As Fig. 8.18 shows, the production rate and accumulative production both decrease with the stress sensitivity increasing. According to Eq. (8.29), when pressure declines, a higher stress sensitivity leads to a greater decrease in the natural fracture permeability. If we assume that the stress sensitivity coefficient is 0.01, 0.05, and 0.1, after 3000 days, the recovery is 52%, 49%, and 47%, respectively.

The porosity of the fracture system mainly affects the volume of free gas in a shale gas reservoir, and its influence is obvious in the stage of production. As the porosity of

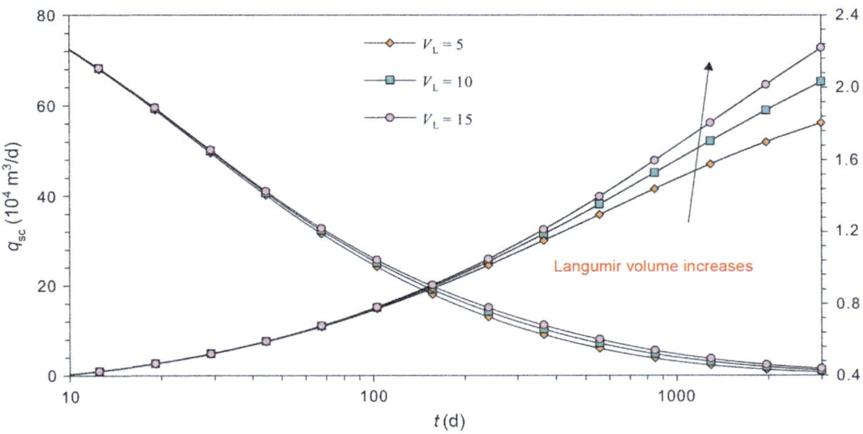

Fig. 8.17 Effect of Langmuir volume on production performance of a MFHW in a shale gas reservoir.

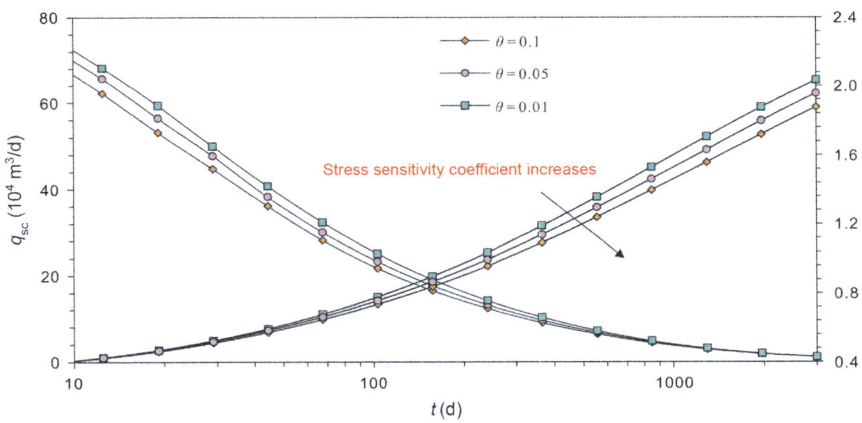

Fig. 8.18 Effect of stress sensitivity coefficient on production performance of a MFHW in shale gas reservoir.

the fracture system increases, the production rate and cumulative production increase. As shown in Fig. 8.19, the orange line represents the conventional gas reservoir production. The plot comparison shows that due to adsorbed gas, shale gas has a high production period at the later period, which can also be maintained for a long time.

The influence of the fracture system permeability on a MFWH in a shale gas reservoir is shown in Fig. 8.20. The fracture system permeability mainly affects the early period of free gas production. The higher the permeability, the higher the early production rate. In the late period, the production decline curves almost become equal.

A larger matrix porosity means that more free gas is stored in the matrix system. However, the gas stored in this system can only be produced after flowing into the

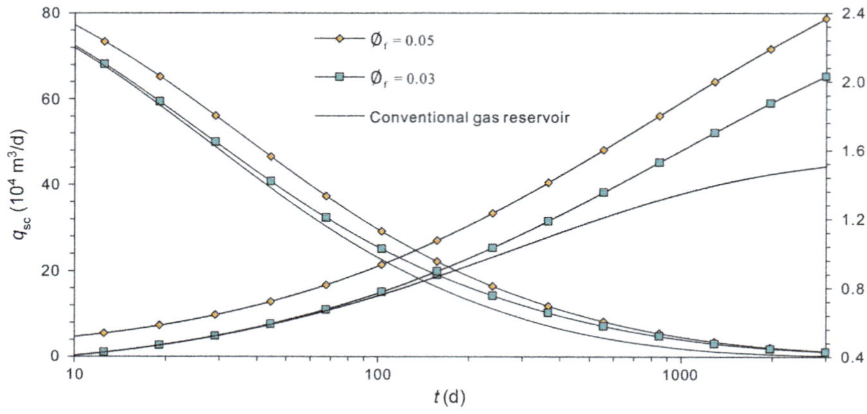

Fig. 8.19 Effect of microfracture system porosity on production performance of a MFHW in shale gas reservoir.

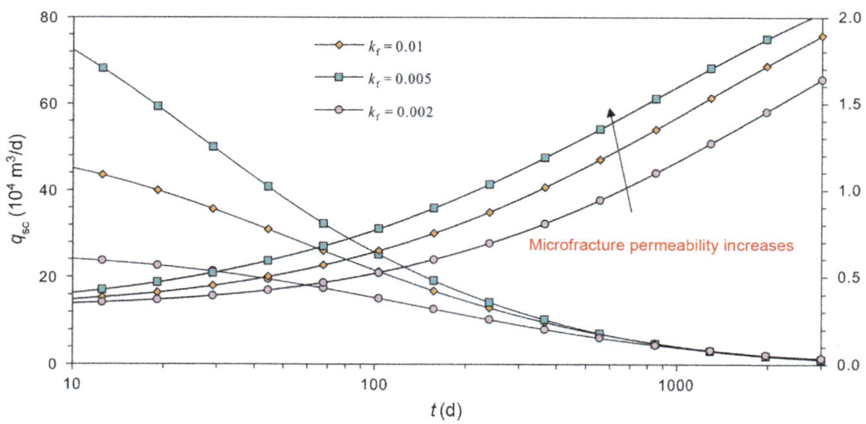

Fig. 8.20 Effect of microfracture system permeability on production performance of a MFHW in a shale gas reservoir.

fracture system. As shown in Fig. 8.21, the higher the matrix porosity, the higher the production rate during the late period; the influence of the matrix porosity on early production is not obvious.

The interporosity flow coefficient α reflects the ability of flowing from matrix to a fracture system, which is mainly related to the properties of rock matrix. Its effect on the productivity of horizontal wells in a shale gas reservoir is shown in Fig. 8.22. This figure shows part of free gas and almost all adsorbed gas being stored in the matrix system. Therefore, the interporosity flow from matrix to a fracture system includes free gas flow and desorption of adsorbed gas. At the early production, the interporosity flow gas is mainly the free gas, and the influence of the interporosity flow coefficient on gas supply is greater. After a few days, the free gas in the reservoir is depleted, and

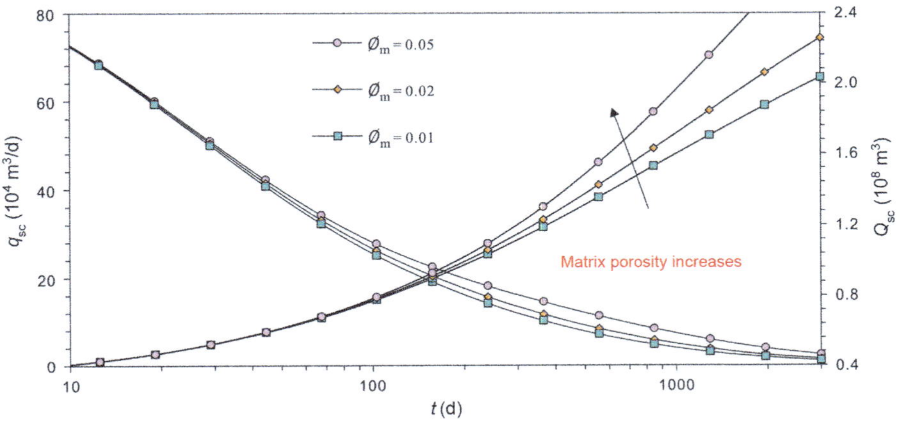

Fig. 8.21 Effect of matrix porosity on production performance of a MFHW in a shale gas reservoir.

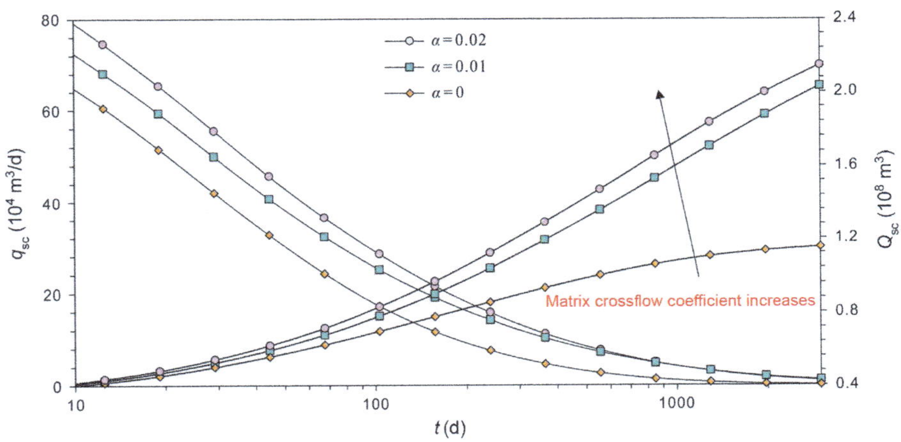

Fig. 8.22 Effect of matrix interporosity flow coefficient on production performance of a MFHW.

desorption gas becomes dominate in the flow. Since the adsorption gas production depends not only on the interporosity flow but also on the desorption gas volume from adsorption, the effect of the coefficient becomes weaker.

8.3.4 Effects of a fracture network on production

A shale gas reservoir has a lot of brittle mineral content, tectonic fractures, overpressure fractures, and contract fractures developed by diagenesis and overlying strata pressure in its geological evolution. Field practices show that the conductivity of these fracture networks can be improved through hydraulic fracturing (a stimulated rock volume). It is generally believed that the technique of drilling multi-stage fractured horizontal wells is the key to the efficient development of shale gas reservoirs. Although a fracture network system (natural and SRV) is very complex in shale, branch fractures models, rectangular composite models, and local fracture network models are widely used to represent fracture network systems (Xu et al., 2015; Zhao et al., 2013; Zhang et al., 2016; Zhang et al., 2015).

8.3.4.1 Branch fracture models

During hydraulic fracturing, if a net pressure for fracture extension is greater than the sum of the difference of two horizontal principle stresses and rock tensile strength, branch fractures are generated during main fracture extension and form networks after certain branch numbers and density are achieved, as shown in Fig. 8.23.

In this section, a network of branch fractures is simplified into a 1D line source and discretized into unstructured triangular grids. Results and calculation parameters are shown in Fig. 8.24 and Table 8.4.

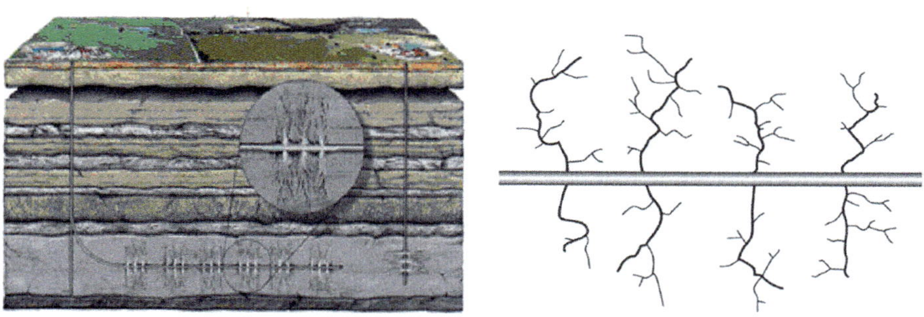

Fig. 8.23 Network of branch fractures of a MFHW.

Fig. 8.24 Triangular grid discretization of a MFHW with branch fractures.

Table 8.4 Model properties of a MFHW with branching fractures

Reservoir property	Value	Reservoir property	Value
Formation thickness, h, m	50	Initial reservoir pressure, p_i, MPa	20
Gas specific gravity, r_g, fraction	0.6	Reservoir temperature, T, °C	100
Formation porosity, Φ, %	5	Formation permeability, k, mD	0.005
Langmuir volume, V_L, m^3/m^3	10	Langmuir pressure, P_L, MPa	4

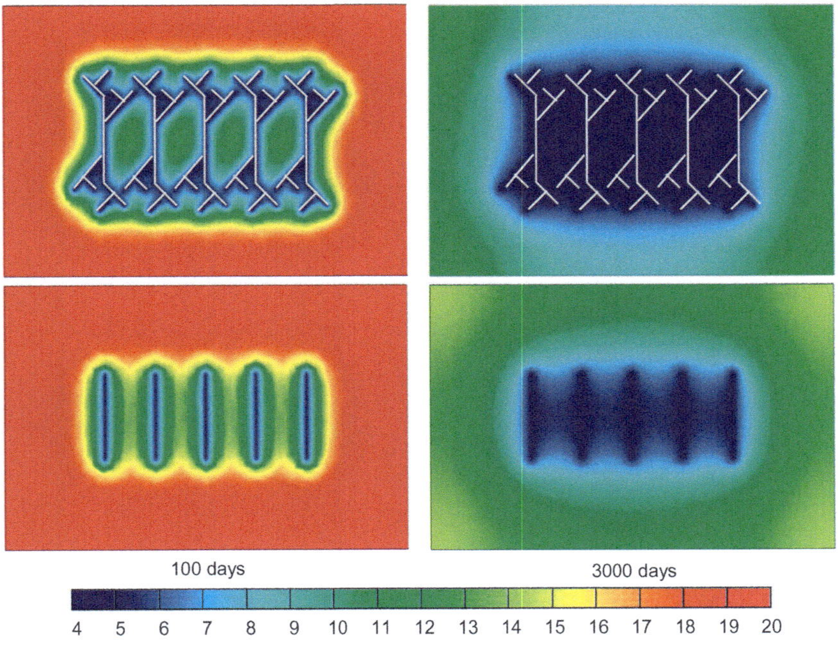

Fig. 8.25 Pressure distribution for branch fractures and conventional models.

Fig. 8.25 shows pressure distributions at different times for the branch fracture model and the conventional MFHW model without induced fractures. Due to the induced fracture networks, after 100 days, the linear and elliptic flow periods occur quickly, and the flow regime is dominated by the formation radial flow; however, after the same time, the elliptic flow and interference between fractures just begin to occur for the conventional well model. Moreover, in the late period, a low-pressure region around the branch fracture well is obviously larger than that from the conventional model, which eventually leads to greater production. Fig. 8.26 shows a 3D pressure projection graph of the branch fractures model in the late period. We can conclude that in the late period, the flow regime is dominated by the formation radial flow; the low pressure region around the induced fracture networks is similar to the shape of a basin floor, and gas desorbs from the matrix of this region as a supplement to gas production.

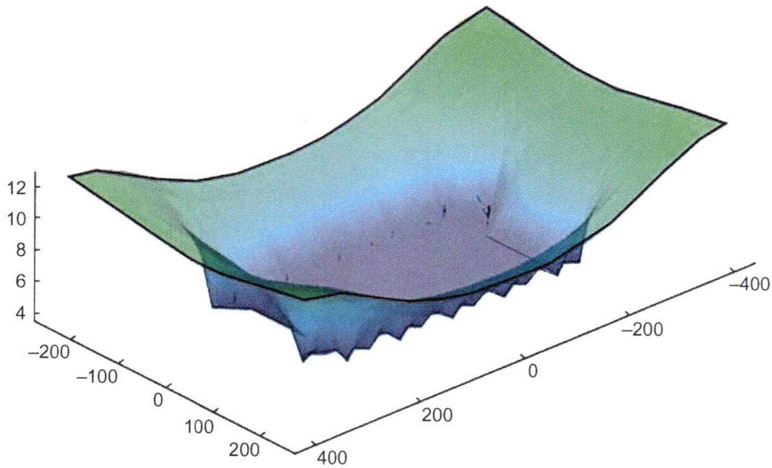

Fig. 8.26 3D projection graph of branch fractures model.

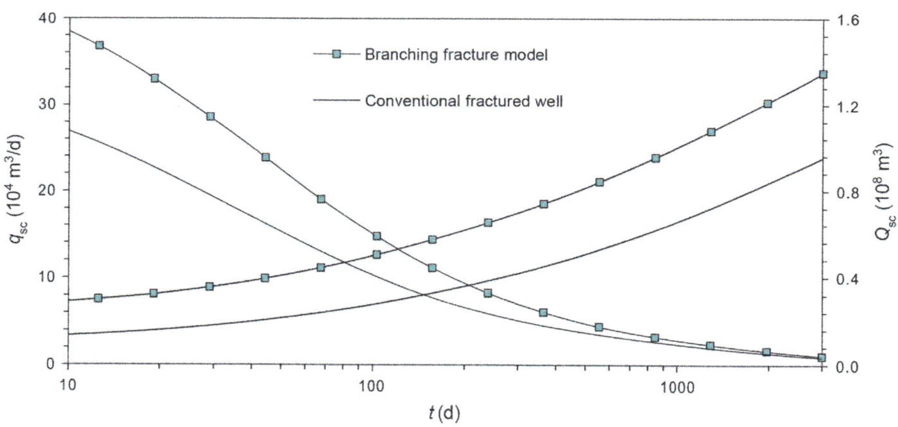

Fig. 8.27 Effect of branch fractures on production.

The OGIP is $2.27 \times 10^8 \text{m}^3$, and the volume of free gas is $1.17 \times 10^8 \text{m}^3$. The influence of branch fractures on a production rate and accumulative production is shown in Fig. 8.27. It can be seen from this figure that a branch fracture network can increase the flow paths of free gas in the early period and thus increase the low pressure region for gas desorption for a higher production rate in the middle and late periods. After 3000 days, the recovery of the branch fracture network model is 61.3%, which is 19.4% higher than the conventional fractured horizontal well model.

8.3.4.2 Rectangular composite models

According to the characteristics of a stimulated rock volume, a composite model can be used to represent a MFHW in a shale gas reservoir. As shown in Fig. 8.28, the shale gas reservoir is treated as three regions: an outer region, a stimulated region, and a

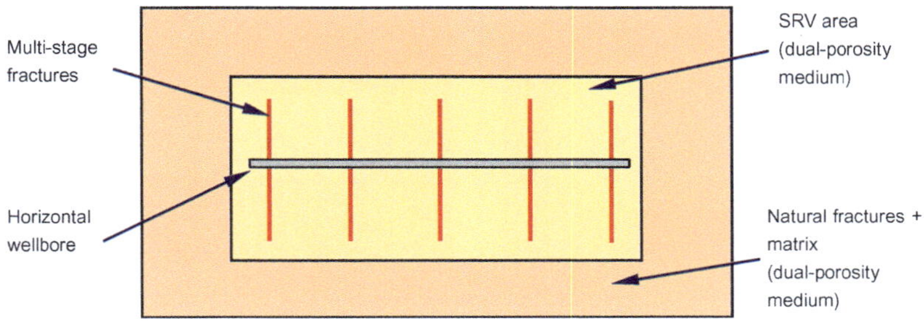

Fig. 8.28 Physical model of a MFHW in a closed composite rectangle shale gas reservoir.

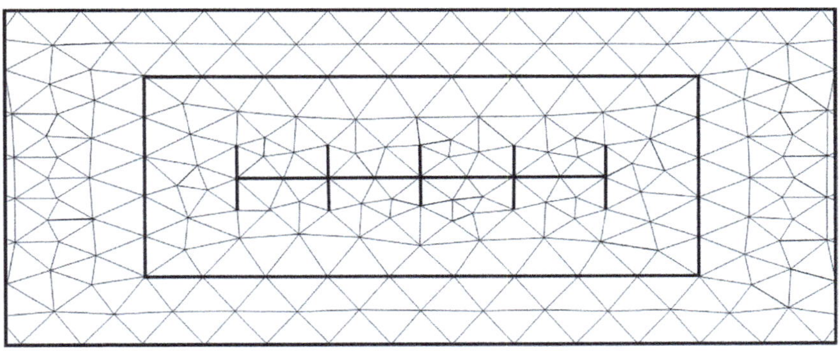

Fig. 8.29 Grid discretization of a MFHW in closed composite rectangle shale gas reservoir.

Table 8.5 Shale reservoir properties

Reservoir property	Value
Inner region fracture porosity, Φ_{f1}, %	3
Outer region fracture porosity, Φ_{f2}, %	1
Matrix porosity, Φ_m, %	1
Inner region microfracture permeability, k_{f1}, mD	0.01
Outer region microfracture permeability, k_{f2}, mD	0.001
Matrix permeability, k_m, mD	0.0001

hydraulic fracture region. Each of them has distinct reservoir properties. Both the inner and outer regions are dual-porosity and dual-permeability, but the inner region, which is stimulated through hydraulic fracturing, has better properties than the outer region.

Fig. 8.29 is a diagram of a MFHW in a closed rectangular composite reservoir. The dimensions of the outer and inner regions are 1600×800 and 1200×400 m, respectively. The main fractures are perpendicular to the horizontal wellbore with equal spacing. The relevant parameters are shown in Table 8.5.

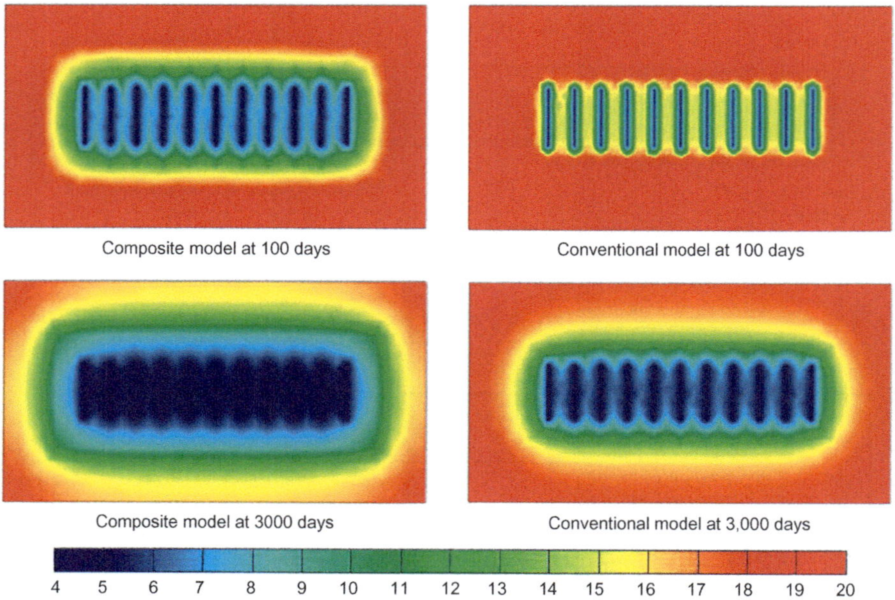

Fig. 8.30 Pressure distribution in a MFHW in a closed composite rectangular gas reservoir.

The calculations show that the geologic OGIPs of this reservoir are $10.6 \times 10^8 \, m^3$, and the volume of adsorption gas is $4.9 \times 10^8 \, m^3$. Fig. 8.30 shows the pressure distribution of the MFHW in the composite shale gas reservoir at different days. The influence of the fracture network on the drainage area is greater than a natural reservoir model without a fracture system. Due to high permeability, the velocity of flow in the inner region is quicker than that in the outer region; at the same time, the pressure propagation is faster. In the middle and late production periods, the fracture network system forms a broader low pressure area than the area in conventional fractured horizontal wells. Fig. 8.31 is a 3D projection graph of a MFHW in a composite or conventional reservoir without SRV. It can be seen that the bottom of the low-pressure region in the composite model is like a basin shape, while the bottom is similar to a gully shape for a conventional model; the low-pressure region in this composite model is obviously bigger than that in the conventional model, and thus the production is larger.

The curves of a production rate and accumulative production of a MFHW in a composite reservoir are shown as Fig. 8.32. At the beginning of production, a high conductivity fracture network means that much free gas is produced, which results in two to three times production from this fracture network system than from a conventional fractured horizontal well. At the late part of production, the high conductive fracture network can form a large low-pressure region. This situation improves the desorption gas production, while it also reduces the flow resistance to the well. It is noted that there is a rapid decline rate period on the production curve, which corresponds to the time at which pressure propagates to the boundary between the inner and outer regions.

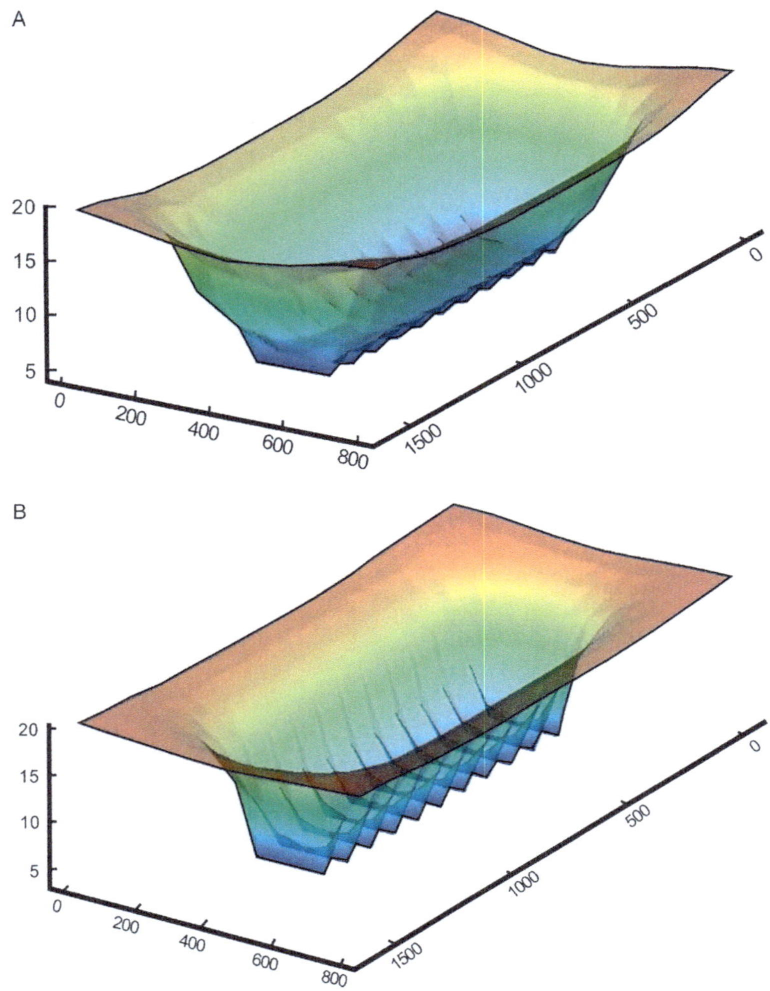

Fig. 8.31 3D projection graph of a MFHW in a composite or conventional reservoir.

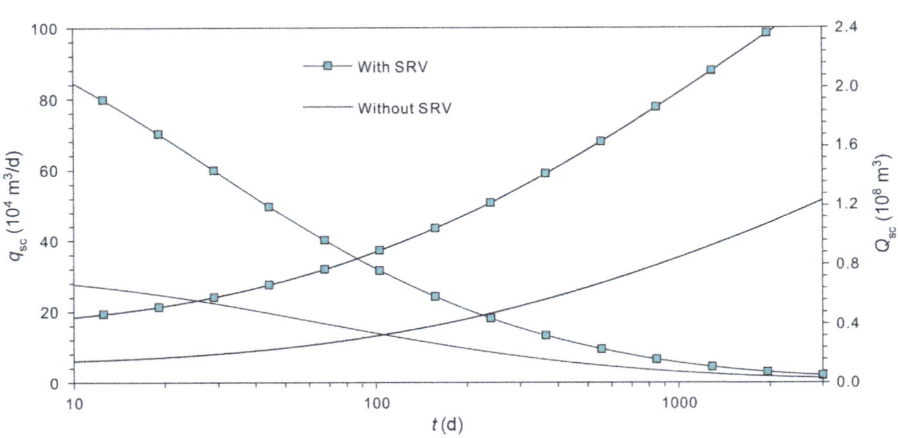

Fig. 8.32 Effect of SRV on production.

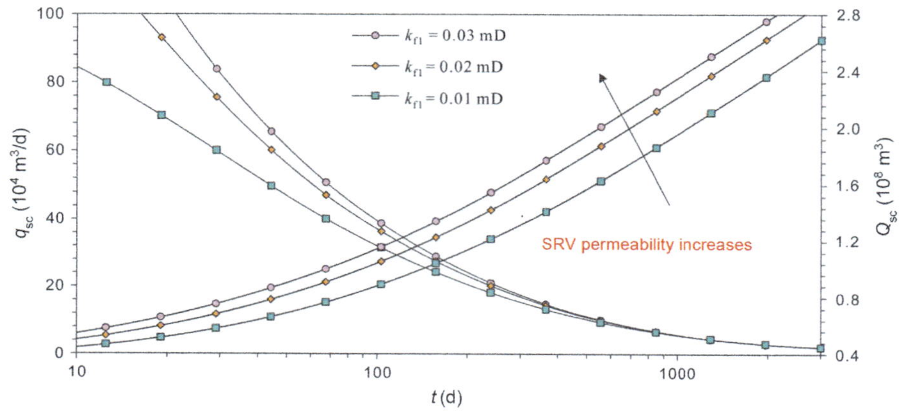

Fig. 8.33 Effect of SRV permeability on production.

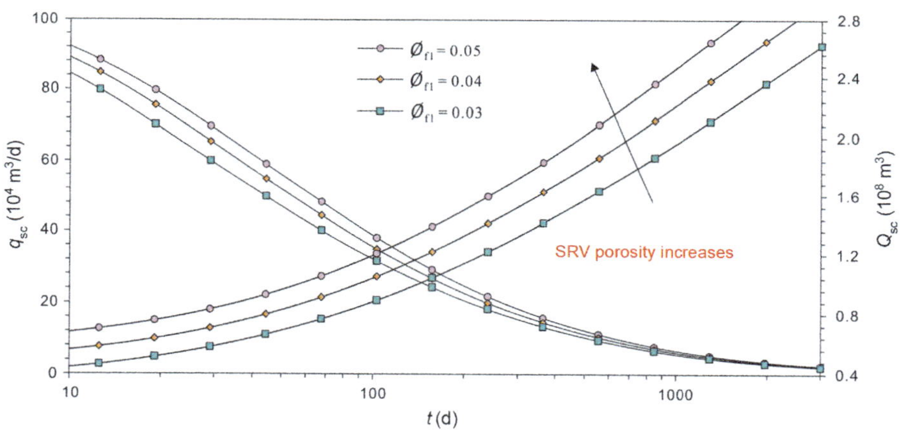

Fig. 8.34 Effect of SRV porosity on production.

Fig. 8.33 shows the effect of SRV permeability on the production performance of a MFHW. The SRV permeability mainly affects the early period. The more the permeability, the larger the early production rate. However, in the middle and late periods, gas resources are mainly from desorption gas on the surface of the outer region matrix, and the effect of the SRV permeability decreases.

As Fig. 8.34 shows, a larger SRV porosity reflects a better hydraulic fracturing stimulation improvement. We can conclude that when the SRV porosity is increased from 3% to 5%, the total volume of free gas will be increased by 66%. Thus the improvement of the SRV porosity will obviously improve the production rate and accumulative production.

Figs. 8.35 and 8.36 show the effects of initial reservoir properties (permeability and porosity) on the production performance. We can conclude that these properties

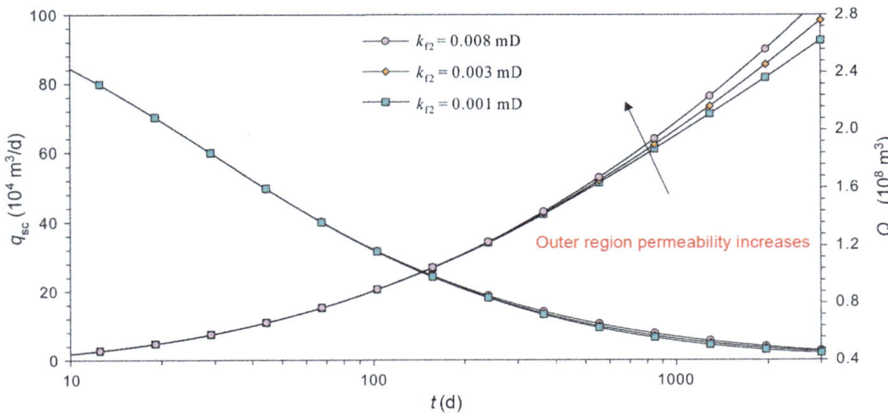

Fig. 8.35 Effect of outer region permeability on production.

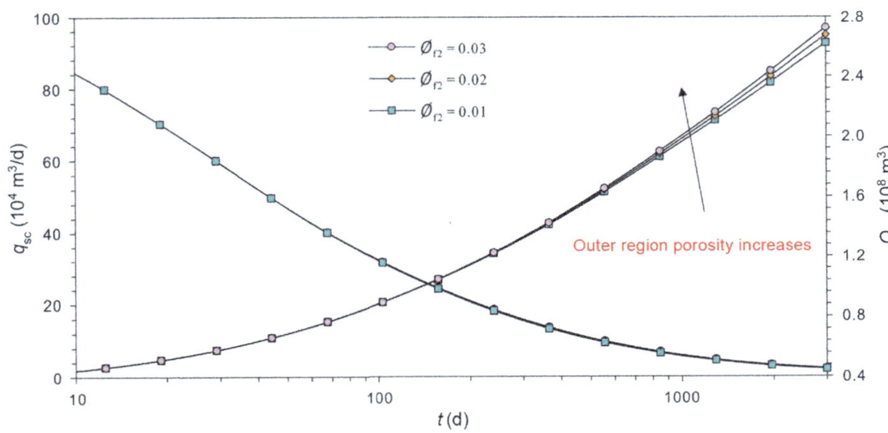

Fig. 8.36 Effect of outer region porosity on production.

mainly affect the middle and late periods. However, the impact of permeability and porosity increasing is minimal. Due to the extremely low initial permeability, the production mainly depend on the SRV properties.

8.3.4.3 Local SRV models

In the previous section, it is introduced that an inner SRV region can be formed around a MFHW in a reservoir with well-developed natural fractures and high-index brittleness. However, in field practice, subject to a fracture spacing, a fracturing degree, and poor reservoir properties, an actual SRV region cannot fully communicate with adjacent fractures, but form an independent local SRV region around the main fractures, as shown in Fig. 8.37.

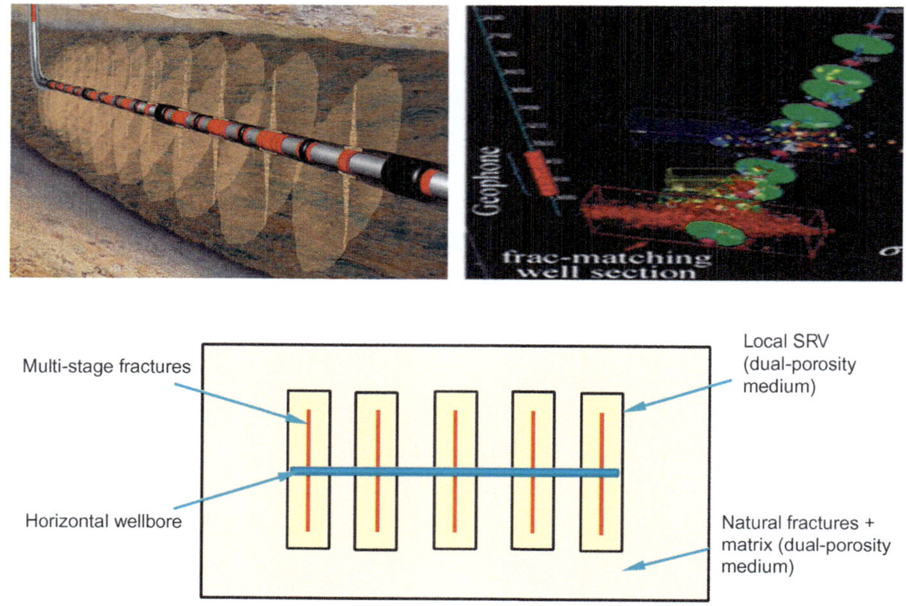

Fig. 8.37 Local SRV model of a MFHW in a closed rectangular composite reservoir.

Table 8.6 Shale reservoir properties

Reservoir property	Value	Reservoir property	Value
Formation thickness, h, m	50	Initial reservoir pressure, p_i, MPa	20
Fracture half-length, x_f, m	80	Reservoir boundary, m	800×500
Local SRV region, m	200×30	Hydraulic fracture number, N	5

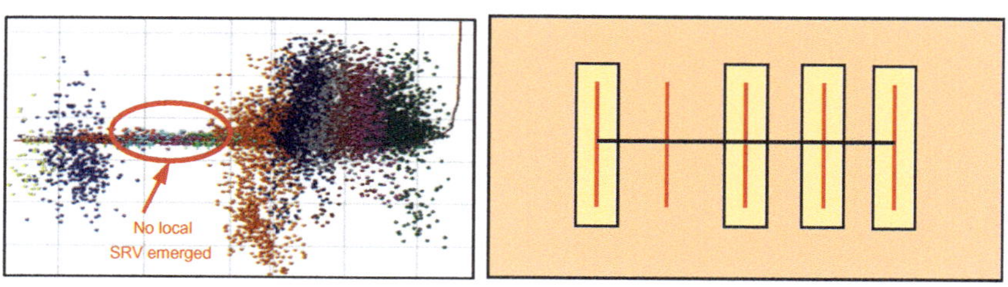

Fig. 8.38 Local SRV model of a MFHW in a closed rectangle composite reservoir (four local SRVs).

According to the basic parameters given in Table 8.6, a local SRV model is assumed as in Fig. 8.38. A pressure distribution and a 3D projection graph are shown in Fig. 8.39. In Fig. 8.40, the production of a local fracture network system is compared to a conventional model without a fracture network system. The influence of the fracture

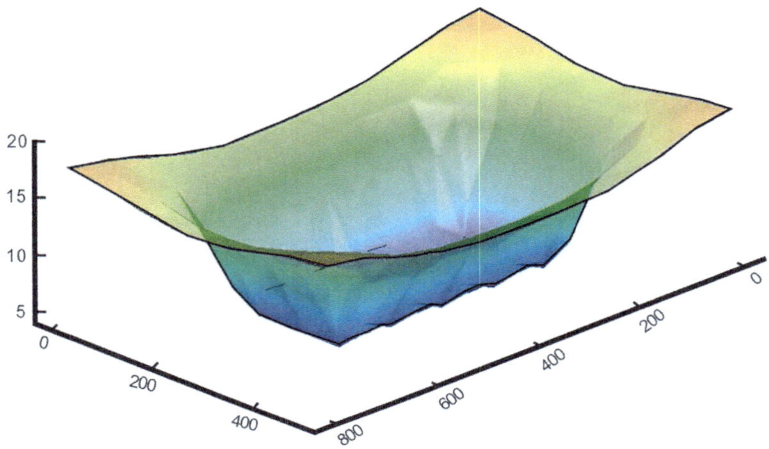

Fig. 8.39 Pressure distribution and 3D projection graph of a local SRV model.

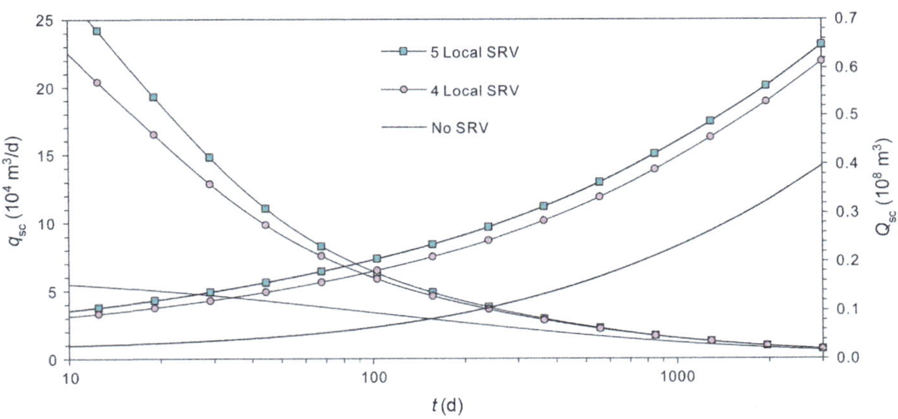

Fig. 8.40 Production performance of a local SRV model of a MFHW in a rectangular reservoir.

network system on the shale gas reservoir is very obvious; even just a local SRV regions can significantly improve the production rate and cumulative production.

8.4 Tri-porosity models in shale gas reservoirs

This section proposes a tri-porosity model that consists of a dual-porosity/dual-permeability medium and discrete fractures to simulate a finite conductivity MFHW in a rectangular composite shale gas reservoir (Fig. 8.41). Nonlinearity of gas properties with pressure, multiple flow mechanisms including ad-desorption, diffusion and stress sensitivity of a natural fracture network, and a stimulated rock volume are all taken into account. The CVFEM with fully implicit and sequential iterative algorithms, based on unstructured 3D tetrahedral meshes, is applied.

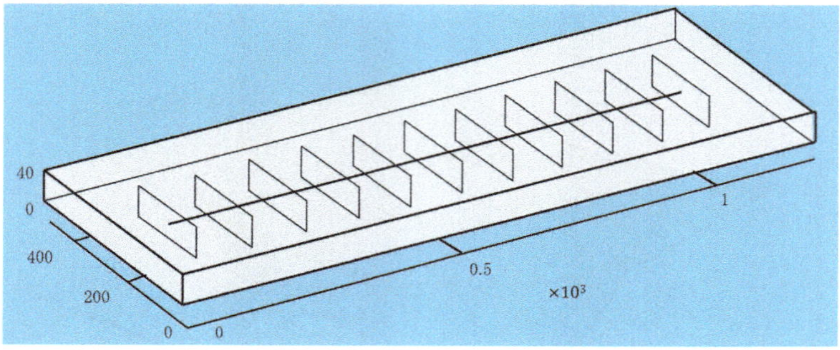

Fig. 8.41 3D physical model of a MFHW in a rectangular reservoir (1200 × 500 × 50 m).

8.4.1 Model assumptions and mathematical model

To make the mathematical model more tractable and easier to understand, the following assumptions and descriptions are applied: ① flow is isothermal and in two phases: gas and water; ② a finite conductivity MFHW in a composite shale gas reservoir is simplified as a tri-porosity medium, in which there are two sets of reservoir properties in the natural and SRV regions, respectively; ③ considering the micro–nanoscale mechanisms in shale matrix, gas, and water flow from matrix into a fracture system, and then into hydraulic fractures.

Based on the tri-porosity model concept, the diffusivity equations in shale gas reservoirs can be derived by combining mass conservation equation, equation of state, and equation of motion as:

$$\nabla\cdot\left(\frac{k_g k_{\mathrm{rgm}}}{\mu_g B_g}\nabla p_{\mathrm{gm}}\right)-\alpha\frac{k_g k_{\mathrm{rgm}}}{\mu_g B_g}(p_{\mathrm{gm}}-p_{\mathrm{gf}})+q^*=\frac{\partial(\phi_{\mathrm{m}}/B_g)}{\partial t}-\frac{\phi_{\mathrm{m}}\partial(S_{\mathrm{wm}}/B_g)}{\partial t} \tag{8.44}$$

$$\nabla\cdot\left[\frac{k_{\mathrm{m}} k_{\mathrm{rwm}}}{\mu_{\mathrm{w}} B_w}\nabla(p_{\mathrm{gm}}-p_{\mathrm{cm}})\right]-\alpha\frac{k_{\mathrm{m}} k_{\mathrm{rwm}}}{\mu_{\mathrm{w}} B_w}(p_{\mathrm{gm}}-p_{\mathrm{cm}}-p_{\mathrm{gf}})=\frac{\partial(S_{\mathrm{wm}}\phi_{\mathrm{m}}/B_{\mathrm{w}})}{\partial t} \tag{8.45}$$

$$\nabla\cdot\left(\frac{k_{\mathrm{fe}} k_{\mathrm{rgf}}}{\mu_g B_g}\nabla p_{\mathrm{gf}}\right)+\alpha\frac{k_{\mathrm{app}} k_{\mathrm{rgm}}}{\mu_g B_g}(p_{\mathrm{gm}}-p_{\mathrm{gf}})=\frac{\partial(\phi_{\mathrm{f}}/B_g)}{\partial t}-\frac{\phi_{\mathrm{f}}\partial(S_{\mathrm{wf}}/B_g)}{\partial t} \tag{8.46}$$

$$\nabla\cdot\left(\frac{k_{\mathrm{fe}} k_{\mathrm{rwf}}}{\mu_{\mathrm{w}} B_w}\nabla p_{\mathrm{gf}}\right)+\alpha\frac{k_{\mathrm{m}} k_{\mathrm{rwm}}}{\mu_{\mathrm{w}} B_w}(p_{\mathrm{gm}}-p_{\mathrm{cm}}-p_{\mathrm{gf}})=\frac{\partial(S_{\mathrm{wf}}\phi_{\mathrm{f}}/B_{\mathrm{w}})}{\partial t} \tag{8.47}$$

The linear flow in hydraulic fractures can be expressed without a compressibility item as:

$$\nabla \cdot \left[\frac{k_F k_{rl}}{\mu_l B_l} \nabla p\right] + q = 0 \quad (l = g, w) \tag{8.48}$$

where k_{me}, k_{fe}, and k_F are the permeability in the matrix, fracture system, and hydraulic fractures, respectively.

8.4.2 Numerical solutions

As Figs. 8.42 and 8.43 show, the thick line represents a horizontal well. The whole reservoir is discretized into unstructured 3D tetrahedral meshes. The 2D embedded surfaces are decomposed into triangular elements that are the faces of the tetrahedrons surrounding the reservoir–hydraulic fracture interface. For variables, an interpolation function is constructed on each discrete tetrahedron as:

$$\phi(x, y, z) = \sum_{l=1}^{4} N_l \phi_l \tag{8.49}$$

$$\sum_{l=1}^{4} N_l = 1 \tag{8.50}$$

$$\frac{\partial N_1}{\partial \beta} + \frac{\partial N_2}{\partial \beta} + \frac{\partial N_3}{\partial \beta} + \frac{\partial N_4}{\partial \beta} = 0, \quad (\beta = x, y, z) \tag{8.51}$$

where:

$$N_l = \frac{1}{6V}(a_l + b_l x + c_l y + d_l z), \quad m = 1,2,3,4$$

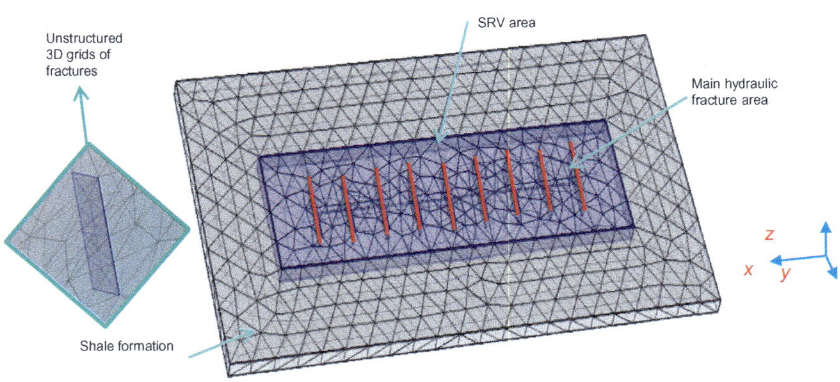

Fig. 8.42 Typical unstructured 3D grids of a MFHW in a shale gas reservoir (2200 × 1000 × 50 m).

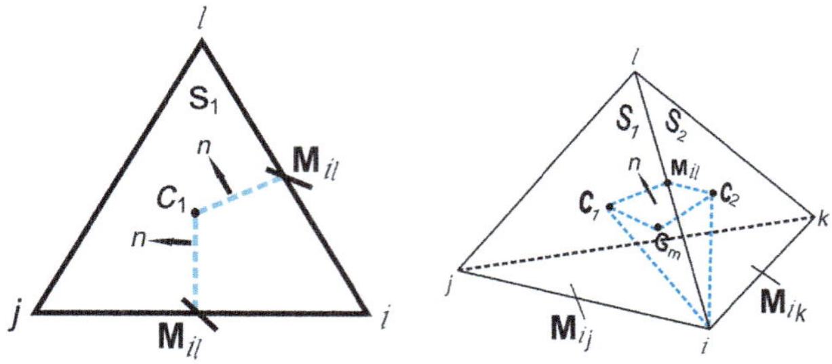

Fig. 8.43 Unstructured element.

$$V=\frac{1}{6}\begin{vmatrix}1 & x_1 & y_1 & z_1\\ 1 & x_2 & y_2 & z_2\\ 1 & x_3 & y_3 & z_3\\ 1 & x_4 & y_4 & z_4\end{vmatrix};\quad \begin{cases} a_1=\begin{vmatrix}x_2 & y_2 & z_2\\ x_3 & y_3 & z_3\\ x_4 & y_4 & z_4\end{vmatrix}, & b_1=-\begin{vmatrix}1 & y_2 & z_2\\ 1 & y_3 & z_3\\ 1 & y_4 & z_4\end{vmatrix},\\ c_1=-\begin{vmatrix}x_2 & 1 & z_2\\ x_3 & 1 & z_3\\ x_4 & 1 & z_4\end{vmatrix}, & d_1=-\begin{vmatrix}x_2 & y_2 & 1\\ x_3 & y_3 & 1\\ x_4 & y_4 & 1\end{vmatrix}\end{cases}$$

$$\begin{cases} a_2=\begin{vmatrix}x_3 & y_3 & z_3\\ x_1 & y_1 & z_1\\ x_4 & y_4 & z_4\end{vmatrix}, & b_2=-\begin{vmatrix}1 & y_3 & z_3\\ 1 & y_1 & z_1\\ 1 & y_4 & z_4\end{vmatrix},\\ c_2=-\begin{vmatrix}x_3 & 1 & z_3\\ x_1 & 1 & z_1\\ x_4 & 1 & z_4\end{vmatrix}, & d_2=-\begin{vmatrix}x_3 & y_3 & 1\\ x_1 & y_1 & 1\\ x_4 & y_4 & 1\end{vmatrix}\end{cases};\quad \begin{cases} a_3=\begin{vmatrix}x_1 & y_1 & z_1\\ x_2 & y_2 & z_2\\ x_4 & y_4 & z_4\end{vmatrix}, & b_3=-\begin{vmatrix}1 & y_1 & z_1\\ 1 & y_2 & z_2\\ 1 & y_4 & z_4\end{vmatrix},\\ c_3=-\begin{vmatrix}x_1 & 1 & z_1\\ x_2 & 1 & z_2\\ x_4 & 1 & z_4\end{vmatrix}, & d_3=-\begin{vmatrix}x_1 & y_1 & 1\\ x_2 & y_2 & 1\\ x_4 & y_4 & 1\end{vmatrix}\end{cases}$$

$$\begin{cases} a_4=\begin{vmatrix}x_1 & y_1 & z_1\\ x_3 & y_3 & z_3\\ x_2 & y_2 & z_2\end{vmatrix}, & b_4=-\begin{vmatrix}1 & y_1 & z_1\\ 1 & y_3 & z_3\\ 1 & y_2 & z_2\end{vmatrix},\\ c_4=-\begin{vmatrix}x_1 & 1 & z_1\\ x_3 & 1 & z_3\\ x_2 & 1 & z_2\end{vmatrix}, & d_4=-\begin{vmatrix}x_1 & y_1 & 1\\ x_3 & y_3 & 1\\ x_2 & y_2 & 1\end{vmatrix}\end{cases}$$

According to the Gauss–Green formula and the finite volume method, when a trail function equals 1, an equivalent integral of the diffusion term on volume V_1 becomes:

$$\iiint_{V_1}\nabla\cdot\left(\frac{\overline{\overline{k}}k_{rl}}{\mu_l B_l}\cdot\nabla p\right)\mathrm{d}V=\iint_{\Omega_1}\left(\frac{\overline{\overline{k}}k_{rl}}{\mu_l B_l}\cdot\nabla p\right)\cdot\boldsymbol{n}d\Omega=F_{1l}\quad(l=\mathrm{g},\mathrm{w}) \tag{8.52}$$

The permeability tensor considering stress sensitivity can be expressed as:

$$\overline{\overline{k}} = \begin{bmatrix} k_{ex} & & \\ & k_{ey} & \\ & & k_{ez} \end{bmatrix} \tag{8.53}$$

Substituting Eq. (8.49) into Eq. (8.52), there is:

$$F_{1l} = \iint_{\Omega_1} \left(\frac{\overline{\overline{k}} k_{rl}}{\mu_l B_l} \cdot \nabla p \right) \cdot \boldsymbol{n} d\Omega = \sum_{v=1,2,3,4} T_{1vl}(p_1 - p_v) \tag{8.54}$$

where:

$$T_{1vl} = \frac{1}{36V} \left(\frac{k_{ex} k_{rl}}{\mu_l B_l} b_1 b_v + \frac{k_{ey} k_{rl}}{\mu_l B_l} c_1 c_v + \frac{k_{ez} k_{rl}}{\mu_l B_l} d_1 d_v \right) \tag{8.55}$$

Similarity, we can calculate the fluxes F_{2l}, F_{3l}, and F_{4l}, and the stiffness matrix on the finite volume V_1 is established as:

$$\boldsymbol{T}_\zeta^{(e)} \boldsymbol{p}_\zeta^{(e)} = \begin{pmatrix} T_{12}+T_{13}+T_{14} & -T_{12} & -T_{13} & -T_{14} \\ -T_{21} & T_{21}+T_{23}+T_{24} & -T_{23} & -T_{24} \\ -T_{31} & -T_{32} & T_{31}+T_{32}+T_{34} & -T_{34} \\ -T_{41} & -T_{42} & -T_{43} & T_{41}+T_{42}+T_{43} \end{pmatrix}_\zeta \begin{pmatrix} p_1 \\ p_1 \\ p_3 \\ p_4 \end{pmatrix}_\zeta \tag{8.56}$$

where:

$$\zeta = \mathrm{m, f}$$

We consider an upstream weighting strategy for transmissivity and transfer coefficients to avoid numerical oscillations (Chen, 2011). Moreover, the stress sensitivity coefficients are also upstream weighted:

$$T_{12l} = T_{upl} = \begin{cases} \dfrac{k_e(p_1) k_{rl}(S_{w1})}{\mu_l(p_{l1}) B_l(p_{l1})} & p_1 \geq p_2 \\ \dfrac{k_e(p_2) k_{rl}(S_{w2})}{\mu_l(p_{l2}) B_l(p_{l2})} & p_1 < p_2 \end{cases} \tag{8.57}$$

In this section, the transfer matrix in the fracture system is the same as that in the matrix based on the upstream weighting concept, which considers that gas and water in the fracture system always are supplemented from the matrix system.

$$w_1 = \left(\frac{\alpha k_{app}}{\mu_{v1} B_{v1}} \right)_{up} = \left(\frac{\alpha k_{app}}{\mu_v B_v} \right)_{p_{ml}} \quad (v = \mathrm{g, w}) \tag{8.58}$$

Applying Eq. (8.56) for each element, we can establish the discrete equations for the matrix, fracture, and hydraulic fracture systems as follows:

For the fracture system,

$$\boldsymbol{R}_{\mathrm{gf}}^{k}=\begin{cases}\left(\mathbf{T}_{\mathrm{gf}}+\mathbf{T}_{\mathrm{gf\_dp}}+\mathbf{W}_{\mathrm{gup}}-\dfrac{\mathbf{N}_{\mathrm{gf}}\_\mathrm{d}p}{\Delta t}+\dfrac{\mathbf{N}_{\mathrm{gf\_swf}}\_\mathrm{d}p}{\Delta t}\right)\delta p_{\mathrm{gf}}-W_{\mathrm{gup}}\_\mathrm{d}p\delta p_{\mathrm{gm}}\\+\left(\mathbf{T}_{\mathrm{gf}}\_\mathrm{d}S_{\mathrm{wf}}+\dfrac{\mathbf{N}_{\mathrm{gf}}}{\Delta t}\right)\delta S_{\mathrm{wf}}-\mathrm{W}_{\mathrm{gup}}\_\mathrm{d}S_{\mathrm{w}}\delta S_{\mathrm{wm}}\end{cases}$$

$$\mathbf{R}_{\mathrm{wf}}^{k}=\begin{cases}\left(\mathbf{T}_{\mathrm{wf}}-\mathrm{W}_{\mathrm{wup}}\right)\delta p_{\mathrm{gf}}+\mathrm{W}_{\mathrm{wup}}\delta p_{\mathrm{gm}}\\+\left(\mathbf{T}_{\mathrm{wf}}\_\mathrm{d}S_{\mathrm{w}}+\dfrac{\mathrm{N}_{\mathrm{gf}}}{\Delta t}\right)\delta S_{\mathrm{wf}}-\mathbf{W}_{\mathrm{gup}}\_\mathrm{d}S_{\mathrm{w}}\delta S_{\mathrm{wm}}\end{cases}$$

(8.59)

where:

$$\boldsymbol{R}_{\mathrm{gf}}^{k}=\frac{\left(\mathbf{N}_{\mathrm{gf}}^{k}-\mathbf{N}_{\mathrm{gf}}^{n}\right)}{\Delta t}-\frac{\left(\mathbf{N}_{\mathrm{gf\_swf}}^{k}-\mathbf{N}_{\mathrm{gf\_swf}}^{n}\right)}{\Delta t}-\mathbf{T}_{\mathrm{gf}}^{k}\mathbf{P}_{\mathrm{gf}}^{k}-\mathbf{W}_{\mathrm{gup}}^{k}\left(\mathbf{P}_{\mathrm{gm}}^{k}-\mathbf{P}_{\mathrm{gf}}^{k}\right)$$

$$\boldsymbol{R}_{\mathrm{wf}}^{k}=\frac{\left(\mathbf{N}_{\mathrm{wf}}^{k}-\mathbf{N}_{\mathrm{wf}}^{n}\right)}{\Delta t}-\mathbf{T}_{\mathrm{wf}}^{k}\left(\mathbf{P}_{\mathrm{gf}}^{k}-\mathbf{P}_{\mathrm{c}}^{k}\right)-\mathbf{W}_{\mathrm{gup}}^{k}\left(\mathbf{P}_{\mathrm{gm}}^{k}-\mathbf{P}_{\mathrm{gf}}^{k}-\mathbf{P}_{\mathrm{c}}^{k}\right)$$

(8.60)

For the matrix system:

$$\boldsymbol{R}_{\mathrm{gm}}^{k}=\begin{cases}\left(\mathbf{T}_{\mathrm{gm}}+\mathbf{T}_{\mathrm{gm\_dp}}-\mathbf{W}_{\mathrm{gup}}-\mathbf{W}_{\mathrm{gup}}\_\mathrm{d}p-\dfrac{\mathbf{N}_{\mathrm{gm}}\_\mathrm{d}p}{\Delta t}+\dfrac{\mathbf{N}_{\mathrm{gm\_swm}}\_\mathrm{d}p}{\Delta t}+\dfrac{V_{\mathrm{E}}^{k}\_\mathrm{d}p}{\Delta t}\right)\delta p_{\mathrm{gm}}\\+\mathbf{W}_{\mathrm{gup}}\delta p_{\mathrm{gf}}+\left(\mathbf{T}_{\mathrm{gm}}\_\mathrm{d}S_{\mathrm{w}}+\dfrac{\mathbf{N}_{\mathrm{gm}}}{\Delta t}-\mathbf{W}_{\mathrm{gup}}\_\mathrm{d}S_{\mathrm{w}}\right)\delta S_{\mathrm{wm}}\end{cases}$$

$$\boldsymbol{R}_{\mathrm{wm}}^{k}=\begin{cases}\left(\mathbf{T}_{\mathrm{wm}}-\mathbf{W}_{\mathrm{wup}}\right)\delta p_{\mathrm{gm}}+\mathbf{W}_{\mathrm{wup}}\delta p_{\mathrm{gf}}\\+\left(\mathbf{T}_{\mathrm{wm}}\_\mathrm{d}S_{\mathrm{w}}+\dfrac{\mathbf{N}_{\mathrm{gm}}}{\Delta t}-\mathbf{T}_{\mathrm{wm\_dp_c}}-\mathbf{W}_{\mathrm{wup}}\_\mathrm{d}S_{\mathrm{w}}\right)\delta S_{\mathrm{wm}}\end{cases}$$

(8.61)

where:

$$\boldsymbol{R}_{\mathrm{gm}}^{k}=\frac{\left(\mathbf{N}_{\mathrm{gm}}^{k}-\mathbf{N}_{\mathrm{gm}}^{n}\right)}{\Delta t}+\frac{\left(\mathbf{V}_{\mathrm{E}}^{k}-\mathbf{V}_{\mathrm{E}}^{n}\right)}{\Delta t}-\frac{\left(\mathbf{N}_{\mathrm{gm\_swm}}^{k}-\mathbf{N}_{\mathrm{gm\_swm}}^{n}\right)}{\Delta t}$$

$$-\mathbf{T}_{\mathrm{gm}}^{k}\mathbf{P}_{\mathrm{gm}}^{k}+\mathbf{W}_{\mathrm{gup}}^{k}\left(\mathbf{P}_{\mathrm{gm}}^{k}-\mathbf{P}_{\mathrm{gf}}^{k}\right)$$

$$\boldsymbol{R}_{\mathrm{wm}}^{k}=\frac{\left(\mathbf{N}_{\mathrm{wm}}^{k}-\mathbf{N}_{\mathrm{wm}}^{\mathrm{n}}\right)}{\Delta t}-\mathbf{T}_{\mathrm{wm}}^{k}\left(\mathbf{P}_{\mathrm{gm}}^{k}-\mathbf{P}_{\mathrm{c}}^{k}\right)+\mathbf{W}_{\mathrm{gup}}^{k}\left(\mathbf{P}_{\mathrm{gm}}^{k}-\mathbf{P}_{\mathrm{gf}}^{k}-\mathbf{P}_{\mathrm{c}}^{k}\right)$$

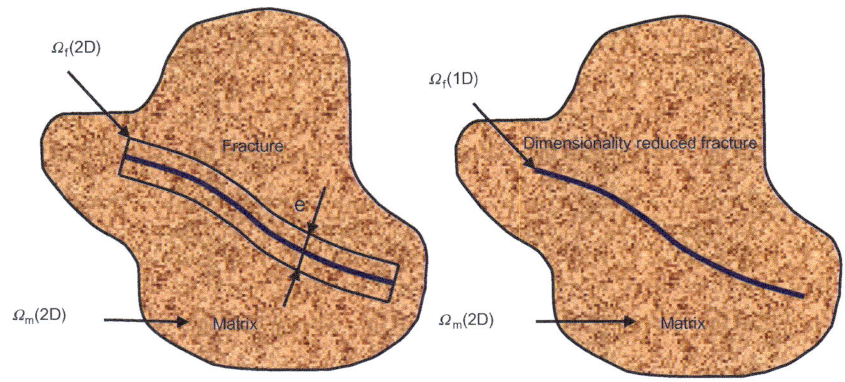

Fig. 8.44 Dimensionality reduction of discretized pseudo 3D fractures.

As Fig. 8.44 shows, based on the discrete fracture theory (Karimi-Fard and Firoozabadi, 2001; Karimi-Fard et al., 2003; Kim and Deo, 2000; Lange et al., 2004; Moinfar et al., 2014), an equivalent integral weak form over a control volume region containing discrete fractures is:

$$\int_{\Omega_d} \text{PDEs}\, d\Omega = \int_{\Omega_m} \text{PDEs}\, d\Omega_m + w_F \times \int_{\Omega_f} \text{PDEs}\, d\Omega_f \tag{8.62}$$

Assuming equivalent pressure and continuous flow at the joint surfaces of hydraulic fractures and the reservoir and ignoring minor flow from the reservoir matrix to hydraulic fractures, the element discretization forms in the reservoir and the hydraulic fracture system can be combined:

$$\boldsymbol{R}_{gf}^{k} + \boldsymbol{q} = \begin{cases} \left(\mathbf{T}_{gf+F} + \mathbf{T}_{gf+F\_dp} + \mathbf{W}_{gup} - \dfrac{\mathbf{N}_{gf+F\_dp}}{\Delta t} + \dfrac{\mathbf{N}_{gf+F\_swf\_dp}}{\Delta t}\right)\delta p_{gf} - \mathbf{W}_{gup\_dp}\delta p_{gm} \\ +\left(\mathbf{T}_{gf+F\_dS_{wf}} + \dfrac{\mathbf{N}_{gf+F}}{\Delta t}\right)\delta S_{wf} - \mathbf{W}_{gup\_dS_w}\delta S_{wm} \end{cases}$$

$$\boldsymbol{R}_{wf}^{k} = \begin{cases} \left(\mathbf{T}_{wf+F} - \mathbf{W}_{wup}\right)\delta p_{gf} + \mathbf{W}_{wup}\delta p_{gm} \\ +\left(\mathbf{T}_{wf+F\_dS_w} + \dfrac{\mathbf{N}_{gf}}{\Delta t}\right)\delta S_{wf} - \mathbf{W}_{gup\_dS_w}\delta S_{wm} \end{cases} \tag{8.63}$$

where:

$$\boldsymbol{T}_{f+F}^{(e)} = \sum_{\eta=f,F} \boldsymbol{T}_{\eta}^{(e)}$$

The term $\boldsymbol{T}\_\mathrm{d}p$ is the transmissibility partial derivative matrix in the matrix system, and the term $\boldsymbol{T}_{\mathrm{f+F}}\mathrm{d}p = \sum_{\eta=\mathrm{f,F}} \boldsymbol{T}_\eta \mathrm{d}p$ is the combined transmissibility partial derivative matrix in the microfracture and hydraulic fracture systems:

$$T_{\mathrm{f}}\_\mathrm{d}p = \begin{pmatrix} \begin{bmatrix} \frac{\partial T_{12}}{\partial p_1}(p_1-p_2) \\ +\frac{\partial T_{13}}{\partial p_1}(p_1-p_3) \\ +\frac{\partial T_{14}}{\partial p_1}(p_1-p_4) \end{bmatrix} & \frac{\partial T_{12}}{\partial p_2}(p_1-p_2) & \frac{\partial T_{13}}{\partial p_3}(p_1-p_3) & \frac{\partial T_{14}}{\partial p_4}(p_1-p_4) \\ \frac{\partial T_{12}}{\partial p_1}(p_2-p_1) & \begin{bmatrix} \frac{\partial T_{12}}{\partial p_2}(p_2-p_1) \\ +\frac{\partial T_{23}}{\partial p_2}(p_2-p_3) \\ +\frac{\partial T_{24}}{\partial p_2}(p_2-p_4) \end{bmatrix} & \frac{\partial T_{23}}{\partial p_3}(p_2-p_3) & \frac{\partial T_{24}}{\partial p_4}(p_2-p_4) \\ \frac{\partial T_{13}}{\partial p_1}(p_3-p_1) & \frac{\partial T_{23}}{\partial p_2}(p_3-p_2) & \begin{bmatrix} \frac{\partial T_{13}}{\partial p_3}(p_3-p_1) \\ +\frac{\partial T_{23}}{\partial p_3}(p_3-p_2) \\ +\frac{\partial T_{34}}{\partial p_4}(p_3-p_4) \end{bmatrix} & \frac{\partial T_{34}}{\partial p_4}(p_3-p_4) \\ \frac{\partial T_{14}}{\partial p_1}(p_4-p_1) & \frac{\partial T_{24}}{\partial p_2}(p_4-p_2) & \frac{\partial T_{34}}{\partial p_4}(p_4-p_3) & \begin{bmatrix} \frac{\partial T_{14}}{\partial p_4}(p_4-p_1) \\ +\frac{\partial T_{24}}{\partial p_4}(p_4-p_2) \\ +\frac{\partial T_{34}}{\partial p_4}(p_4-p_3) \end{bmatrix} \end{pmatrix}_{\zeta} \tag{8.64}$$

$$T_{\mathrm{F}}\_\mathrm{d}p = \begin{pmatrix} \begin{bmatrix} \frac{\partial T_{ij}}{\partial p_i}\left(p_i-p_j\right) \\ +\frac{\partial T_{ik}}{\partial p_i}(p_i-p_k) \end{bmatrix} & \frac{\partial T_{ij}}{\partial p_j}\left(p_i-p_j\right) & \frac{\partial T_{ik}}{\partial p_k}(p_i-p_k) \\ \frac{\partial T_{ij}}{\partial p_i}\left(p_j-p_i\right) & \begin{bmatrix} \frac{\partial T_{ij}}{\partial p_j}\left(p_j-p_i\right) \\ +\frac{\partial T_{jk}}{\partial p_j}\left(p_j-p_k\right) \end{bmatrix} & \frac{\partial T_{jk}}{\partial p_k}\left(p_j-p_k\right) \\ \frac{\partial T_{ik}}{\partial p_{ii}}(p_k-p_i) & \frac{\partial T_{jk}}{\partial p_2}\left(p_k-p_j\right) & \begin{bmatrix} \frac{\partial T_{ik}}{\partial p_k}(p_k-p_i) \\ +\frac{\partial T_{jk}}{\partial p_k}\left(p_k-p_j\right) \end{bmatrix} \end{pmatrix}_{F} \tag{8.65}$$

The quantity $\boldsymbol{N}$_dp is the partial derivative matrix of the accumulation term:

$$\boldsymbol{N}\_\mathrm{d}p=\frac{V}{4}\begin{pmatrix} \frac{\partial}{\partial p}\left(\frac{\phi_\zeta}{B_\mathrm{g}}\right)\Big|_{p_{\zeta1}^k} & 0 & 0 & 0 \\ 0 & \frac{\partial}{\partial p}\left(\frac{\phi_\zeta}{B_\mathrm{g}}\right)\Big|_{p_{\zeta2}^k} & 0 & 0 \\ 0 & 0 & \frac{\partial}{\partial p}\left(\frac{\phi_\zeta}{B_\mathrm{g}}\right)\Big|_{p_{\zeta3}^k} & 0 \\ 0 & 0 & 0 & \frac{\partial}{\partial p}\left(\frac{\phi_\zeta}{B_\mathrm{g}}\right)\Big|_{p_{\zeta4}^k} \end{pmatrix} \tag{8.66}$$

The quantity $\boldsymbol{W}\_\mathrm{d}p$ is the partial derivative matrix of the element interporosity flow term:

$$\boldsymbol{W}\_\mathrm{d}p=\frac{\alpha V}{4}\begin{pmatrix} \frac{\partial}{\partial p}\left(\frac{k}{\mu B_\mathrm{g}}\right)\Big|_{p=p_1} & 0 & 0 & 0 \\ 0 & \frac{\partial}{\partial p}\left(\frac{k}{\mu B_\mathrm{g}}\right)\Big|_{p=p_2} & 0 & 0 \\ 0 & 0 & \frac{\partial}{\partial p}\left(\frac{k}{\mu B_\mathrm{g}}\right)\Big|_{p=p_3} & 0 \\ 0 & 0 & 0 & \frac{\partial}{\partial p}\left(\frac{k}{\mu B_\mathrm{g}}\right)\Big|_{p=p_4} \end{pmatrix}_\mathrm{m} \begin{pmatrix} p_1 \\ p_2 \\ p_3 \\ p_4 \end{pmatrix}_\zeta \tag{8.67}$$

Finally, the term $\boldsymbol{V}_\mathrm{E}\_\mathrm{d}p$ is the partial derivative matrix of the element adsorption term:

$$\boldsymbol{V}_\mathrm{E}\_\mathrm{d}p=\frac{V}{4}\begin{pmatrix} \frac{\partial[(1-\phi_\mathrm{t})V_\mathrm{E}]}{\partial p}\Big|_{p_{\mathrm{m}1}^k} & 0 & 0 & 0 \\ 0 & \frac{\partial[(1-\phi_\mathrm{t})V_\mathrm{E}]}{\partial p}\Big|_{p_{\mathrm{m}2}^k} & 0 & 0 \\ 0 & 0 & \frac{\partial[(1-\phi_\mathrm{t})V_\mathrm{E}]}{\partial p}\Big|_{p_{\mathrm{m}3}^k} & 0 \\ 0 & 0 & 0 & \frac{\partial[(1-\phi_\mathrm{t})V_\mathrm{E}]}{\partial p}\Big|_{p_{\mathrm{m}4}^k} \end{pmatrix} \tag{8.68}$$

Eqs. (8.61) and (8.63) are the implicit iteration matrices on one 3D regular tetrahedral element among N tetrahedral elements in the whole reservoir. The general matrix can be acquired in the solution domain by a finite element matrix assembly method. By applying Eqs. (8.61) and (8.63) at each vertex, we obtain a $4N \times 4N$ linear system of equations, which can be solved for the $4N$ unknowns $\delta p_{\mathrm{gf}1}, \ldots, \delta p_{\mathrm{gf}N}, \delta p_{\mathrm{gm}1}, \ldots, \delta p_{\mathrm{gm}N}, \delta S_{\mathrm{wf}1}, \ldots, \delta S_{\mathrm{wf}N}, \delta S_{\mathrm{wm}1}, \ldots, \delta S_{\mathrm{wm}N}$. The resulting whole matrix can be expressed as:

$$[\mathbf{k}]_{4N*4N}[\delta\mathbf{X}]_{4N*1} = \left[\mathbf{R}^k\right]_{4N*1} \tag{8.69}$$

In this section, for a finite conductivity MFHW, a modified Peaceman's well model is established based on unstructured meshes (Chen et al., 2006; Li et al., 2014; Moinfar et al., 2013). The production term by a point source calculation formula can be obtained as:

$$q_{scl}^{n+1} = \mathrm{PI}_l^{n+1}(p_{\mathrm{bh}} - p_{\mathrm{ave}})^{n+1} \tag{8.70}$$

where:

$$\mathrm{PI}_l = \frac{0.543 k_{\mathrm{F}} k_{\mathrm{r}l} w_0}{\mu_l B_l\left[\ln\left(\dfrac{r_0}{r_w}\right) + s_{\mathrm{c}}\right]}$$
$$r_0 = \sqrt{\frac{A_{\mathrm{fv}}}{\pi}} \tag{8.71}$$

For the situation of producing at constant bottom-hole pressure (BHP), when neglecting the flow resistance in the horizontal well, the BHP at each intersection is equal. Eq. (8.70) is added to Eq. (8.69) to obtain iterative solutions of δp and δS_{w}, and the pressure, saturation, and production at the next time level can be calculated. However, in the situation of constant production, we supply an unknown δp_{bh} to Eq. (8.69); the whole matrix can be expressed as:

$$\begin{bmatrix}
kk_{11} & kk_{12} & \cdots & \cdots & \cdots\ \cdots & kk_{14N} & 0 \\
kk_{i1} & kk_{i2} & \cdots & (kk_{ii}+a_g) & \cdots & kk_{i4N} & \mathrm{PI}_{\mathrm{g}}^k \\
kk_{N+11} & kk_{N+12} & \cdots & \cdots & \cdots\ \cdots & kk_{N+14N} & 0 \\
kk_{N+i1} & \cdots & (kk_{N+iN+i}+a_{\mathrm{w}}) & \cdots & & kk_{N+i4N} & \mathrm{PI}_{\mathrm{w}}^k \\
kk_{2N1} & kk_{2N2} & \cdots & kk_{2NN} & \cdots & kk_{2N4N} & 0 \\
\vdots & \vdots & \vdots & \vdots & \vdots & & \vdots \\
kk_{2N+i1} & kk_{2N+i2} & \cdots & \cdots & \cdots & kk_{2N+i4N} & 0 \\
kk_{4N1} & kk_{4N2} & \cdots & \cdots & \cdots\ \cdots & kk_{4N4N} & 0 \\
0 & 0 & \cdots & a_g + a_{\mathrm{w}} & \cdots\ \cdots & 0_{(4N4N)} & \mathrm{PI}_{\mathrm{g}}^k + \mathrm{PI}_{\mathrm{w}}^k
\end{bmatrix}
\begin{bmatrix}
\delta p_{\mathrm{f}1}^k \\ \delta p_{\mathrm{f}i}^k \\ \delta p_{\mathrm{f}N}^k \\ \delta p_{\mathrm{m}i}^k \\ \delta p_{\mathrm{m}N}^k \\ \vdots \\ \delta S_{\mathrm{wf}i}^k \\ \delta S_{\mathrm{wm}N}^k \\ \delta p_{\mathrm{bh}}^k
\end{bmatrix}
=
\begin{bmatrix}
R_1^k \\ R_i^k \\ R_N^k \\ R_{N+i}^k \\ R_{2N}^k \\ \vdots \\ R_{3N+i}^k \\ R_{4N}^k \\ q_{\mathrm{t}} - c
\end{bmatrix} \tag{8.72}$$

where:

$$a_l = \delta \mathrm{PI}_l^k \left(p_{\mathrm{bh}}^k - p_{\mathrm{gf}}^k\right) - \mathrm{PI}_l^k \quad (l = \mathrm{g}, \mathrm{w})$$

$$c = \sum \mathrm{PI}_l^k \left(p_{\mathrm{bh}}^k - p_{\mathrm{gf}}^k\right)$$

8.4.3 Production performance analysis

In this section, the effects of some key parameters on pressure responses and production performance are investigated. The basic parameters are given in Table 8.7.

Fig. 8.45 shows the pressure wave diffusion of a MFHW in a shale gas reservoir. In the initial period, a low-pressure region is formed at the interface between the horizontal well and the fractures, and the pressure wave expands as a wedge. As production continues, the pressure wave extends to the ends of the fractures, and the flow regime transitions to the linear and elliptic flow. When the pressure wave continues to expand, the long elliptical pressure drop zones in each fracture are superimposed with interference, and the flow regime is dominated by the flow from the reservoir to the fractured horizontal well. Finally, the pressure wave continues to propagate to the boundary, and the MFHW region eventually forms the whole low-pressure area, which is similar to a long rectangle.

Fig. 8.46 shows a production rate distribution in a finite conductivity MFHW in a shale gas reservoir. The distribution of the production rate shows a "U" shape. In detail, according to the interference between hydraulic fractures, the production rate of the fractures lying at the toe and heel of the horizontal well is bigger than that in the fractures initiating at the middle section of the well. Moreover, as production progresses, the interference between fractures becomes more serious; the production

Table 8.7 Shale reservoir properties

Reservoir property	Value	Reservoir property	Value
Formation thickness, h, m	50	Rectangular boundary, $X \times Y$, m	1200×500
Initial reservoir pressure, p_i, MPa	20	Reservoir temperature, T, °C	100
Gas specific gravity, r_g, fraction	0.6	Horizontal wellbore length, L, m	1000
Hydraulic fracture number, N	11	Fracture half-length, x_f, m	100
Hydraulic fracture width, m	0.001	Stress sensitive coefficient, θ, MPa^{-1}	0.01
Microfracture porosity, Φ, %	3	Microfracture permeability, k, mD	0.01
Matrix porosity, Φ_m, %	1	Matrix permeability, k_m, mD	0.001
Langmuir volume, V_L, m^3/m^3	10	Langmuir pressure, P_L, MPa	4
Irreducible water saturation, S_{wr}	0.2	Residual gas saturation, S_{gr}	0.2

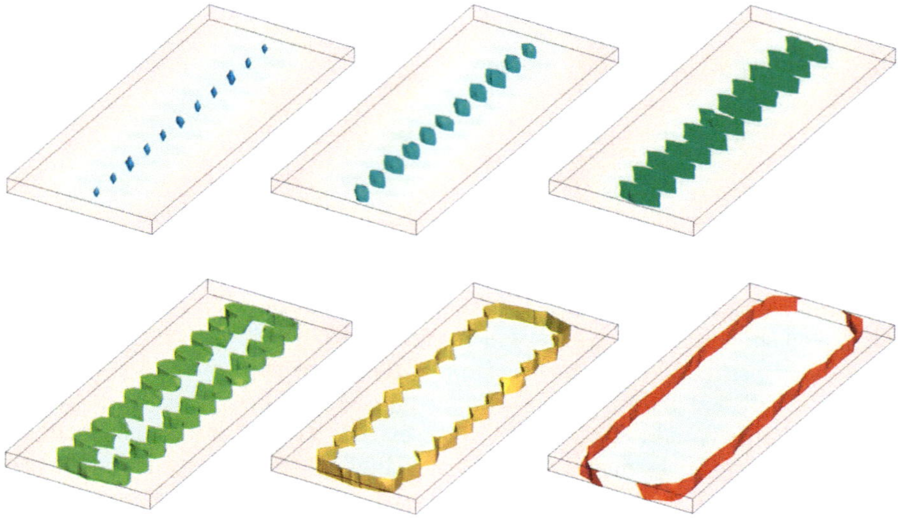

Fig. 8.45 Pressure distribution in a finite conductivity MFHW.

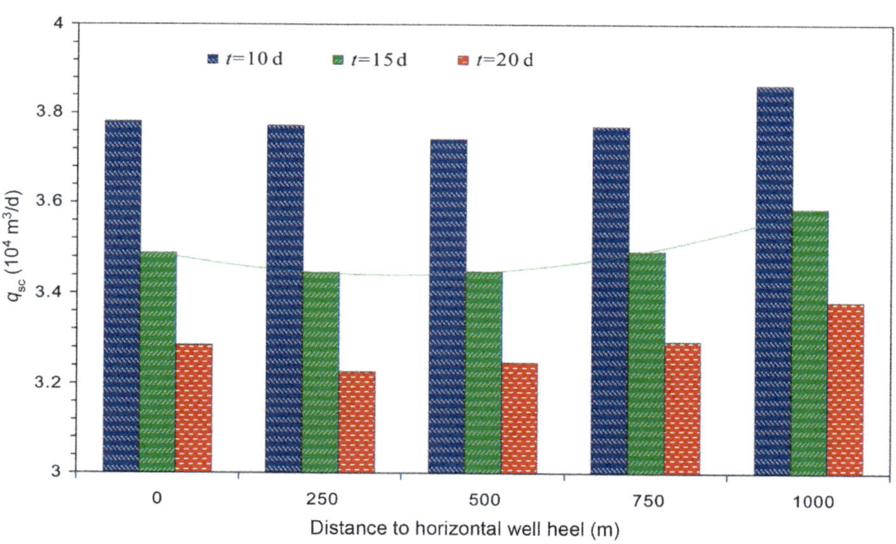

Fig. 8.46 Production rate distribution in a finite conductivity MFHW in a shale gas reservoir.

difference between the end and middle sections becomes larger, and the concave part of the "U" shape becomes deeper.

As Fig. 8.47 shows, we assume three groups of the initial saturation to study the impact of saturation on production performance. We can conclude that the smaller the initial water saturation, the greater the production rate and accumulative production. In detail, a smaller water saturation leads to a longer stable production period and

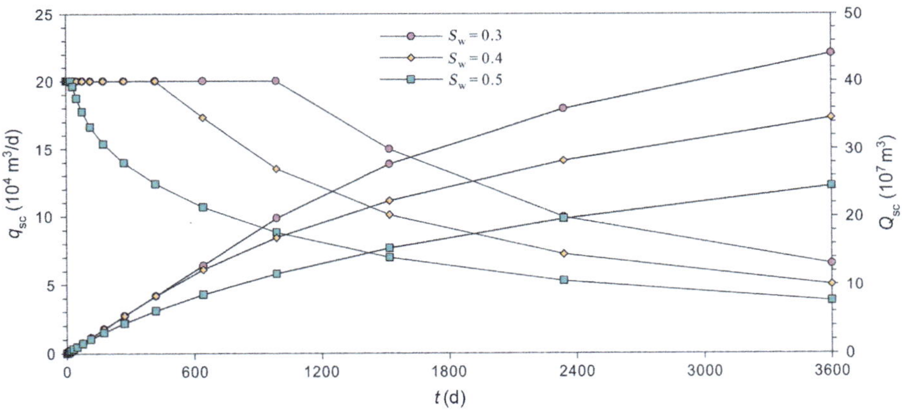

Fig. 8.47 Effect of initial water saturation on production.

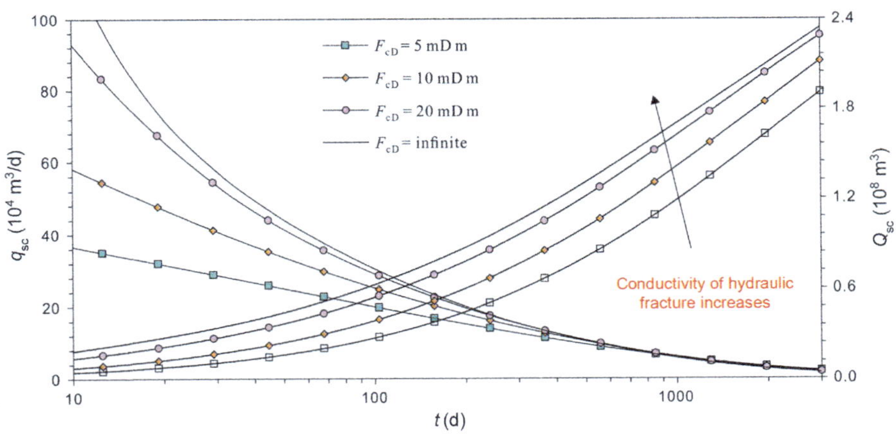

Fig. 8.48 Effect of hydraulic fracture conductivity on production of a MFHW in a shale gas reservoir.

a greater production rate at the constant BHP stage for the MFHW. This is because a smaller water saturation means a larger initial gas saturation, which represents more fluid mobility and a larger volume of gas at initial conditions.

The production decline and cumulative production curves are shown in Fig. 8.48. Compared to an infinite conductive model, the early production of a horizontal well with finite conductive hydraulic fractures is lower. This is because the early time flow regimes are fracture linear flow and elliptical flow, which both happen around the hydraulic fractures. Higher fracture conductivity means less flow resistance, and, therefore, higher production under the same BHP. After flow enters into the system linear flow and boundary dominated flow, the surrounding region of hydraulic fractures is not the main gas supply and production from both models tends to become equalized. Moreover, it can be seen that the production rate and cumulative production of the fractured horizontal well increase with an increase in fracture conductivity. This effect is more obvious during the early time flow period and diminishes with production time. Note that the increase in early production with fracture conductivity tends to maximize at a certain level. As the plot shows, for hydraulic fracture conductivity of 20 mD m, the production rate and cumulative production are almost close to those of a horizontal well with infinite conductive fractures.

According to the above analysis, there is an optimal value of fracture conductivity for a hydraulic fracturing design. For fracture conductivity of 20 mD m, the calculated fracture permeability is 20 D for a fracture width of 0.001 m. Obviously, it is easily achieved during normal fracturing operation. Therefore, it is not worth trying to improve well production through increasing hydraulic fracture conductivity.

As Fig. 8.49 shows, we define a fracturing degree coefficient β, which is the ratio of hydraulic fracture height to reservoir thickness:

$$\beta = \frac{h_F}{h} \tag{8.73}$$

Fig. 8.50 shows the influence of a hydraulic fracturing degree on production of a MFHW in a shale gas reservoir. A fracture opening degree directly affects the early production. After some days, an approximate rectangular low pressure region is created surrounding the multistage fractured horizontal well. At this time, the effect of the fracturing degree is minimal.

We assume a dimensionless inter-porosity coefficient λ_{m-f} ($\lambda_{m-f} = \alpha * L^2 * k_m/k_f$), which reflects the capacity of the fluid transfer from the matrix to small scale fractures. First, choosing different inter-porosity flow coefficients $\lambda_{mf} = (0.1, 1, 10)$ and keeping the other parameters the same, the production rate and accumulative production curves are shown in Fig. 8.51. Next, the production performance of the tri-porosity media and a conventional single porosity fracture system is compared to demonstrate the significant effect of inter-porosity flow on a production improvement. It can be seen that the bigger the λ_{m-f}, the greater the production rate

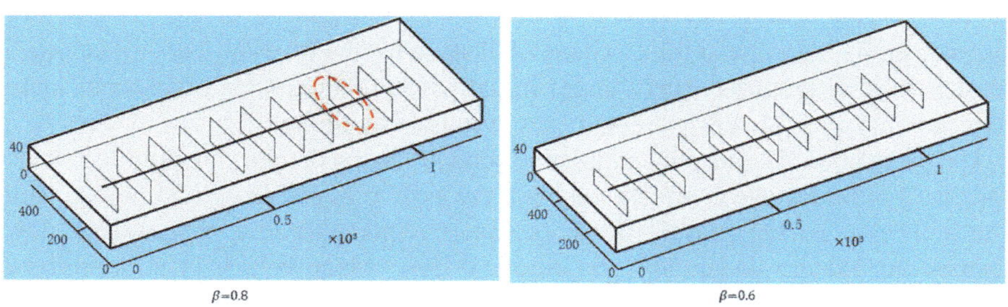

Fig. 8.49 Models with different fracturing degree of a MFHW in a shale gas reservoir.

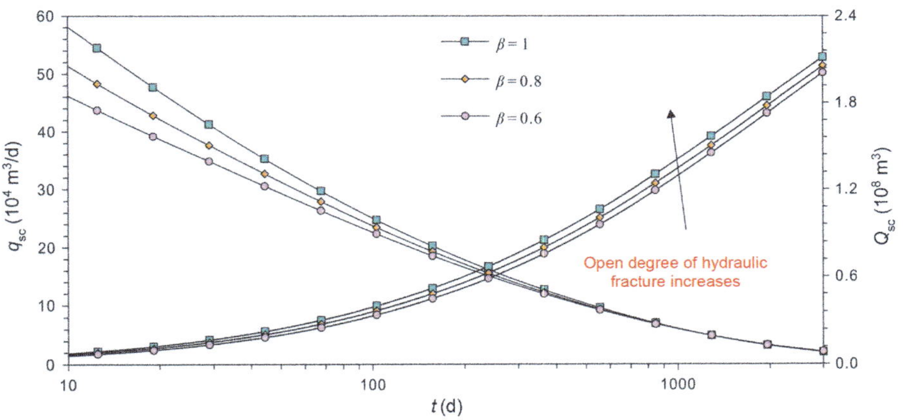

Fig. 8.50 Effect of hydraulic fracturing degree on production of a MFHW in a shale gas reservoirs.

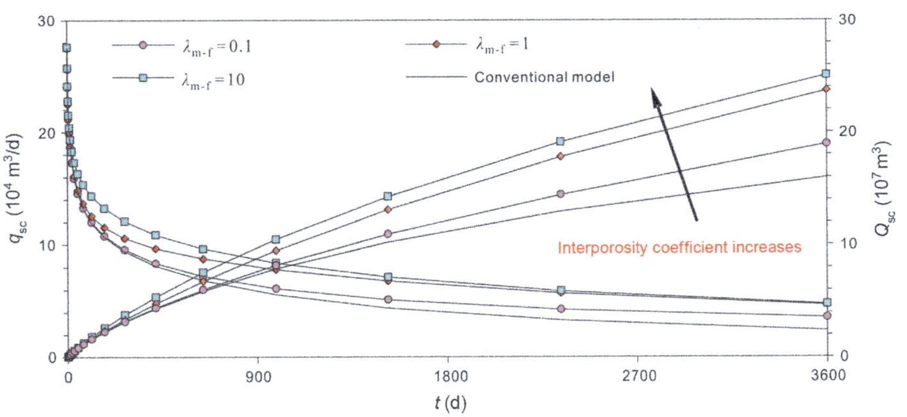

Fig. 8.51 Effect of inter-porosity coefficient on production of a MFHW in a shale gas reservoir.

and accumulative production, the production declines more slowly, and the longer the stable production period. After a little amount of gas stored in the fracture system is produced in the early period, under a pressure difference, the gas stored in the extremely tight shale matrix begins transferring into the fracture system. A bigger λ_{m-f} means more gas from the matrix as a supplement for the fracture system, which ultimately leads to an improvement of production.

In field practice, a MFHW well pad model is used to develop an extremely low permeability shale gas reservoir. As Fig. 8.52 shows, four MFHWs are represented by unstructured elements. From the pressure distribution obtained by the simulator introduced in this book, we can conclude that a well pad model can expand a stimulated reservoir volume, which finally improves production.

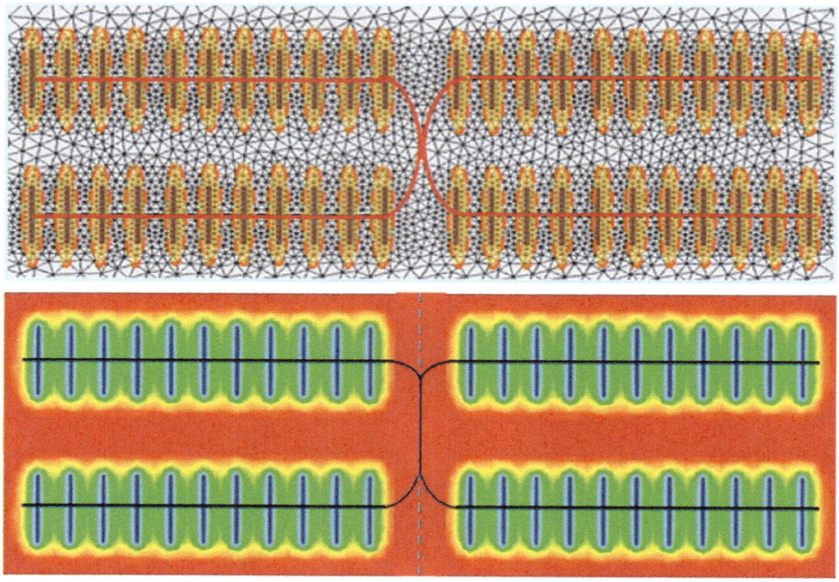

Fig. 8.52 A four-well pad model in a shale gas reservoir.

Case studies

9

Chapter Outline

9.1 Application of a well test analysis model 287
9.1.1 Pressure buildup test analysis of well H2-2 287
9.1.2 Pressure buildup test analysis of well H6-4 289
9.1.3 Pressure buildup test analysis of well H6-6 290
9.1.4 Pressure buildup test analysis of well H9-6 294
9.2 Application of numerical simulation 299
9.2.1 Overview of target regions and well 299
9.2.2 Hydraulic fracturing operation for well H9-1 300
9.2.3 Hydraulic fracture simulation results 303
9.2.4 Numerical simulation study 304

9.1 Application of a well test analysis model

Pressure buildup test data from four shale gas wells from Changning, Shanghai, and Weiyuan, Sichuan, shale gas reservoirs is interpreted by using the multistage fracturing horizontal well test analysis models established in the previous chapters. From the characteristics of a pseudo-pressure derivative curve, the flow feature of a fractured well is obvious, which shows linear flow or bilinear flow in a later flow period. The fitting curves of the tested data and the theoretical curves are plotted in log–log and semi-log graphs. According to the diagnosis curves and the fitting curves, it can be concluded that our models established in the previous chapters are reasonable and reliable for the interpretation of shale gas wells.

9.1.1 Pressure buildup test analysis of well H2-2

The production time of well H2-2 before the pressure buildup test is 1445.68 h, and the shut-in time is 282.77 h. A log–log diagnosis curve of the bottom-hole pressure is shown in Fig. 9.1. It can be seen from Fig. 9.1 that the wellbore storage effect is obvious. In this flow period, the pseudo-pressure difference and its derivative curves are straight lines with a slope of 1; in the late period, the linear flow happens, with the derivative curve exhibiting a straight line with a slope of 1/2. The linear flow period is a very important feature for fractured horizontal wells in unconventional gas reservoirs, which can last a long time since the pressure wave in such reservoirs is very strong.

Developments in Petroleum Science, Vol. 66. https://doi.org/10.1016/B978-0-444-64315-5.00009-7

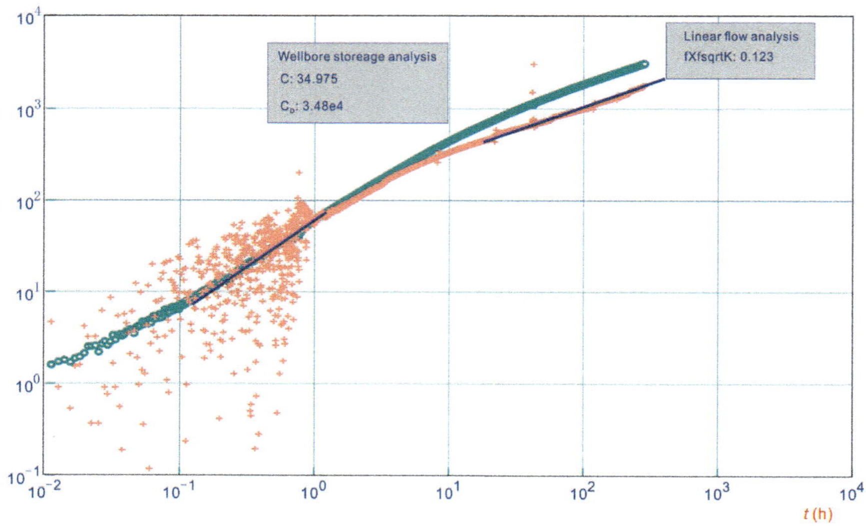

Fig. 9.1 Type curves overlay including the derivative for well H2-2.

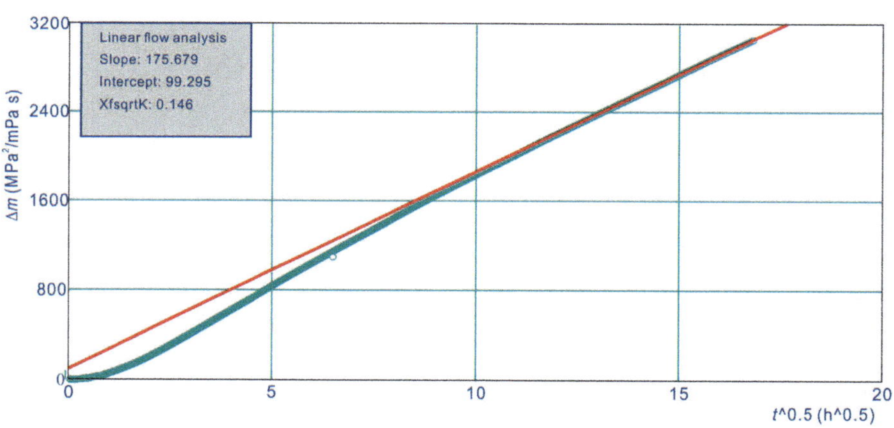

Fig. 9.2 Linear flow analysis for well H2-2.

Fig. 9.2 is a linear flow analysis diagram of well H2-2. The value of $x_f.k^{0.5}$ can be solved by using the slope of linear flow in a Cartesian coordinate diagram of $\Delta\psi$ and $t^{0.5}$. The value of $x_f.k^{0.5}$ for well H2-2 is 0.146 m μm.

In the matching process, both the pseudo-pressure and the corresponding derivative are plotted in a log–log graph with a compatible scale to a dimensionless type curve, as shown in Fig. 9.3. Through repeated adjustments of parameter simulation, a log–log plot, a semi-log plot, and pressure history fit maps with good fitting effects can be obtained (Figs. 9.3–9.5). The final interpretation results are given in Table 9.1.

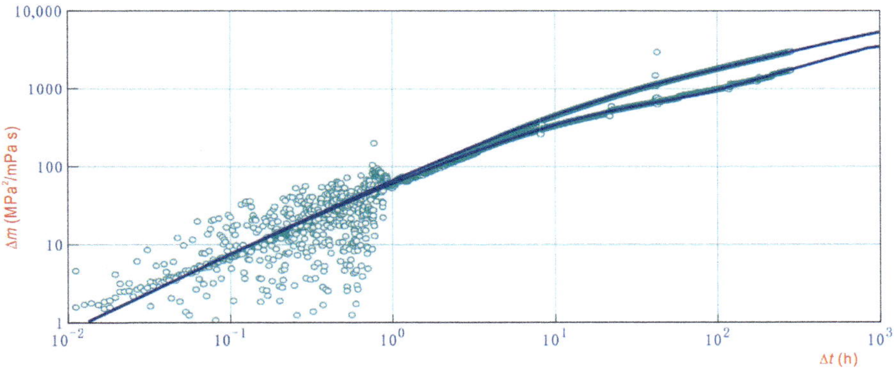

Fig. 9.3 Log–log plot of matching result for well H2-2.

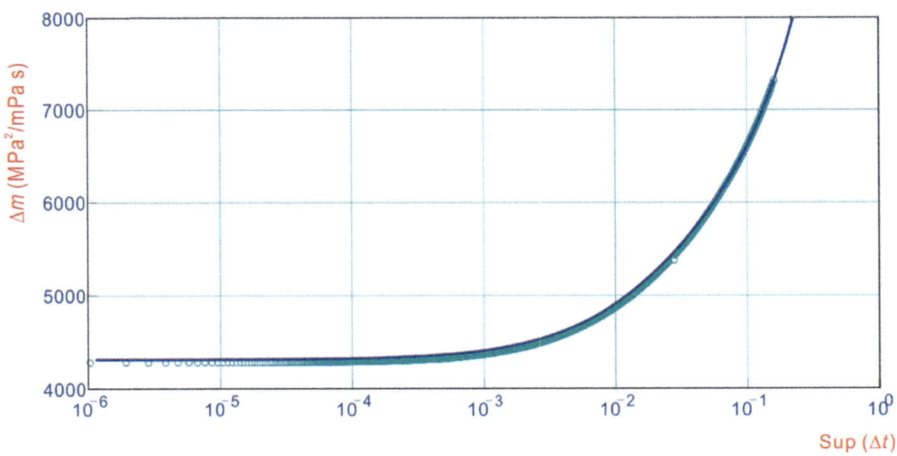

Fig. 9.4 Semi-log plot of matching result for well H2-2.

9.1.2 Pressure buildup test analysis of well H6-4

The production time of well H6-4 before the pressure buildup test is 2227.00 h, and the shut-in time is 494.5 h. A log–log diagnosis curve of the bottom-hole pressure is shown in Fig. 9.6. It can be seen from Fig. 9.6 that the wellbore storage effect is clear. In this flow period, the pseudo-pressure difference and its derivative curves are straight lines with a slope of 1; in the late period, the linear flow happens with a derivative curve exhibiting a straight line with a slope of 1/2. The linear flow period is a very important feature for fractured horizontal wells in unconventional gas reservoirs, which can last a long time since the pressure wave in such reservoirs is very strong.

Fig. 9.7 is a linear flow analysis diagram of well H6-4. The value of $x_f.k^{0.5}$ can be solved by using the slope of linear flow in a Cartesian coordinate diagram of $\Delta\psi$ and $t^{0.5}$. The value of $x_f.k^{0.5}$ for well H2-2 is 0.02 m μm.

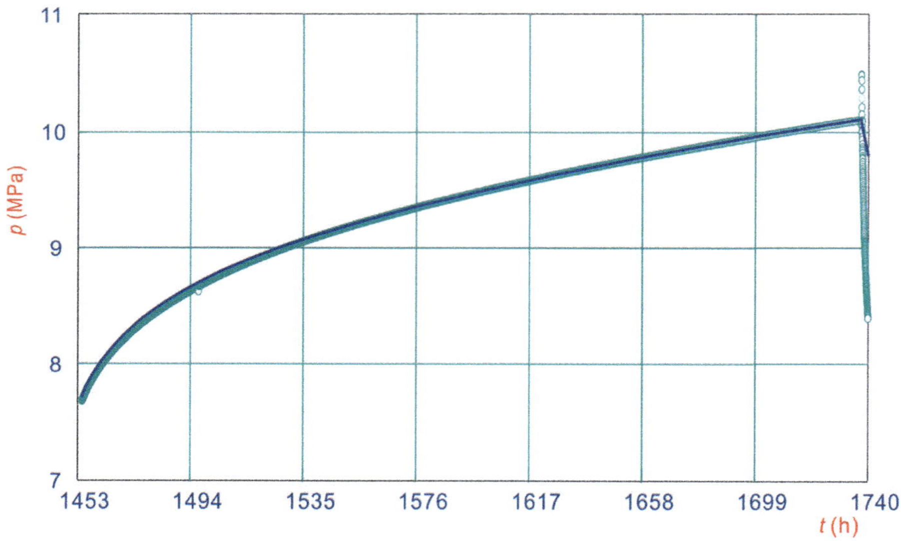

Fig. 9.5 The history of pressure buildup for well H2-2.

Table 9.1 Interpretation results of well H2-2

Parameters	Value	Parameters	Value
Formation pressure (MPa)	14.64	Dimensionless fracture conductivity	220.00
Reservoir permeability (mD)	8.33×10^{-3}	Fracture half length (m)	135.00
Storativity ratio of internal fracture	0.0244	Wellbore storage coefficient (m^3/MPa)	25.40
Flow coefficient of internal fracture	2.28×10^{-8}	Formation permeability (mD)	4.45×10^{-4}

In the matching process, both the pseudo-pressure and the corresponding derivative are plotted in a log–log graph with a compatible scale to a dimensionless type curve, as shown in Fig. 9.8. Through repeated adjustments of parameter simulation, a log–log plot, a semi-log plot, and pressure history fit maps with good fitting effects can be obtained (Figs. 9.8–9.10). The final interpretation results are given in Table 9.2.

9.1.3 Pressure buildup test analysis of well H6-6

The production time of well H6-6 before the pressure buildup test is 2202.95 h, and the shut-in pressure buildup time is 495.46 h. The log–log diagnosis curves of the shut-in pressure buildup are shown in Figs. 9.11 and 9.12. It can be seen from Fig. 9.13 that, in

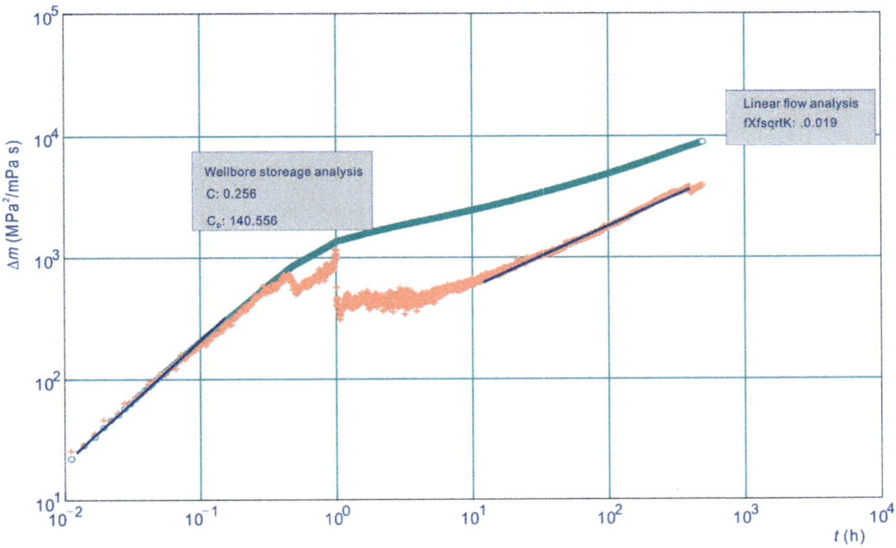

Fig. 9.6 Type curves overlay including the derivative for well H6-4.

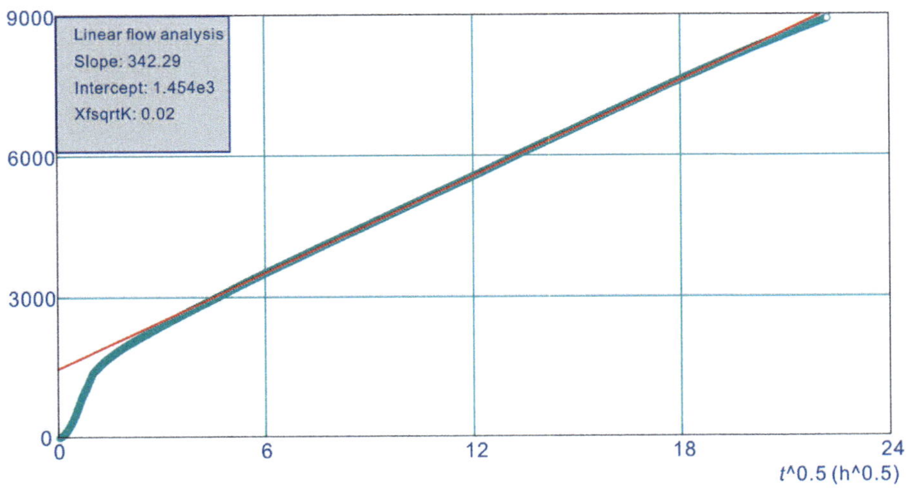

Fig. 9.7 Linear flow analysis diagram for well H6-4.

the early stage of shut-in, the pure wellbore reservoir effect phase, the pseudo-pressure difference, and its derivative curve exhibit straight lines with a slope of 1. After the transition section of the wellbore reservoir effect, the derivative curve approximately presents a straight line with a slope of 1/4, describing the bilinear flow characteristics of the fracture crack, and the gas in the inner zone formation, indicating that the conductivity of the fracture crack is relatively low.

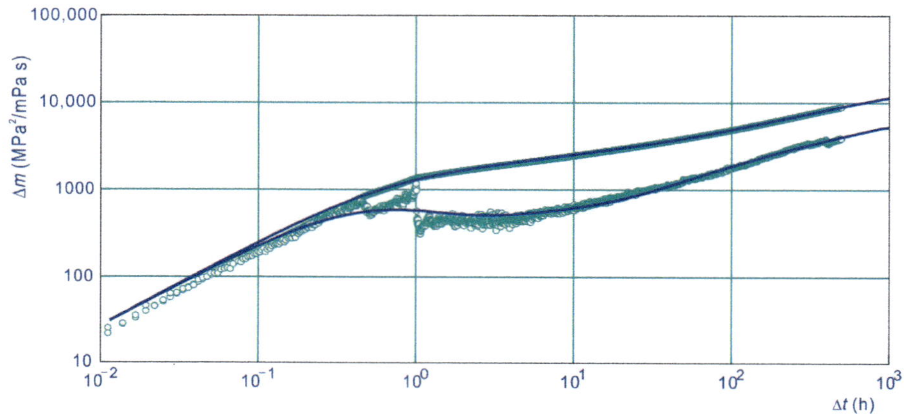

Fig. 9.8 Log–log plot of matching result for well H6-4.

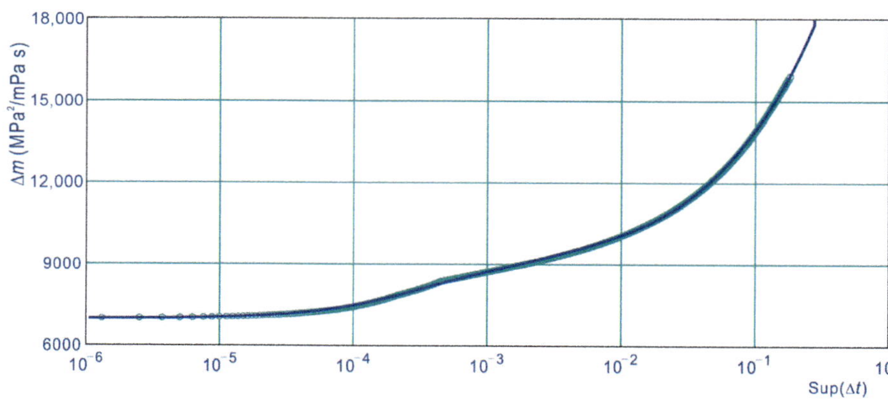

Fig. 9.9 Semi-log plot of matching result for well H6-4.

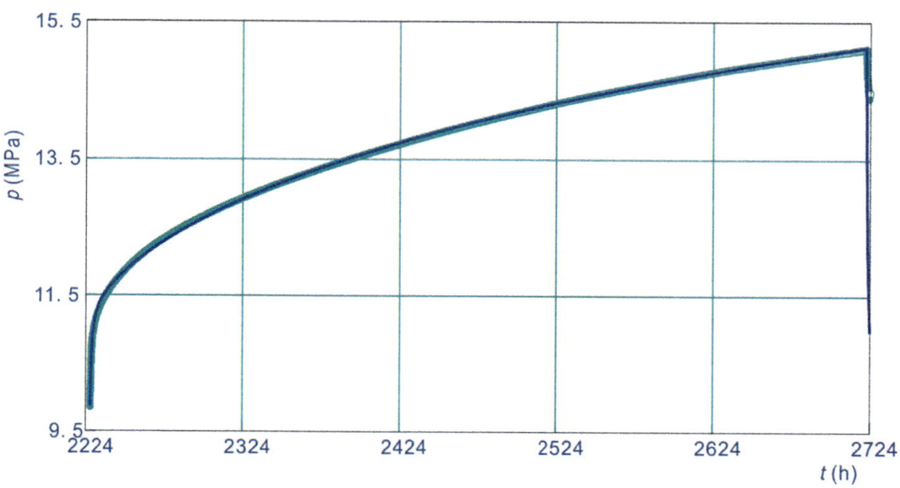

Fig. 9.10 The history of pressure buildup for well H6-4.

Table 9.2 Interpretation results of well H6-4

Parameter	Value	Parameter	Value
Formation pressure (MPa)	22.09	Dimensionless fracture conductivity	7.36
Reservoir permeability (mD)	3.42×10^{-3}	Fracture half length (m)	65.00
Storativity ratio of internal fracture	0.012	Wellbore storage coefficient (m^3/MPa)	0.192
Flow coefficient of internal fracture	2.55×10^{-8}	Formation permeability (mD)	3.03×10^{-5}

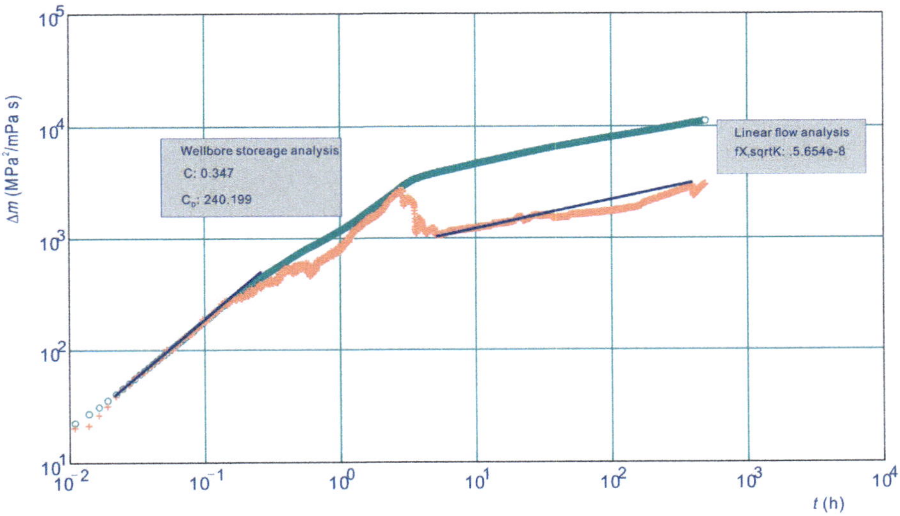

Fig. 9.11 Type curves overlay including the derivative for well H6-6(1).

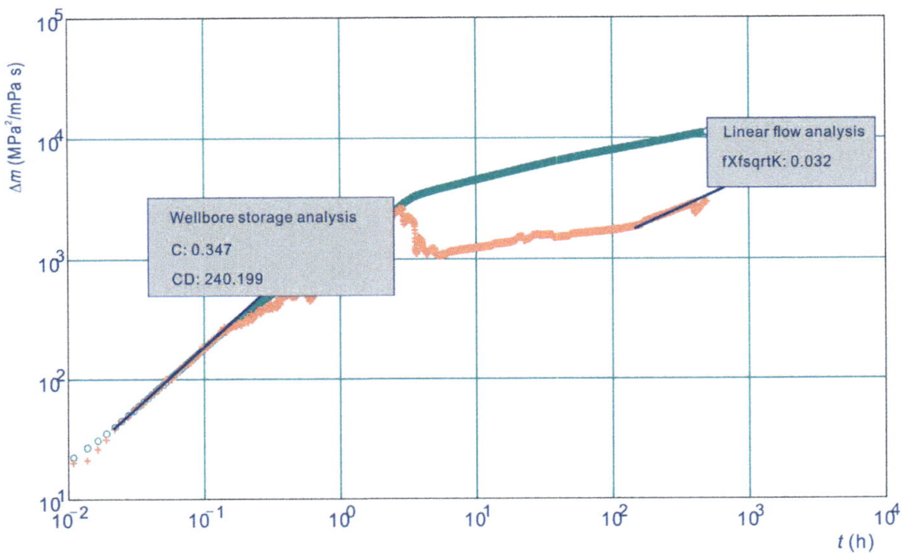

Fig. 9.12 Log–log plot of matching result for well H6-6(2).

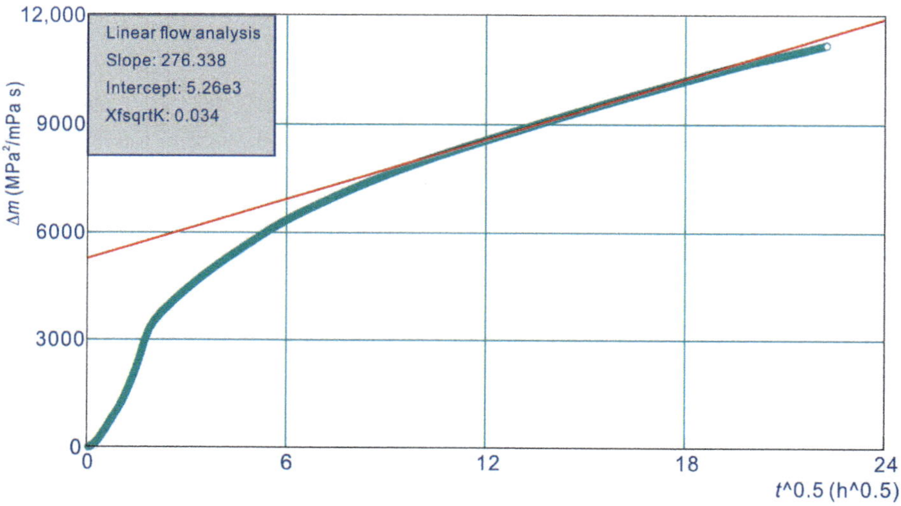

Fig. 9.13 Linear flow analysis diagram of well H6-6.

It can be seen from Fig. 9.12 that, in the late stage of the test, the pseudo-pressure difference and its derivative curve both approximate a straight line with a slope of 1/2, describing the flow of gas in the reservoir as a characteristic of linear flow. The linear flow stage has not ended at the end of the test.

Fig. 9.13 is a linear flow analysis diagram of well H6-6. The value of $x_{f}.k^{0.5}$ can be solved by using the slope of linear flow in a Cartesian coordinate diagram of $\Delta\psi$ and $t^{0.5}$. The value of $x_{f}.k^{0.5}$ for well H2-2 is 0.034 m μm.

Fig. 9.14 shows the bilinear flow analysis of well H6-6. The value of $k_f w_f \cdot k^{0.5}$ can be solved by using the Cartesian coordinate relationship of $\Delta\psi$ and $t^{0.25}$ and is found to be 5.652×10^{-8} m μm^3.

In the matching process, both the pseudo-pressure and the corresponding derivative are plotted in a log–log graph with a compatible scale to the dimensionless type curve, as shown in Fig. 9.15. Through repeated adjustments of parameter simulation, a log–log plot, a semi-log plot, and pressure history fit maps with good fitting effects can be obtained (Figs. 9.15–9.17). The final interpretation results are given in Table 9.3.

9.1.4 Pressure buildup test analysis of well H9-6

The production time of well H9-6 before the pressure buildup test is 594.6 h, and the shut-in pressure buildup time is 374.34 h. A log–log diagnosis curve of the bottomhole pressure is shown in Fig. 9.18. It can be seen from Fig. 9.18 that the wellbore storage effect is obvious. In this flow period, the pseudo-pressure difference and its derivative curves are straight lines with a slope of 1; in the late period, the linear

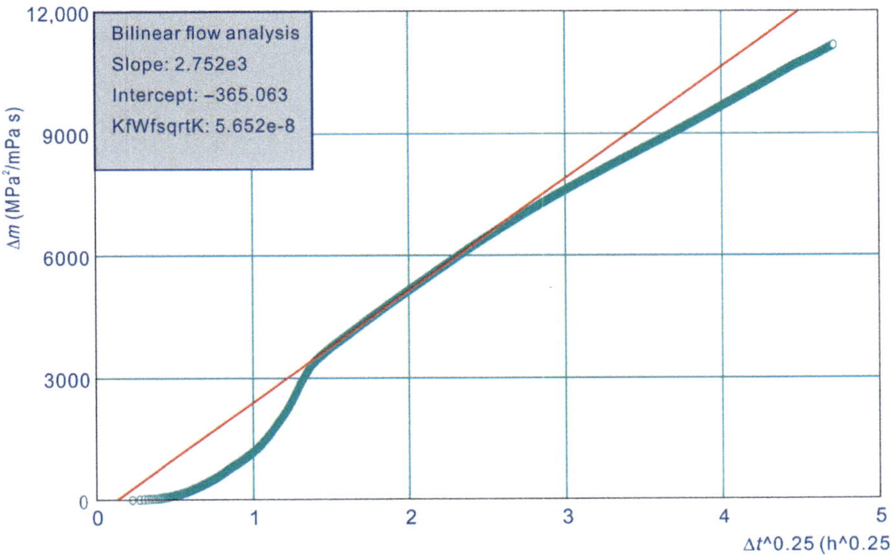

Fig. 9.14 Bilinear flow analysis diagram of well H6-6.

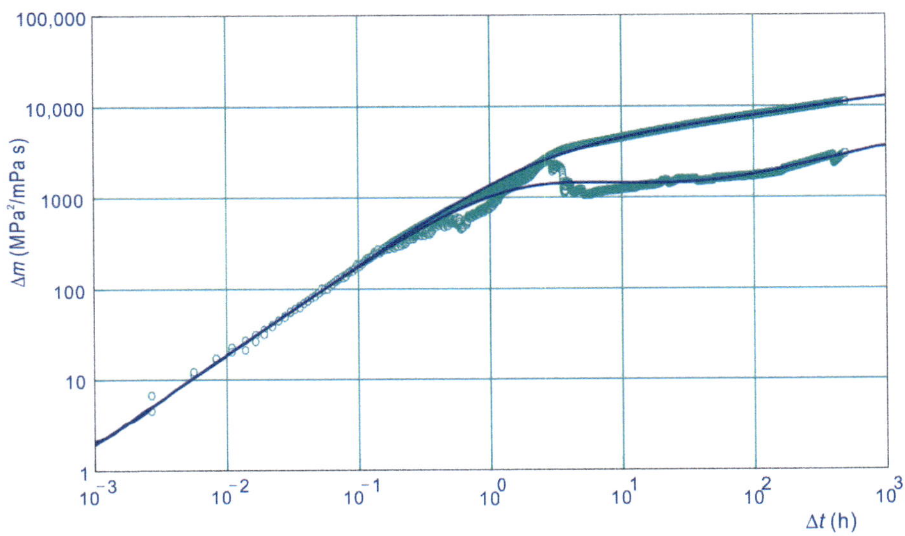

Fig. 9.15 Log–log plot of matching result for well H6-6.

flow happens, whose derivative curve exhibits a straight line with a slope of 1/2. The linear flow period is a very important feature for fractured horizontal wells in unconventional gas reservoirs, which can last a long time because the pressure wave in such reservoirs is very strong.

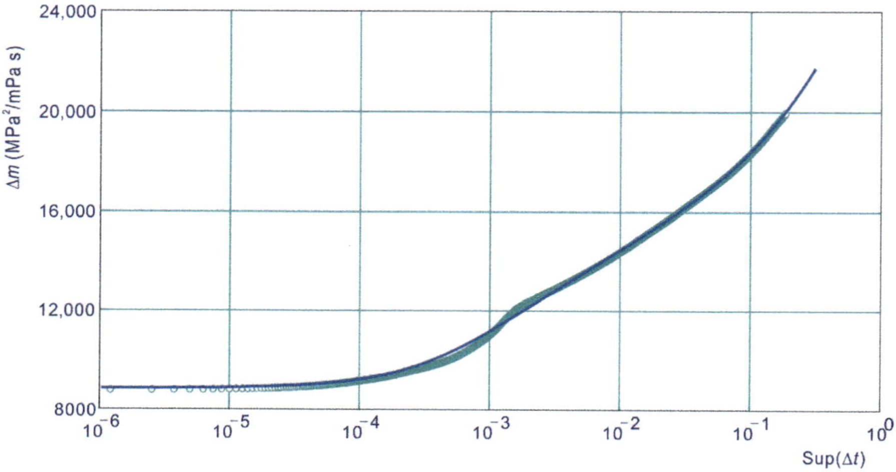

Fig. 9.16 Semi-log plot of matching result for well H6-6.

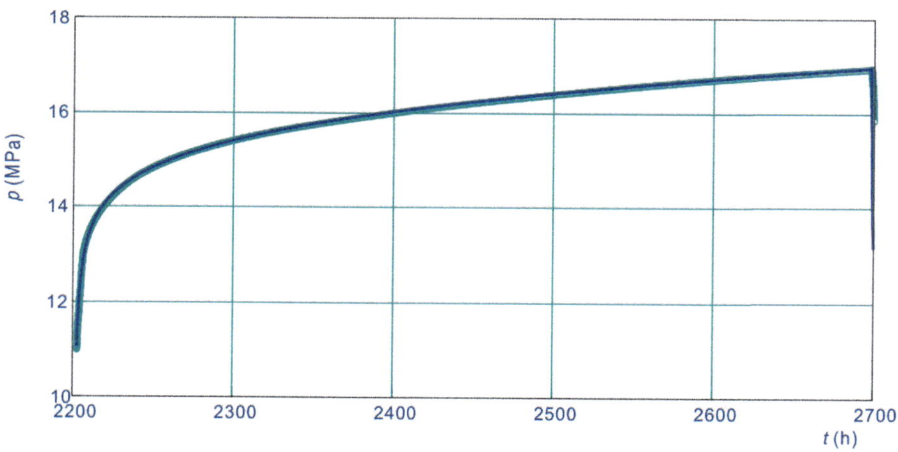

Fig. 9.17 The history of pressure buildup for well H6-6.

Table 9.3 Interpretation results of well H6-6

Parameter	Value	Parameter	Value
Formation pressure (MPa)	22.63	Dimensionless fracture conductivity	25.55
Reservoir permeability (mD)	2.84×10^{-3}	Fracture half length (m)	34.63
Storativity ratio of internal fracture	0.06	Wellbore storage coefficient (m^3/MPa)	0.321
Flow coefficient of internal fracture	7.83×10^{-8}	Formation permeability (mD)	8.00×10^{-5}
Skin factor of fracture surface	0.363		

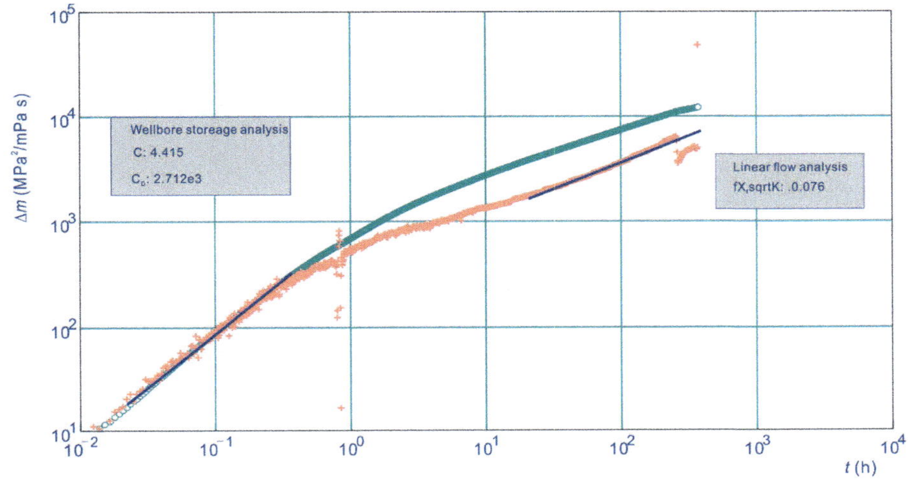

Fig. 9.18 Type curves overlay including the derivative for well H9-6.

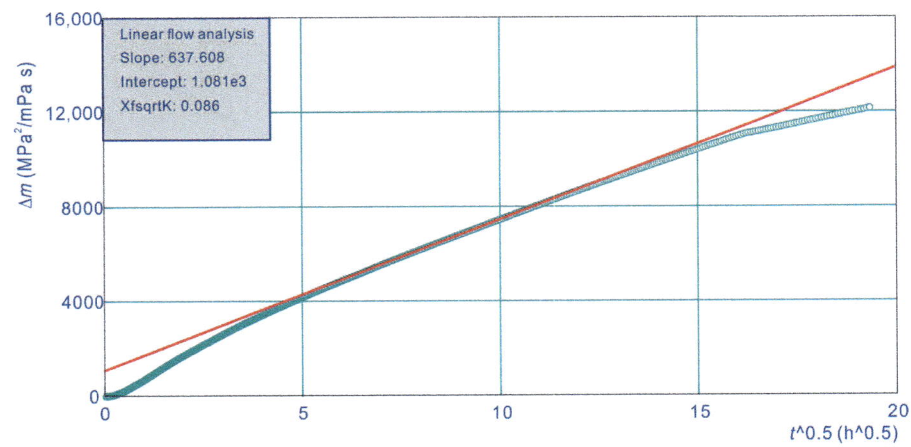

Fig. 9.19 Linear flow analysis diagram of well H9-6.

Fig. 9.19 is a linear flow analysis diagram of well H9-6. The value of $x_f.k^{0.5}$ can be solved by using the slope of linear flow in a Cartesian coordinate diagram of $\Delta\psi$ and $t^{0.5}$. The value of $x_f.k^{0.5}$ for well H2-2 is 0.086 m μm.

Based on the results of the wellbore reservoir effect and the linear flow analysis, the interpretation and calculation of the interpretation model are performed. Through repeated adjustments of parameter simulation, a log–log plot, a semi-log plot, and pressure history fit maps with good fitting effects can be obtained (Figs. 9.20–9.22). The final interpretation results are given in Table 9.4.

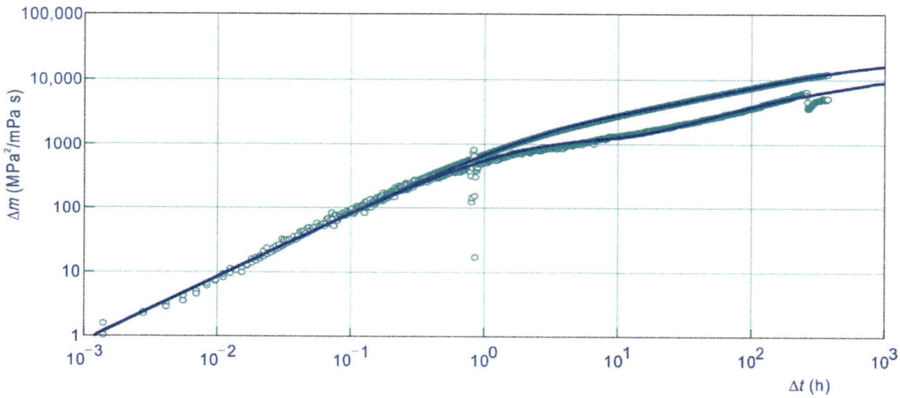

Fig. 9.20 Log–log plot of matching result for well H9-6.

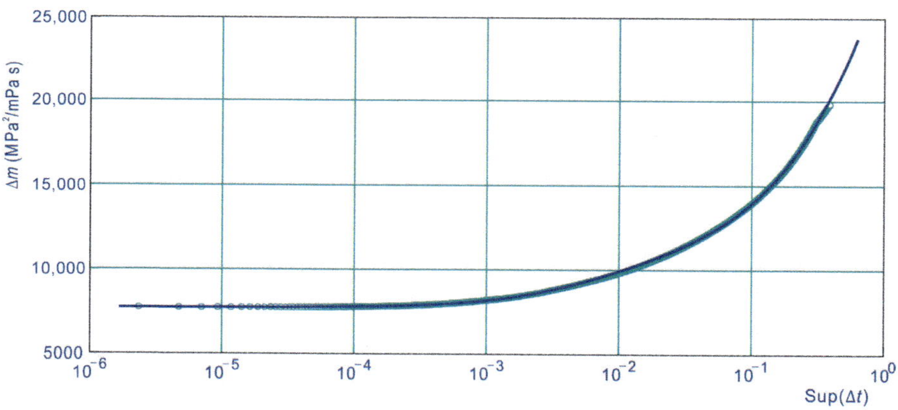

Fig. 9.21 Semi-log plot of matching result for well H9-6.

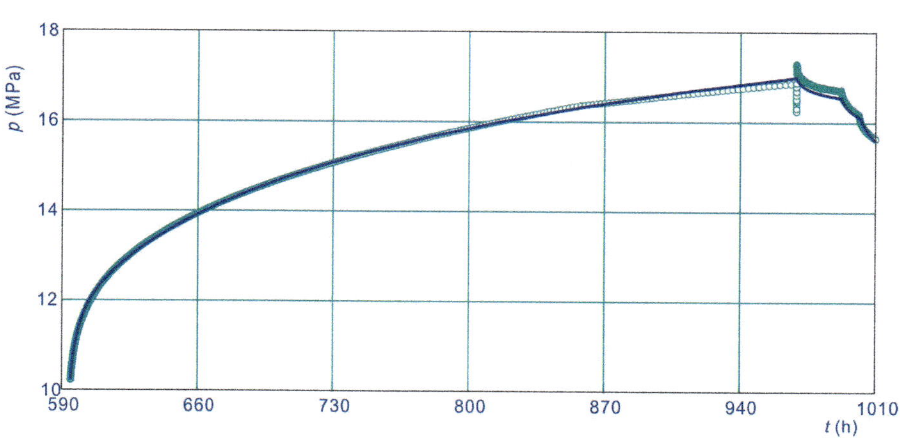

Fig. 9.22 The history of pressure buildup for well H9-6.

Table 9.4 Interpretation results of well H9-6

Parameter	Interpretation result	Interpretation parameter	Interpretation result
Formation pressure (MPa)	22.84	Dimensionless fracture conductivity	10.07
Reservoir permeability (mD)	0.03	Fracture half length (m)	40.00
Storativity ratio of internal fracture	0.065	Wellbore storage coefficient (m^3/MPa)	4.36
Flow coefficient of internal fracture	3.56×10^{-8}	Formation permeability (mD)	1.49×10^{-3}

9.2 Application of numerical simulation

9.2.1 Overview of target regions and well

The region in which block N-201 is located belongs to the Changning–Weiyuan National Shale Gas Demonstration Development area, southwest of the Sichuan Basin, China. The area covers the Changning County, Gong County, Xingwen County, and Qilian County of Yibin City in Sichuan Province. More precisely, the region belongs to the Shuifu–Suiyong Development Zone; it is mountainous with elevations from 400 to 1300 m, being dominated by low-lying mountains and hills.

In terms of its geological structure, it is located in the south of the Changning anticline. The reservoirs for shale gas exploration and development are organic-rich shale formations, which are mainly located in the upper part of the Wufeng Formation of the Ordovician and the lower part of the Longmaxi Formation of the Silurian. The total thickness is between 30 and 50 m, and the buried depth is between 2300 and 3200 m.

The Wufeng–Longmaxi Formation is a continental-shelf sedimentation, forming an organic-rich mudstone, which has characteristics of single lithology, thick grain layers, extensive distribution, and richness in fossils. The sedimentary stratum can be subdivided into two subfacies: an inland formation and an outer formation, and seven microfacies, such as an organic siliceous mud formation and an organic silty mud formation. The Wufeng–Long 1^1 subsection is an organic-rich siliceous mud formation, which is the most favorable sedimentary microfacies for shale gas enrichment and accumulation.

The shale gas reservoir is dominated by black carbonaceous shale, black shale, siliceous shale, black mudstone, black silty mudstone, and grayish black silty mudstone. The brittle minerals are relatively high in mineral components, with an average content of >70%, are mainly composed of siliceous minerals and have good compressibility. The clay minerals are composed of illite (52.2%), chlorite (25%), illite–montmorillonite, and few expansive minerals. The organic components are mainly saprolite, whose main type is kerogen-I. The thermal maturity

(Ro) is generally >2.5%, reaching the stage of over-maturation. Therefore, dry gas is mainly produced. The organic total content (TOC) in the Wufeng Formation and the Long 1^1 subsection is generally higher, with the measured TOC being 3.0%–4.2% in the laboratory and 2.7%–4.5% as determined by logging.

The maximum horizontal principal stress direction in the shale gas reservoir mainly runs southeast–northwest, with some areas being in the northeast direction. The triaxial compressive strength of the core is 181.73–321.74 MPa, with an average of 254.04 MPa; Young's modulus is $(1.548–5.599) \times 10^4$ MPa, with an average of 3.52×10^4 MPa; and Poisson's ratio is 0.158–0.331, with an average of 0.225.

There are various pore structures in the reservoir, including pores and microfractures. The pore types include organic pores, intercrystalline pores, intragranular dissolved pores, and intergranular pores. The types of fractures include structural, diagenetic, dissolution, and hydrocarbon-generating joints. The measured porosity is 2.0%–6.8% as measured in the laboratory and 3.6%–7.3% as determined by logging. The matrix permeability is low, being $(0.714–1.48) \times 10^{-4}$ mD, with the average of 1.02×10^{-4} mD. The correlation between porosity and permeability (nitrogen measurement) is poor, and the shale layering fractures and microfractures have an important influence on shale permeability.

The shale gas is mainly methane, accounting for more than 97%, with few hydrocarbon contents above C_3, no hydrogen sulfide, and 0.22%–0.54% CO_2. The gas has high maturity and the drying coefficient (C_1/C_2+) is 134.65–282.98. The gas saturation is high, and the average gas saturation of a single well is 50%–70%. The total gas content is relatively high, being 2.0–3.5 m^3/t as measured in the laboratory and 2.9–7.4 m^3/t as determined by logging, with an average of 4.8 m^3/t.

The N-201 block is in a relatively stable geological structural zone, and the upper and lower strata of Longmaxi are dense barrier layers. Therefore, the gas preservation state is good and the TOC is high. At present, it is considered that the abundance of organic matter in the N-201 block is one of the key factors controlling the gas properties of the Wufeng–Longmaxi Formation shale reservoir.

Well H9-1 is a horizontal well located in the H9 platform in the N-201 block. The well was drilled in June 2015, and the drilling depth is 4520 m. The maximum well deviation depth of the whole well is 3669.80 m, the well inclination is 95.97 degrees, the azimuth is 11.36 degrees, the closing distance is 1325.76 m, and the closing orientation is 55.66 degrees. The basic information of the well is given in Table 9.5, and the well structure is shown in Fig. 9.23.

9.2.2 Hydraulic fracturing operation for well H9-1

From November 7 to 29, 2015, the hydraulic fracturing operation in the 18 sections of well H9-1 was completed. Well H9-1 uses large bore bridge plugs as the segmentation tool, a slick water fracturing fluid system, and 100 mesh silt + 40/70 mesh ceramsite for slug-type sand fracturing.

The first section of H9-1 uses coiled tubing perforation (in a conventional helical shot pattern), divided into two clusters of perforation. Each cluster perforation section length is 1.5 m, the perforation density is 16 holes/m, and the phase angle is 60 degrees.

Table 9.5 Basic information of well H9-1

Basic data			
Geographic location	**Group 7 of Qixing Village, Shangluo Town, Yi County, Yibin City, Sichuan Province**		
Geological structure location	South of Ordovician top structure in Changning anticline structure		
Elevation (m)	879.317	Kelly bushings (m)	888.737
Start drilling date	2015.04.27	Finish drilling date	2015.06.24
Total depth (m)	4560	Target formation	Baota
Artificial bottom (m)	4520	Completion method	Casing perforation
Maximum deviation (degree)	95.97	Depth of depth at maximum deviation (m)	3669.80
Pressure of water test (MPa)	85		

Well structure

Drill size × depth (mm × m)	Casing size × depth (mm × m)	Cement top (m)	Pressure test (MPa (min MPa))
444.5 × 513.00	339.7 × 510.55	0	17.430 (17.2)
311.2 × 1742.00	244.5 × 1740.03	0	17.430 (17.2)
215.9 × 4560.00	139.7 × 4553.50	1777.00	35.5630 (35.39)

Continued

Table 9.5 Continued

Parameters of drilling fluid					
Formation name	**Well section (m)**	**Drilling fluid system**	**Density (g/cm^3)**	**Viscosity (mPa s)**	**Chloride (mg/L)**
Jialingjiang–Feixianguan	0.00–513.00	Water	—	—	—
Feixianguan–Shiniulan	513.00–2404.25	Water-based polymer	1.03–1.50	32–63	2840–16,330
Shiniulan–Baota	2404.25–4560.00	Oil-based drilling fluid	1.78–1.95	65–100	—

Cementing quality

The well cementing quality pass rate standard: The first interface interpretation conclusion is mainly based on the 3 ft. sound amplitude.

The conventional density cement cementing interpretation standard: the sound amplitude value is <15% for good cementation; the sound amplitude value is between 15% and 30% for middle cementation; the sound amplitude value is >30% for poor cementation.

The evaluation of section 10.0–4472.5 m for cementing logging: The well cementing section is 82.57%, the medium cementing section is 11.62%, and the poor cementing section is 5.81%. The well logging for cementing evaluation is qualified.

Fracturing parameters						
Formation	**Section (m)**	**Length (m)**	**Effective porosity (%)**	**Mineral brittleness index (%)**	**Pressure coefficient**	**Temperature (°C)**
Longmaxi	3160–4520	1360	3.8–6.7	44.4–59.1	2.03 (prediction)	93.1 (4560 m)

Production casing parameters						
Outer diameter (mm)	**Landing depth (m)**	**Steel grade**	**Wall thickness (mm)**	**Internal diameter (mm)**	**Resistance to internal pressure (MPa)**	**Resistance to external pressure (MPa)**
139.7	4520.22	BG125V	12.7	114.3	137.2	156.7

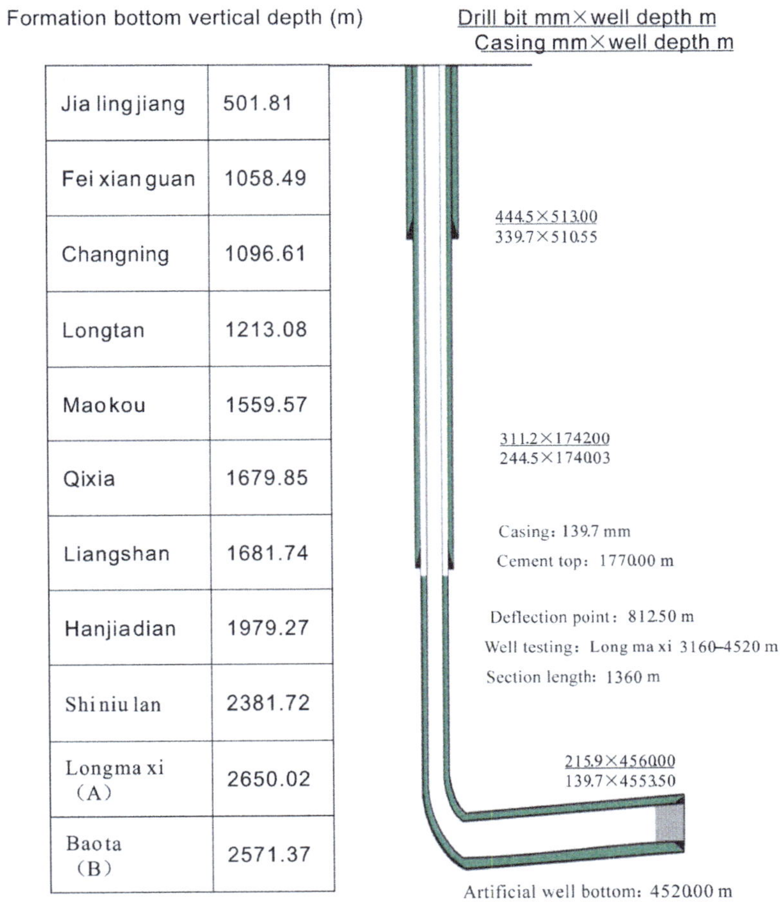

Fig. 9.23 H9-1 casing program.

The total number of holes is 48. The rest of the sections use wireline conveyed perforation (in a conventional helical shot pattern), divided into three clusters of perforation. Each cluster perforation section length is 1.0 m; the perforation density is 16 holes/m; the phase angle is 60 degrees, and the total number of holes is 48. Table 9.6 lists the fluid and sand addition data of each fracturing section of well H9-1.

9.2.3 *Hydraulic fracture simulation results*

In the process of the hydraulic fracturing simulation, the fracturing geometry is fitted and calibrated by using the actual downhole monitoring data. At the same time, the pump data is used for history fitting, and then parameters of the hydraulic fractures are used to correct the hydraulic parameters of the fracture distribution, as shown in Fig. 9.25. Table 9.7 and Fig. 9.24 show the hydraulic fracturing simulation results for H9-1.

Table 9.6 Fluid and sand addition data of each fracturing section of well H9-1

Fracturing section	100 mesh (t)	40/70 mesh (t)	Total sand addition (t)	Total liquid volume (m^3)
1	18.41	32.73	51.14	1838.31
2	20.63	80.67	101.30	1872.99
3	13.00	107.87	120.87	1904.99
4	13.89	108.30	122.19	1850.86
5	12.70	108.75	121.45	1873.56
6	10.67	109.65	120.32	1885.43
7	11.18	108.99	120.17	1863.36
8	11.90	109.82	121.72	1804.46
9	10.56	50.05	60.61	1310.5
10	20.73	54.38	75.11	1931.93
11	20.91	69.30	90.21	1891.4
12	18.31	57.50	75.81	1922.3
13	17.29	47.89	65.18	1898.54
14	15.50	54.82	70.32	1934.91
15	11.32	52.68	64.00	1909.78
16	21.47	68.74	90.21	1956.7
17	30.07	91.10	121.17	1897.88
18	18.25	112.30	130.55	1919.4
Total	296.8	1425.5	1722.3	33,467.3

From the simulation of well H9-1, the average hydraulic fracturing length is determined to be 225–233 m, the average supporting fracturing length is determined to be 185–189 m; the average hydraulic fracturing height is determined to be 50–62 m, and the average supporting fracturing height is determined to be 12–18 m (as shown in Fig. 9.25). The proppants match well with the hydraulic seam length on the plane, but the vertical coverage of the proppants needs to be improved.

9.2.4 Numerical simulation study

Figs. 9.26 and 9.27 show the production history of gas and water production rates and tubing pressure of well H9-1.

By introducing the fracture simulation results into the numerical simulator, we can use this simulator to obtain the production data to fit the production history and the pressure distribution at different times, as shown in Fig. 9.28.

Table 9.7 Hydraulic fracturing simulation results for H9-1

Fracturing section	Perforation cluster	Hydraulic fracture length (m)	Supporting fracture length (m)	Hydraulic fracture height (m)	Supporting fracture height (m)	Average conductivity (mD m)
1	1	391.7	324.9	40	16	64.3
	2	184.2	159.1	47	13	132.3
2	1	241.0	226.6	69	31	196.8
	2	133.9	101.9	64	14	671.1
	3	170.2	56.0	56	16	195.4
3	1	312.0	290.3	69	15	320.6
	2	212.0	107.9	86	8	563.2
	3	145.0	83.8	80	13	426.6
4	1	183.5	174.2	71	22	428.7
	2	93.2	64.5	35	16	653.6
	3	298.3	258.7	58	16	155.9
5	1	325.7	323.2	69	18	250.1
	2	343.8	338.3	51	26	180.2
	3	173.0	154.7	38	22	120.1
6	1	453.0	419.4	64	24	171.4
	2	425.2	411.8	40	19	147.1
	3	197.3	146.9	75	13	439.0
7	1	304.6	302.2	68	18	294.9
	2	377.0	366.3	39	17	170.5
	3	129.9	116.0	42	18	241.9
8	1	269.5	265.7	35	17	192.4
	2	339.7	325.4	50	24	207.1
	3	336.2	323.7	44	28	116.4
9	1	211.7	203.0	43	29	78.5
	2	134.4	129.6	26	22	100.3
	3	287.7	187.2	24	19	121.9
10	1	220.8	215.0	85	27	150.8
	2	71.9	68.7	107	33	254.6
	3	270.3	261.8	28	17	129.4
11	1	267.9	251.7	79	18	149.6
	2	164.8	158.9	70	30	168.7
	3	260.4	236.6	41	15	66.0
12	1	274.0	257.6	75	26	412.9
	2	71.5	27.0	47	8	64.0
	3	223.1	193.1	47	8	64.0

Continued

Table 9.7 Continued

Fracturing section	Perforation cluster	Hydraulic fracture length (m)	Supporting fracture length (m)	Hydraulic fracture height (m)	Supporting fracture height (m)	Average conductivity (mD m)
13	1	309.4	287.0	100	17	135.0
	2	72.5	62.5	106	15	507.9
	3	156.6	102.3	89	12	487.6
14	1	199.4	152.6	86	11	542.7
	2	132.4	74.5	88	12	1211.5
	3	186.3	113.3	84	11	531.4
15	1	113.3	98.2	55	11	646.7
	2	195.2	104.2	64	8	476.4
	3	201.8	122.8	73	8	310.5
16	1	72.9	50.8	75	14	771.6
	2	290.9	188.3	88	15	428.8
	3	221.0	71.6	54	23	251.4
17	1	79.2	69.4	55	23	282.7
	2	199.2	171.3	60	13	476.7
	3	382.1	313.8	35	12	350.4
18	1	397.7	307.2	60	20	269.6
	2	141.0	134.3	80	19	664.1
	3	68.8	62.7	66	24	617.3
Average		224.9	189.0	62	18	321.9

We can see that the fracture network greatly improves the flow ability of a stimulated reservoir volume (SRV), and the pressure decline spreads along the fractures.

Because the production well only records the wellhead pressure, a wellbore multiphase flow theory is used to calculate the pressure at the bottom of the well, as shown in Fig. 9.29. Based on the simulation theory and the method, we can match the production history of well H9-1, as shown in Figs. 9.30 and 9.31. By adjusting parameters, we find that the average fracture permeability is 200 mD and that the reservoir permeability near the fractured area is 0.001 mD. Then we can predict production in 5 years. Well H9-1's cumulative production is 1.4375 billion m^2 (as given in Table 9.8; Fig. 9.32).

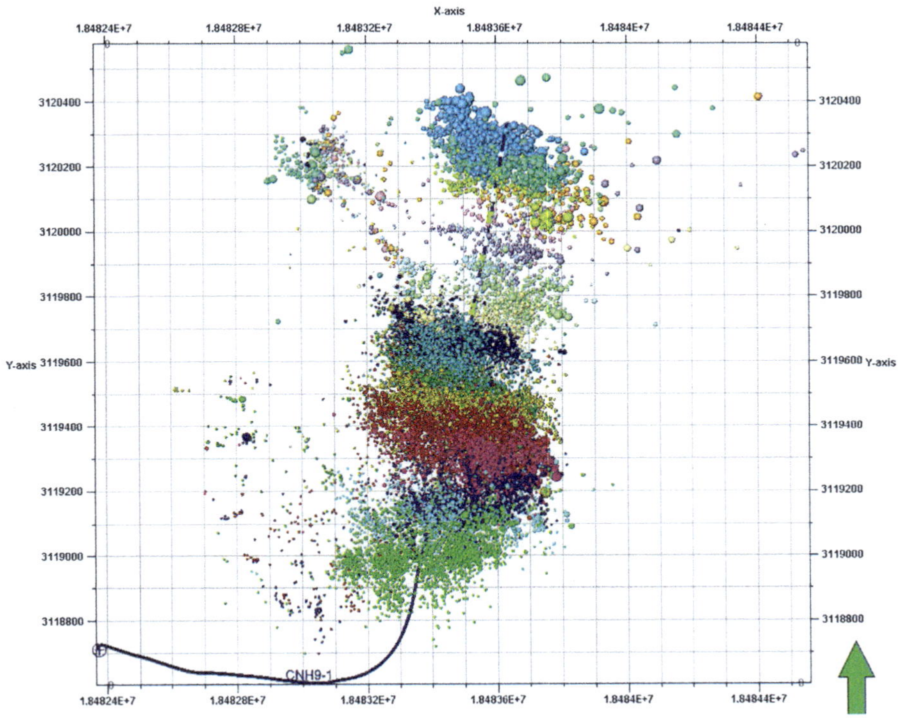

Fig. 9.24 Hydraulic fracturing simulation of H9-1.

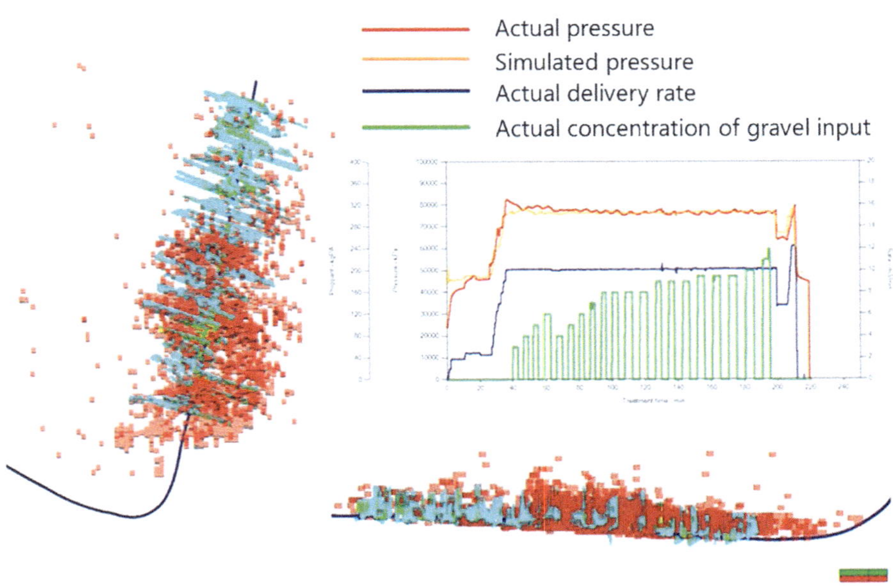

Fig. 9.25 Average conductivity of fractures.

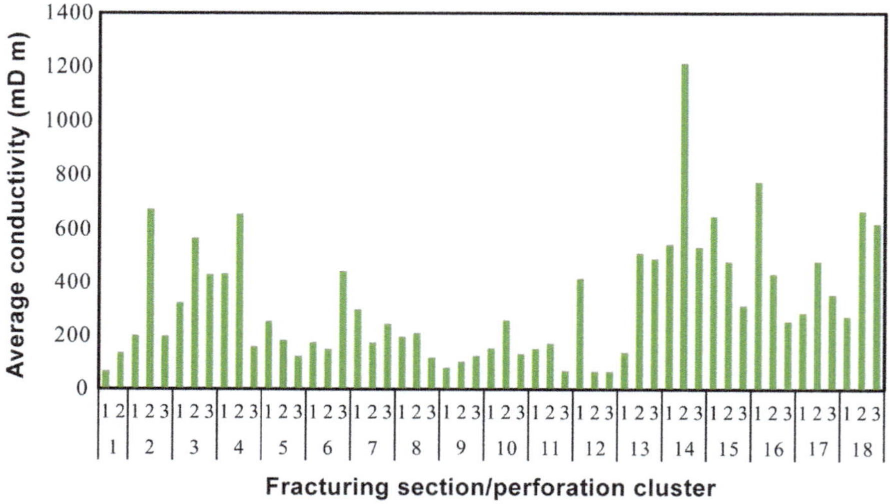

Fig. 9.26 Production history of H9-1.

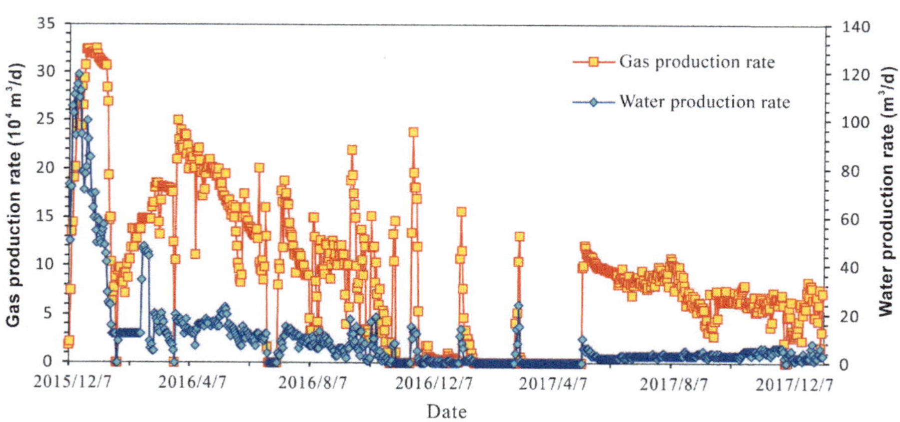

Fig. 9.27 Tubing pressure of H9-1.

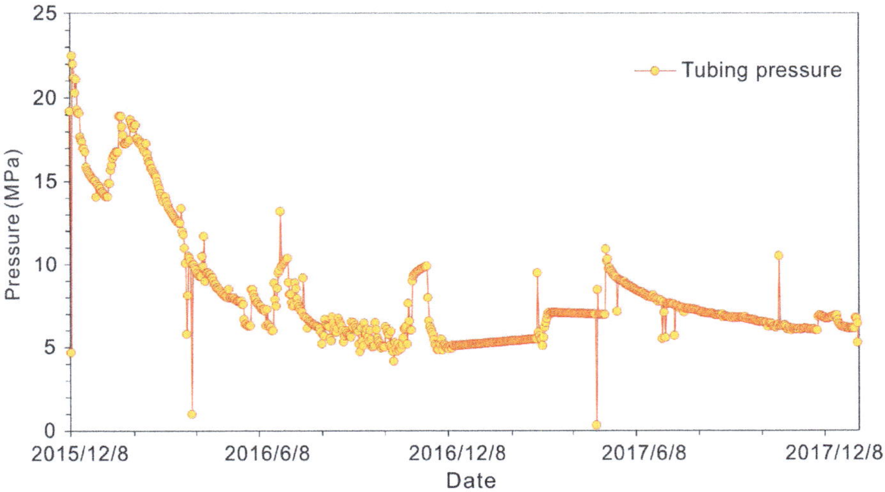

Fig. 9.28 Pressure field distribution of well H9-1 under different production times.

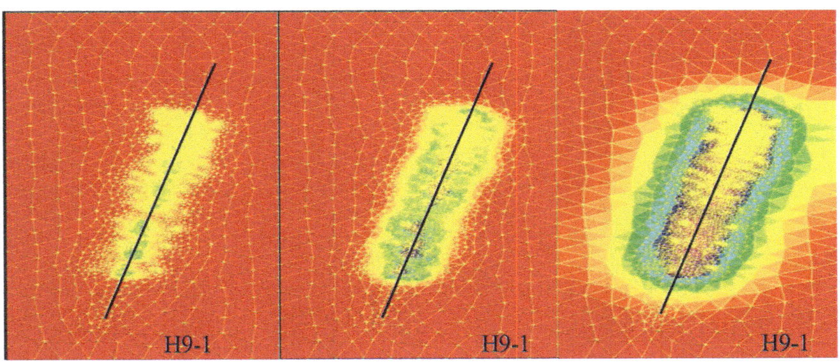

Fig. 9.29 History of well pressure.

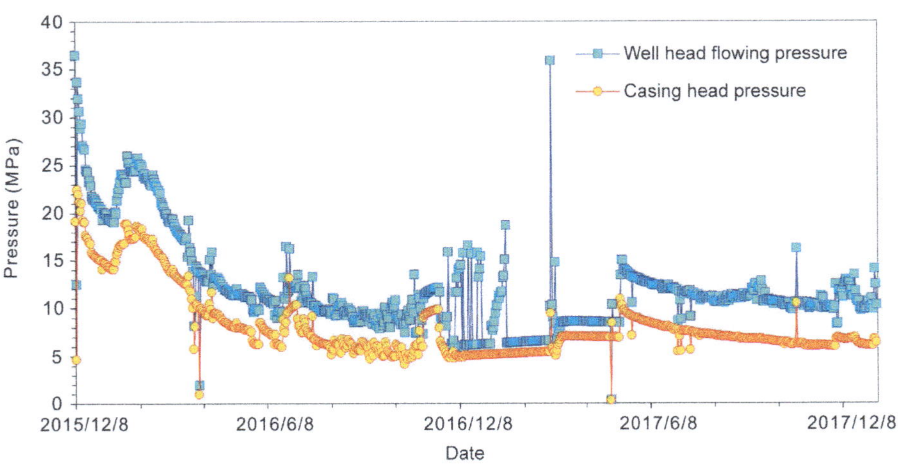

Fig. 9.30 History match of well bottom pressure.

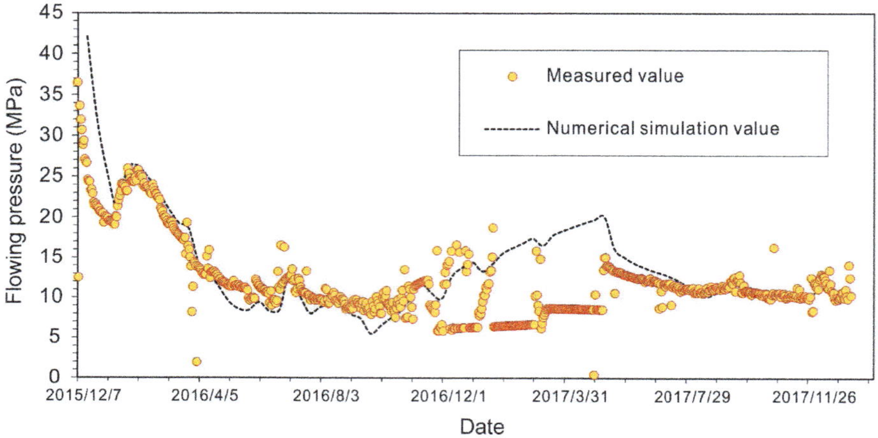

Fig. 9.31 History match and prediction of production.

Table 9.8 Prediction result

Platform	**Well name**	**Production date**	**Gas rate (10^4 m^3/day)**	**Cumulative production (10^4 m^3)**	**Prediction of gas rate (10^4 m^3/day)**	**Prediction of cumulative production (10^4 m^3)**
			2018/1/8		**2022/12/8**	
Changing H9 Platform	H9-1	2015/12/7	7.19	6742.697	3.70	14,375.02

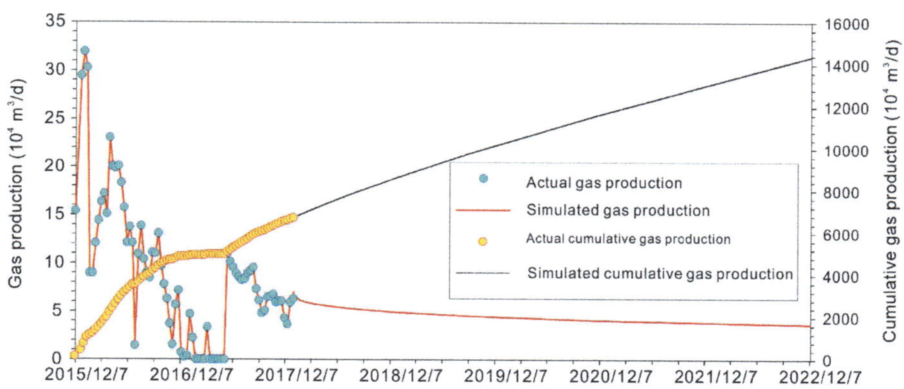

Fig. 9.32 Production forecast for well H9-1.

References

Aguilera, R., 1995. Naturally Fractured Reservoirs, second ed. PennWell Books, Tulsa, Oklahoma.

Ahmed, D.U., Meehan, N., 2016. Unconventional Oil and Gas Resources: Exploitation and Development. CRC Press, Taylor & Francis Group.

Al-Ahmadi, H.A., Almarzooq, A.M., Watenbargen, R.A., 2010. Application of linear flow analysis to shale gas wells-field cases. In: Paper SPE 130370 Presented at the SPE Unconventional Gas Conference, Pittsburgh, Pennsylvania, USA.

Arthur, J.D., Langhus, B.P.G., Alleman, D., 2004. An overview of modern shale gas development in the United States. Chem. Technol. Fuels Oils 36 (2), 82–88.

Askarieh, M.M., Chambers, A.V., Daniel, F.B.D., et al., 2000. The chemical and microbial degradation of cellulose in the near field of a repository for radioactive wastes. Waste Manag. 20, 93–106.

Ayala, L.F., Ertekin, T., Adewumi, M.A., 2005. Compositional modeling of retrograde gas-condensate reservoirs in multimechanistic flow domains. In: Paper SPE 94856 Presented at the SPE Latin American and Caribbean Petroleum Engineering Conference, Rio de Janeiro, Brazil.

Bello, R.O., Watenbargen, R.A., 2010. Multi-stage hydraulically fractured horizontal shale gas well rate transient analysis. In: Paper SPE 126754 Presented at the North Arica Technical Conference and Exhibition, Cairo, Egypt.

Bernabe, Y., 1987. The effective pressure law for permeability during pore pressure and confining pressure cycling of several crystalline rocks. J. Geophys. Res. Solid Earth 92, 649–657.

Brohi, I., Pooladi-Darvish, M., Aguilera, R., 2011. In: Modeling fractured horizontal wells as dual porosity composite reservoirs-application to tight gas, shale gas and tight oil cases. Paper SPE 144057 Presented at the SPE Western North American Region Meeting, Anchorage, Alaska, USA.

Brown, M., 2009. Analytical Trilinear Pressure Transient Model for Multiply Fractured Horizontal Wells in Tight Shale Reservoirs. Colorado School of Mines, Golden, Colorado, USA.

Brown, M., Ozkan, E., Ragahavan, R., et al., 2011. Practical solutions for pressure-transient responses of fractured horizontal wells in unconventional shale reservoirs. SPE Reserv. Eval. Eng. 14 (6), 663–676.

Brunauer, S., Emmett, P.H., Teller, E., 1938. Adsorption of gases in multimolecular layers. J. Am. Chem. Soc. 60 (2), 309–319.

Brunauer, S., Deming, L.S., Edwards, W., 1940. On a theory of the der Waals adsorption of gases. J. Am. Chem. Soc. 62 (7), 1723–1732.

Bumb, A.C., McKee, C.R., 1988. Gas-well testing in the presence of desorption for coalbed methane and Devonian shale. SPE Form. Eval. 3 (1), 179–185.

Carlson, E.S., Mercer, J.C., 1991. Devonian shale gas production: mechanisms and simple models. J. Petrol. Technol. 43 (4), 476–482.

Carslaw, H.S., Jaeger, J.C., 1959. Conduction of Heat in Solids. Oxford University Press, Oxford, pp. 255–277.

Chawathe, A., Ertekin, T., Grader, A., 1996. In: Numerical simulation of multimechanistic gas-water flow in fractured reservoirs.Paper SPE 35186 Presented at the Permian Basin Oil and Gas Recovery Conference, Midland, Texas, USA.

Chen, Z., 2005. Finite Element Methods and Their Applications. Springer-Verlag, Heidelberg and New York.

Chen, Z., 2006. The control volume finite element methods and their applications to multiphase flow. Netw. Heterogeneous Media 1 (2), 689–706.

Chen, Z., 2011. The Finite Element Method: Its Fundamentals and Applications in Engineering. World Scientific, New York.

Chen, C.C., Raghavan, R.A., 1996. A multiply-fractured horizontal well in a rectangular drainage region. In: Paper SPE 37072 Presented at the International Conference on Horizontal Well Technology, Calgary, Canada.

Chen, Z., Huan, G.R., Li, B.Y., 2003. Modeling 2D and 3D horizontal wells using CVFA. Commun. Math. Sci. 1 (1), 30–43.

Chen, Z., Huan, G.R., Ma, Y., 2006. Computational Methods for Multiphase Flows in Porous Media. Computational Science and Engineering Series, vol. 2. SIAM, Philadelphia.

Chen, G.S., Dong, D.Z., Wang, S.Q., et al., 2009. A preliminary study on accumulation mechanism and enrichment pattern of shale gas. Nat. Gas Ind. 29 (5), 17–21 (in Chinese).

Chin, L.Y., Raghavan, R., Thomas, L.K., et al., 2000. In: Fully coupled analysis of well responses in stress-sensitive reservoirs.Paper SPE 66222 Presented at ATCE, New Orleans, LA, USA.

Cho, Y., Ozkan, E., Apaydin, O.G., 2013. Pressure-dependent natural-fracture permeability in shale and its effect on shale-gas well production. SPE Reserv. Eval. Eng. 16 (2), 216–228.

Civan, F., 2010. Effective correlation of apparent gas permeability in tight porous media. Transp. Porous Media 82 (2), 375–384.

Civan, F., Rai, C.S., Sondergeld, C.H., 2011. Shale-gas permeability and diffusivity inferred by improved formulation of relevant retention and transport mechanisms. Transp. Porous Media 86 (3), 925–944.

Clarkson, C.R., Bustin, R.M., Seidel, J.P., 2007. Production-data analysis of single-phase (gas) coalbed-methane wells. SPE Reserv. Eval. Eng. 10 (3), 312–331.

Clarkson, C.R., Nobakht, M., Kaviani, D., et al., 2012. Production analysis of tight-gas and shale-gas reservoirs using the dynamic-slippage concept. SPE J. 17 (1), 230–242.

Curtis, J.B., 2002. Fractured shale-gas systems. AAPG Bull. 86 (11), 1921–1938.

Darabi, H., Ettehad, A., Javadpour, F., et al., 2012. Gas flow in ultra-tight shale strata. J. Fluid Mech. 710, 641–658.

Dehghanpour, H., Shirdel, M., 2011. A triple porosity model for shale gas reservoirs. In: Paper SPE 149501 Presented at the Canadian Unconventional Resources Conference, Calgary, Alberta, Canada.

Deng, J., Zhu, W.Y., Ma, Q., 2014. A new seepage model for shale gas reservoir and productivity analysis of fractured well. Fuel 124, 232–240.

Ekeroth, E., Roth, O., Jonsson, M., 2006. The relative impact of radiolysis products in radiation induced oxidative dissolution of UO2. J. Nucl. Mater. 355, 38–46.

Ertekin, T., King, G.R., Schwerer, F.C., 1986. Dynamic gas slippage: a unique dual-mechanism approach to the flow of gas in tight formations. SPE Form. Eval. 1 (1), 43–52.

Everdingen, V., Hust, A.F., 1949. The application of the Laplace transformation to flow problems in reservoirs. J. Petrol. Technol. 1 (12), 305–324.

Fan, D.Y., 2013. Well Test Theory and Interpretation Method of Multi-Fractured Horizontal Wells Based on the Discrete Fracture Model. China University of Petroleum, Dongying, China (in Chinese).

Florence, F.A., Rushing, J.A., Newsham, K.E., et al., 2007. Improved permeability prediction relations for low-permeability sands. In: Paper SPE 107954 Presented at SPE Rocky Mountain Oil & Gas Technology Symposium, Denver, Colorado, USA.

Gao, C., Lee, W.J., Spivey, J.P., et al., 1994. Modeling multilayer gas reservoirs including sorption effects. In: Paper SPE 29173 presented at the SPE Eastern regional Meeting, Charleston, West Virginia.

Gao, H.M., He, Y.F., Jiang, H.Q., et al., 2007. Pressure transient analysis of arbitrarily shaped fractured reservoirs. Petrol. Sci. 4 (2), 66–70.

Gasparik, M., Ghanizadeh, A., Bertier, P., et al., 2013. Geological controls on the methane storage capacity in organic-rich shales. Int. J. Coal Geol. 123 (SI), 34–51.

Gasparik, M., Gensterblum, Y., Ghanizadeh, A., et al., 2014. High-pressure/High-temperature methane sorption measurements on carbonaceous shales using manometric method: experimental and data evaluation considerations for improved accuracy. SPE J. 20 (4), 790–809.

Gensterblum, Y., Ghanizadeh, A., Cuss, R.L., et al., 2015. Gas transport and storage capacity in shale gas reservoirs-Areview. Part A:Transport processes. J. Unconv. Oil Gas Resour. 12, 87–122.

Gringarten, A.C., Ramey, J.R., Henry, J., 1973. The use of source and Green's functions in solving unsteady-flow problems in reservoirs. SPE J. 13, 285–296.

Guo, C.H., Bai, B.J., Wei, M.Z., et al., 2013. Study on gas permeability in nano pores of shale gas reservoirs. In: Paper SPE 167179 Presented at the SPE Unconventional Resources Conference, Calgary, Alberta, Canada.

He, Y.S., Li, Z., Xi, H.X., et al., 2004. Research progress of gas-solid adsorption isotherms. Ion Exchange Adsorp. 20 (4), 376–384 (In Chinese).

Heller, R., Vermylen, J., Zoback, M., 2014. Experimental investigation of matrix permeability of gas shales. AAPG Bull. 98, 975–995.

Horne, R.N., Temeng, K.O., 1995. Relative productivities and pressure transient modeling of horizontal wells with multiple fractures. In: Paper SPE 29891 Presented at SPE Middle East Oil Show, Bahrain, Bahrain.

Hu, J., Zhang, C., Rui, Z., Yu, Y., Chen, Z., 2017. Fractured horizontal well productivity prediction in tight oil reservoirs. J. Petrol. Sci. Eng. 151, 159–168.

Huang, J.L., Ju, C.N., Li, J.Z., et al., 2012. Shale gas accumulation conditions and favorable zones of Silurian Longmaxi Formation in south Sichuan Basin, China. J. Chin. Coal Soc. 37 (5), 782–787 (in Chinese).

Igwe, G.J.I., 1987. Gas transport mechanism and slippage phenomenon in porous media. In: Paper SPE 16479 Presented at the SPE Conference, Richardson, Texas, USA.

Javadpour, F., 2009. Nanopores and apparent permeability of gas flow in mudrocks (shales and siltstone). J. Gas Can. Petrol. Technol. 48 (8), 16–21.

Javadpour, F., Fisher, D., Unsworth, M., 2007. Nanoscale gas flow in shale gas sediments. J. Gas Can. Petrol. Technol. 46 (10), 55–61.

Jiang, J.M., Rami, M.Y., 2015. A multimechanistic multicontinuum model for simulating shale gas reservoir with complex fractured system. Fuel 161, 333–344.

Jizba, D.L., 1991. Mechanical and Acoustical Properties of Sandstones and Shales. Department of Geophysics, Stanford University.

Jones, F.O., Owens, W.W., 1980. A laboratory study of low-permeability gas sands. J. Petrol. Technol. 32 (9), 1–10.

Jongkittinarukorn, K., Tiab, D., 1998. Development of the boundary element method for a horizontal well in multilayer reservoir. In: Paper SPE 39939 Presented at the SPE Rocky Mountain regional/Low-Permeability Reservoirs Symposium, Denver, Colorado.

Karimi-Fard, M., Firoozabadi, A., 2001. Numerical simulation of water injection in 2D fractured media using discrete-fracture model. In: SPE71615 Presented at SPE Annual Technical Conference and Exhibition, New Orleans, Louisiana.

Karimi-Fard, M., Durlofsky, L.J., Aziz, K., 2003. An efficient discrete fracture model applicable for general purpose reservoir simulators. In: SPE 79699 Presented at SPE Reservoir Simulation Symposium, Houston, Texas.

Kikani, J., Horne, R.N., 1989. Application of boundary element method to reservoir engineering problems. J. Petrol. Sci. Eng. 3, 229–241.

Kim, J.G., Deo, M.D., 2000. Finite element, discrete-fracture model for multiphase flow in porous media. AICHE J. 46 (6), 1120–1130.

King, G.R., 1990. Material-balance techniques for coal-seam and Devonian shale gas reservoirs with limited water influx. In: Paper SPE 20730 Presented at the SPE Annual Technical Conference and Exhibition, New Orleans, Louisiana.

Klinkenberg, L.J., 1941. The Permeability of Porous Media to Liquids and Gases. API Drilling and Production Practice, NY, New York.

Kryuchkov, S., Sanger, S., 2004. Asymptotic description of vertically fractured wells within the boundary element method. J. Can. Petrol. Technol. 43 (3), 31–36.

Kuuskraa, V.A., Wicks, D.E., Thruber, J.L., et al., 1992. Geologic and reservoir mechanisms controlling gas recovery from the Antrim shale. In: Paper SPE 24883 Presented at the SPE Annual Technical Conference and Exhibition, Washington, DC, USA.

Kwon, O., Kronenberg, A.K., Gangi, A.F., et al., 2001. Permeability of Wilcox shale and its effective pressure law. J. Geophys. Res. Solid Earth 106, 19339–19353.

Lane, H.S., Watson, A.T., Lancaster, D.E., 1989. Identifying and estimating desorption from Devonian shale gas production data. In: Paper SPE 19794 Presented at the SPE Annual Technical Conference and Exhibition, San Antonio, Texas.

Lange, A., Basquet, R., Bourbiaux, B., 2004. Hydraulic characterization of faults and fractures using a dual medium discrete fracture network simulator. In: SPE88675 Presented at 11th Abu Dhabi International Conference and Exhibition, Abu Dhabi.

Langmuir, I., 1918. The desorption of gases on plane surfaces of glass, mica and platinum. J. Am. Chem. Soc. 40 (9), 1361–1403.

Li, D.M., Jiang, H.Q., Li, J.J., 2014. The impact of diffusion type on multiscale discrete fracture model numerical simulation for shale gas. Nat. Gas. Sci. Eng. 20, 74–81.

Loucks, R.G., Reed, R.M., Ruppel, S.C., et al., 2012. Spectrum of pore types and networks in mudrocks and a descriptive classification for matrix-related mudrock pores. AAPG Bull. 96 (6), 1071–1098.

Marschall, P., Horseman, S., Gimmi, T., 2005. Characterisation of gas transport properties of the Opalinus Clay, a potential host rock formation for radioactive waste disposal. Oil Gas Sci. Technol. – Rev. IFP. 60, 121–139.

Mayerhofer, M.J., Lolon, E., Warpomski, N.R., et al., 2008. What is stimulated rock volume? SPE Prod. Operations 25 (1), 89–98 (SPE-119890-PA).

Medeiros, F., Kurtoglu, B., Ozkan, E., et al., 2007. Pressure-transient performances of hydraulically fractured horizontal wells in locally and globally naturally fractured formations. In: Paper IPTC-11781 Presented at the International Petroleum Technology Conference, Dubai.

Medeiros, F.J., Ozkan, E., Kazemi, H.A., 2008. Productivity and drainage area of fractured horizontal wells in tight gas reservoirs. SPE Reserv. Eval. Eng. 11 (5), 902–911.

Medeiros, F.J., Ozkan, E., Kazemi, H.A., 2010. Semianalytical approach to model pressure transients in heterogeneous reservoirs. SPE Reserv. Eval. Eng. 13 (2), 341–358.

Michel, G.G., Sigal, R.F., Civan, F., et al., 2011. Parametric investigation of shale gas production considering nano-scale pore size distribution, formation factor, and non-Darcy flow mechanisms. In: Paper SPE 147438 Presented at SPE Annual Technical Conference and Exhibition, Denver, Colorado, USA.

Mohamed, G.H., 1999. The fluid mechanics of micro devices—the freeman scholar lecture. J. Fluids Eng. 121, 5–33.

Moinfar, A., Varavei, A., Sepehrnoori, K., et al., 2013. Development of a Coupled Dual Continuum and Discrete Fracture Model for the Simulation of Unconventional Reservoirs. In: Paper SPE 163647 Presented at the SPE Reservoir Simulation Symposium. Woodlands, Texas, USA.

Moinfar, A., Varavei, A., Sepehrnoori, K., et al., 2014. Development of an efficient embedded discrete fracture model for 3D compositional reservoir simulation in fractured reservoirs. SPE J. 19 (2), 289–303.

Monteagudo, E.P.J., Firoozabadi, A., 2004. Control-volume method for numerical simulation of two-phase immiscible flow in two- and three-dimensional discrete-fractured media. Water Resour. Res. 40, 1–20.

Newman, A.B., 1936. Heating and cooling rectangular and cylindrical solids. Ind. Eng. Chem. 28 (1988).

Nobakht, M., Clarkson, C.R., 2012. A new analytical method for analyzing linear flow in tight/shale gas reservoirs: constant-flowing-pressure boundary condition. SPE Reserv. Eval. Eng. 15 (3), 370–384 (SPE-143989).

Nobakht, M., Mattar, L., Moghadam, S., et al., 2012. Simplified forecasting of tight, shale-gas production in linear flow. J. Can. Petrol. Technol. 51 (6), 476–486 (SPE-133615).

Nobakht, M., Clarkson, C.R., Kaviani, D., 2013. New type curves for analyzing horizontal well with multiple fractures in shale gas reservoirs. J. Nat. Gas Sci. Eng. 10, 99–112.

Ozkan, E., Raghavan, R., 1991a. New solutions for well-test-analysis problems: part I—analytical considerations. SPE Form. Eval. 6 (3), 359–368 (SPE 19615).

Ozkan, E., Raghavan, R., 1991b. New solutions for well-test-analysis problems: part II—computational considerations and applications. SPE Form. Eval. 6 (3), 369–378 (SPE 18616).

Ozkan, E., Raghavan, R., 1994. New Solutions for well-test-analysis problems: part III—additional algorithms. In: Paper SPE 28424 Presented at the SPE 69th Annual Technical Conference and Exhibition, New Orleans, LA, USA.

Ozkan, E., Raghavan, R., Apaydin, O.G., 2010a. Modeling of fluid transfer from shale matrix to fracture network. In: Paper SPE 134830 Presented at the SPE Annual Technical Conference and Exhibition, Florence, Italy.

Ozkan, E., Raghavan, R., Apaydin, O.G., 2010b. Modeling of fluid transfer from shale matrix to fracture network. In: Paper SPE 134830 Presented at the SPE Annual Technical and Exhibition, Florence, Italy.

Ozkan, E., Brown, M., Raghavan, R., et al., 2011. Comparison of fractured-horizontal-well performance in tight sand and shale reservoirs. SPE Reserv. Eval. Eng. 14 (2), 248–259.

Pecher, R., Stanislav, J.F., 1997. Boundary element techniques in petroleum reservoir simulation. J. Petrol. Sci. Eng. 17, 353–366.

Rouquerol, J., Avnir, D., Fairbridge, C.W., et al., 1994. Recommendations for the characterization of porous solids. Pure Appl. Chem. 66 (8), 1739–1758.

Roy, S., Raju, R., 2003. Modeling gas flow through microchannels and nanopores. J. Appl. Phys. 93 (8), 4870–4878.

Ruthven, D.M., 1984. Principles of Adsorption and Adsorption Processes. John Wiley and Sons, pp. 80–109.

Sampath, C.W., Keighin, K., 1982. Factors affecting gas slippage in tight sandstones. J. Petrol. Technol. 34 (11), 2715–2720.

Sato, K., Horne, R.N., 1993. Perturbation boundary element method for heterogeneous reservoirs: part 2—transient-flow problems. SPE Form. Eval. 8 (4), 1–8.

Schepers, K.C., Gonzalez, R.J., Koperna, G.J., et al., 2009. Reservoir modeling in support of shale gas exploration. In: Paper SPE 123057 Presented at the Latin American and Caribbean Petroleum Engineering Conference, Cartagena de Indias, Colombia.

Shabro, V., Javadpour, F., Torres, V.C., 2009. A generalized finite-difference diffusive advective (FDDA) model for gas flow in micro- and nano-porous media. World J. Eng. 6 (3), 7–15.

Shabro, V., Torres, V.C., Javadpour, F., 2011a. Pore-scale quantification of apparent permeability and electrical resistivity of hydrocarbon-bearing shale in the presence of gas desorption. In: Paper Presented at the SPWLA 52nd Annual Logging Symposium, Colorado Springs, CO, USA.

Shabro, V., Torres, V.C., Javadpour, F., 2011b. Numerical simulation of shale-gas production: from pore-scale modeling of slip-flow, Knudsen diffusion, and Langmuir desorption to reservoir modeling of compressible fluid. In: Paper SPE 144355 Presented at the SPE North American Unconventional Gas Conference and Exhibition, Woodlands, Texas, USA.

Shabro, V., Torres-Verdin, C., Sepehrnoori, K., 2012. Forecasting gas production in organic shale with the combined numerical simulation of gas diffusion in kerogen, Langmuir desorption from surfaces, and advection in nanopores. In: Paper SPE 159250 Presented at the 2012 SPE Annual Technical Conference and Exhibition, San Antonio, Texas, USA.

Singh, H., Javadpour, F., Ettehadtavakkol, A., et al., 2013. Nonempirical apparent permeability of shale. SPE Reserv. Eval. Eng. 7 (3), 414–424.

Slatt, E.M., O'Neal, N.R., 2011. Pore types in the Barrett and Woodford gas shales: contribution to understanding gas storage and migration pathways in fine-grained rocks. AAPG Bull. 95 (12), 2017–2030.

Smith, M.B., Montgomery, C., 2015. Hydraulic Fracturing. CRC Press, Taylor & Francis Group.

Song, B., 2010. Pressure Transient Analysis and Production Analysis for New Albany Shale Gas Wells. Master Thesis. Texas A&M University, pp. 1–40.

Stalgorova, E., Mattar, L., 2012. Practical analytical model to simulate production of horizontal wells with branch fractures. In: Paper SPE 162515 Presented at the SPE Canadian Unconventional Resources Conference, Calgary, Alberta, Canada.

Stalgorova, E., Mattar, L., 2013. Analytical model for unconventional multifractured composite systems. SPE Reserv. Eval. Eng. 16 (3), 246–256 (SPE 162516-PA).

Stehfest, H., 1970. Algorithm 368: numerical inversion of Laplace transforms. Commun. ACM 13 (1), 47–49.

Sun, H., 2014. Multi-Scale Simulation Theory and Method of Gas Transport in Shale Gas Reservoirs. China University of Petroleum, Dongying, China.

Swami, V., 2012. Shale gas reservoir modeling: from nanopores to laboratory. In: Paper SPE 163065-STU Presented at the SPE International Student Paper Contest at the SPE Annual Technical Conference and Exhibition, San Antonio, Texas, USA.

Swami, V., Settari, A., 2012. A pore scale gas flow model for shale gas reservoir. In: Paper SPE 155756 Presented at the Americas Unconventional Resources Conference, Pittsburgh, Pennsylvania, USA.

Swami, V., Settari, A.T., Javadpour, F., 2013. A numerical model for multi-mechanism flow in shale gas reservoirs with application to laboratory scale testing. In: Paper SPE 164840 Presented at the EAGE Annual Conference & Exhibition Incorporating SPE Europec, London, United Kingdom.

Tian, L., Xiao, C., Liu, M.J., et al., 2014. Well testing model for multi-fractured horizontal well for shale gas reservoirs with consideration of dual diffusion in matrix. J. Nat. Gas Sci. Eng. 21, 283–295.

Wan, J., Aziz, K., 1999. Multiple hydraulic fractures in horizontal wells. In: Paper SPE 54627 Presented at the SPE Western Regional Meeting, Anchorage, Alaska, USA.

Wan, J., Aziz, K., 2002. Semi-analytical well model of horizontal wells with multiple hydraulic fractures. SPE J. 7, 437–445.

Wang, H.T., Zhang, L.H., 2009. A boundary element method applied to pressure transient analysis of geometrically complex gas reservoirs. In: Paper SPE 122055 Presented at the Latin American and Caribbean Petroleum Engineering Conference, Cartagena de Indias, Colombia.

Wang, C., Didier, D., Wu, Y.S., 2013. Characterizing hydraulic fractures in shale gas reservoirs using transient pressure tests. In: Paper SPE 163819 Presented at SPE Hydraulic Fracturing Technology Conference, Woodlands, TX, USA.

Wang, J., Luo, H.S., Liu, H.Q., et al., 2015. Variations of gas flow regimes and petro-physical properties during gas production considering volume consumed by adsorbed gas and stress dependence effect in shale gas reservoirs. In: Paper SPE 174996 Presented at the SPE Annual Technical Conference and Exhibition, Houston, Texas, USA.

Warren, J.E., Root, P.J., 1963. The behavior of naturally fractured reservoirs. SPE J. 3 (3), 245–255.

Wei, C.J., 2013. Formation Evaluation and Numerical Modeling on Hydraulic Fracturing for an Emerging Marine Shale Gas Reservoir. PhD Dissertation. University of Wyoming, p. 65.

Wu, K., Chen, Z., 2016. Review of gas transport through nanopores of shale gas reservoirs. Petrol. Sci. Bull. 1, 91–127 (in Chinese).

Xu, M.Y., Liao, X.W., He, Y.F., 2012. A new model of fractured horizontal well coupling with the tight gas reservoir at unsteady state. Petrol. Sci. Technol. 30, 2604–2612.

Xu, B.X., Haghighi, M., Li, X.F., et al., 2013. Development of new type curves for production analysis in naturally fractured shale gas/tight gas reservoirs. J. Petrol. Sci. Eng. 105, 107–115.

Xu, J.C., Guo, C.H., Wei, M.Z., et al., 2015. Production performance analysis for composite shale gas reservoir considering multiple transport mechanisms. Nat. Gas. Sci. Eng. 26, 382–395.

Yao, J., Sun, H., Huang, Z.Q., et al., 2013a. Key mechanical problems in the development of shale gas reservoirs. Sci. Sin. Phys. Mech. Astron. 43 (12), 1527–1547 (in Chinese).

Yao, S.S., Zeng, F.H., Liu, H., et al., 2013b. A semi-analytical model for multi-stage fractured horizontal wells. J. Hydrol. 507, 201–212.

Yin, H.J., He, Y.F., Fu, C.Q., 2005. Pressure transient analysis of heterogeneous reservoirs with impermeability barriers using perturbation boundary element method. J. Hydrodyn. B 17 (1), 102–109.

Zeng, X.L., Liu, S.G., Huang, W.M., et al., 2011. Comparison of Silurian Longmaxi Formation shale of shale of Sichuan Basin in China and Carboniferous Barnett Formation shale of Fort Worth Basin in United States. Geol. Bull. Chin. 30 (2/3), 372–384 (in Chinese).

Zerzar, A., Tiab, D., Bettam, Y., 2004. Interpretation of multiple hydraulically fractured horizontal wells. In: Paper SPE 88707-MS Abu Dhabi International Conference and Exhibition, Abu Dhabi, United Arab Emirates.

Zhang, J.C., Jin, Z.J., Yuan, M.S., 2004. Reservoir mechanism of shale gas and its distribution. Nat. Gas Ind. 24 (7), 15–18 (in Chinese).

Zhang, D.L., Zhang, L.H., Zhao, Y.L., et al., 2015. Composite model to analyze the decline performance of a multiple fractured horizontal well in shale reservoirs. J. Nat. Gas Sci. Eng. 26, 999–1010.

Zhang, R.H., Zhang, L.H., Wang, R.H., et al., 2016. Research on transient flow theory of a multiple fractured horizontal well in a composite shale gas reservoir based on the finite-element method. J. Nat. Gas Sci. Eng. 33, 587–598.

Zhao, Y.L., Zhang, L.H., Zhao, J.Z., Luo, J.X., Zhang, B.N., 2013. "Triple porosity" modeling of transient well test and rate decline analysis for multi-fractured horizontal well in shale gas reservoirs. J. Petrol. Sci. Eng. 110, 253–262.

Zhao, Y.L., Zhang, L.H., Luo, J.X., et al., 2014. Performance of fractured horizontal well with stimulated reservoir volume in unconventional gas reservoir. J. Hydrol. 512, 447–456.

Zhao, Y.L., Zhang, L.H., Liu, Y.H., et al., 2015. Transient pressure analysis of fractured well in bi-zonal gas reservoirs. J. Hydrol. 524, 89–99.

Zhou, L., Zhou, Y.P., Bai, S.P., et al., 2002. Studies on the transition behavior of physical adsorption from the sub- to the supercritical region: experiments on silica gel. J. Colloid Interface Sci. 253 (1), 9–15.

Zoback, M.D., Byerlee, J.D., 1975. Permeability and effective stress. AAPG Bull. 59, 154–158.

Zou, C.N., Dong, D.Z., Wang, S.J., et al., 2010. Geological characteristics, formation mechanism and resource potential of shale gas in China. Petrol. Explor. Dev. 37 (6), 641–653 (in Chinese).

Zwillinger, D., 1996. Standard Mathematical Tables and Formulae. CRC Press, New York, pp. 100–150. Section 6.19.

Nomenclatures

A	the area of each triangle (m^2)
b	constant (dimensionless)
B	Langmuir adsorption equilibrium constant (1/Pa)
B_g	gas volume factor at bottomhole pressure (sm^3/m^3)
b_k	slippage factor or coefficient (Pa)
C	wellbore and fracture system storage coefficient (m^3/Pa)
C_E	gas molar concentration (mol/m^3)
C_{gm}	gas compressibility (Pa^{-1})
C_m	molar mass of gas in matrix (mol/m^3)
d	molecular collision diameter (nm)
D_F	Fick's diffusion coefficient (m^2/s)
$f(s)$	parameter group for different mechanism flow models (dimensionless)
f, mc	fractures and macropore media (dimensionless)
F_g	geometry factor ($1/m^2$)
F_s	shape factor ($1/m^2$)
G	gas adsorption amount under equilibrium status (m^3/m^3)
G_d	molar concentration of gas dissolved in water (mol/m^3)
G_L	Langmuir adsorption volume of shale gas (m^3/m^3)
G_m	limited gas adsorption amount (m^3/m^3)
h	effective thickness of the gas reservoir (m)
J_{ads}	gas adsorption amount on the unit area of rock surface (m/s)
J_{des}	gas desorption amount on the unit area of rock surface (m/s)
J_F	Fick's mass diffusion flux ($kg/(m^2 s)$)
J_K	gas mass flux of Knudsen diffusion ($kg/(m^2 s)$)
J_l	mass velocity of gas in medium l ($kg/(m^2 s)$)
K	empirical constant (dimensionless)
$\mathbf{k_{app}}$	apparent or effective permeability (m^2)
k_a	gas equilibrium adsorption rate (m/(Pa s))
k_{app}	apparent permeability (dimensionless)
k_B	Boltzmann constant (1.3805×10^{-23} J/K)
K_c	Henry constant (m^3 Pa/mol)
k_d	gas equilibrium desorption rate (m/s)
k_F	hydraulic fracture permeability (mD)
k_f	permeability of microfracture system (m^2)
k_{f0}	microfracture permeability at the reference pressure (m^2)
k_{fe}	fracture permeability (mD)
k_{fh}	horizontal permeability of the fracture system (m^2)
k_{fz}	vertical permeability of the fracture system (m^2)
k_l	permeability of medium l (m^2)
k_m	matrix permeability (equivalent liquid permeability) (m^2)

k_{me}	matrix permeability (mD)
k_z	vertical permeability of the gas reservoir (m^2)
L_{ref}	reference length (m)
M	gas mass adsorbed by unit mass adsorbent (kg/kg)
M_g	gas molecule mass (kg/mol)
$m(p)$	gas pseudo-pressure (Pa/s)
$m(p_L)$	Langmuir pseudo-pressure (Pa/s)
m_{wD1}	pseudo-pressure difference excluding skin effect (dimensionless)
m_{wD2}	pseudo-pressure difference including skin effect (dimensionless)
m_{wD3}	bottomhole pressure with wellbore storage and skin effect (dimensionless)
n	the outer normal vector
N	number of gas moles (mol)
p	gas pressure (Pa)
$\bar{p}$	average pressure of core plug between outlet and inlet (Pa)
p_b	gas partial pressure (Pa)
p_f	system pressure of microfractures (Pa)
p_{f0}	reference pressure, usually original reservoir pressure (Pa)
p_i	initial reservoir pore pressure (MPa)
p_L	Langmuir pressure (Pa)
p_l	pressure in medium l (Pa)
p_m	pore pressure of matrix (Pa)
p_o	reference pressure (Pa)
p_{wf}	flowing bottomhole pressure (Pa)
q	the term of source and sink (dimensionless)
q^*	desorption flux
$q_D(t_D)$	well production rate, $q_D = q/q_{sc}$ (dimensionless)
q_{des}	mass flow rate of gas desorption from reservoir of volume V_b (kg/s)
q_F	Fick's mass flow rate (mass of gas passing through volume V_b in unit time) (kg/s)
q_m	interporosity flux rate from matrix to natural fracture system (m^3/s)
q_{sc}	surface production rate under standard conditions (m^3/s)
q_{sf}	subsurface production rate (m^3/s)
R	gas constant (8.314 J/(mol K))
r/z	radial and vertical coordinates (m)
R_h	average hydraulic radius of pore media (nm)
r_m	inner diameter of sphere matrix element (m)
R_m	radius of sphere matrix (m)
s	Laplace variable (dimensionless)
S_{kin}	skin factor (dimensionless)
T	gas absolute temperature (K)
t	time (s)
V	gas volume (m^3)
V_b	shale matrix volume (m^3)
v_{Fm}	Fick's diffusion velocity (m/s)
v_{km}	Knudsen diffusion velocity (m/s)
v_r	gas radial flow velocity (m/s)
v_z	gas vertical flow velocity (m/s)
w_F	width of hydraulic fracture (m)
x, y, z	coordinates in Cartesian coordinate system (m)
Z	gas compressibility factor (dimensionless)

$\boldsymbol{\Gamma}$	the boundary of V_{i}
$\boldsymbol{\Lambda}$	parameter group, different for different mechanism models (dimensionless)
$\boldsymbol{\alpha}$	matrix shape factor, $(1/m^2)$
$\boldsymbol{\gamma}$	stress sensitivity coefficient (1/Pa)
$\boldsymbol{\delta}$	collision diameter of gas molecule (nm)
$\boldsymbol{\theta}$	gas coverage of the porous rock surface (dimensionless)
$\boldsymbol{\lambda}$	gas molecular mean free path of gas (nm)
$\boldsymbol{\mu}_{\mathbf{g}}$	gas viscosity (Pa s)
$\boldsymbol{\rho}_{\mathbf{g}}$	gas density at the given conditions (kg/m^3)
$\boldsymbol{\rho}_{\mathbf{gf}}$	gas density at the fracture conditions (m)
$\boldsymbol{\rho}_{\mathbf{gm}}$	gas density at matrix pressure condition (kg/m^3)
$\boldsymbol{\rho}_{\mathbf{gs}}$	shale gas density at standard conditions (kg/m^3)
$\boldsymbol{\varnothing}_{\mathbf{f}}$	porosity of microfracture system (dimensionless)
$\boldsymbol{\varnothing}_{\mathbf{m}}$	porosity of matrix macropores (dimensionless)
$\boldsymbol{\Delta m}_{\mathbf{s}}$	additional pseudo-pressure drop caused by skin effect (Pa/s)
$\boldsymbol{\nabla}$	gradient operator $\nabla = \frac{\partial}{\partial x} i + \frac{\partial}{\partial y} j + \frac{\partial}{\partial z} k$ (dimensionless)

Appendices

Appendix A Solution derivation in shale gas reservoirs under different transport mechanisms

A.1 Microfractures + steady state adsorption/desorption and diffusion model (model 1)

For a natural fracture system, the diffusivity equation in spherical coordinates can be expressed as follows:

$$\frac{1}{r^2}\frac{\partial}{\partial r}\left(\frac{k_\mathrm{f}}{\mu_\mathrm{g}}\rho_\mathrm{g} r^2 \frac{\partial p_\mathrm{f}}{\partial r}\right)=\frac{\partial(\phi_\mathrm{f}\rho_\mathrm{g})}{\partial t}+q_\mathrm{des} \tag{A.1}$$

For steady-state desorption and diffusion, according to the Langmuir isothermal adsorption equation, the mass flow rate of desorption gas from a matrix particle surface is:

$$q_\mathrm{des}=\rho_\mathrm{gsc}(1-\phi_\mathrm{f})\frac{G_\mathrm{L} m(p_\mathrm{L})}{[m(p_\mathrm{L})+m(p_\mathrm{f})]^2}\frac{\partial m(p_\mathrm{f})}{\partial t} \tag{A.2}$$

The equation most commonly used to model real gas pressure/volume/temperature behavior is the real gas law given by:

$$\rho_\mathrm{g}=\frac{p_\mathrm{f} M_\mathrm{g}}{ZRT} \tag{A.3}$$

For compressible gas, the pseudo-pressure, $m(p_\mathrm{f})$, can be used to linearize the diffusivity equation, which is defined as:

$$m(p_\mathrm{f})=2\int_{p_\mathrm{o}}^{p_\mathrm{f}}\frac{p}{\mu_\mathrm{g}(p_\mathrm{f})Z(p_\mathrm{f})}dp \tag{A.4}$$

Recall the definition of compressibility of gas:

$$c_g=\frac{1}{\rho_\mathrm{g}}\frac{\partial \rho_\mathrm{g}}{\partial p_\mathrm{f}} \tag{A.5}$$

Define the formation compressibility:

$$c_{\mathrm{f}} = \frac{1}{\phi_{\mathrm{f}}} \frac{\partial \phi_{\mathrm{f}}}{\partial p_{\mathrm{f}}} \tag{A.6}$$

As the gas compressibility (c_{g}) is much larger than the formation compressibility (c_{f}), the porosity of the fracture system (ϕ_{f}) can be treated as a constant. Eqs. (A.1) and (A.2) can be written as:

$$\frac{1}{r^2}\frac{\partial}{\partial r}\left(r^2 \frac{\partial m(p_{\mathrm{f}})}{\partial r}\right) = \frac{\phi_{\mathrm{f}}\mu_{\mathrm{g}}c_{\mathrm{fg}}}{k_{\mathrm{f}}}\frac{\partial m(p_{\mathrm{f}})}{\partial t} + \frac{1}{k_{\mathrm{f}}}\frac{2RT}{M_{\mathrm{g}}} q_{\mathrm{des}} \tag{A.7}$$

$$q_{\mathrm{des}} = \frac{M_{\mathrm{g}} p_{\mathrm{sc}}}{RT_{\mathrm{sc}}} \frac{(1-\phi_{\mathrm{f}})G_{\mathrm{L}} m(p_{\mathrm{L}})}{[m(p_{\mathrm{L}}) + m(p_{\mathrm{f}})]^2} \frac{\partial m(p_{\mathrm{f}})}{\partial t} \tag{A.8}$$

Substitution of Eq. (A.8) into Eq. (A.7) and setting $m_{\mathrm{f}} = m(p_{\mathrm{f}})$, we have:

$$\frac{1}{r^2}\frac{\partial}{\partial r}\left(r^2 \frac{\partial m_{\mathrm{f}}}{\partial r}\right) = \frac{\phi_{\mathrm{f}}\mu_{\mathrm{g}}c_{\mathrm{mg}}}{k_{\mathrm{f}}}\frac{\partial m_{\mathrm{f}}}{\partial t} + \frac{\phi_{\mathrm{f}}\mu_{\mathrm{gi}}}{k_{\mathrm{f}}} \frac{2Tp_{\mathrm{sc}}}{\phi_{\mathrm{f}}\mu_{\mathrm{gi}}T_{\mathrm{sc}}} \frac{(1-\phi_{\mathrm{f}})G_{\mathrm{L}} m(p_{\mathrm{L}})}{[m(p_{\mathrm{L}}) + m(p_{\mathrm{f}})]^2} \frac{\partial m_{\mathrm{f}}}{\partial t} \tag{A.9}$$

Following a similar technique as in Bumb and McKee (1988), the desorption compressibility can be introduced to take into account the contribution of desorption to the effective compressibility, which is defined as:

$$c_{\mathrm{d}} = \frac{2Tp_{\mathrm{sc}}}{\phi_{\mathrm{f}}\mu_{\mathrm{gi}}T_{\mathrm{sc}}} \frac{(1-\phi_{\mathrm{f}})G_{\mathrm{L}} m(p_{\mathrm{L}})}{[m(p_{\mathrm{L}}) + m(p_{\mathrm{f}})]^2} \tag{A.10}$$

Using Eq. (A.10), Eq. (A.9) can be rewritten as:

$$\frac{1}{r^2}\frac{\partial}{\partial r}\left(r^2 \frac{\partial m_{\mathrm{f}}}{\partial r}\right) = \frac{\phi_{\mathrm{f}}\mu_{\mathrm{g}}c_{\mathrm{fg}}}{k_{\mathrm{f}}}\frac{\partial m_{\mathrm{f}}}{\partial t} + \frac{\phi_{\mathrm{f}}\mu_{\mathrm{gi}}c_{\mathrm{d}}}{k_{\mathrm{f}}}\frac{\partial m_{\mathrm{f}}}{\partial t} \tag{A.11}$$

Note that the parameters of gas compressibility and gas viscosity in Eq. (A.11) make the problem nonlinear, which are the functions of reservoir pressure in both the fracture and matrix systems. In order to simplify and linearize the problem, these two parameters are always treated as constants under the initial condition; that is: $\mu_{\mathrm{g}} = \mu_{\mathrm{gi}}$ and $c_{\mathrm{g}} = c_{\mathrm{gi}}$. Thus Eq. (A.11) becomes:

$$\frac{1}{r^2}\frac{\partial}{\partial r}\left(r^2 \frac{\partial m_{\mathrm{f}}}{\partial t}\right) = \frac{\phi_{\mathrm{f}}\mu_{\mathrm{gi}}c_{\mathrm{fgi}}}{k_{\mathrm{f}}}\frac{\partial m_{\mathrm{f}}}{\partial t} + \frac{\phi_{\mathrm{f}}\mu_{\mathrm{gi}}c_{\mathrm{d}}}{k_{\mathrm{f}}}\frac{\partial m_{\mathrm{f}}}{\partial t} \tag{A.12}$$

Define the pseudo-pressure difference:

$$\Delta m_{\mathrm{f}} = m(p_{\mathrm{i}}) - m(p_{\mathrm{f}}) = 2\int_{p_{\mathrm{f}}}^{p_{\mathrm{i}}} \frac{p}{\mu_{\mathrm{g}}(p_{\mathrm{f}})Z(p_{\mathrm{f}})}\mathrm{d}p \tag{A.13}$$

Also, define the ratio of adsorption compressibility to gas compressibility under the initial condition as:

$$\omega = \frac{c_{\mathrm{d}}}{c_{\mathrm{gi}}} \tag{A.14}$$

The dimensionless radius and dimensionless time are defined as:

$$r_D = \frac{r}{L_{\mathrm{ref}}} \tag{A.15}$$

$$t_{\mathrm{D}} = \frac{k_{\mathrm{f}}t}{\phi_{\mathrm{f}}\mu_{\mathrm{gi}}c_{\mathrm{fgi}}L_{\mathrm{ref}}^2} \tag{A.16}$$

Substituting Eqs. (A.15) and (A.16) into Eq. (A.9) yields:

$$\frac{1}{r_{\mathrm{D}}^2}\frac{\partial}{\partial r_{\mathrm{D}}}\left(r_{\mathrm{D}}^2\frac{\partial \Delta m_{\mathrm{f}}}{\partial r_{\mathrm{D}}}\right) = (1+\omega)\frac{\partial \Delta m_{\mathrm{f}}}{\partial t_{\mathrm{D}}} \tag{A.17}$$

Introduce the flowing Laplace transform:

$$\Delta \overline{m}_{\mathrm{f}}| = \int_0^{\infty} \Delta m_{\mathrm{f}} e^{-st_{\mathrm{D}}}\mathrm{d}t_{\mathrm{D}} \tag{A.18}$$

After taking the Laplace transformation with Eq. (A.17), the governing equations can be reduced to:

$$\frac{1}{r_{\mathrm{D}}^2}\frac{\partial}{\partial r_{\mathrm{D}}}\left(r_{\mathrm{D}}^2\frac{\partial \Delta \overline{m}_{\mathrm{f}}}{\partial r_{\mathrm{D}}}\right) = f(s)\Delta \overline{m}_{\mathrm{f}} \tag{A.19}$$

where the expression of $f(s)$ is:

$$f(s) = (1+\omega)s \tag{A.20}$$

Eq. (A.19) is the general dimensionless diffusivity equation in the micro-fracture system in spherical coordinates.

A.2 Microfractures + matrix macropores + steady state adsorption/desorption and diffusion model (model 2)

A.2.1 Transient matrix flow

In spherical coordinates, the continuity equation of shale gas flow in the fracture system is:

$$\frac{1}{r^2}\frac{\partial}{\partial r}\left(\frac{k_f}{\mu_g}\rho_g r^2 \frac{\partial p_f}{\partial r}\right) = \frac{\partial(\phi_f \rho_g)}{\partial t} - q_m \tag{A.21}$$

The diffusivity equation in the matrix system is:

$$\frac{1}{r_m^2}\frac{\partial}{\partial r_m}\left(\frac{k_m}{\mu_g}\rho_g r_m^2 \frac{\partial p_m}{\partial r_m}\right) = \frac{\partial(\phi_m \rho_g)}{\partial t} + q_{des} \tag{A.22}$$

The initial condition is:

$$p_m(t=0, r_m) = p_i \tag{A.23}$$

The inner boundary condition is:

$$\frac{\partial p_m}{\partial r_m}(t, r_m = 0) = 0 \tag{A.24}$$

Due to the outer boundary of the matrix connected with the fracture system, the pressure in them at this point is the same:

$$p_m(t, r_m = R_m) = p_f \tag{A.25}$$

For transient flow, q_m can be expressed as:

$$q_m = -\frac{3\rho_{gm}}{R_m}\frac{k_m}{\mu_g}\frac{\partial p_m}{\partial r_m}\bigg|_{r_m = R_m} \tag{A.26}$$

According to the Langmuir isotherm adsorption equation, the desorption term of the shale gas, q_{des}, can be expressed as follows:

$$q_{des} = \rho_g(1 - \phi_f - \phi_m)\frac{G_L m(p_L)}{[m(p_L) + m(p_m)]^2}\frac{\partial m(p_m)}{\partial t} \tag{A.27}$$

Setting $m_f = m(p_f)$ and $m_m = m(p_m)$, the diffusivity equation in the fracture system can be rewritten as follows by the substitution of pseudo-pressure and gas density:

$$\frac{1}{r^2}\frac{\partial}{\partial r}\left(r^2\frac{\partial m_f}{\partial r}\right) = \frac{\mu_{gi}\phi_f c_{fgi}}{k_f}\frac{\partial m_f}{\partial t} + \frac{3}{R_m}\frac{k_m}{k_f}\frac{\partial m_m}{\partial r_m}\bigg|_{r_m = R_m} \tag{A.28}$$

For a spherical matrix, the shape factor is:

$$\alpha = \frac{15}{R_m^2} \tag{A.29}$$

Using the shape factor, Eq. (A.28) can be changed to:

$$\frac{1}{r^2}\frac{\partial}{\partial r}\left(r^2\frac{\partial m_f}{\partial r}\right) = \frac{\mu_{gi}\phi_f c_{fgi}}{k_f}\frac{\partial m_f}{\partial t} + \frac{\alpha R_m}{5}\frac{k_m}{k_f}\frac{\partial m_m}{\partial r_m}\bigg|_{r_m=R_m} \tag{A.30}$$

Substituting Eq. (A.27) into the diffusivity equation in the matrix system yields:

$$\frac{1}{r_m^2}\frac{\partial}{\partial r_m}\left(r_m^2\frac{\partial m_m}{\partial r_m}\right) = \frac{\phi_m \mu_{gi} c_{mgi}}{k_m}\frac{\partial m_m}{\partial t} + \frac{\phi_m \mu_{gi} c_d}{k_m}\frac{\partial m_m}{\partial t} \tag{A.31}$$

where c_d is the desorption compressibility, which can be formulated as:

$$c_d = \frac{2Tp_{sc}}{\phi_m \mu_{gi} T_{sc}}\frac{(1-\phi_f-\phi_m)G_L m(p_L)}{[m(p_L)+m(p_m)]^2} \tag{A.32}$$

The initial condition in the spherical matrix system is:

$$m_m|(t=0, r_m) = m(p_i) \tag{A.33}$$

The inner boundary condition is given by:

$$\frac{\partial m_m}{\partial r_m}\bigg|(t, r_m=0) = 0 \tag{A.34}$$

The outer boundary condition in the matrix system is:

$$m_m|(t, r_m=R_m) = m_f \tag{A.35}$$

Define the following variables:

$$r_{mD} = \frac{r_m}{R_m},\quad r_D = \frac{r}{L_{ref}},\quad t_D = \frac{k_f t}{(\phi_m c_{mgi} + \phi_f c_{fgi})\mu_{gi} L_{ref}^2},\quad \omega_f = \frac{\phi_f c_{fgi}}{\phi_m c_{mgi} + \phi_f c_{fgi}},$$
$$\omega_d = \frac{\phi_m c_d}{\phi_m c_{mgi} + \phi_f c_{fgi}},\quad \lambda = \alpha\frac{k_m}{k_f}L_{ref}^2,\quad \Delta m_f = m(p_i) - m(p_f),\quad \Delta m_m = m(p_i) - m(p_m)$$

Substituting the above variables into the mathematical model, we see that:

$$\frac{1}{r_D^2}\frac{\partial}{\partial r_D}\left(r_D^2\frac{\partial \Delta m_f}{\partial r_D}\right) = \omega_f\frac{\partial \Delta m_f}{\partial t_D} + \frac{\lambda}{5}\frac{\partial \Delta m_m}{\partial r_{mD}}\bigg|_{r_{mD}=1} \tag{A.36}$$

$$\frac{1}{r_{\mathrm{mD}}^{2}}\frac{\partial}{\partial r_{\mathrm{mD}}}\left(r_{\mathrm{mD}}^{2}\frac{\partial \Delta m_{\mathrm{m}}}{\partial r_{\mathrm{mD}}}\right)=\frac{15(1-\omega_{\mathrm{f}})}{\lambda}\frac{\partial \Delta m_{\mathrm{m}}}{\partial t_{\mathrm{D}}}+\frac{15\omega_{\mathrm{d}}}{\lambda}\frac{\partial \Delta m_{\mathrm{m}}}{\partial t_{\mathrm{D}}} \tag{A.37}$$

$$\Delta m_{\mathrm{m}}|(t_{\mathrm{D}}=0, r_{\mathrm{mD}})=0 \tag{A.38}$$

$$\left.\frac{\partial \Delta m_{\mathrm{m}}}{\partial r_{\mathrm{mD}}}\right|(t_{\mathrm{D}}, r_{\mathrm{mD}}=0)=0 \tag{A.39}$$

$$\Delta m_{\mathrm{m}}|(t_{\mathrm{D}}, r_{\mathrm{mD}}=1)=\Delta m_{\mathrm{f}} \tag{A.40}$$

Introducing the following Laplace transform:

$$\Delta \overline{m}|=\int_{0}^{\infty}\Delta m e^{-st_{\mathrm{D}}}\mathrm{d}t_{\mathrm{D}} \tag{A.41}$$

the models become:

$$\frac{1}{r_{\mathrm{D}}^{2}}\frac{\partial}{\partial r_{\mathrm{D}}}\left(r_{\mathrm{D}}^{2}\frac{\partial \Delta \overline{m}_{\mathrm{f}}}{\partial r_{\mathrm{D}}}\right)=\omega_{\mathrm{f}}s\Delta \overline{m}_{\mathrm{f}}+\frac{\lambda}{5}\left.\frac{\partial \Delta \overline{m}_{\mathrm{m}}}{\partial r_{\mathrm{mD}}}\right|_{r_{\mathrm{mD}}=1} \tag{A.42}$$

$$\frac{1}{r_{\mathrm{mD}}^{2}}\frac{\partial}{\partial r_{\mathrm{mD}}}\left(r_{\mathrm{mD}}^{2}\frac{\partial \Delta \overline{m}_{\mathrm{m}}}{\partial r_{\mathrm{mD}}}\right)=\frac{15s(1-\omega_{\mathrm{f}}+\omega_{\mathrm{d}})}{\lambda}\Delta \overline{m}_{\mathrm{m}} \tag{A.43}$$

$$\Delta \overline{m}_{\mathrm{m}}|(t_{\mathrm{D}}=0, r_{\mathrm{mD}})=0 \tag{A.44}$$

$$\left.\frac{\partial \Delta \overline{m}_{\mathrm{m}}}{\partial r_{\mathrm{mD}}}\right|(s, r_{\mathrm{mD}}=0)=0 \tag{A.45}$$

$$\Delta \overline{m}_{\mathrm{m}}|(s, r_{\mathrm{mD}}=1)=\Delta \overline{m}_{\mathrm{f}} \tag{A.46}$$

Define the following variable substitution:

$$W=r_{\mathrm{mD}}\Delta \overline{m}_{\mathrm{m}} \tag{A.47}$$

Then Eq. (A.43) reduces to:

$$\frac{\partial^{2}W}{\partial r_{\mathrm{mD}}^{2}}=\frac{15s(1-\omega_{\mathrm{f}}+\omega_{\mathrm{d}})}{\lambda}W \tag{A.48}$$

The general solution of the above equation is:

$$W=A\sinh\left(\sqrt{g}r_{\mathrm{mD}}\right)+B\cosh\left(\sqrt{g}r_{\mathrm{mD}}\right) \tag{A.49}$$

where:

$$g = \frac{15(1-\omega_{\mathrm{f}}+\omega_{\mathrm{d}})s}{\lambda} \tag{A.50}$$

Substituting Eq. (A.49) back into Eq. (A.47) yields:

$$\Delta\overline{m}_{\mathrm{m}} = \frac{A\sinh\left(\sqrt{g}r_{\mathrm{mD}}\right)+B\cosh\left(\sqrt{g}r_{\mathrm{mD}}\right)}{r_{\mathrm{mD}}} \tag{A.51}$$

According to the inner boundary condition, we can derive at $B=0$, and then the other coefficient A can be obtained by the outer boundary equation, which is:

$$A = \frac{\Delta\overline{m}_{\mathrm{f}}}{\sinh\left(\sqrt{g}\right)} \tag{A.52}$$

Substituting A and B into Eq. (A.51) yields:

$$\Delta\overline{m}_{\mathrm{m}} = \frac{\Delta\overline{m}_{\mathrm{f}}}{\sinh\left(\sqrt{g}\right)}\frac{\sinh\left(\sqrt{g}r_{\mathrm{mD}}\right)}{r_{\mathrm{mD}}} \tag{A.53}$$

Taking the derivative of Eq. (A.53) on the inner boundary has:

$$\left.\frac{\partial\Delta\overline{m}_{\mathrm{m}}}{\partial r_{\mathrm{mD}}}\right|_{r_{\mathrm{mD}}=1} = \left[\sqrt{g}\coth\left(\sqrt{g}-1\right)\right]\Delta\overline{m}_{\mathrm{f}} \tag{A.54}$$

Hence the general diffusivity equation in the micro-fracture system can be obtained by substituting Eq. (A.54) into Eq. (A.42), which is:

$$\frac{1}{r_{\mathrm{D}}^2}\frac{\partial}{\partial r_{\mathrm{D}}}\left(r_{\mathrm{D}}^2\frac{\partial\Delta\overline{m}_{\mathrm{f}}}{\partial r_{\mathrm{D}}}\right) = f(s)\Delta\overline{m}_{\mathrm{f}} \tag{A.55}$$

where the parameter group $f(s)$ is expressed as:

$$f(s) = \omega_{\mathrm{f}}s + \frac{\lambda}{5}\left[\sqrt{\frac{15(1-\omega_{\mathrm{f}}+\omega_{\mathrm{d}})s}{\lambda}}\coth\sqrt{\frac{15(1-\omega_{\mathrm{f}}+\omega_{\mathrm{d}})s}{\lambda}}-1\right] \tag{A.56}$$

A.2.2 Pseudo-steady state interporosity flow

When gas flow from macro pores to microfractures is pseudo-steady state, the diffusivity equation in the microfracture system can be derived by substituting the interporosity flow equation and the formula of pseudo pressure into the diffusivity equation mentioned above, which can be written as follows:

$$\frac{1}{r^2}\frac{\partial}{\partial r}\left(r^2\frac{\partial m_\mathrm{f}}{\partial r}\right)=\frac{\phi_\mathrm{f}\mu_\mathrm{gi}c_\mathrm{fgi}}{k_\mathrm{f}}\frac{\partial m_\mathrm{f}}{\partial t}-\frac{\alpha k_\mathrm{m}}{k_\mathrm{f}}[m_\mathrm{m}-m_\mathrm{f}] \tag{A.57}$$

The gas flow equation in the macropores in the matrix system is:

$$-\alpha\frac{k_\mathrm{m}}{k_\mathrm{f}}[m_\mathrm{m}-m_\mathrm{f}]=\frac{\phi_\mathrm{m}\mu_\mathrm{gi}c_\mathrm{mgi}}{k_\mathrm{f}}\frac{\partial m_\mathrm{m}}{\partial t}+\frac{\phi_\mathrm{m}\mu_\mathrm{gi}c_\mathrm{d}}{k_\mathrm{f}}\frac{\partial m_\mathrm{m}}{\partial t} \tag{A.58}$$

where c_d is the desorption compressibility, which is:

$$c_\mathrm{d}=\frac{2Tp_\mathrm{sc}}{\phi_\mathrm{m}\mu_\mathrm{gi}T_\mathrm{sc}}\frac{(1-\phi_\mathrm{f}-\phi_\mathrm{m})G_\mathrm{L}m(p_\mathrm{L})}{[m(p_\mathrm{L})+m(p_\mathrm{m})]^2} \tag{A.59}$$

Define the following dimensionless variables:

$$r_\mathrm{D}=\frac{r}{L_\mathrm{ref}},\quad t_\mathrm{D}=\frac{k_\mathrm{f}t}{\left(\phi_\mathrm{m}c_\mathrm{mgi}+\phi_\mathrm{f}c_\mathrm{mgi}\right)\mu_\mathrm{gi}L_\mathrm{ref}^2},\quad \omega_\mathrm{f}=\frac{\phi_\mathrm{f}c_\mathrm{fgi}}{\phi_\mathrm{m}c_\mathrm{mgi}+\phi_\mathrm{f}c_\mathrm{fgi}},\quad \omega_\mathrm{d}=\frac{\phi_\mathrm{m}c_\mathrm{d}}{\phi_\mathrm{m}c_\mathrm{mgi}+\phi_\mathrm{f}c_\mathrm{fgi}},$$

$$\lambda=\alpha\frac{k_\mathrm{m}}{k_\mathrm{f}}L_\mathrm{ref}^2$$

The dimensionless diffusivity equation in the fracture system can be obtained by substituting the above dimensionless variables into the corresponding equations:

$$\frac{1}{r_\mathrm{D}^2}\frac{\partial}{\partial r_\mathrm{D}}\left(r_\mathrm{D}^2\frac{\partial \Delta m_\mathrm{f}}{\partial r_\mathrm{D}}\right)=\omega_\mathrm{f}\frac{\partial \Delta m_\mathrm{f}}{\partial t_\mathrm{D}}-\lambda[\Delta m_\mathrm{m}-\Delta m_\mathrm{f}] \tag{A.60}$$

$$-\lambda[\Delta m_\mathrm{m}-\Delta m_\mathrm{f}]=(1-\omega_\mathrm{f})\frac{\partial \Delta m_\mathrm{m}}{\partial t_\mathrm{D}}+\omega_\mathrm{d}\frac{\partial \Delta m_\mathrm{m}}{\partial t_\mathrm{D}} \tag{A.61}$$

Taking the Laplace transform for Eqs. (A.60) and (A.61) yields:

$$\frac{1}{r_\mathrm{D}^2}\frac{\partial}{\partial r_\mathrm{D}}\left(r_\mathrm{D}^2\frac{\partial \Delta \overline{m}_\mathrm{f}}{\partial r_\mathrm{D}}\right)=\omega_\mathrm{f}s\Delta\overline{m}_\mathrm{f}-\lambda[\Delta\overline{m}_\mathrm{m}-\Delta\overline{m}_\mathrm{f}] \tag{A.62}$$

$$-\lambda[\Delta\overline{m}_\mathrm{m}-\Delta\overline{m}_\mathrm{f}]=(1-\omega_\mathrm{f}+\omega_\mathrm{d})s\Delta\overline{m}_\mathrm{m} \tag{A.63}$$

Combining Eqs. (A.62) and (A.63), the diffusivity equation in the fracture system becomes:

$$\frac{1}{r_\mathrm{D}^2}\frac{\partial}{\partial r_\mathrm{D}}\left(r_\mathrm{D}^2\frac{\partial \Delta \overline{m}_\mathrm{f}}{\partial r_\mathrm{D}}\right)=f(s)\Delta\overline{m}_\mathrm{f} \tag{A.64}$$

where $f(s)$ can be expressed as:

$$f(s)=\frac{\lambda(1+\omega_\mathrm{d})+\omega_\mathrm{f}(1-\omega_\mathrm{f}+\omega_\mathrm{d})s}{\lambda+(1-\omega_\mathrm{f}+\omega_\mathrm{d})s}s \tag{A.65}$$

A.3 Microfractures + gas adsorption/desorption + matrix Fick's diffusion model (model 3)

When the diffusion of the adsorbed gas from the matrix to microfractures satisfies Fick's diffusion law, the diffusivity equation in the fractures system in spherical coordinates is:

$$\frac{1}{r^2}\frac{\partial}{\partial r}\left(\frac{k_{\rm f}}{\mu_{\rm g}}\rho_{\rm g}r^2\frac{\partial p_{\rm f}}{\partial r}\right)=\frac{\partial(\phi_{\rm f}\rho_{\rm g})}{\partial t}+q_{\rm F} \tag{A.66}$$

where $q_{\rm F}$ is the gas diffusion from the matrix to the fractures in a unit reservoir volume, which can be expressed as:

$$q_{\rm F}=M_{\rm g}(1-\phi_{\rm f})\frac{{\rm d}C_{\rm m}}{{\rm d}t} \tag{A.67}$$

For pseudo-steady state and unsteady state, gas diffusion from the matrix to the fractures, there is:

$$\frac{{\rm d}C_{\rm m}}{{\rm d}t}=\begin{cases}\left.\dfrac{3D_{\rm F}}{R_{\rm m}}\dfrac{\partial C_{\rm m}}{\partial r_{\rm m}}\right|_{r_{\rm m}=R_{\rm m}} & \text{unsteady state diffusion}\\[2ex] \dfrac{6D_{\rm F}\pi^2}{R_{\rm m}^2}\left[C_{\rm E}(p_{\rm f})-C_{\rm m}\right] & \text{pseudo-steady state diffusion}\end{cases} \tag{A.68}$$

Substituting the expressions of the gas density into Eqs. (A.66) and (A.67) yields:

$$\frac{1}{r^2}\frac{\partial}{\partial r}\left(r^2\frac{\partial m_{\rm f}}{\partial r}\right)=\frac{\phi_{\rm f}\mu_{\rm gi}c_{\rm fgi}}{k_{\rm f}}\frac{\partial m_{\rm f}}{\partial t}+\frac{2RT}{k_{\rm f}}(1-\phi_{\rm f})\frac{{\rm d}C_{\rm m}}{{\rm d}t} \tag{A.69}$$

The following dimensionless variables are defined as:

$$r_D=\frac{r}{L_{\rm ref}},\quad r_{\rm mD}=\frac{r_{\rm m}}{R_{\rm m}},\quad t_{\rm D}=\frac{k_{\rm f}t}{\Lambda L_{\rm ref}^2},\quad \omega=\frac{\phi_{\rm f}\mu_{\rm gi}c_{\rm fgi}}{\Lambda},$$

$$\Lambda=\begin{cases}\phi_{\rm f}\mu_{\rm gi}c_{\rm fgi}+\dfrac{6k_{\rm f}h}{q_{\rm sc}} & \text{unsteady-state diffusion}\\[2ex] \phi_{\rm f}\mu_{\rm gi}c_{\rm fgi}+\dfrac{2k_{\rm f}h}{q_{\rm sc}} & \text{pseudo-steady-state diffusion}\end{cases},$$

$$\lambda=\begin{cases}\dfrac{k_{\rm f}\tau}{\Lambda L_{\rm ref}^2} & \text{unsteady-state diffusion}\\[2ex] \dfrac{k_{\rm f}\tau}{6\Lambda L_{\rm ref}^2} & \text{pseudo-steady-state diffusion}\end{cases},$$

$$\tau=\begin{cases}\dfrac{R_{\mathrm{m}}^{2}}{D_{\mathrm{F}}} & \text{unsteady-state diffusion}\\ \dfrac{R_{\mathrm{m}}^{2}}{\pi^{2}D_{\mathrm{F}}} & \text{pseudo-steady-state diffusion}\end{cases}$$

Substituting the above dimensionless variables and Eq. (A.68) into Eq.(A.69) and then taking the Laplace transformation, the diffusivity equation in the Laplace domain can be obtained. For unsteady state diffusion, it is:

$$\frac{1}{r_{\mathrm{D}}^{2}}\frac{\partial}{\partial r_{\mathrm{D}}}\left(r_{\mathrm{D}}^{2}\frac{\partial\Delta\overline{m}_{\mathrm{f}}}{\partial r_{\mathrm{D}}}\right)=\omega s\Delta\overline{m}_{\mathrm{f}}-\frac{M_{\mathrm{g}}}{\rho_{\mathrm{sc}}}\frac{(1-\omega)(1-\phi_{\mathrm{f}})}{\lambda}\frac{q_{\mathrm{sc}}p_{\mathrm{sc}}T}{k_{\mathrm{f}}hT_{\mathrm{sc}}}\frac{\partial\overline{C}_{\mathrm{m}}}{\partial r_{\mathrm{mD}}}\bigg|_{r_{\mathrm{mD}}=1} \tag{A.70}$$

For pseudo-steady-state-diffusion, it is:

$$\frac{1}{r_{\mathrm{D}}^{2}}\frac{\partial}{\partial r_{\mathrm{D}}}\left(r_{\mathrm{D}}^{2}\frac{\partial\Delta\overline{m}_{\mathrm{f}}}{\partial r_{\mathrm{D}}}\right)=\omega s\Delta\overline{m}_{\mathrm{f}}-\frac{M_{\mathrm{g}}}{\rho_{\mathrm{sc}}}\frac{(1-\omega)(1-\phi_{\mathrm{f}})}{\lambda}\frac{q_{\mathrm{sc}}p_{\mathrm{sc}}T}{k_{\mathrm{f}}hT_{\mathrm{sc}}}\left[\overline{C}_{\mathrm{E}}(p_{\mathrm{f}})-\overline{C}_{\mathrm{m}}\right] \tag{A.71}$$

According to the gas diffusion described in Section 1.6.3, two situations below will be discussed.

A.3.1 Unsteady state diffusion

The micropore diffusivity equation, which describes the transport of shale gas in the shale matrix, can be written in concentration by employing mass balance on a spherical element volume and defining the velocity term using Fick's law of diffusion:

$$\frac{\partial C_{\mathrm{m}}}{\partial t}=\frac{1}{r_{\mathrm{m}}^{2}}\frac{\partial}{\partial r_{\mathrm{m}}}\left(D_{\mathrm{F}}r_{\mathrm{m}}^{2}\frac{\partial C_{\mathrm{m}}}{\partial r_{\mathrm{m}}}\right) \tag{A.72}$$

Using the existing symmetry condition, the center of the element can be treated as a no-flow boundary:

$$C_{\mathrm{m}}(t=0,r_{\mathrm{m}})=C_{\mathrm{m}}(p_{\mathrm{i}}) \tag{A.73}$$

The inner boundary condition is:

$$\frac{\partial C_{\mathrm{m}}(t,r_{\mathrm{m}}=0)}{\partial r_{\mathrm{m}}}=0 \tag{A.74}$$

The concentration of the shale gas on the external surface of the matrix element is evaluated at the gas pressure in the fracture system:

$$C_{\mathrm{m}}(t,r_{\mathrm{m}}=R_{\mathrm{m}})=C_{\mathrm{m}}(p_{\mathrm{f}}) \tag{A.75}$$

The following dimensionless gas concentration, C_{mD} and C_{ED}, are defined as:

$$C_{\mathrm{mD}}=C_{\mathrm{m}}(p_{\mathrm{m}})-C_{\mathrm{m}}(p_{\mathrm{i}}),\quad C_{\mathrm{ED}}=C_{\mathrm{E}}(p_{\mathrm{f}})-C_{\mathrm{m}}(p_{\mathrm{i}}) \tag{A.76}$$

The dimensionless forms in the matrix system can be obtained by substituting the dimensionless gas concentration and dimensionless variables into the above model and then taking the Laplace transformation, which becomes:

$$\frac{1}{r_{\mathrm{mD}}^2}\frac{\partial}{\partial r_{\mathrm{mD}}}\left(r_{\mathrm{mD}}^2\frac{\partial\overline{C}_{\mathrm{mD}}}{\partial r_{\mathrm{mD}}}\right)=\lambda s\overline{C}_{\mathrm{mD}} \tag{A.77}$$

$$\frac{\partial\overline{C}_{\mathrm{mD}}(s,r_{\mathrm{mD}}=0)}{\partial r_{\mathrm{mD}}}=0 \tag{A.78}$$

$$\overline{C}_{\mathrm{mD}}(s,r_{\mathrm{mD}}=1)=\overline{C}_{\mathrm{ED}}[m(p_{\mathrm{f}})] \tag{A.79}$$

Define the following group:

$$W_{\mathrm{D}}=\overline{C}_{\mathrm{mD}}r_{\mathrm{mD}} \tag{A.80}$$

Substituting the above equation into Eq. (A.77) yields:

$$\frac{\partial^2 W_{\mathrm{D}}}{\partial r_{\mathrm{mD}}^2}=\lambda s W_{\mathrm{D}} \tag{A.81}$$

The general solution is:

$$W_{\mathrm{D}}=A\sinh\left(\sqrt{\lambda s}r_{\mathrm{mD}}\right)+B\cosh\left(\sqrt{\lambda s}r_{\mathrm{mD}}\right) \tag{A.82}$$

Substituting Eq. (A.82) into Eq. (A.80) yields:

$$\overline{C}_{\mathrm{mD}}=\frac{A\sinh\left(\sqrt{\lambda s}r_{\mathrm{mD}}\right)+B\cosh\left(\sqrt{\lambda s}r_{\mathrm{mD}}\right)}{r_{\mathrm{mD}}} \tag{A.83}$$

According to the inner boundary condition of matrix system, we have:

$$\lim_{r_{\mathrm{mD}}\to 0}\overline{C}_{\mathrm{mD}}=\lim_{r_{\mathrm{mD}}\to 0}\frac{A\sinh\left(\sqrt{\lambda s}r_{\mathrm{mD}}\right)+B\cosh\left(\sqrt{\lambda s}r_{\mathrm{mD}}\right)}{r_{\mathrm{mD}}}=\text{finite value} \tag{A.84}$$

According to the L'Hospital rule, the solution of parameter B can be derived from Eq. (A.84):

$$B=0 \tag{A.85}$$

Then the following equation can be obtained from the inner boundary condition:

$$\lim_{r_{\mathrm{mD}}\to 0} A\frac{\sqrt{\lambda s}\cosh\left(\sqrt{\lambda s}r_{\mathrm{mD}}\right)-\sinh\left(\sqrt{\lambda s}r_{\mathrm{mD}}\right)}{r_{\mathrm{mD}}^2}=0 \tag{A.86}$$

Similarly, the above equation can be simplified to:

$$\frac{A\lambda s}{2}\lim_{r_{\mathrm{mD}}\to 0}\sinh\left(\sqrt{\lambda s}r_{\mathrm{mD}}\right)=0 \tag{A.87}$$

Combining with the outer boundary condition and substituting B into the corresponding equation, parameter A can be obtained as:

$$A=\frac{\overline{C}_{\mathrm{ED}}[m(p_{\mathrm{f}})]}{\sinh\left(\sqrt{\lambda s}\right)} \tag{A.88}$$

Substituting Eqs. (A.85) and (A.88) into Eq. (A.83), the solution of the gas concentration is:

$$\overline{C}_{\mathrm{mD}}=\frac{\overline{C}_{\mathrm{ED}}[m(p_{\mathrm{f}})]}{\sinh\left(\sqrt{\lambda s}\right)}\frac{\sinh\left(\sqrt{\lambda s}r_{\mathrm{mD}}\right)}{r_{\mathrm{mD}}} \tag{A.89}$$

According to the Langmuir isotherm desorption equation, the following equation is established:

$$\overline{C}_{\mathrm{ED}}[m(p_{\mathrm{f}})]=\frac{\rho_{\mathrm{sc}}}{M_{\mathrm{g}}}\mathbf{L}\left[\frac{G_{\mathrm{L}}m(p_{\mathrm{f}})}{m(p_{\mathrm{L}})+m(p_{\mathrm{f}})}-\frac{G_{\mathrm{L}}m(p_{\mathrm{i}})}{m(p_{\mathrm{L}})+m(p_{\mathrm{i}})}\right] \tag{A.90}$$

Define the following desorption coefficient σ:

$$\sigma=\frac{G_{\mathrm{L}}m(p_{\mathrm{L}})}{[m(p_{\mathrm{L}})+m(p_{\mathrm{f}})][m(p_{\mathrm{L}})+m(p_{\mathrm{i}})]}\frac{q_{\mathrm{sc}}p_{\mathrm{sc}}T}{k_{\mathrm{f}}hT_{\mathrm{sc}}} \tag{A.91}$$

Eq. (A.90) becomes:

$$\overline{C}_{\mathrm{ED}}[m(p_{\mathrm{f}})]=\frac{\rho_{\mathrm{sc}}}{M_{\mathrm{g}}}\sigma[\overline{m}(p_{\mathrm{f}})-\overline{m}(p_{\mathrm{i}})]=-\frac{\rho_{\mathrm{sc}}}{M_{\mathrm{g}}}\frac{k_{\mathrm{f}}hT_{\mathrm{sc}}}{q_{\mathrm{sc}}p_{\mathrm{sc}}T}\sigma\Delta\overline{m}_{\mathrm{f}} \tag{A.92}$$

Substituting Eq. (A.92) into Eq. (A.89) yields:

$$\overline{C}_{\mathrm{mD}}=-\frac{\rho_{\mathrm{sc}}}{M_{\mathrm{g}}}\frac{k_{\mathrm{f}}hT_{\mathrm{sc}}}{q_{\mathrm{sc}}p_{\mathrm{sc}}T}\frac{\sigma}{\sinh\left(\sqrt{\lambda s}\right)}\frac{\sinh\left(\sqrt{\lambda s}r_{\mathrm{mD}}\right)}{r_{\mathrm{mD}}}\Delta\overline{m}_{\mathrm{f}} \tag{A.93}$$

Taking the derivative of the above equation yields:

$$\left.\frac{\partial\overline{C}_{\mathrm{mD}}}{\partial r_{\mathrm{mD}}}\right|_{r_{\mathrm{mD}}=1}=-\frac{\rho_{\mathrm{sc}}}{M_{\mathrm{g}}}\frac{k_{\mathrm{f}}hT_{\mathrm{sc}}}{q_{\mathrm{sc}}p_{\mathrm{sc}}T}\sigma\left[\sqrt{\lambda s}\coth\left(\sqrt{\lambda s}\right)-1\right]\Delta\overline{m}_{\mathrm{f}} \tag{A.94}$$

Substituting Eq. (A.94) into Eq. (A.70), the diffusivity equation in the fracture system with the consideration of unsteady state diffusion is:

$$\frac{1}{r_{\mathrm{D}}^{2}}\frac{\partial}{\partial r_{\mathrm{D}}}\left(r_{\mathrm{D}}^{2}\frac{\partial\Delta\overline{m}_{\mathrm{f}}}{\partial r_{\mathrm{D}}}\right)=f(s)\Delta\overline{m}_{\mathrm{f}} \tag{A.95}$$

where the expression of $f(s)$ is:

$$f(s)=\omega s+\frac{(1-\omega)(1-\phi_{\mathrm{f}})\sigma}{\lambda}\left[\sqrt{\lambda s}\coth\left(\sqrt{\lambda s}\right)-1\right] \tag{A.96}$$

A.3.2 Pseudo-steady state diffusion

For pseudo-steady state diffusion, the dimensionless expression of Eq. (A.68) becomes:

$$\lambda\frac{dC_{\mathrm{mD}}}{dt_{\mathrm{D}}}=C_{\mathrm{ED}}-C_{\mathrm{mD}} \tag{A.97}$$

The Laplace transformation of the above equation can be reduced to:

$$\overline{C}_{\mathrm{mD}}=\frac{1}{\lambda s+1}\overline{C}_{\mathrm{ED}} \tag{A.98}$$

According to the dimensionless concentration and Langmuir isotherm desorption equation, we have:

$$\overline{C}_{\mathrm{ED}}=\frac{\rho_{\mathrm{sc}}}{M_{\mathrm{g}}}\mathbf{L}\left[\frac{G_{\mathrm{L}}m(p_{\mathrm{f}})}{m(p_{\mathrm{L}})+m(p_{\mathrm{f}})}-\frac{G_{\mathrm{L}}m(p_{\mathrm{i}})}{m(p_{\mathrm{L}})+m(p_{\mathrm{i}})}\right] \tag{A.99}$$

Define the desorption coefficient σ as follows:

$$\sigma=\frac{G_{\mathrm{L}}m(p_{\mathrm{L}})}{[m(p_{\mathrm{L}})+m(p_{\mathrm{f}})][m(p_{\mathrm{L}})+m(p_{\mathrm{i}})]}\frac{q_{\mathrm{sc}}p_{\mathrm{sc}}T}{k_{\mathrm{f}}hT_{\mathrm{sc}}} \tag{A.100}$$

Thus we see that:

$$\overline{C}_{\mathrm{ED}}=-\frac{\rho_{\mathrm{sc}}}{M_{\mathrm{g}}}\frac{k_{\mathrm{f}}hT_{\mathrm{sc}}}{q_{sc}p_{sc}T}\sigma\Delta\overline{m}_{\mathrm{f}} \tag{A.101}$$

Combining Eqs. (A.101) and (A.98) yields:

$$\overline{C}_{\mathrm{ED}}-\overline{C}_{\mathrm{mD}}=-\frac{\rho_{\mathrm{sc}}}{M_{\mathrm{g}}}\frac{k_{\mathrm{f}}hT_{\mathrm{sc}}}{q_{\mathrm{sc}}p_{\mathrm{sc}}T}\frac{\sigma\lambda s}{\lambda s+1}\Delta\overline{m}_{\mathrm{f}} \tag{A.102}$$

The dimensionless expression in the right-hand side is:

$$\frac{1}{r_D^2}\frac{\partial}{\partial r_D}\left(r_D^2\frac{\partial\Delta\overline{m}_f}{\partial r_D}\right)=\omega s\Delta\overline{m}_f-\frac{M_g}{\rho_{sc}}\frac{q_{sc}p_{sc}T}{k_f h T_{sc}}\frac{(1-\omega)}{\lambda}\left(\overline{C}_{ED}-\overline{C}_{mD}\right) \tag{A.103}$$

Substituting Eq. (A.102) into the above equation yields:

$$\frac{1}{r_D^2}\frac{\partial}{\partial r_D}\left(r_D^2\frac{\partial\Delta\overline{m}_f}{\partial r_D}\right)=f(s)\Delta\overline{m}_f \tag{A.104}$$

The expression of $f(s)$ is:

$$f(s)=\omega s+\frac{\sigma(1-\omega)(1-\phi_f)s}{\lambda s+1} \tag{A.105}$$

A.4 Microfractures + matrix macropores + gas adsorption/desorption + nanopore Fick's diffusion model (model 4)

The diffusivity equation in the fracture system in spherical coordinates is:

$$\frac{1}{r^2}\frac{\partial}{\partial r}\left(r^2\frac{\partial m_f}{\partial r}\right)=\frac{\phi_f\mu_{gi}c_{fgi}}{k_f}\frac{\partial m_f}{\partial t}-\frac{1}{k_f}\frac{2RT}{M_g}q_m \tag{A.106}$$

A.4.1 Transient interporosity flow

When the gas flow from macropores into microfractures is transient flow, the interporosity flow term is:

$$q_m=-\frac{3\rho_g}{R_{mac}}\frac{k_m}{\mu}\frac{\partial p_m}{\partial r_m}\bigg|_{r_m=R_{mac}} \tag{A.107}$$

The diffusivity equation in the matrix system is:

$$\frac{1}{r_m^2}\frac{\partial}{\partial r_m}\left(r_m^2\frac{\partial m_m}{\partial r_m}\right)=\frac{\phi_m\mu_{gi}c_{mgi}}{k_m}\frac{\partial m_m}{\partial t}+\frac{1}{k_m}\frac{2RT}{M_g}q_F \tag{A.108}$$

The gas diffusion from matrix to fractures in a unit reservoir volume is:

$$q_F=M_g(1-\phi_f-\phi_m)\frac{dC_m}{dt} \tag{A.109}$$

The derivative of gas concentration in time has the following relationship with a different diffusion model:

$$\frac{dC_m}{dt} = \begin{cases} \left.\dfrac{3D_F}{R_m}\dfrac{\partial C_m}{\partial r_m}\right|_{r_m=R_m} & \text{unsteady state diffusion} \\ \dfrac{6D_F\pi^2}{R_m^2}[C_E(p_f) - C_m] & \text{pseudo-steady state diffusion} \end{cases} \tag{A.110}$$

Steady state diffusion. When the gas diffusion follows the pseudo-steady diffusion, the model is simplified to:

$$\frac{1}{r^2}\frac{\partial}{\partial r}\left(r^2\frac{\partial m_f}{\partial r}\right) = \frac{\phi_f\mu_{gi}c_{fgi}}{k_f}\frac{\partial m_f}{\partial t} + \left.\frac{3}{R_{mac}}\frac{k_m}{k_f}\frac{\partial m_m}{\partial r_m}\right|_{r_m=R_{mac}} \tag{A.111}$$

$$\frac{1}{r_m^2}\frac{\partial}{\partial r_m}\left(r_m^2\frac{\partial m_m}{\partial r_m}\right) = \frac{\phi_m\mu_{gi}c_{mgi}}{k_m}\frac{\partial m_m}{\partial t} + \frac{2RT(1-\phi_f-\phi_m)}{k_m}\frac{6D_F\pi^2}{R_m^2}[C_E(p_m) - C_m] \tag{A.112}$$

The shape factor for the spherical matrix is:

$$\alpha = \frac{15}{R_{mac}^2} \tag{A.113}$$

Eq. (A.111) becomes:

$$\frac{1}{r^2}\frac{\partial}{\partial r}\left(r^2\frac{\partial m_f}{\partial r}\right) = \frac{\phi_f\mu_{gi}c_{fgi}}{k_f}\frac{\partial m_f}{\partial t} + \left.\frac{\alpha R_{mac}}{5}\frac{k_m}{k_f}\frac{\partial m_m}{\partial r_m}\right|_{r_m=R_{mac}} \tag{A.114}$$

The following variables are defined as:

$$\lambda_{mf} = \alpha\frac{k_m}{k_f}L_{ref}^2,\quad r_D = \frac{r}{L_{ref}},\quad r_{mD} = \frac{r_m}{R_{mac}},\quad t_D = \frac{k_f t}{\Lambda L_{ref}^2},\quad \omega_f = \frac{\phi_f\mu_{gi}c_{fgi}}{\Lambda},$$

$$\omega_m = \frac{\phi_m\mu_{gi}c_{mgi}}{\Lambda},\quad \Lambda = \phi_m\mu_{gi}c_{mgi} + \phi_f\mu_{gi}c_{fgi} + \frac{2k_f h}{q_{sc}}\frac{R_{mac}^2}{L_{ref}^2},\quad \lambda = \frac{k_m\tau}{6\Lambda L_{ref}^2},$$

$$\tau = \frac{R_m^2}{\pi^2 D_F},\quad \phi_{mf} = \frac{k_f}{k_m}$$

Substituting the above variables into the governing equation, we have:

$$\frac{1}{r_D^2}\frac{\partial}{\partial r_D}\left(r_D^2\frac{\partial\Delta m_f}{\partial r_D}\right) = \omega_f\frac{\partial\Delta m_f}{\partial t_D} + \left.\frac{\lambda_{mf}}{5}\frac{\partial\Delta m_m}{\partial r_{mD}}\right|_{r_{mD}=1} \tag{A.115}$$

$$\frac{1}{r_{mD}^2}\frac{\partial}{\partial r_{mD}}\left(r_{mD}^2\frac{\partial\Delta m_m}{\partial r_{mD}}\right) = \frac{15\omega_m}{\lambda_{mf}}\frac{\partial\Delta m_m}{\partial t_D} - \frac{M_g}{\rho_{sc}}\frac{q_{sc}p_{sc}T\beta}{k_f h T_{sc}\lambda}[C_{ED} - C_{mD}] \tag{A.116}$$

where $\beta=(1-\omega_{\rm f}-\omega_{\rm m})(1-\phi_{\rm f}-\phi_{\rm m})$. Taking the Laplace transformation with the above continuity equations, we have:

$$\frac{1}{r_{\rm D}^2}\frac{\partial}{\partial r_{\rm D}}\left(r_{\rm D}^2\frac{\partial\Delta\overline{m}_{\rm f}}{\partial r_{\rm D}}\right)=\omega_{\rm f}s\Delta\overline{m}_{\rm f}+\frac{\lambda_{\rm mf}}{5}\frac{\partial\Delta\overline{m}_{\rm m}}{\partial r_{\rm mD}}\bigg|_{r_{\rm mD}=1} \tag{A.117}$$

$$\frac{1}{r_{\rm mD}^2}\frac{\partial}{\partial r_{\rm mD}}\left(r_{\rm mD}^2\frac{\partial\Delta\overline{m}_{\rm m}}{\partial r_{\rm mD}}\right)=\frac{15\omega_{\rm m}}{\lambda_{\rm mf}}s\Delta\overline{m}_{\rm m}-\frac{M_{\rm g}}{\rho_{\rm sc}}\frac{q_{\rm sc}p_{\rm sc}T}{k_{\rm f}hT_{\rm sc}}\frac{\beta}{\lambda}\left[\overline{C}_{\rm ED}-\overline{C}_{\rm mD}\right] \tag{A.118}$$

For pseudo-steady state diffusion, the diffusion equation becomes:

$$\lambda\theta_{\rm mf}\frac{dC_{\rm mD}}{dt_{\rm D}}=C_{\rm ED}-C_{\rm mD} \tag{A.119}$$

According to the result in Section A.3.2, for pseudo-steady state diffusion, the following formula can be obtained:

$$\overline{C}_{\rm ED}-\overline{C}_{\rm mD}=-\frac{\rho_{\rm sc}}{M_{\rm g}}\frac{k_{\rm f}hT_{\rm sc}}{q_{\rm sc}p_{\rm sc}T}\frac{\sigma\lambda\theta_{\rm mf}s}{\lambda\theta_{\rm mf}s+1}\Delta\overline{m}_{\rm m} \tag{A.120}$$

where the expression of the adsorption/desorption coefficient, σ, is:

$$\sigma=\frac{G_{\rm L}m(p_{\rm L})}{[m(p_{\rm L})+m(p_{\rm f})][m(p_{\rm L})+m(p_{\rm i})]}\frac{q_{\rm sc}p_{\rm sc}T}{k_{\rm f}hT_{\rm sc}} \tag{A.121}$$

Substituting Eq. (A.120) into Eq. (A.118) yields:

$$\frac{1}{r_{\rm mD}^2}\frac{\partial}{\partial r_{\rm mD}}\left(r_{\rm mD}^2\frac{\partial\Delta\overline{m}_{\rm m}}{\partial r_{\rm mD}}\right)=g(s)\Delta\overline{m}_{\rm m} \tag{A.122}$$

where $g(s)=\left(\frac{15\omega_{\rm m}}{\lambda_{\rm mf}}+\frac{\beta\sigma\theta_{\rm mf}}{\lambda\theta_{\rm mf}s+1}\right)s$.

According to the derivation in Section A.2.1 regarding unsteady state diffusion, if we add the boundary conditions into the continuity equation in the matrix system, the following equation can be obtained:

$$\frac{\partial\Delta\overline{m}_{\rm m}}{\partial r_{\rm mD}}\bigg|_{r_{\rm mD}=1}=\left[\sqrt{g(s)}\coth\left(\sqrt{g(s)}\right)-1\right]\Delta\overline{m}_{\rm f} \tag{A.123}$$

Substituting Eq. (A.123) into Eq. (A.117), we have:

$$\frac{1}{r_{\rm D}^2}\frac{\partial}{\partial r_{\rm D}}\left(r_{\rm D}^2\frac{\partial\Delta\overline{m}_{\rm f}}{\partial r_{\rm D}}\right)=f(s)\Delta\overline{m}_{\rm f} \tag{A.124}$$

where the expression of $f(s)$ is:

$$f(s)=\omega_{\mathrm{f}}s+\frac{\lambda_{\mathrm{mf}}}{5}\left[\sqrt{g(s)}\coth\left(\sqrt{g(s)}\right)-1\right] \tag{A.125}$$

Note that the expression of parameter Λ can be expressed with the interporosity coefficient according to its definition, which is: $\Lambda=\phi_{\mathrm{m}}\mu_{\mathrm{gi}}c_{\mathrm{mgi}}+\phi_{\mathrm{f}}\mu_{\mathrm{gi}}c_{\mathrm{fgi}}+\frac{2k_{\mathrm{f}}h}{q_{\mathrm{sc}}}\frac{15}{\lambda_{\mathrm{mf}}\theta_{\mathrm{mf}}}$.

Unsteady state diffusion. When the desorption gas obeys the unsteady state diffusion from nanopores into micropores, the continuity equation in the micropores in the matrix system is:

$$\frac{1}{r_{\mathrm{m}}^{2}}\frac{\partial}{\partial r_{\mathrm{m}}}\left(r_{\mathrm{m}}^{2}\frac{\partial m_{\mathrm{m}}}{\partial r_{\mathrm{m}}}\right)=\frac{\phi_{\mathrm{m}}\mu_{\mathrm{gi}}c_{\mathrm{mgi}}}{k_{\mathrm{m}}}\frac{\partial m_{\mathrm{m}}}{\partial t}+\frac{2RT}{k_{\mathrm{m}}}\frac{3D_{\mathrm{F}}}{R_{\mathrm{m}}}\frac{\partial C_{\mathrm{m}}}{\partial r_{\mathrm{m}}}\bigg|_{r_{\mathrm{m}}=R_{\mathrm{m}}} \tag{A.126}$$

Define the following variables:

$$\lambda_{\mathrm{mf}}=\alpha\frac{k_{\mathrm{m}}}{k_{\mathrm{f}}}L_{\mathrm{ref}}^{2},\quad r_{D}=\frac{r}{L_{\mathrm{ref}}},\quad r_{\mathrm{mD}}=\frac{r_{\mathrm{m}}}{R_{\mathrm{mac}}},\quad t_{\mathrm{D}}=\frac{k_{\mathrm{f}}t}{\Lambda L_{\mathrm{ref}}^{2}},\quad \omega_{\mathrm{f}}=\frac{\phi_{\mathrm{f}}\mu_{\mathrm{gi}}c_{\mathrm{fgi}}}{\Lambda},\quad \omega_{\mathrm{m}}=\frac{\phi_{\mathrm{m}}\mu_{\mathrm{gi}}c_{\mathrm{mgi}}}{\Lambda},$$

$$\Lambda=\phi_{\mathrm{m}}\mu_{\mathrm{gi}}c_{\mathrm{mgi}}+\phi_{\mathrm{f}}\mu_{\mathrm{gi}}c_{\mathrm{fgi}}+\frac{6k_{\mathrm{f}}hR_{\mathrm{mac}}^{2}}{q_{\mathrm{sc}}\ L_{\mathrm{ref}}^{2}},\quad \lambda=\frac{k_{\mathrm{m}}\tau}{\Lambda L_{\mathrm{ref}}^{2}},\quad \tau=\frac{R_{\mathrm{m}}^{2}}{D_{\mathrm{F}}},\quad \theta_{\mathrm{mf}}=\frac{k_{\mathrm{f}}}{k_{\mathrm{m}}}$$

Substituting the above variables into Eq. (A.126) and then taking the Laplace transformation with it, we have:

$$\frac{\partial}{\partial r_{\mathrm{mD}}}\left(r_{\mathrm{mD}}^{2}\frac{\partial\Delta\overline{m}_{\mathrm{m}}}{\partial r_{\mathrm{mD}}}\right)=\frac{15\omega_{\mathrm{m}}}{\lambda_{\mathrm{mf}}}s\Delta\overline{m}_{\mathrm{m}}-\frac{M_{\mathrm{g}}}{\rho_{sc}}\frac{q_{\mathrm{sc}}p_{\mathrm{sc}}T}{k_{\mathrm{f}}hT_{\mathrm{sc}}}\frac{\beta}{\lambda}\frac{\partial\overline{C}_{\mathrm{mD}}}{\partial r_{\mathrm{mD}}}\bigg|_{r_{\mathrm{mD}}=1} \tag{A.127}$$

where: $\beta=(1-\omega_{\mathrm{f}}-\omega_{\mathrm{m}})(1-\phi_{\mathrm{f}}-\phi_{\mathrm{m}})$.

Similarly, the unsteady diffusion equation in the matrix system is:

$$\frac{1}{r_{\mathrm{mD}}^{2}}\frac{\partial}{\partial r_{\mathrm{mD}}}\left(r_{\mathrm{mD}}^{2}\frac{\partial\overline{C}_{\mathrm{mD}}}{\partial r_{\mathrm{mD}}}\right)=\lambda\theta_{\mathrm{mf}}s\overline{C}_{\mathrm{mD}} \tag{A.128}$$

With the same method as above, the following equation can be obtained:

$$\frac{\partial\overline{C}_{\mathrm{mD}}}{\partial r_{\mathrm{mD}}}\bigg|_{r_{\mathrm{mD}}=1}=-\frac{\rho_{\mathrm{sc}}}{M_{\mathrm{g}}}\frac{k_{\mathrm{f}}hT_{\mathrm{sc}}}{q_{\mathrm{sc}}p_{\mathrm{sc}}T}\sigma\left[\sqrt{\lambda\theta_{\mathrm{mf}}s}\coth\left(\sqrt{\lambda\theta_{\mathrm{mf}}s}\right)-1\right]\Delta\overline{m}_{\mathrm{f}} \tag{A.129}$$

Substituting Eq. (A.129) into Eq. (A.127), we have:

$$\frac{1}{r_{\mathrm{mD}}^{2}}\frac{\partial}{\partial r_{\mathrm{mD}}}\left(r_{\mathrm{mD}}^{2}\frac{\partial\Delta\overline{m}_{\mathrm{m}}}{\partial r_{\mathrm{mD}}}\right)=g(s)\Delta\overline{m}_{\mathrm{m}} \tag{A.130}$$

where: $g(s) = \left(\frac{15\omega_m}{\lambda_{mf}} + \frac{\beta}{\lambda}\sigma\left[\sqrt{\lambda\theta_{mf}s}\coth\left(\sqrt{\lambda\theta_{mf}s}\right) - 1\right]\right)s$.

Hence, inserting Eq. (A.130) into Eq. (A.124), the following dimensionless continuity equation in the natural fracture system with unsteady diffusion can be obtained:

$$\frac{1}{r_D^2}\frac{\partial}{\partial r_D}\left(r_D^2\frac{\partial\Delta\overline{m}_f}{\partial r_D}\right) = f(s)\Delta\overline{m}_f \tag{A.131}$$

where the expression of $f(s)$ is:

$$f(s) = \omega_f s + \frac{\lambda_{mf}}{5}\left[\sqrt{g(s)}\coth\left(\sqrt{g(s)}\right) - 1\right] \tag{A.132}$$

A.4.2 Pseudo-steady state interporosity flow

When gas from micropores into natural fractures follows the pseudo-steady state interporosity flow, the following continuity equation in the natural fracture system can be obtained:

$$\frac{1}{r^2}\frac{\partial}{\partial r}\left(r^2\frac{\partial m_f}{\partial r}\right) = \frac{\phi_f\mu_{gi}c_{fgi}}{k_f}\frac{\partial m_f}{\partial t} - \frac{\alpha k_m}{k_f}(m_m - m_f) \tag{A.133}$$

Steady state diffusion. The continuity equation in micropores in the matrix system is:

$$-\alpha\frac{k_m}{k_f}(m_m - m_f) = \frac{\phi_m\mu_{gi}c_{mgi}}{k_f}\frac{\partial m_m}{\partial t} + \frac{2RT}{k_f}\frac{6D_F\pi^2}{R_m^2}[C_E(p_m) - C_m] \tag{A.134}$$

Define the following variables:

$$\lambda_{mf} = \alpha\frac{k_m}{k_f}L_{ref}^2,\quad r_D = \frac{r}{L_{ref}},\quad t_D = \frac{k_f t}{\Lambda L_{ref}^2},\quad \omega_f = \frac{\phi_f\mu_{gi}c_{fgi}}{\Lambda},\quad \omega_m = \frac{\phi_m\mu_{gi}c_{mgi}}{\Lambda},$$
$$\Lambda = \phi_m\mu_{gi}c_{mgi} + \phi_f\mu_{gi}c_{fgi} + \frac{2k_f h}{q_{sc}},\quad \lambda = \frac{k_f\tau}{6\Lambda L_{ref}^2},\quad \tau = \frac{R_m^2}{\pi^2 D_F}$$

Thus, substituting the above variables into Eqs. (A.133) and (A.134), we have:

$$\frac{1}{r_D^2}\frac{\partial}{\partial r_D}\left(r_D^2\frac{\partial\Delta m_f}{\partial r_D}\right) = \omega_f\frac{\partial\Delta m_f}{\partial t_D} - \lambda_{mf}(\Delta m_m - \Delta m_f) \tag{A.135}$$

The continuity equation in micropores in the matrix system is:

$$-\lambda_{mf}(\Delta m_m - \Delta m_f) = -\omega_m\frac{\partial\Delta m_m}{\partial t_D} - \frac{M_g}{\rho_{sc}}\frac{q_{sc}p_{sc}T\beta}{k_f hT_{sc}\lambda}[C_{ED} - C_{mD}] \tag{A.136}$$

where $\beta = (1 - \omega_f - \omega_m)(1 - \phi_f - \phi_m)$. Taking the Laplace transformation with the above equation yields:

$$\frac{1}{r_{\mathrm{D}}^{2}}\frac{\partial}{\partial r_{\mathrm{D}}}\left(r_{\mathrm{D}}^{2}\frac{\partial\Delta\overline{m}_{\mathrm{f}}}{\partial r_{\mathrm{D}}}\right)=\omega_{\mathrm{f}}s\Delta\overline{m}_{\mathrm{f}}-\lambda_{\mathrm{mf}}(\Delta\overline{m}_{\mathrm{m}}-\Delta\overline{m}_{\mathrm{f}}) \tag{A.137}$$

$$-\lambda_{\mathrm{mf}}(\Delta\overline{m}_{\mathrm{m}}-\Delta\overline{m}_{\mathrm{f}})=\omega_{\mathrm{m}}s\Delta\overline{m}_{\mathrm{m}}-\frac{M_{\mathrm{g}}}{\rho_{\mathrm{sc}}}\frac{q_{\mathrm{sc}}p_{\mathrm{sc}}T}{k_{\mathrm{f}}hT_{\mathrm{sc}}}\frac{\beta}{\lambda}\left[\overline{C}_{\mathrm{ED}}-\overline{C}_{\mathrm{mD}}\right] \tag{A.138}$$

According to the pseudo-steady state diffusion model discussed before, the following formula can be obtained:

$$\overline{C}_{\mathrm{ED}}-\overline{C}_{\mathrm{mD}}=-\frac{\rho_{\mathrm{sc}}}{M_{\mathrm{g}}}\frac{k_{\mathrm{f}}hT_{\mathrm{sc}}}{q_{\mathrm{sc}}p_{\mathrm{sc}}T}\frac{\sigma\lambda s}{\lambda s+1}\Delta\overline{m}_{\mathrm{m}} \tag{A.139}$$

where the expression of adsorption/desorption coefficient is:

$$\sigma=\frac{G_{\mathrm{L}}m(p_{\mathrm{L}})}{[m(p_{\mathrm{L}})+m(p_{\mathrm{f}})][m(p_{\mathrm{L}})+m(p_{\mathrm{i}})]}\frac{q_{\mathrm{sc}}p_{\mathrm{sc}}T}{k_{\mathrm{f}}hT_{\mathrm{sc}}} \tag{A.140}$$

Substituting Eq. (A.139) into Eq. (A.138) yields:

$$-\lambda_{\mathrm{mf}}(\Delta\overline{m}_{\mathrm{m}}-\Delta\overline{m}_{\mathrm{f}})=g(s)\Delta\overline{m}_{\mathrm{m}} \tag{A.141}$$

where $g(s)=\omega_{\mathrm{m}}s+\frac{\sigma\beta s}{\lambda s+1}$. The solution of Eq. (A.141) is

$$\Delta\overline{m}_{\mathrm{m}}=\frac{\lambda_{\mathrm{mf}}}{[g(s)+\lambda_{\mathrm{mf}}]}\Delta\overline{m}_{\mathrm{f}} \tag{A.142}$$

Substituting Eq. (A.142) into Eq. (A.137) yields:

$$\frac{1}{r_{\mathrm{D}}^{2}}\frac{\partial}{\partial r_{\mathrm{D}}}\left(r_{\mathrm{D}}^{2}\frac{\partial\Delta\overline{m}_{\mathrm{f}}}{\partial r_{\mathrm{D}}}\right)=f(s)\Delta\overline{m}_{\mathrm{f}} \tag{A.143}$$

where the expression of $f(s)$ is:

$$f(s)=\omega_{\mathrm{f}}s+\frac{\lambda_{\mathrm{mf}}g(s)}{g(s)+\lambda_{\mathrm{mf}}} \tag{A.144}$$

Unsteady state diffusion. For unsteady state diffusion, the continuity equation in microspores in the matrix system is:

$$-\alpha\frac{k_{\mathrm{m}}}{k_{\mathrm{f}}}(m_{\mathrm{m}}-m_{\mathrm{f}})=\frac{\phi_{\mathrm{m}}\mu_{\mathrm{gi}}c_{\mathrm{mgi}}}{k_{\mathrm{f}}}\frac{\partial m_{\mathrm{m}}}{\partial t}+\frac{2RT}{k_{\mathrm{f}}}\frac{3D_{\mathrm{F}}}{R_{\mathrm{m}}}\frac{\partial C_{\mathrm{m}}}{\partial r_{\mathrm{m}}}\bigg|_{r_{\mathrm{m}}=R_{\mathrm{m}}} \tag{A.145}$$

Define the following variables:

$$\lambda_{\mathrm{mf}}=\alpha\frac{k_{\mathrm{m}}}{k_{\mathrm{f}}}L_{\mathrm{ref}}^{2},\quad r_{D}=\frac{r}{L_{\mathrm{ref}}},\quad t_{\mathrm{D}}=\frac{k_{\mathrm{f}}t}{\Lambda L_{\mathrm{ref}}^{2}},\quad \omega_{\mathrm{f}}=\frac{\phi_{\mathrm{f}}\mu_{\mathrm{gi}}c_{\mathrm{fgi}}}{\Lambda},\quad \omega_{\mathrm{m}}=\frac{\phi_{\mathrm{m}}\mu_{\mathrm{gi}}c_{\mathrm{mgi}}}{\Lambda},$$

$$\Lambda=\phi_{\mathrm{m}}\mu_{\mathrm{gi}}c_{\mathrm{mgi}}+\phi_{\mathrm{f}}\mu_{\mathrm{gi}}c_{\mathrm{fgi}}+\frac{6k_{\mathrm{f}}h}{q_{\mathrm{sc}}},\quad \lambda=\frac{k_{\mathrm{f}}\tau}{\Lambda L_{\mathrm{ref}}^{2}},\quad \tau=\frac{R_{\mathrm{m}}^{2}}{D_{\mathrm{F}}}$$

The dimensionless continuity equation in the micropores in the matrix system is:

$$-\lambda_{\mathrm{mf}}(\Delta m_{\mathrm{m}}-\Delta m_{\mathrm{f}})=\omega_{\mathrm{m}}\frac{\partial\Delta m_{\mathrm{m}}}{\partial t_{\mathrm{D}}}-\frac{M_{\mathrm{g}}}{\rho_{\mathrm{sc}}}\frac{q_{\mathrm{sc}}p_{\mathrm{sc}}T}{k_{\mathrm{f}}hT_{\mathrm{sc}}}\frac{\beta}{\lambda}\frac{\partial\overline{C}_{\mathrm{mD}}}{\partial r_{\mathrm{mD}}}\bigg|_{r_{\mathrm{mD}}=1} \tag{A.146}$$

where: $\beta=(1-\omega_{\mathrm{f}}-\omega_{\mathrm{m}})(1-\phi_{\mathrm{f}}-\phi_{\mathrm{m}})$.

According to the unsteady state diffusion model for spherical matrix, we have:

$$\frac{\partial\overline{C}_{\mathrm{mD}}}{\partial r_{\mathrm{mD}}}\bigg|_{r_{\mathrm{mD}}=1}=-\frac{M_{\mathrm{g}}}{\rho_{\mathrm{sc}}}\frac{k_{\mathrm{f}}hT_{\mathrm{sc}}}{q_{\mathrm{sc}}p_{\mathrm{sc}}T}\sigma\left[\sqrt{\lambda s}\coth\left(\sqrt{\lambda s}\right)-1\right]\Delta\overline{m}_{\mathrm{f}} \tag{A.147}$$

Taking the Laplace transformation for Eq. (A.146) and then combining Eq. (A.147) with it, we have:

$$-\lambda_{\mathrm{mf}}(\Delta\overline{m}_{\mathrm{m}}-\Delta\overline{m}_{\mathrm{f}})=g(s)\Delta\overline{m}_{\mathrm{m}} \tag{A.148}$$

where: $g(s)=\omega_{\mathrm{m}}s+\frac{\sigma\beta}{\lambda}\left[\sqrt{\lambda s}\coth\left(\sqrt{\lambda s}\right)-1\right]$.

Combining Eq. (A.148) with Eq. (A.135), we have:

$$\frac{1}{r_{\mathrm{D}}^{2}}\frac{\partial}{\partial r_{\mathrm{D}}}\left(r_{\mathrm{D}}^{2}\frac{\partial\Delta\overline{m}_{\mathrm{f}}}{\partial r_{\mathrm{D}}}\right)=f(s)\Delta\overline{m}_{\mathrm{f}} \tag{A.149}$$

where the expression of $f(s)$ is:

$$f(s)=\omega_{\mathrm{f}}s+\frac{\lambda_{\mathrm{mf}}g(s)}{g(s)+\lambda_{\mathrm{mf}}} \tag{A.150}$$

A.5 Microfractures + gas adsorption/desorption + nanopore Knudsen diffusion model (model 5)

In a spherical coordinate system, the continuity equation in the natural system is:

$$\frac{1}{r^{2}}\frac{\partial}{\partial r}\left(\frac{k_{f}}{\mu}\rho_{\mathrm{g}}r^{2}\frac{\partial p_{\mathrm{f}}}{\partial r}\right)+q_{\mathrm{m}}=\frac{\partial(\phi_{\mathrm{f}}\rho_{\mathrm{g}})}{\partial t} \tag{A.151}$$

A.5.1 Transient interporosity flow

According to the previous analysis, the permeability of the matrix system can be presented by an apparent permeability, k_{app}, to take into account the effects of Knudsen diffusion and Darcy's flow. Thus the continuity equation in the matrix system is:

$$\frac{1}{r_{\mathrm{m}}^{2}}\frac{\partial}{\partial r_{\mathrm{m}}}\left(\frac{k_{\mathrm{app}}}{\mu}\rho_{\mathrm{g}}r_{\mathrm{m}}^{2}\frac{\partial p_{\mathrm{m}}}{\partial r_{\mathrm{m}}}\right)=\frac{\partial(\phi_{\mathrm{m}}\rho_{\mathrm{g}})}{\partial t}+q_{\mathrm{des}} \tag{A.152}$$

For the transient interporosity flow model, the interporosity flux can be expressed as:

$$q_{\rm m} = -\frac{3\rho_{\rm g}}{R_{\rm m}}\frac{k_{\rm app}}{\mu_{\rm g}}\frac{\partial p_{\rm m}}{\partial r_{\rm m}}\bigg|_{r_{\rm m}=R_{\rm m}} \tag{A.153}$$

For the adsorption/desorption gas, this item can be expressed as:

$$q_{\rm des} = \rho_{\rm gsc}(1-\phi_{\rm f})G_{\rm L}\frac{p_{\rm L}}{(p_{\rm L}+p_{\rm m})^2}\frac{\partial p_{\rm m}}{\partial t} \tag{A.154}$$

Introducing the pseudo pressure and gas density expressions, the continuity equation in the natural fracture system becomes:

$$\frac{1}{r^2}\frac{\partial}{\partial r}\left(r^2\frac{\partial m(p_{\rm f})}{\partial r}\right) = \frac{\mu_{\rm gi}\phi_{\rm f}c_{\rm fgi}}{k_{\rm f}}\frac{\partial m(p_{\rm f})}{\partial t} + \frac{3}{R_{\rm m}}\frac{k_{\rm app}}{k_{\rm f}}\frac{\partial m(p_{\rm m})}{\partial r_{\rm m}}\bigg|_{r_{\rm m}=R_{\rm m}} \tag{A.155}$$

By expressing the Langmuir isotherm adsorption equation with the pseudo pressure, the continuity equation in the matrix system is:

$$\frac{1}{r_{\rm m}^2}\frac{\partial}{\partial r_{\rm m}}\left(r_{\rm m}^2\frac{\partial m(p_{\rm m})}{\partial r_{\rm m}}\right) = \frac{\mu_{\rm gi}\phi_{\rm m}c_{\rm mgi}}{k_{\rm app}}\frac{\partial m(p_{\rm m})}{\partial t} + \frac{\phi_{\rm m}\mu_{\rm gi}c_{\rm d}}{k_{\rm app}}\frac{\partial m(p_{\rm m})}{\partial t} \tag{A.156}$$

where $c_{\rm d}$ is the additional compressibility coefficient, whose expression is:

$$c_{\rm d} = \frac{2Tp_{\rm sc}}{\phi_{\rm m}\mu_{\rm gi}T_{\rm sc}}\frac{(1-\phi_{\rm f}-\phi_{\rm m})G_{\rm L}m(p_{\rm L})}{[{\rm m}(p_{\rm L})+m(p_{\rm m})]^2} \tag{A.157}$$

Define the following variables:

$$r_{\rm mD} = \frac{r_{\rm m}}{R_{\rm m}},\quad r_{\rm D} = \frac{r}{L_{\rm ref}},\quad t_{\rm D} = \frac{k_{\rm f}t}{(\phi_{\rm m}c_{\rm mgi}+\phi_{\rm f}c_{\rm fgi})\mu_{\rm gi}L_{\rm ref}^2},\quad \omega_{\rm f} = \frac{\phi_{\rm f}c_{\rm fgi}}{\phi_{\rm m}c_{\rm mgi}+\phi_{\rm f}c_{\rm fgi}},$$

$$\omega_{\rm d} = \frac{\phi_{\rm m}c_{\rm d}}{\phi_{\rm m}c_{\rm mgi}+\phi_{\rm f}c_{\rm fgi}},\quad \lambda = \alpha\frac{k_{\rm m}}{k_{\rm f}}L_{\rm ref}^2,\quad \theta_{\text{ka-m}} = \frac{k_{\rm app}}{k_{\rm m}}$$

Substituting the above dimensionless variables into the corresponding continuity equations in the matrix and natural fracture systems, we have:

$$\frac{1}{r_{\rm D}^2}\frac{\partial}{\partial r_{\rm D}}\left(r_{\rm D}^2\frac{\partial\Delta m_{\rm f}}{\partial r_{\rm D}}\right) = \omega_{\rm f}\frac{\partial\Delta m_{\rm f}}{\partial t_{\rm D}} + \frac{\lambda\theta_{\text{ka-m}}}{5}\frac{\partial\Delta m_{\rm m}}{\partial r_{\rm mD}}\bigg|_{r_{\rm mD}=1} \tag{A.158}$$

$$\frac{1}{r_{\rm mD}^2}\frac{\partial}{\partial r_{\rm mD}}\left(r_{\rm mD}^2\frac{\partial\Delta m_{\rm f}}{\partial r_{\rm m\,D}}\right) = \frac{15(1-\omega_{\rm f})}{\lambda\theta_{\text{ka-m}}}\frac{\partial\Delta m_{\rm m}}{\partial t_{\rm D}} + \frac{15\omega_{\rm d}}{\lambda\theta_{\text{ka-m}}}\frac{\partial\Delta m_{\rm m}}{\partial t_{\rm D}} \tag{A.159}$$

Taking the Laplace transformation with the above equations yields:

$$\frac{1}{r_D^2}\frac{\partial}{\partial r_D}\left(r_D^2\frac{\partial \Delta \overline{m}_f}{\partial r_D}\right)=\omega_f s\Delta \overline{m}_f+\frac{\lambda\theta_{\text{ka-m}}}{5}\frac{\partial \Delta \overline{m}_m}{\partial t_{mD}}\bigg|_{r_{mD}=1} \tag{A.160}$$

$$\frac{1}{r_{mD}^2}\frac{\partial}{\partial r_{mD}}\left(r_{mD}^2\frac{\partial \Delta \overline{m}_m}{\partial r_{mD}}\right)=\frac{15(1-\omega_f)}{\lambda\theta_{\text{ka-m}}}s\overline{m}_m+\frac{15\omega_d}{\lambda\theta_{\text{ka-m}}}s\Delta \overline{m}_m \tag{A.161}$$

According to the derivation method before, combining Eq. (A.161) with its essential definiteness condition equations, we have:

$$\frac{\partial \Delta \overline{m}_m}{\partial r_{mD}}\bigg|_{r_{mD}=1}=\left[\sqrt{g}\coth\left(\sqrt{g}\right)-1\right]\Delta \overline{m}_f \tag{A.162}$$

where the expression of g is:

$$g=\frac{15(1-\omega_f+\omega_d)s}{\lambda\theta_{\text{ka-m}}} \tag{A.163}$$

Inserting Eq. (A.162) into Eq. (A.160) yields:

$$\frac{1}{r_D^2}\frac{\partial}{\partial r_D}\left(r_D^2\frac{\partial \Delta \overline{m}_f}{\partial r_D}\right)=f(s)\Delta \overline{m}_f \tag{A.164}$$

where the expression of $f(s)$ is:

$$f(s)=\omega_f s+\frac{\lambda\theta_{\text{ka-m}}}{5}\left[\sqrt{g}\coth\left(\sqrt{g}\right)-1\right] \tag{A.165}$$

A.5.2 Pseudo-steady state interporosity flow

When gas flows from the matrix system to the natural fractures system, the interporosity flux is:

$$q_m=\frac{\alpha k_{app}}{\mu}\left(\rho_{gm}p_m-\rho_{gf}p_f\right) \tag{A.166}$$

Substituting the above interporosity flux equation into Eq. (A.151), the continuity equation in the natural fracture system is:

$$\frac{1}{r^2}\frac{\partial}{\partial r}\left(\frac{k_f}{\mu}\rho_g r^2\frac{\partial p_f}{\partial r}\right)=\frac{\partial(\phi_f\rho_g)}{\partial t}-\frac{\alpha k_{app}}{\mu}\left(\rho_{gm}p_m-\rho_{gf}p_f\right) \tag{A.167}$$

The continuity equation in the matrix system is:

$$-\frac{\alpha k_{app}}{\mu}\left(\rho_{gm}p_m-\rho_{gf}p_f\right)=\frac{\partial(\phi_m\rho_g)}{\partial t}+q_{des} \tag{A.168}$$

Substituting the gas density and desorption equations into the above equation, the continuity equation in the natural fracture system is:

$$\frac{1}{r^2}\frac{\partial}{\partial r}\left(r^2\frac{\partial m(p_\mathrm{f})}{\partial r}\right)=\frac{\mu_\mathrm{gi}\phi_\mathrm{f}c_\mathrm{fgi}}{k_\mathrm{f}}\frac{\partial m(p_\mathrm{f})}{\partial t}-\frac{\alpha k_\mathrm{app}}{k_\mathrm{f}}[m(p_\mathrm{m})-m(p_\mathrm{f})] \quad (A.169)$$

The continuity equation in nanopores is:

$$-\alpha\frac{k_\mathrm{app}}{k_\mathrm{f}}[m(p_\mathrm{m})-m(p_\mathrm{f})]=\frac{\mu_\mathrm{gi}\phi_\mathrm{m}c_\mathrm{mgi}}{k_\mathrm{f}}\frac{\partial m(p_\mathrm{m})}{\partial t}+\frac{\phi_\mathrm{m}\mu_\mathrm{gi}c_\mathrm{d}}{k_\mathrm{f}}\frac{\partial m(p_\mathrm{m})}{\partial t} \quad (A.170)$$

where c_d is the additional compressibility:

$$c_\mathrm{d}=\frac{2Tp_\mathrm{sc}}{\phi_\mathrm{m}\mu_\mathrm{gi}T_\mathrm{sc}}\frac{(1-\phi_\mathrm{f}-\phi_\mathrm{m})G_\mathrm{L}m(p_\mathrm{L})}{[m(p_\mathrm{L})+m(p_\mathrm{m})]^2} \quad (A.171)$$

Define the following variables:

$$r_\mathrm{D}=\frac{r}{L_\mathrm{ref}},\quad t_\mathrm{D}=\frac{k_\mathrm{f}t}{\left(\phi_\mathrm{m}c_\mathrm{mgi}+\phi_\mathrm{f}c_\mathrm{fgi}\right)\mu_\mathrm{gi}L_\mathrm{ref}^2},\quad \omega_\mathrm{f}=\frac{\phi_\mathrm{f}c_\mathrm{fgi}}{\phi_\mathrm{m}c_\mathrm{mgi}+\phi_\mathrm{f}c_\mathrm{fgi}},\quad \omega_\mathrm{d}=\frac{\phi_\mathrm{m}c_\mathrm{d}}{\phi_\mathrm{m}c_\mathrm{mgi}+\phi_\mathrm{f}c_\mathrm{fgi}},$$

$$\lambda=\alpha\frac{k_\mathrm{m}}{k_\mathrm{f}}L_\mathrm{ref}^2,\quad \theta_\text{ka-m}=\frac{k_\mathrm{app}}{k_\mathrm{m}}$$

Substituting the above dimensionless variables into the continuity equations in the matrix and natural fracture systems, we have:

$$\frac{1}{r_\mathrm{D}^2}\frac{\partial}{\partial r_\mathrm{D}}\left(r_\mathrm{D}^2\frac{\partial\Delta m_\mathrm{f}}{\partial r_\mathrm{D}}\right)=\omega_\mathrm{f}\frac{\partial\Delta m_\mathrm{f}}{\partial t_\mathrm{D}}-\lambda\theta_\text{ka-m}[\Delta m_\mathrm{m}-\Delta m_\mathrm{f}] \quad (A.172)$$

$$-\lambda\theta_\text{ka-m}[\Delta m_\mathrm{m}-\Delta m_\mathrm{f}]=(1-\omega_\mathrm{f})\frac{\partial\Delta m_\mathrm{m}}{\partial t_\mathrm{D}}+\omega_\mathrm{d}\frac{\partial\Delta m_\mathrm{m}}{\partial t_D} \quad (A.173)$$

Taking the Laplace transformation for them, there is:

$$\frac{1}{r_\mathrm{D}^2}\frac{\partial}{\partial r_\mathrm{D}}\left(r_\mathrm{D}^2\frac{\partial\Delta\overline{m}_\mathrm{f}}{\partial r_\mathrm{D}}\right)=\omega_\mathrm{f}s\Delta\overline{m}_\mathrm{f}-\lambda\theta_\text{ka-m}(\Delta\overline{m}_\mathrm{m}-\Delta\overline{m}_\mathrm{f}) \quad (A.174)$$

$$-\lambda\theta_\text{ka-m}(\Delta\overline{m}_\mathrm{m}-\Delta\overline{m}_\mathrm{f})=(1-\omega_\mathrm{f})s\Delta\overline{m}_\mathrm{m}+\omega_\mathrm{d}s\Delta\overline{m}_\mathrm{m} \quad (A.175)$$

Combining the above two equations, we have:

$$\frac{1}{r_\mathrm{D}^2}\frac{\partial}{\partial r_\mathrm{D}}\left(r_\mathrm{D}^2\frac{\partial\Delta\overline{m}_\mathrm{f}}{\partial r_\mathrm{D}}\right)=f(s)\Delta\overline{m}_\mathrm{f} \quad (A.176)$$

where the expression of $f(s)$ is:

$$f(s)=\frac{\lambda\theta_{\text{ka-m}}(1+\omega_{\text{d}})+\omega_{\text{f}}(1-\omega_{\text{f}}+\omega_{\text{d}})s}{\lambda\theta_{\text{ka-m}}+(1-\omega_{\text{f}}+\omega_{\text{d}})s}s \tag{A.177}$$

Appendix B Solution derivation for a continuous line source in a composite model

B.1 Microfractures + steady state adsorption/desorption and diffusion model (model 1)

Assuming that a fully penetrated continuous linear source with a constant production rate in the center of a reservoir, the continuity equation of the linear source that takes into account the gas adsorption/desorption and diffusion can be established according to the mass conservation theory.

For the inner region in the natural fracture system, it is:

$$\frac{1}{r}\frac{\partial}{\partial r}\left(\frac{k_{\text{f1}}}{\mu_{\text{g}}}\frac{p_{\text{f1}}M_{\text{g}}}{ZRT}r\frac{\partial p_{\text{f1}}}{\partial r}\right)=\phi_{\text{f1}}c_{\text{fg1}}\frac{p_{\text{f1}}M_{\text{g}}}{ZRT}\frac{\partial p_{\text{f1}}}{\partial t}+q_{\text{des1}} \tag{B.1}$$

For the outer region in the natural fracture system, it is:

$$\frac{1}{r}\frac{\partial}{\partial r}\left(\frac{k_{\text{f2}}}{\mu_{\text{g}}}\frac{p_{\text{f2}}M_{\text{g}}}{ZRT}r\frac{\partial p_{\text{f2}}}{\partial r}\right)=\phi_{\text{f2}}c_{\text{fg2}}\frac{p_{\text{f2}}M_{\text{g}}}{ZRT}\frac{\partial p_{\text{f2}}}{\partial t}+q_{\text{des2}} \tag{B.2}$$

Because the pseudo-steady state adsorption/desorption diffusion model is used in this case, the desorption item in the above equations can be described as:

$$q_{\text{des1,2}}=\frac{M_{\text{g}}p_{\text{sc}}}{RT_{\text{sc}}}\frac{(1-\phi_{\text{f1,2}})G_{\text{L}}m(p_{\text{L}})}{[m(p_{\text{L}})+m(p_{\text{f1,2}})]^2}\frac{\partial m(p_{\text{f1,2}})}{\partial t} \tag{B.3}$$

Substituting Eq. (A.8) into Eqs. (A.1) and (B.2) yields:

$$\frac{1}{r}\frac{\partial}{\partial r}\left(r\frac{\partial m_{\text{f1}}}{\partial r}\right)=\frac{\phi_{\text{f1}}c_{\text{fg1}}}{k_{\text{f1}}}\frac{\partial m_{\text{f1}}}{\partial t}+\frac{p_{\text{sc}}T}{T_{\text{sc}}}\frac{(1-\phi_{\text{f1}})G_{\text{L}}m(p_{\text{L}})}{[m(p_{\text{L}})+m(p_{\text{f1}})]^2}\frac{\partial m_{\text{f1}}}{\partial t} \tag{B.4}$$

$$\frac{1}{r}\frac{\partial}{\partial r}\left(r\frac{\partial m_{\text{f2}}}{\partial r}\right)=\frac{\phi_{\text{f2}}c_{\text{fg2}}}{k_{\text{f2}}}\frac{\partial m_{\text{f2}}}{\partial t}+\frac{p_{\text{sc}}T}{T_{\text{sc}}}\frac{(1-\phi_{\text{f2}})G_{\text{L}}m(p_{\text{L}})}{[m(p_{\text{L}})+m(p_{\text{f2}})]^2}\frac{\partial m_{\text{f2}}}{\partial t} \tag{B.5}$$

Define the following additional adsorption/desorption compressibility:

$$c_{\text{d1}}=\frac{2Tp_{\text{sc}}}{\phi_{\text{f1}}\mu_{\text{gi}}T_{\text{sc}}}\frac{(1-\phi_{\text{f1}})G_{\text{L}}m(p_{\text{L}})}{[m(p_{\text{L}})+m(p_{\text{f1}})]^2} \tag{B.6}$$

$$c_{d2} = \frac{2Tp_{sc}}{\phi_{f2}\mu_{gi}T_{sc}} \frac{(1-\phi_{f2})G_L m(p_L)}{[m(p_L)+m(p_{f2})]^2} \tag{B.7}$$

Substituting the above two equations into Eqs. (B.4) and (B.5), the continuity equation in the natural fracture system in the inner region is:

$$\frac{1}{r}\frac{\partial}{\partial r}\left(r\frac{\partial m_{f1}}{\partial r}\right) = \frac{\phi_{f1}\mu_{gi}c_{fg1}}{k_{f1}}\frac{\partial m_{f1}}{\partial t} + \frac{\phi_{f1}\mu_{gi}c_{d1}}{k_{f1}}\frac{\partial m_{f1}}{\partial t} \tag{B.8}$$

For the natural fracture system in the outer region, there is:

$$\frac{1}{r}\frac{\partial}{\partial r}\left(r\frac{\partial m_{f2}}{\partial r}\right) = \frac{\phi_{f2}\mu_{gi}c_{fg2}}{k_{f2}}\frac{\partial m_{f2}}{\partial t} + \frac{\phi_{f2}\mu_{gi}c_{d2}}{k_{f2}}\frac{\partial m_{f2}}{\partial t} \tag{B.9}$$

The following dimensionless variables are defined:

$$r_D = \frac{r}{L_{ref}},\ M_{12} = \frac{k_{f1}/\mu_{gi}}{k_{f2}/\mu_{gi}},\ t_D = \frac{k_{f2}t}{\left(\phi_f\mu_{gi}c_{fgi}\right)_{1+2}L_{ref}^2},\ \omega_1 = \frac{\left(\phi_f\mu_{gi}c_{fgi}\right)_1}{\left(\phi_f\mu_{gi}c_{fgi}\right)_{1+2}},\ \omega_1 = \frac{\left(\phi_f\mu_{gi}c_{fgi}\right)_2}{\left(\phi_f\mu_{gi}c_{fgi}\right)_{1+2}},$$

$$\alpha_{d1} = \frac{c_{d1}}{c_{gi}},\ \alpha_{d2} = \frac{c_{d2}}{c_{gi}}$$

Substituting the above dimensionless variables into the continuity equations in the inner and outer regions, we have:

$$\frac{1}{r_D}\frac{\partial}{\partial r_D}\left(r_D\frac{\partial m_{f1}}{\partial r_D}\right) = \frac{\omega_1}{M_{12}}\frac{\partial m_{f1}}{\partial t_D} + \frac{\omega_1\alpha_{d1}}{M_{12}}\frac{\partial m_{f1}}{\partial t_D} \tag{B.10}$$

For the outer natural fracture system, we see that:

$$\frac{1}{r_D}\frac{\partial}{\partial r_D}\left(r_D\frac{\partial m_{f2}}{\partial r_D}\right) = \omega_2\frac{\partial m_{f2}}{\partial t_D} + \omega_2\alpha_{d2}\frac{\partial m_{f2}}{\partial t_D} \tag{B.11}$$

Substituting the pseudo pressure difference into the above continuity equations and then taking the Laplace transformation for them, we have:

$$\frac{1}{r_D}\frac{\partial}{\partial r_D}\left(r_D\frac{\partial \Delta\overline{m}_{f1}}{\partial r_D}\right) = f_1(s)\Delta\overline{m}_{f1} \tag{B.12}$$

$$\frac{1}{r_D}\frac{\partial}{\partial r_D}\left(r_D\frac{\partial \Delta\overline{m}_{f2}}{\partial r_D}\right) = f_2(s)\Delta\overline{m}_{f2} \tag{B.13}$$

where $f_1(s) = \omega s\frac{1+\alpha_{d1}}{M_{12}}$ and $f_2(s) = \omega_2 s(1+\alpha_{d2})$.

B.2 Microfractures + matrix macropores + steady state adsorption/desorption and diffusion (model 2)

B.2.1 Transient matrix flow

The continuity equation in the inner natural fracture system is:

$$\frac{1}{r}\frac{\partial}{\partial r}\left(\frac{k_{\text{f1}}}{\mu_{\text{g}}}\frac{p_{\text{f1}}M_{\text{g}}}{ZRT}r\frac{\partial p_{\text{f1}}}{\partial r}\right)=\phi_{\text{f1}}c_{\text{fg1}}\frac{p_{\text{f1}}M_{\text{g}}}{ZRT}\frac{\partial p_{\text{f1}}}{\partial t}-q_{\text{m1}} \tag{B.14}$$

The interporosity flux, q_{m1}, from the matrix system into the fracture system is:

$$q_{\text{m1}}=-\frac{3\rho_{\text{gm}}}{R_{\text{m1}}}\frac{k_{\text{m1}}}{\mu_{\text{g}}}\frac{\partial p_{\text{m1}}}{\partial r_{\text{m1}}}\bigg|_{r_{\text{m1}}=R_{\text{m1}}} \tag{B.15}$$

The continuity equation in the matrix system is:

$$\frac{1}{r}\frac{\partial}{\partial r}\left(\frac{k_{\text{m1}}}{\mu_{\text{g}}}\frac{p_{\text{m1}}M_{\text{g}}}{ZRT}r\frac{\partial p_{\text{m1}}}{\partial r}\right)=\phi_{\text{m1}}c_{\text{mg1}}\frac{p_{\text{m1}}M_{\text{g}}}{ZRT}\frac{\partial p_{\text{m1}}}{\partial t}+q_{\text{des1}} \tag{B.16}$$

The desorption flux is:

$$q_{\text{des1}}=\frac{M_{\text{g}}p_{\text{sc}}}{RT_{\text{sc}}}\frac{(1-\phi_{\text{f1}}-\phi_{\text{m1}})G_{\text{L}}m(p_{\text{L}})}{\left[m(p_{\text{L}})+m(p_{\text{f1}})\right]^2}\frac{\partial m(p_{\text{f1}})}{\partial t} \tag{B.17}$$

The continuity equation in the outer natural fracture system is:

$$\frac{1}{r}\frac{\partial}{\partial r}\left(\frac{k_{\text{f2}}}{\mu_{\text{g}}}\frac{p_{\text{f2}}M_{\text{g}}}{ZRT}r\frac{\partial p_{\text{f2}}}{\partial r}\right)=\phi_{\text{f2}}c_{\text{fg2}}\frac{p_{\text{f2}}M_{\text{g}}}{ZRT}\frac{\partial p_{\text{f2}}}{\partial t}-q_{\text{m2}} \tag{B.18}$$

The interporosity flux, q_{m2}, is:

$$q_{\text{m2}}=-\frac{3\rho_{\text{gm}}}{R_{\text{m2}}}\frac{k_{\text{m2}}}{\mu_{\text{g}}}\frac{\partial p_{\text{m2}}}{\partial r_{\text{m2}}}\bigg|_{r_{\text{m2}}=R_{\text{m2}}} \tag{B.19}$$

The continuity equation in the outer region in the matrix system is:

$$\frac{1}{r}\frac{\partial}{\partial r}\left(\frac{k_{\text{m2}}}{\mu_{\text{g}}}\frac{p_{\text{m2}}M_{\text{g}}}{ZRT}r\frac{\partial p_{\text{m2}}}{\partial r}\right)=\phi_{\text{m2}}c_{\text{mg2}}\frac{p_{\text{m2}}M_{\text{g}}}{ZRT}\frac{\partial p_{\text{m2}}}{\partial t}+q_{\text{des2}} \tag{B.20}$$

The desorption flux, q_{des2}, is:

$$q_{\mathrm{des2}}=\frac{M_{\mathrm{g}}p_{\mathrm{sc}}}{RT_{\mathrm{sc}}}\frac{(1-\phi_{\mathrm{f2}}-\phi_{\mathrm{m2}})G_{\mathrm{L}}m(p_{\mathrm{L}})}{[m(p_{\mathrm{L}})+m(p_{\mathrm{f2}})]^2}\frac{\partial m(p_{\mathrm{f2}})}{\partial t} \tag{B.21}$$

Define the adsorption/desorption compressibility in the inner and outer regions:

$$c_{\mathrm{d1}}=\frac{2Tp_{\mathrm{sc}}}{\phi_{\mathrm{m1}}\mu_{\mathrm{gi}}T_{\mathrm{sc}}}\frac{(1-\phi_{\mathrm{f1}}-\phi_{\mathrm{m1}})G_{\mathrm{L}}m(p_{\mathrm{L}})}{[m(p_{\mathrm{L}})+m(p_{\mathrm{m1}})]^2} \tag{B.22}$$

$$c_{\mathrm{d2}}=\frac{2Tp_{\mathrm{sc}}}{\phi_{\mathrm{m2}}\mu_{\mathrm{gi}}T_{\mathrm{sc}}}\frac{(1-\phi_{\mathrm{f2}}-\phi_{\mathrm{m2}})G_{\mathrm{L}}m(p_{\mathrm{L}})}{[m(p_{\mathrm{L}})+m(p_{\mathrm{m2}})]^2} \tag{B.23}$$

Introducing the pseudo pressure and adsorption/desorption compressibility into Eqs. (B.20) and (B.21), the continuity equations in the inner region are:

$$\frac{1}{r}\frac{\partial}{\partial r}\left(r\frac{\partial m_{\mathrm{f1}}}{\partial r}\right)=\frac{\phi_{\mathrm{f1}}\mu_{\mathrm{gi}}c_{\mathrm{fg1}}}{k_{\mathrm{f1}}}\frac{\partial m_{\mathrm{f1}}}{\partial t}+\frac{3}{R_{\mathrm{m1}}}\frac{k_{\mathrm{m1}}}{k_{\mathrm{f1}}}\left.\frac{\partial m_{\mathrm{m1}}}{\partial r_{\mathrm{m1}}}\right|_{r_{\mathrm{m1}}=R_{\mathrm{m1}}} \tag{B.24}$$

$$\frac{1}{r_{\mathrm{m1}}^2}\frac{\partial}{\partial r_{\mathrm{m1}}}\left(r_{\mathrm{m1}}^2\frac{\partial m_{\mathrm{m1}}}{\partial r_{\mathrm{m1}}}\right)=\frac{\phi_{\mathrm{m1}}\mu_{\mathrm{gi}}c_{\mathrm{mg1}}}{k_{\mathrm{m1}}}\frac{\partial m_{\mathrm{m1}}}{\partial t}+\frac{\phi_{\mathrm{m1}}\mu_{\mathrm{gi}}c_{\mathrm{d1}}}{k_{\mathrm{m1}}}\frac{\partial m_{\mathrm{m1}}}{\partial t} \tag{B.25}$$

In the outer region, they are:

$$\frac{1}{r}\frac{\partial}{\partial r}\left(r\frac{\partial m_{\mathrm{f2}}}{\partial r}\right)=\frac{\phi_{\mathrm{f2}}\mu_{\mathrm{gi}}c_{\mathrm{fg2}}}{k_{\mathrm{f2}}}\frac{\partial m_{\mathrm{f2}}}{\partial t}+\frac{3}{R_{\mathrm{m2}}}\frac{k_{\mathrm{m2}}}{k_{\mathrm{f2}}}\left.\frac{\partial m_{\mathrm{m2}}}{\partial r_{\mathrm{m1}}}\right|_{r_{\mathrm{m1}}=R_{\mathrm{m2}}} \tag{B.26}$$

$$\frac{1}{r_{\mathrm{m2}}^2}\frac{\partial}{\partial r_{\mathrm{m2}}}\left(r_{\mathrm{m2}}^2\frac{\partial m_{\mathrm{m2}}}{\partial r_{\mathrm{m2}}}\right)=\frac{\phi_{\mathrm{m2}}\mu_{\mathrm{gi}}c_{\mathrm{mg2}}}{k_{\mathrm{m2}}}\frac{\partial m_{\mathrm{m2}}}{\partial t}+\frac{\phi_{\mathrm{m2}}\mu_{\mathrm{gi}}c_{\mathrm{d2}}}{k_{\mathrm{m2}}}\frac{\partial m_{\mathrm{m2}}}{\partial t} \tag{B.27}$$

Define the following variables:

$$r_{\mathrm{mD1}}=\frac{r_{\mathrm{m1}}}{R_{\mathrm{m1}}},\quad r_{\mathrm{mD2}}=\frac{r_{\mathrm{m2}}}{R_{\mathrm{m2}}},\quad r_{\mathrm{D}}=\frac{r}{L_{\mathrm{ref}}},\quad t_{\mathrm{D}}=\frac{k_{\mathrm{f2}}t}{(\phi_{\mathrm{m}}c_{\mathrm{mgi}}+\phi_{\mathrm{f}}c_{\mathrm{fgi}})_{1+2}\mu_{\mathrm{gi}}L_{\mathrm{ref}}^2},$$

$$\omega_1=\frac{(\phi_{\mathrm{m}}c_{\mathrm{mgi}}+\phi_{\mathrm{f}}c_{\mathrm{fgi}})_1}{(\phi_{\mathrm{m}}c_{\mathrm{mgi}}+\phi_{\mathrm{f}}c_{\mathrm{fgi}})_{1+2}},\quad \omega_{\mathrm{f1}}=\frac{(\phi_{\mathrm{f}}c_{\mathrm{fgi}})_1}{(\phi_{\mathrm{m}}c_{\mathrm{mgi}}+\phi_{\mathrm{f}}c_{\mathrm{fgi}})_1},\quad \omega_2=\frac{(\phi_{m}c_{\mathrm{mgi}}+\phi_{\mathrm{f}}c_{\mathrm{fgi}})_2}{(\phi_{\mathrm{m}}c_{\mathrm{mgi}}+\phi_{\mathrm{f}}c_{\mathrm{fgi}})_{1+2}},$$

$$\omega_{\mathrm{f2}}=\frac{(\phi_{\mathrm{f}}c_{\mathrm{fgi}})_2}{(\phi_{\mathrm{m}}c_{\mathrm{mgi}}+\phi_{\mathrm{f}}c_{\mathrm{fgi}})_2},\quad \alpha_{\mathrm{d1}}=\frac{\phi_{\mathrm{m1}}c_{\mathrm{d1}}}{(\phi_{\mathrm{m}}c_{\mathrm{mgi}}+\phi_{\mathrm{f}}c_{\mathrm{fgi}})_1},\quad \alpha_{\mathrm{d2}}=\frac{\phi_{\mathrm{m2}}c_{\mathrm{d2}}}{(\phi_{\mathrm{m}}c_{\mathrm{mgi}}+\phi_{\mathrm{f}}c_{\mathrm{fgi}})_2},$$

$$\lambda_1=\alpha_1\frac{k_{\mathrm{m1}}}{k_{\mathrm{f1}}}L_{\mathrm{ref}}^2,\quad \lambda_2=\alpha_2\frac{k_{\mathrm{m2}}}{k_{\mathrm{f2}}}L_{\mathrm{ref}}^2,\quad M_{12}=\frac{k_{\mathrm{f1}}/\mu_{\mathrm{gi}}}{k_{\mathrm{f2}}/\mu_{\mathrm{gi}}},\quad \alpha_1=\frac{15}{R_{\mathrm{m1}}^2},\quad \alpha_2=\frac{15}{R_{\mathrm{m2}}^2}$$

Substituting the above dimensionless variables into Eqs. (B.24)–(B.27), the dimensionless equations in the inner region are:

$$\frac{1}{r_D}\frac{\partial}{\partial r_D}\left(r_D\frac{\partial \Delta m_{f1}}{\partial r_D}\right)=\frac{\omega_1\omega_{f1}}{M_{12}}\frac{\partial \Delta m_{f1}}{\partial t_D}+\frac{\lambda_1}{5}\frac{\partial \Delta m_{m1}}{\partial r_{mD1}}\bigg|_{r_{mD1}=1} \tag{B.28}$$

$$\frac{1}{r_{mD1}^2}\frac{\partial}{\partial r_{mD1}}\left(r_{mD1}^2\frac{\partial \Delta m_{m1}}{\partial r_{mD1}}\right)=\frac{15\omega_1(1-\omega_{f1})}{\lambda_1 M_{12}}\frac{\partial \Delta m_{m1}}{\partial t_D}+\frac{15\alpha_{d1}\omega_1}{\lambda_1 M_{12}}\frac{\partial \Delta m_{m1}}{\partial t_D} \tag{B.29}$$

In the outer region, they are:

$$\frac{1}{r_D}\frac{\partial}{\partial r_D}\left(r_D\frac{\partial \Delta m_{f2}}{\partial r_D}\right)=\omega_2\omega_{f2}\frac{\partial \Delta m_{f2}}{\partial t_D}+\frac{\lambda_2}{5}\frac{\partial \Delta m_{m2}}{\partial r_{mD1}}\bigg|_{r_{mD1}=1} \tag{B.30}$$

$$\frac{1}{r_{mD2}^2}\frac{\partial}{\partial r_{mD2}}\left(r_{mD2}^2\frac{\partial \Delta m_{m2}}{\partial r_{mD2}}\right)=\frac{15\omega_2(1-\omega_{f2})}{\lambda_2}\frac{\partial \Delta m_{m2}}{\partial t_D}+\frac{15\alpha_{d2}\omega_2}{\lambda_2}\frac{\partial \Delta m_{m2}}{\partial t_D} \tag{B.31}$$

By adding the corresponding inner and outer boundary conditions in the matrix system, which are the same as in the homogeneous models described in the previous sections, the solutions in the matrix system in the Laplace space can be obtained. Thereafter, through taking the derivation for them, the following equations can be obtained:

$$\frac{\partial \Delta \overline{m}_{m1}}{\partial r_{mD1}}\bigg|_{r_{mD1}=1}=\Delta \overline{m}_{f1}\left[\sqrt{g_1}\coth\left(\sqrt{g_1}\right)-1\right] \tag{B.32}$$

$$\frac{\partial \Delta \overline{m}_{m2}}{\partial r_{mD2}}\bigg|_{r_{mD2}=1}=\Delta \overline{m}_{f2}\left[\sqrt{g_2}\coth\left(\sqrt{g_2}\right)-1\right] \tag{B.33}$$

where:

$$g_1=\frac{15\omega_1(1-\omega_{f1}+\alpha_{d1})s}{\lambda_1 M_{12}} \tag{B.34}$$

$$g_2=\frac{15\omega_2(1-\omega_{f2}+\alpha_{d2})s}{\lambda_2} \tag{B.35}$$

Substituting Eqs. (B.32) and (B.35) into Eqs. (B.28) and (B.30), the dimensionless diffusivity equations in the inner and outer regions in the natural fracture system can be solved in the Laplace domain. In the inner fracture system, there is:

$$\frac{1}{r_D}\frac{\partial}{\partial r_D}\left(r_D\frac{\partial \Delta \overline{m}_{f1}}{\partial r_D}\right)=f_1(s)\Delta \overline{m}_{f1} \tag{B.36}$$

In the outer natural fracture system, there is:

$$\frac{1}{r_D}\frac{\partial}{\partial r_D}\left(r_D\frac{\partial \Delta \overline{m}_{f2}}{\partial r_D}\right)=f_2(s)\Delta \overline{m}_{f2} \tag{B.37}$$

where:

$$f_1(s) = \frac{\omega_1 \omega_{f1} s}{M_{12}} + \frac{\lambda_1}{5}\left[\sqrt{g_1}\coth\left(\sqrt{g_1}\right) - 1\right] \tag{B.38}$$

$$f_2(s) = \omega_2 \omega_{f2} s + \frac{\lambda_1}{5}\left[\sqrt{g_2}\coth\left(\sqrt{g_2}\right) - 1\right] \tag{B.39}$$

B.2.2 Pseudo-steady state matrix flow

For the pseudo-steady state interporosity flow model, the dimensionless diffusivity equation in the inner region can be obtained with the similar approach described before, which is:

$$\frac{1}{r_D}\frac{\partial}{\partial r_D}\left(r_D \frac{\partial \Delta m_{f1}}{\partial r_D}\right) = \frac{\omega_1 \omega_{f1}}{M_{12}}\frac{\partial \Delta m_{f1}}{\partial t_D} - \lambda_1[\Delta m_{m1} - \Delta m_{f1}] \tag{B.40}$$

$$-\lambda_1[\Delta m_{m1} - \Delta m_{f1}] = \frac{\omega_1(1-\omega_{f1})}{M_{12}}\frac{\partial \Delta m_{m1}}{\partial t_D} + \frac{\alpha_{d1}\omega_1}{M_{12}}\frac{\partial \Delta m_{m1}}{\partial t_D} \tag{B.41}$$

Similarly, the following equations in the outer region can also be obtained:

$$\frac{1}{r_D}\frac{\partial}{\partial r_D}\left(r_D \frac{\partial \Delta m_{f2}}{\partial r_D}\right) = \omega_2 \omega_{f2}\frac{\partial \Delta m_{f2}}{\partial t_D} - \lambda_2[\Delta m_{m2} - \Delta m_{f2}] \tag{B.42}$$

$$-\lambda_2[\Delta m_{m2} - \Delta m_{f2}] = \omega_2(1-\omega_{f2})\frac{\partial \Delta m_{m2}}{\partial t_D} + \alpha_{d2}\omega_2\frac{\partial \Delta m_{m2}}{\partial t_D} \tag{B.43}$$

The corresponding dimensionless variables used in the above equations are defined as follows:

$$r_D = \frac{r}{L_{ref}}, \quad t_D = \frac{k_{f2} t}{\left(\phi_m c_{mgi} + \phi_f c_{fgi}\right)_{1+2}\mu_{gi} L_{ref}^2}, \quad \omega_1 = \frac{\left(\phi_m c_{mgi} + \phi_f c_{fgi}\right)_1}{\left(\phi_m c_{mgi} + \phi_f c_{fgi}\right)_{1+2}},$$

$$\omega_2 = \frac{\left(\phi_m c_{mgi} + \phi_f c_{fgi}\right)_2}{\left(\phi_m c_{mgi} + \phi_f c_{fgi}\right)_{1+2}}, \quad \omega_{f1} = \frac{\left(\phi_f c_{fgi}\right)_1}{\left(\phi_m c_{mgi} + \phi_f c_{fgi}\right)_1}, \quad \lambda_1 = \alpha_1 \frac{k_{m1}}{k_{f1}} L_{ref}^2,$$

$$\lambda_2 = \alpha_2 \frac{k_{m2}}{k_{f2}} L_{ref}^2, \quad \omega_{f2} = \frac{\left(\phi_f c_{fgi}\right)_2}{\left(\phi_m c_{mgi} + \phi_f c_{fgi}\right)_2}, \quad \alpha_{d1} = \frac{\phi_{m1} c_{d1}}{\left(\phi_m c_{mgi} + \phi_f c_{fgi}\right)_1},$$

$$\alpha_{d2} = \frac{\phi_{m2} c_{d2}}{\left(\phi_m c_{mgi} + \phi_f c_{fgi}\right)_2}, \quad c_{d1} = \frac{2Tp_{sc}}{\phi_{m1}\mu_{gi} T_{sc}} \frac{(1-\phi_{f1}-\phi_{m1}) G_L m(p_L)}{[m(p_L) + m(p_{m1})]^2},$$

$$c_{d2} = \frac{2Tp_{sc}}{\phi_{m2}\mu_{gi} T_{sc}} \frac{(1-\phi_{f2}-\phi_{m2}) G_L m(p_L)}{[m(p_L) + m(p_{m2})]^2}, \quad M_{12} = \frac{k_{f1}/\mu_{gi}}{k_{f2}/\mu_{gi}}$$

Taking the Laplace transform with respect to dimensionless time from Eqs. (B.40) to (B.43) and then substituting the solution in the matrix system into the corresponding natural fracture system, the dimensionless equations in the inner and outer regions natural fractures system can be simplified to the following expressions:

$$\frac{1}{r_D}\frac{\partial}{\partial r_D}\left(r_D\frac{\partial \Delta \overline{m}_{f1}}{\partial r_D}\right)=f_1(s)\Delta \overline{m}_{f1} \tag{B.44}$$

$$\frac{1}{r_D}\frac{\partial}{\partial r_D}\left(r_D\frac{\partial \Delta \overline{m}_{f2}}{\partial r_D}\right)=f_2(s)\Delta \overline{m}_{f2} \tag{B.45}$$

where:

$$f_1(s)=\left[\frac{\omega_{f1}}{M_{12}}+\frac{(1-\omega_{f1}+\alpha_{d1})\lambda_1}{\lambda_1 M_{12}+\omega_1(1-\omega_{f1}+\alpha_{d1})s}\right]\omega_1 s \tag{B.46}$$

$$f_2(s)=\left[\omega_{f2}+\frac{(1-\omega_{f2}+\alpha_{d2})\lambda_2}{\lambda_2+\omega_2(1-\omega_{f2}+\alpha_{d2})s}\right]\omega_2 s \tag{B.47}$$

B.3 Microfractures + gas adsorption/desorption + matrix Fick's diffusion (model 3)

When the gas diffusion follows Fick's diffusion, with a similar approach as before, the diffusivity equations in the inner region are:

$$\frac{1}{r}\frac{\partial}{\partial r}\left(r\frac{k_{f1}p_{f1}M_g}{\mu_g ZRT}\frac{\partial p_{f1}}{\partial r}\right)=\phi_{f1}c_{fg1}\frac{p_{f1}M_g}{ZRT}\frac{\partial p_{f1}}{\partial t}+q_{F1} \tag{B.48}$$

$$q_{F1}=M_g(1-\phi_{f1})\frac{dC_{m1}}{dt} \tag{B.49}$$

In the outer natural fracture system, it is:

$$\frac{1}{r}\frac{\partial}{\partial r}\left(r\frac{k_{f2}p_{f2}M_g}{\mu_g ZRT}\frac{\partial p_{f2}}{\partial r}\right)=\phi_{f2}c_{fg2}\frac{p_{f2}M_g}{ZRT}\frac{\partial p_{f2}}{\partial t}+q_{F2} \tag{B.50}$$

$$q_{F2}=M_g(1-\phi_{f2})\frac{dC_{m2}}{dt} \tag{B.51}$$

According to Fick's diffusion theory, there are two types of diffusion equations, which are:

$$\frac{dC_m}{dt}=\begin{cases}\left.\dfrac{3D_F}{R_m}\dfrac{\partial C_m}{\partial r_m}\right|_{r_m=R_m} & \text{unsteady state diffusion}\\[2ex] \dfrac{6D_F\pi^2}{R_m^2}[C_E(p_f)-C_m] & \text{pseudo-steady state diffusion}\end{cases} \tag{B.52}$$

Substituting Eqs. (B.49) and (B.51) into Eqs. (A.66) and (B.50) and introducing the pseudo pressure, the governing equations in the natural fractures become:

$$\frac{1}{r}\frac{\partial}{\partial r}\left(r\frac{\partial m_{\mathrm{f1}}}{\partial r}\right)=\frac{\phi_{\mathrm{f1}}\mu_{\mathrm{gi}}c_{\mathrm{fg1}}}{k_{\mathrm{f1}}}\frac{\partial m_{\mathrm{f1}}}{\partial t}+\frac{2RT}{k_{\mathrm{f1}}}(1-\phi_{\mathrm{f1}})\frac{\mathrm{d}C_{\mathrm{m1}}}{\mathrm{d}t} \quad \text{(B.53)}$$

$$\frac{1}{r}\frac{\partial}{\partial r}\left(r\frac{\partial m_{\mathrm{f2}}}{\partial r}\right)=\frac{\phi_{\mathrm{f2}}\mu_{\mathrm{gi}}c_{\mathrm{fg2}}}{k_{\mathrm{f2}}}\frac{\partial m_{\mathrm{f2}}}{\partial t}+\frac{2RT}{k_{\mathrm{f2}}}(1-\phi_{\mathrm{f2}})\frac{\mathrm{d}C_{\mathrm{m2}}}{\mathrm{d}t} \quad \text{(B.54)}$$

Define the following dimensionless variables:

$$t_{\mathrm{D}}=\frac{k_{\mathrm{f2}}t}{\Lambda L_{\mathrm{ref}}^2},\ \omega_1=\frac{\left(\phi_1\mu_{\mathrm{gi}}c_{\mathrm{g1}}\right)_{\mathrm{f}}}{\Lambda},\ \omega_2=\frac{\left(\phi_2\mu_{\mathrm{gi}}c_{\mathrm{g2}}\right)_{\mathrm{f}}}{\Lambda},\ M_{12}=\frac{k_{\mathrm{f1}}/\mu_{\mathrm{gi}}}{k_{\mathrm{f2}}/\mu_{\mathrm{gi}}},\ r_{\mathrm{Dm}}=\frac{r_{\mathrm{m}}}{R_{\mathrm{m}}},\ r_{\mathrm{D}}=\frac{r}{L_{\mathrm{ref}}},$$

$$\Lambda=\begin{cases}\left(\phi_{\mathrm{f}}\mu_{\mathrm{gi}}c_{\mathrm{fg}}\right)_{1+2}+\dfrac{6k_{\mathrm{f2}}h}{q_{\mathrm{sc}}} & \text{unsteady state diffusion}\\ \left(\phi_{\mathrm{f}}\mu_{\mathrm{gi}}c_{\mathrm{fg}}\right)_{1+2}+\dfrac{2k_{\mathrm{f2}}h}{q_{\mathrm{sc}}} & \text{pseudo-steady state diffusion}\end{cases},$$

$$\lambda=\begin{cases}\dfrac{k_{\mathrm{f2}}\tau}{\Lambda L_{\mathrm{ref}}^2} & \text{unsteady state diffusion}\\ \dfrac{k_{\mathrm{f2}}\tau}{6\Lambda L_{\mathrm{ref}}^2} & \text{pseudo-steady state diffusion}\end{cases},$$

$$\tau=\begin{cases}\dfrac{R_{\mathrm{m}}^2}{D_{\mathrm{F}}} & \text{unsteady state diffusion}\\ \dfrac{R_{\mathrm{m}}^2}{\pi^2 D_{\mathrm{F}}} & \text{pseudo-steady state diffusion}\end{cases}$$

B.3.1 *Unsteady state diffusion*

For unsteady state diffusion, inserting the expression of the diffusion model in Eq. (A.68) into the diffusivity equations Eqs. (B.53) and (B.54), we have:

$$\frac{1}{r_{\mathrm{D}}}\frac{\partial}{\partial r_{\mathrm{D}}}\left(r_{\mathrm{D}}\frac{\partial\Delta m_{\mathrm{f1}}}{\partial r_{\mathrm{D}}}\right)=\frac{\omega_1}{M_{12}}\frac{\partial\Delta m_{\mathrm{f1}}}{\partial t_{\mathrm{D}}}-\frac{M_{\mathrm{g}}q_{\mathrm{sc}}p_{\mathrm{sc}}T}{\rho_{\mathrm{sc}}k_{\mathrm{f2}}hT_{\mathrm{sc}}}\frac{(1-\omega_1-\omega_2)(1-\phi_{\mathrm{f1}})}{\lambda M_{12}}\left.\frac{\partial C_{\mathrm{m1}}}{\partial r_{\mathrm{mD}}}\right|_{r_{\mathrm{mD}}=1} \quad \text{(B.55)}$$

$$\frac{1}{r_{\mathrm{D}}}\frac{\partial}{\partial r_{\mathrm{D}}}\left(r_{\mathrm{D}}\frac{\partial\Delta m_{\mathrm{f2}}}{\partial r_{\mathrm{D}}}\right)=\omega_2\frac{\partial\Delta m_{\mathrm{f2}}}{\partial t_{\mathrm{D}}}-\frac{M_{\mathrm{g}}q_{\mathrm{sc}}p_{\mathrm{sc}}T}{\rho_{\mathrm{sc}}k_{\mathrm{f2}}hT_{\mathrm{sc}}}\frac{(1-\omega_1-\omega_2)(1-\phi_{\mathrm{f2}})}{\lambda}\left.\frac{\partial C_{\mathrm{m2}}}{\partial r_{\mathrm{mD}}}\right|_{r_{\mathrm{mD}}=1} \quad \text{(B.56)}$$

Introducing the following dimensionless concentration into Eqs. (B.55) and (B.56):

$$C_{\mathrm{mD}}=C_{\mathrm{m}}(p_{\mathrm{m}})-C_{\mathrm{m}}(p_{\mathrm{i}}),C_{\mathrm{ED}}=C_{\mathrm{E}}(p_{\mathrm{f}})-C_{\mathrm{m}}(p_{\mathrm{i}}) \quad \text{(B.57)}$$

and then taking the Laplace transformation for the inner and outer natural fracture systems, the following equations can be obtained:

$$\frac{1}{r_D}\frac{\partial}{\partial r_D}\left(r_D\frac{\partial \Delta\overline{m}_{f1}}{\partial r_D}\right)=\frac{\omega_1 s}{M_{12}}\Delta\overline{m}_{f1}-\frac{M_g}{\rho_{sc}}\frac{q_{sc}p_{sc}T}{k_{f2}hT_{sc}}\frac{\beta_1}{\lambda M_{12}}\frac{\partial \overline{C}_{mD1}}{\partial r_{mD}}\bigg|_{r_{mD}=1} \tag{B.58}$$

$$\frac{1}{r_D}\frac{\partial}{\partial r_D}\left(r_D\frac{\partial \Delta\overline{m}_{f2}}{\partial r_D}\right)=\omega_2 s\Delta\overline{m}_{f2}-\frac{M_g}{\rho_{sc}}\frac{q_{sc}p_{sc}T}{k_{f2}hT_{sc}}\frac{\beta_2}{\lambda}\frac{\partial \overline{C}_{mD2}}{\partial r_{mD}}\bigg|_{r_{mD}=1} \tag{B.59}$$

where: $\beta_1=(1-\omega_1-\omega_2)(1-\phi_{f1})$ and $\beta_2=(1-\omega_1-\omega_2)(1-\phi_{f2})$.

According to the above results, the following formulas can be obtained:

$$\frac{\partial \overline{C}_{mD1}}{\partial r_{mD}}\bigg|_{r_{mD}=1}=-\frac{\rho_{sc}}{M_g}\frac{k_{f2}hT_{sc}}{q_{sc}p_{sc}T}\sigma\left[\sqrt{\lambda s}\coth(\lambda s)-1\right]\Delta\overline{m}_{f1} \tag{B.60}$$

$$\frac{\partial \overline{C}_{mD2}}{\partial r_{mD}}\bigg|_{r_{mD}=1}=-\frac{\rho_{sc}}{M_g}\frac{k_{f2}hT_{sc}}{q_{sc}p_{sc}T}\sigma\left[\sqrt{\lambda s}\coth(\lambda s)-1\right]\Delta\overline{m}_{f2} \tag{B.61}$$

Substituting Eqs. (A.94) and (B.61) into Eqs. (B.58) and (B.59), respectively, we have:

$$\frac{1}{r_D}\frac{\partial}{\partial r_D}\left(r_D\frac{\partial \Delta\overline{m}_{f1}}{\partial r_D}\right)=f_1(s)\Delta\overline{m}_{f1} \tag{B.62}$$

$$\frac{1}{r_D}\frac{\partial}{\partial r_D}\left(r_D\frac{\partial \Delta\overline{m}_{f2}}{\partial r_D}\right)=f_2(s)\Delta\overline{m}_{f2} \tag{B.63}$$

where:

$$f_1(s)=\frac{\omega_1 s}{M_{12}}+\frac{\beta_1\sigma}{\lambda M_{12}}\left[\sqrt{\lambda s}\coth\left(\sqrt{\lambda s}\right)-1\right] \tag{B.64}$$

$$f_2(s)=\omega_2 s+\frac{\beta_2\sigma}{\lambda}\left[\sqrt{\lambda s}\coth\left(\sqrt{\lambda s}\right)-1\right] \tag{B.65}$$

$$\sigma=\frac{G_L m(p_L)}{[m(p_L)+m(p_f)][m(p_L)+m(p_i)]}\frac{q_{sc}p_{sc}T}{k_{f2}hT_{sc}} \tag{B.66}$$

B.3.2 Pseudo-steady state diffusion

For pseudo-steady state diffusion, the following diffusivity equations can be derived by substituting Eq. (A.68) into Eqs. (A.66) and (B.50):

$$\frac{1}{r_D}\frac{\partial}{\partial r_D}\left(r_D\frac{\partial \Delta m_{f1}}{\partial r_D}\right)=\frac{\omega_1}{M_{12}}\frac{\partial \Delta m_{f1}}{\partial t_D}-\frac{M_g}{\rho_{sc}}\frac{q_{sc}p_{sc}T}{k_{f2}hT_{sc}}\frac{(1-\omega_1-\omega_2)(1-\phi_{f1})}{\lambda M_{12}}[C_E(p_{f1})-C_{m1}] \tag{B.67}$$

$$\frac{1}{r_{\mathrm{D}}}\frac{\partial}{\partial r_{\mathrm{D}}}\left(r_{\mathrm{D}}\frac{\partial \Delta m_{\mathrm{f2}}}{\partial r_{\mathrm{D}}}\right)=\omega_2\frac{\partial \Delta m_{\mathrm{f2}}}{\partial t_{\mathrm{D}}}-\frac{M_{\mathrm{g}}\,q_{\mathrm{sc}}p_{\mathrm{sc}}T}{\rho_{\mathrm{sc}}\,k_{\mathrm{f2}}hT_{\mathrm{sc}}}\frac{(1-\omega_1-\omega_2)(1-\phi_{\mathrm{f2}})}{\lambda}\left[C_{\mathrm{E}}(p_{\mathrm{f2}})-C_{\mathrm{m2}}\right] \tag{B.68}$$

According to the pseudo-steady diffusion equation, the following equations can be obtained:

$$\overline{C}_{\mathrm{ED}}-\overline{C}_{\mathrm{mD1}}=-\frac{M_{\mathrm{g}}\,k_{\mathrm{f2}}hT_{\mathrm{sc}}}{\rho_{\mathrm{sc}}\,q_{\mathrm{sc}}p_{\mathrm{sc}}T}\frac{\sigma\lambda s}{\lambda s+1}\Delta\overline{m}_{\mathrm{f1}} \tag{B.69}$$

$$\overline{C}_{\mathrm{ED}}-\overline{C}_{\mathrm{mD2}}=-\frac{M_{\mathrm{g}}\,k_{\mathrm{f2}}hT_{\mathrm{sc}}}{\rho_{\mathrm{sc}}\,q_{\mathrm{sc}}p_{\mathrm{sc}}T}\frac{\sigma\lambda s}{\lambda s+1}\Delta\overline{m}_{\mathrm{f2}} \tag{B.70}$$

With a similar approach as described above, the following equations can be obtained:

$$\frac{1}{r_{\mathrm{D}}}\frac{\partial}{\partial r_{\mathrm{D}}}\left(r_{\mathrm{D}}\frac{\partial \Delta\overline{m}_{\mathrm{f1}}}{\partial r_{\mathrm{D}}}\right)=f_1(s)\Delta\overline{m}_{\mathrm{f1}} \tag{B.71}$$

$$\frac{1}{r_{\mathrm{D}}}\frac{\partial}{\partial r_{\mathrm{D}}}\left(r_{\mathrm{D}}\frac{\partial \Delta\overline{m}_{\mathrm{f2}}}{\partial r_{\mathrm{D}}}\right)=f_2(s)\Delta\overline{m}_{\mathrm{f2}} \tag{B.72}$$

where:

$$f_1(\mathrm{s})=\frac{\omega_1 s}{M_{12}}+\frac{\beta_1}{M_{12}}\frac{\sigma s}{\lambda s+1} \tag{B.73}$$

$$f_2(\mathrm{s})=\omega_2 s+\beta_2\frac{\sigma s}{\lambda s+1} \tag{B.74}$$

B.4 Microfractures + matrix macropores + gas adsorption/desorption + nanopore Fick's diffusion (model 4)

B.4.1 Transient interporosity flow

When the gas flow from macropores into the natural fracture system follows the transient interporosity, the inner diffusivity equations are:

$$\frac{1}{r}\frac{\partial}{\partial r}\left(r\frac{k_{\mathrm{f1}}p_{\mathrm{f1}}M_{\mathrm{g}}}{\mu_{\mathrm{g}}\,ZRT}\frac{\partial p_{\mathrm{f1}}}{\partial r}\right)=\phi_{\mathrm{f1}}c_{\mathrm{fg1}}\frac{p_{\mathrm{f1}}M_{\mathrm{g}}}{ZRT}\frac{\partial p_{\mathrm{f1}}}{\partial t}-q_{\mathrm{m1}} \tag{B.75}$$

$$\frac{1}{r_{\mathrm{m}}^2}\frac{\partial}{\partial r_{\mathrm{m}}}\left(r_{\mathrm{m}}^2\frac{\partial m_{\mathrm{m1}}}{\partial r_{\mathrm{m}}}\right)=\frac{\phi_{\mathrm{m1}}\mu_{\mathrm{gi}}c_{\mathrm{mg1}}}{k_{\mathrm{m1}}}\frac{\partial m_{\mathrm{m1}}}{\partial t}+\frac{1}{k_{\mathrm{m1}}}\frac{2RT}{M_{\mathrm{g}}}q_{\mathrm{F}} \tag{B.76}$$

$$q_{\mathrm{m1}}=-\frac{3\rho_{\mathrm{gm}}\,k_{\mathrm{m1}}}{R_{\mathrm{m1}}\,\mu_{\mathrm{g}}}\frac{\partial p_{\mathrm{m1}}}{\partial r_{\mathrm{m1}}}\bigg|_{r_{\mathrm{m1}}=R_{\mathrm{m1}}} \tag{B.77}$$

$$q_{\mathrm{F}}=M_{\mathrm{g}}(1-\phi_{\mathrm{f1}}-\phi_{\mathrm{m1}})\frac{\mathrm{d}C_{\mathrm{m}}}{\mathrm{d}t} \tag{B.78}$$

In the outer region, they are:

$$\frac{1}{r}\frac{\partial}{\partial r}\left(r\frac{k_{\mathrm{f2}}p_{\mathrm{f2}}M_{\mathrm{g}}}{\mu_{\mathrm{g}}ZRT}\frac{\partial p_{\mathrm{f2}}}{\partial r}\right)=\phi_{\mathrm{f2}}c_{\mathrm{fg2}}\frac{p_{\mathrm{f2}}M_{\mathrm{g}}}{ZRT}\frac{\partial p_{\mathrm{f2}}}{\partial t}-q_{\mathrm{m}} \tag{B.79}$$

$$\frac{1}{r_{\mathrm{m}}^{2}}\frac{\partial}{\partial r_{\mathrm{m}}}\left(r_{\mathrm{m}}^{2}\frac{\partial m_{\mathrm{m2}}}{\partial r_{\mathrm{m}}}\right)=\frac{\phi_{\mathrm{m2}}\mu_{\mathrm{gi}}c_{\mathrm{mg2}}}{k_{\mathrm{m2}}}\frac{\partial m_{\mathrm{m2}}}{\partial t}+\frac{1}{k_{\mathrm{m2}}}\frac{2RT}{M_{\mathrm{g}}}q_{\mathrm{F}} \tag{B.80}$$

$$q_{\mathrm{m2}}=-\frac{3\rho_{\mathrm{gm}}}{R_{\mathrm{m2}}}\frac{k_{\mathrm{m2}}}{\mu_{\mathrm{g}}}\frac{\partial p_{\mathrm{m2}}}{\partial r_{\mathrm{m2}}}\bigg|_{r_{\mathrm{m2}}=R_{\mathrm{m2}}} \tag{B.81}$$

$$q_{\mathrm{F}}=M_{\mathrm{g}}(1-\phi_{\mathrm{f2}}-\phi_{\mathrm{m2}})\frac{\mathrm{d}C_{\mathrm{m}}}{\mathrm{d}t} \tag{B.82}$$

By assembling the above equations, the following expressions can be obtained:

$$\frac{1}{r}\frac{\partial}{\partial r}\left(r\frac{\partial m_{\mathrm{f1}}}{\partial r}\right)=\frac{\phi_{\mathrm{f1}}\mu_{\mathrm{gi}}c_{\mathrm{fg1}}}{k_{\mathrm{f1}}}\frac{\partial m_{\mathrm{f1}}}{\partial t}+\frac{k_{\mathrm{m1}}}{k_{\mathrm{f1}}}\frac{3}{R_{\mathrm{m1}}}\frac{\partial m_{\mathrm{m1}}}{\partial r_{\mathrm{m2}}}\bigg|_{r_{\mathrm{m2}}=R_{\mathrm{m1}}} \tag{B.83}$$

$$\frac{1}{r_{\mathrm{m}}^{2}}\frac{\partial}{\partial r_{\mathrm{m}}}\left(r_{\mathrm{m}}^{2}\frac{\partial m_{\mathrm{m1}}}{\partial r_{\mathrm{m}}}\right)=\frac{\phi_{\mathrm{m1}}\mu_{\mathrm{gi}}c_{\mathrm{mg1}}}{k_{\mathrm{m1}}}\frac{\partial m_{\mathrm{m1}}}{\partial t}+\frac{2RT}{k_{\mathrm{m1}}}(1-\phi_{\mathrm{f1}}-\phi_{\mathrm{m1}})\frac{\mathrm{d}C_{\mathrm{m}}}{\mathrm{d}t} \tag{B.84}$$

In the outer region, there are:

$$\frac{1}{r}\frac{\partial}{\partial r}\left(r\frac{\partial m_{\mathrm{f2}}}{\partial r}\right)=\frac{\phi_{\mathrm{f2}}\mu_{\mathrm{gi}}c_{\mathrm{fg2}}}{k_{\mathrm{f2}}}\frac{\partial m_{\mathrm{f2}}}{\partial t}+\frac{3}{R_{\mathrm{m2}}}\frac{k_{\mathrm{m2}}}{k_{\mathrm{f2}}}\frac{\partial m_{\mathrm{m2}}}{\partial r_{\mathrm{m2}}}\bigg|_{r_{\mathrm{m2}}=R_{\mathrm{m2}}} \tag{B.85}$$

$$\frac{1}{r_{\mathrm{m}}^{2}}\frac{\partial}{\partial r_{\mathrm{m}}}\left(r_{\mathrm{m}}^{2}\frac{\partial m_{\mathrm{m2}}}{\partial r_{\mathrm{m}}}\right)=\frac{\phi_{\mathrm{m2}}\mu_{\mathrm{gi}}c_{\mathrm{mg2}}}{k_{\mathrm{m2}}}\frac{\partial m_{\mathrm{m2}}}{\partial t}+\frac{2RT}{k_{\mathrm{m2}}}(1-\phi_{\mathrm{f2}}-\phi_{\mathrm{m2}})\frac{\mathrm{d}C_{\mathrm{m}}}{\mathrm{d}t} \tag{B.86}$$

According to Fick's diffusion theory, we have:

$$\frac{\mathrm{d}C_{\mathrm{m}}}{\mathrm{d}t}=\begin{cases}\dfrac{3D_{\mathrm{F}}}{R_{\mathrm{m}}}\dfrac{\partial C_{\mathrm{m}}}{\partial r_{\mathrm{m}}}\bigg|_{r_{\mathrm{m}}=R_{\mathrm{m}}} & \text{unsteady state diffusion}\\[2ex] \dfrac{6D_{\mathrm{F}}\pi^{2}}{R_{\mathrm{m}}^{2}}[C_{\mathrm{E}}(p_{\mathrm{f}})-C_{\mathrm{m}}] & \text{pseudo-steady state diffusion}\end{cases} \tag{B.87}$$

In a shale gas reservoir, an induced fracture network just changes the properties of the inner fracture system but has no effect on the matrix system. So the assumptions can be made: $k_{\mathrm{m}}=k_{\mathrm{m1}}=k_{\mathrm{m2}}$ and $R_{\mathrm{mac}}=R_{\mathrm{m1}}=R_{\mathrm{m2}}$.

Define the following variables:

$$t_D=\frac{k_{f2}t}{\Lambda L_{ref}^2},\quad M_{12}=\frac{k_{f1}/\mu_{gi}}{k_{f2}/\mu_{gi}},\quad \alpha=\frac{15}{R_{mac}^2},\quad r_{m1D}=\frac{r_{m1}}{R_{mac}},\quad r_{m2D}=\frac{r_{m2}}{R_{mac}},\quad r_D=\frac{r}{L_{ref}},$$

$$\omega_{f1}=\frac{\left(\phi_f\mu_{gi}c_{fg}\right)_1}{\Lambda},\quad \omega_{f2}=\frac{\left(\phi_f\mu_{gi}c_{fg}\right)_2}{\Lambda},\quad \omega_{m1}=\frac{\left(\phi_m\mu_{gi}c_{mg}\right)_1}{\Lambda},\quad \omega_{m2}=\frac{\left(\phi_m\mu_{gi}c_{mg}\right)_2}{\Lambda},$$

$$\lambda_1=\alpha\frac{k_m}{k_{f1}}L_{ref}^2,\quad \lambda_2=\alpha\frac{k_m}{k_{f2}}L_{ref}^2,$$

$$\theta_{fm}=\frac{k_{f2}}{k_m},$$

$$\Lambda=\begin{cases}\left(\phi_f\mu_{gi}c_{fg}+\phi_m\mu_{gi}c_{mg}\right)_{1+2}+\dfrac{6k_{f2}hR_{mac}^2}{q_{sc}\,L_{ref}^2} & \text{for unsteady state diffusion}\\ \left(\phi_f\mu_{gi}c_{fg}+\phi_m\mu_{gi}c_{mg}\right)_{1+2}+\dfrac{2k_{f2}hR_{mac}^2}{q_{sc}\,L_{ref}^2} & \text{for pseudo-steady state diffusion}\end{cases},$$

$$\lambda=\begin{cases}\dfrac{k_m\tau}{\Lambda L_{ref}^2} & \text{for unsteady state diffusion}\\ \dfrac{k_m\tau}{6\Lambda L_{ref}^2} & \text{for pseudo-steady state diffusion}\end{cases},\quad \tau=\begin{cases}\dfrac{R_m^2}{D_F} & \text{for unsteady state diffusion}\\ \dfrac{R_m^2}{\pi^2 D_F} & \text{for pseudo-steady state diffusion}\end{cases}$$

Unsteady state diffusion. Substituting the above variables into the mathematical model, the continuity equations in the inner region are:

$$\frac{1}{r_D}\frac{\partial}{\partial r_D}\left(r_D\frac{\partial m_{f1}}{\partial r_D}\right)=\frac{\omega_{f1}}{M_{12}}\frac{\partial m_{f1}}{\partial t_D}+\frac{\lambda_1}{5}\frac{\partial m_{m1}}{\partial r_{m1D}}\bigg|_{r_{m1D}=1} \tag{B.88}$$

$$\frac{1}{r_{m1D}^2}\frac{\partial}{\partial r_{m1D}}\left(r_{m1D}^2\frac{\partial m_{m1}}{\partial r_{m1D}}\right)=\frac{15\omega_{m1}}{\lambda_1 M_{12}}\frac{\partial m_{m1}}{\partial t_D}+\frac{2RTR_{mac}^2}{k_m}\left(1-\phi_{f1}-\phi_{m1}\right)\frac{dC_m}{dt} \tag{B.89}$$

$$\frac{2RTR_{mac}^2}{k_m}\frac{dC_m}{dt}=\frac{M_g}{\rho_{sc}}\frac{q_{sc}p_{sc}T}{k_{f2}hT_{sc}}\frac{\left(1-\omega_{f1}-\omega_{m1}\right)}{\lambda}\frac{\partial C_m}{\partial r_{m1D}}\bigg|_{r_{m1D}=1} \tag{B.90}$$

In the outer region, they are:

$$\frac{1}{r_D}\frac{\partial}{\partial r_D}\left(r_D\frac{\partial m_{f2}}{\partial r_D}\right)=\omega_{f2}\frac{\partial m_{f2}}{\partial t_D}+\frac{\lambda_2}{5}\frac{\partial m_{m2}}{\partial r_{mD}}\bigg|_{r_{mD}=1} \tag{B.91}$$

$$\frac{1}{r_{\mathrm{m2D}}^2}\frac{\partial}{\partial r_{\mathrm{m2D}}}\left(r_{\mathrm{m2D}}^2\frac{\partial m_{\mathrm{m2}}}{\partial r_{\mathrm{m2D}}}\right)=\frac{15\omega_{\mathrm{m2}}}{\lambda_2}\frac{\partial m_{\mathrm{m2}}}{\partial t_{\mathrm{D}}}+\frac{2RTR_{\mathrm{mac}}^2}{k_{\mathrm{m}}}(1-\phi_{\mathrm{f2}}-\phi_{\mathrm{m2}})\frac{\mathrm{d}C_{\mathrm{m}}}{\mathrm{d}t} \tag{B.92}$$

$$\frac{2RTR_{\mathrm{mac}}^2}{k_{\mathrm{m}}}\frac{\mathrm{d}C_{\mathrm{m}}}{\mathrm{d}t}=\frac{M_{\mathrm{g}}q_{\mathrm{sc}}p_{\mathrm{sc}}T}{\rho_{\mathrm{sc}}k_{\mathrm{f2}}hT_{\mathrm{sc}}}\frac{(1-\omega_{\mathrm{f2}}-\omega_{\mathrm{m2}})}{\lambda}\left.\frac{\partial C_{\mathrm{m}}}{\partial r_{\mathrm{m2D}}}\right|_{r_{\mathrm{m2D}}=1} \tag{B.93}$$

Substituting Eqs. (B.90) and (B.93) into Eqs. (B.89) and (B.92), the continuity equations in the inner region are:

$$\frac{1}{r_{\mathrm{D}}}\frac{\partial}{\partial r_{\mathrm{D}}}\left(r_{\mathrm{D}}\frac{\partial \Delta m_{\mathrm{f1}}}{\partial r_{\mathrm{D}}}\right)=\frac{\omega_{\mathrm{f1}}}{M_{12}}\frac{\partial \Delta m_{\mathrm{f1}}}{\partial t_{\mathrm{D}}}+\frac{\lambda_1}{5}\left.\frac{\partial \Delta m_{\mathrm{m1}}}{\partial r_{\mathrm{mD}}}\right|_{r_{\mathrm{mD}}=1} \tag{B.94}$$

$$\frac{1}{r_{\mathrm{m1D}}^2}\frac{\partial}{\partial r_{\mathrm{m1D}}}\left(r_{\mathrm{m1D}}^2\frac{\partial \Delta m_{\mathrm{m1}}}{\partial r_{m\mathrm{1D}}}\right)=\frac{15\omega_{\mathrm{m1}}}{\lambda_1 M_{12}}\frac{\partial \Delta m_{\mathrm{m1}}}{\partial t_{\mathrm{D}}}-\frac{M_{\mathrm{g}}q_{\mathrm{sc}}p_{\mathrm{sc}}T\beta_1}{\rho_{\mathrm{sc}}k_{\mathrm{f2}}hT_{\mathrm{sc}}}\frac{1}{\lambda}\left.\frac{\partial C_{\mathrm{m}}}{\partial r_{\mathrm{m1D}}}\right|_{r_{\mathrm{m1D}}=1} \tag{B.95}$$

where: $\beta_1=(1-\phi_{\mathrm{f1}}-\phi_{\mathrm{m1}})(1-\omega_{\mathrm{f1}}-\omega_{\mathrm{m1}})$. In the outer region, they are:

$$\frac{1}{r_{\mathrm{D}}}\frac{\partial}{\partial r_{\mathrm{D}}}\left(r_{\mathrm{D}}\frac{\partial \Delta m_{\mathrm{f2}}}{\partial r_{\mathrm{D}}}\right)=\omega_{\mathrm{f2}}\frac{\partial \Delta m_{\mathrm{f2}}}{\partial t_{\mathrm{D}}}+\frac{\lambda_2}{5}\left.\frac{\partial \Delta m_{\mathrm{m2}}}{\partial r_{\mathrm{mD}}}\right|_{r_{\mathrm{mD}}=1} \tag{B.96}$$

$$\frac{1}{r_{\mathrm{m2D}}^2}\frac{\partial}{\partial r_{\mathrm{m2D}}}\left(r_{\mathrm{m2D}}^2\frac{\partial \Delta m_{\mathrm{m2}}}{\partial r_{\mathrm{m2D}}}\right)=\frac{15\omega_{\mathrm{m2}}}{\lambda_2}\frac{\partial \Delta m_{\mathrm{m2}}}{\partial t_{\mathrm{D}}}-\frac{M_{\mathrm{g}}q_{\mathrm{sc}}p_{\mathrm{sc}}T\beta_2}{\rho_{\mathrm{sc}}k_{\mathrm{f2}}hT_{\mathrm{sc}}}\frac{1}{\lambda}\left.\frac{\partial C_{\mathrm{m}}}{\partial r_{\mathrm{m2D}}}\right|_{r_{\mathrm{m2D}}=1} \tag{B.97}$$

where: $\beta_2=(1-\phi_{\mathrm{f2}}-\phi_{\mathrm{m2}})(1-\omega_{\mathrm{f2}}-\omega_{\mathrm{m2}})$.

According to Fick's diffusion model described before, the following relationships can be obtained:

$$\left.\frac{\partial \overline{C}_{\mathrm{mD1}}}{\partial r_{\mathrm{mD}}}\right|_{r_{\mathrm{mD}}=1}=-\frac{\rho_{\mathrm{sc}}k_{\mathrm{f2}}hT_{\mathrm{sc}}}{M_{\mathrm{g}}q_{\mathrm{sc}}p_{\mathrm{sc}}T}\sigma\left[\sqrt{\lambda\theta_{\mathrm{fm}}s}\coth\left(\sqrt{\lambda\theta_{\mathrm{fm}}s}\right)-1\right]\Delta\overline{m}_{\mathrm{m1}} \tag{B.98}$$

$$\left.\frac{\partial \overline{C}_{\mathrm{mD2}}}{\partial r_{\mathrm{mD}}}\right|_{r_{\mathrm{mD}}=1}=-\frac{\rho_{\mathrm{sc}}k_{\mathrm{f2}}hT_{\mathrm{sc}}}{M_{\mathrm{g}}q_{\mathrm{sc}}p_{\mathrm{sc}}T}\sigma\left[\sqrt{\lambda\theta_{\mathrm{fm}}s}\coth\left(\sqrt{\lambda\theta_{\mathrm{fm}}s}\right)-1\right]\Delta\overline{m}_{\mathrm{m2}} \tag{B.99}$$

Substituting Eqs. (B.98) and (B.99) into Eqs. (B.95) and (B.97), respectively, we have:

$$\frac{1}{r_{\mathrm{m1D}}^2}\frac{\partial}{\partial r_{\mathrm{m1D}}}\left(r_{\mathrm{m1D}}^2\frac{\partial \Delta \overline{m}_{\mathrm{m1}}}{\partial r_{\mathrm{m1D}}}\right)=g_1(s)\Delta\overline{m}_{\mathrm{m1}} \tag{B.100}$$

$$\frac{1}{r_{\mathrm{m2D}}^2}\frac{\partial}{\partial r_{\mathrm{m2D}}}\left(r_{\mathrm{m2D}}^2\frac{\partial \Delta \overline{m}_{\mathrm{m2}}}{\partial r_{\mathrm{m2D}}}\right)=g_2(s)\Delta\overline{m}_{\mathrm{m2}} \tag{B.101}$$

where:

$$g_1(s) = \frac{15\omega_{\mathrm{m1}}}{\lambda_1 M_{12}} s + \frac{\beta_1}{\lambda}\sigma\left[\sqrt{\lambda\theta_{\mathrm{fm}} s}\coth\left(\sqrt{\lambda\theta_{\mathrm{fm}} s}\right) - 1\right] \tag{B.102}$$

$$g_2(s) = \frac{15\omega_{\mathrm{m2}}}{\lambda_2} s + \frac{\beta_2}{\lambda}\sigma\left[\sqrt{\lambda\theta_{\mathrm{fm}} s}\coth\left(\sqrt{\lambda\theta_{\mathrm{fm}} s}\right) - 1\right] \tag{B.103}$$

Pseudo-steady state diffusion. According to the pseudo-steady state diffusion model, the following relationships are established:

$$\overline{C}_{\mathrm{ED}} - \overline{C}_{\mathrm{mD1}} = -\frac{M_{\mathrm{g}} k_{\mathrm{f2}} h T_{\mathrm{sc}}}{\rho_{\mathrm{sc}} q_{\mathrm{sc}} p_{\mathrm{sc}} T} \frac{\sigma\theta_{\mathrm{fm}}\lambda s}{\lambda\theta_{\mathrm{fm}} s + 1} \Delta\overline{m}_{\mathrm{m1}} \tag{B.104}$$

$$\overline{C}_{\mathrm{ED}} - \overline{C}_{\mathrm{mD2}} = -\frac{M_{\mathrm{g}} k_{\mathrm{f2}} h T_{\mathrm{sc}}}{\rho_{\mathrm{sc}} q_{\mathrm{sc}} p_{\mathrm{sc}} T} \frac{\sigma\theta_{\mathrm{fm}}\lambda s}{\lambda\theta_{\mathrm{fm}} s + 1} \Delta\overline{m}_{\mathrm{m2}} \tag{B.105}$$

Substituting Eqs. (B.104) and (B.105) into Eqs. (B.95) and (B.97) yields:

$$\frac{1}{r_{\mathrm{m1D}}^2}\frac{\partial}{\partial r_{\mathrm{m1D}}}\left(r_{\mathrm{m1D}}^2 \frac{\partial\Delta\overline{m}_{\mathrm{m1}}}{\partial r_{\mathrm{m1D}}}\right) = g_1(s)\Delta\overline{m}_{\mathrm{m1}} \tag{B.106}$$

$$\frac{1}{r_{\mathrm{m2D}}^2}\frac{\partial}{\partial r_{\mathrm{m2D}}}\left(r_{\mathrm{m2D}}^2 \frac{\partial\Delta\overline{m}_{\mathrm{m2}}}{\partial r_{\mathrm{m2D}}}\right) = g_2(s)\Delta\overline{m}_{\mathrm{m2}} \tag{B.107}$$

where:

$$g_1(s) = \frac{15\omega_{\mathrm{m1}}}{\lambda_1 M_{12}} s + \beta_1 \frac{\sigma\theta_{\mathrm{fm}} s}{\lambda\theta_{\mathrm{fm}} s + 1} \tag{B.108}$$

$$g_2(s) = \frac{15\omega_{\mathrm{m2}}}{\lambda_2} s + \beta_2 \frac{\sigma\theta_{\mathrm{fm}} s}{\lambda\theta_{\mathrm{fm}} s + 1} \tag{B.109}$$

According to the transient interporosity flow equation, the following expressions can be obtained:

$$\left.\frac{\partial\Delta\overline{m}_{\mathrm{m1}}}{\partial r_{\mathrm{mD1}}}\right|_{r_{\mathrm{mD1}}=1} = \Delta\overline{m}_{\mathrm{f1}}\left[\sqrt{g_1}\coth\left(\sqrt{g_1}\right) - 1\right] \tag{B.110}$$

$$\left.\frac{\partial\Delta\overline{m}_{\mathrm{m2}}}{\partial r_{\mathrm{mD2}}}\right|_{r_{\mathrm{mD2}}=1} = \Delta\overline{m}_{\mathrm{f2}}\left[\sqrt{g_2}\coth\left(\sqrt{g_2}\right) - 1\right] \tag{B.111}$$

Substituting Eqs. (B.110) and (B.111) into Eqs. (B.94) and (B.96) and then making some arrangements, the following integrated equations can be obtained: In the inner natural fracture system:

$$\frac{1}{r_D}\frac{\partial}{\partial r_D}\left(r_D\frac{\partial \Delta \overline{m}_{f1}}{\partial r_D}\right)=f_1(s)\Delta \overline{m}_{f1} \tag{B.112}$$

In the outer natural fracture system:

$$\frac{1}{r_D}\frac{\partial}{\partial r_D}\left(r_D\frac{\partial \Delta \overline{m}_{f2}}{\partial r_D}\right)=f_2(s)\Delta \overline{m}_{f2} \tag{B.113}$$

where:

$$f_1(s)=\frac{\omega_{f1}}{M_{12}}s+\frac{\lambda_1}{5}\left[\sqrt{g_1}\coth\left(\sqrt{g_1}\right)-1\right] \tag{B.114}$$

$$f_2(s)=\omega_{f2}s+\frac{\lambda_2}{5}\left[\sqrt{g_2}\coth\left(\sqrt{g_2}\right)-1\right] \tag{B.115}$$

B.4.2 Pseudo-steady state interporosity flow

The gas flow from the matrix system to the natural fracture system follows the pseudo-steady interporosity flow, the continuity equations in the inner region are:

$$\frac{1}{r}\frac{\partial}{\partial r}\left(r\frac{\partial \Delta m_{f1}}{\partial r}\right)=\frac{\phi_{f1}\mu_{gi}c_{fg1}}{k_{f1}}\frac{\partial \Delta m_{f1}}{\partial t}-\frac{\alpha k_{m1}}{k_{f1}}(\Delta m_{m1}-\Delta m_{f1}) \tag{B.116}$$

$$-\frac{\alpha_1 k_{m1}}{k_{f1}}(\Delta m_{m1}-\Delta m_{f1})=\frac{\phi_{m1}\mu_{gi}c_{mg1}}{k_{f1}}\frac{\partial \Delta m_{m1}}{\partial t}-\frac{2RT}{k_{f1}}(1-\phi_{f1}-\phi_{m1})\frac{\mathrm{d}C_m}{dt} \tag{B.117}$$

In the outer region, they are:

$$\frac{1}{r}\frac{\partial}{\partial r}\left(r\frac{\partial \Delta m_{f2}}{\partial r}\right)=\frac{\phi_{f2}\mu_{gi}c_{fg2}}{k_{f2}}\frac{\partial \Delta m_{f2}}{\partial t}-\frac{\alpha k_{m2}}{k_{f2}}(\Delta m_{m2}-\Delta m_{f2}) \tag{B.118}$$

$$-\frac{\alpha_2 k_{m2}}{k_{f2}}(\Delta m_{m2}-\Delta m_{f2})=\frac{\phi_{m2}\mu_{gi}c_{mg2}}{k_{f2}}\frac{\partial \Delta m_{m2}}{\partial t}-\frac{2RT}{k_{f2}}(1-\phi_{f2}-\phi_{m2})\frac{\mathrm{d}C_m}{dt} \tag{B.119}$$

Define the following variables:

$$t_D=\frac{k_{f2}t}{\Lambda L_{ref}^2}, M_{12}=\frac{k_{f1}/\mu_{gi}}{k_{f2}/\mu_{gi}}, r_D=\frac{r_{ref}}{L_{ref}}, \omega_{f1}=\frac{\left(\phi_f\mu_{gi}c_{fg}\right)_1}{\Lambda}, \omega_{f2}=\frac{\left(\phi_f\mu_{gi}c_{fg}\right)_2}{\Lambda},$$

$$\omega_{m1}=\frac{(\phi_m\mu_{gi}c_{mg})_1}{\Lambda},$$

$$\omega_{m2}=\frac{(\phi_m\mu_{gi}c_{mg})_2}{\Lambda},\lambda_1=\alpha\frac{k_m}{k_{f1}}L_{ref}^2,\lambda_2=\alpha\frac{k_m}{k_{f2}}L_{ref}^2,\theta_{fm}=\frac{k_{f2}}{k_m},$$

$$\lambda=\begin{cases}\dfrac{k_{f2}\tau}{\Lambda L_{ref}^2} & \text{for unsteady state diffusion}\\ \dfrac{k_{f2}\tau}{6\Lambda L_{ref}^2} & \text{for pseudo-steady state diffusion}\end{cases},$$

$$\tau=\begin{cases}\dfrac{R_m^2}{D_F} & \text{for unsteady state diffusion}\\ \dfrac{R_m^2}{\pi^2 D_F} & \text{for pseudo-steady state diffusion}\end{cases},$$

$$\Lambda=\begin{cases}(\phi_f\mu_{gi}c_{fg}+\phi_m\mu_{gi}c_{mg})_{1+2}+\dfrac{6k_{f2}h}{q_{sc}} & \text{for unsteady state diffusion}\\ (\phi_f\mu_{gi}c_{fg}+\phi_m\mu_{gi}c_{mg})_{1+2}+\dfrac{2k_{f2}h}{q_{sc}} & \text{for pseudo-steady diffusion}\end{cases}$$

When the gas flow from macropores into the natural fracture system follows the pseudo-steady state interporosity, the diffusivity equation in the inner region is:

$$\frac{1}{r_D}\frac{\partial}{\partial r_D}\left(r_D\frac{\partial\Delta m_{f1}}{\partial r_D}\right)=\frac{\omega_{f1}}{M_{12}}\frac{\partial\Delta m_{f1}}{\partial t_D}-\lambda_1(\Delta m_{m1}-\Delta m_{f1}) \tag{B.120}$$

$$-\lambda_1(\Delta m_{m1}-\Delta m_{f1})=\frac{\omega_{m1}}{M_{12}}\frac{\partial\Delta m_{m1}}{\partial t_D}-\frac{2RTL_{ref}^2}{M_{12}k_{f2}}(1-\phi_{f1}-\phi_{m1})\frac{dC_m}{dt} \tag{B.121}$$

In the outer region, it is:

$$\frac{1}{r_D}\frac{\partial}{\partial r_D}\left(r_D\frac{\partial\Delta m_{f2}}{\partial r_D}\right)=\omega_{f2}\frac{\partial\Delta m_{f2}}{\partial t_D}-\lambda_2(\Delta m_{m2}-\Delta m_{f2}) \tag{B.122}$$

$$-\lambda_2(\Delta m_{m2}-\Delta m_{f2})=\omega_{m2}\frac{\partial\Delta m_{m2}}{\partial t_D}-\frac{2RTL_{ref}^2}{k_{f2}}(1-\phi_{f2}-\phi_{m2})\frac{dC_m}{dt} \tag{B.123}$$

Unsteady state diffusion. For the unsteady state diffusion model, the diffusivity equations in the inner region can be obtained by substituting Eq. (B.87) into Eqs. (B.121) and (B.123), which are:

$$-\lambda_1(\Delta m_{m1}-\Delta m_{f1})=\frac{\omega_{m1}}{M_{12}}\frac{\partial\Delta m_{m1}}{\partial t_D}-\frac{2RTL_{ref}^2}{M_{12}k_{f2}}(1-\phi_{f1}-\phi_{m1})\frac{dC_m}{dt} \tag{B.124}$$

$$\frac{2RTL_{\text{ref}}^2}{M_{12}k_{\text{f2}}}\frac{\text{d}C_{\text{m}}}{\text{d}t}=\frac{M_{\text{g}}q_{\text{sc}}p_{\text{sc}}T}{\rho_{\text{sc}}k_{\text{f2}}hT_{\text{sc}}}\frac{(1-\omega_{\text{f1}}-\omega_{\text{m1}})}{\lambda M_{12}}\frac{\partial C_{\text{m}}}{\partial r_{\text{m1D}}}\bigg|_{r_{\text{m1D}}=1} \tag{B.125}$$

In the outer region, they are:

$$-\lambda_2(\Delta m_{\text{m2}}-\Delta m_{\text{f2}})=\omega_{\text{m2}}\frac{\partial\Delta m_{\text{m2}}}{\partial t_{\text{D}}}-\frac{2RTL_{\text{ref}}^2}{k_{\text{f2}}}(1-\phi_{\text{f2}}-\phi_{\text{m2}})\frac{\text{d}C_{\text{m}}}{\text{d}t} \tag{B.126}$$

$$\frac{2RTL_{\text{ref}}^2}{k_{\text{f2}}}\frac{\text{d}C_{\text{m}}}{\text{d}t}=\frac{M_{\text{g}}q_{\text{sc}}p_{\text{sc}}T}{\rho_{\text{sc}}k_{\text{f2}}hT_{\text{sc}}}\frac{(1-\omega_{\text{f2}}-\omega_{\text{m2}})}{\lambda}\frac{\partial C_{\text{m}}}{\partial r_{\text{m2D}}}\bigg|_{r_{\text{m2D}}=1} \tag{B.127}$$

Substituting Eqs. (B.125) and (B.127) into Eqs. (B.124) and (B.126) and then taking the Laplace transformation, we have:

$$-\lambda_1(\Delta\overline{m}_{\text{m1}}-\Delta\overline{m}_{\text{f1}})=\frac{\omega_{\text{m1}}}{M_{12}}s\Delta\overline{m}_{\text{m1}}-\frac{M_{\text{g}}q_{\text{sc}}p_{\text{sc}}T}{\rho_{\text{sc}}k_{\text{f2}}hT_{\text{sc}}}\frac{\beta_1}{\lambda M_{12}}\frac{\partial\overline{C}_{\text{mD}}}{\partial r_{\text{m1D}}}\bigg|_{r_{\text{m1D}}=1} \tag{B.128}$$

$$-\lambda_2(\Delta\overline{m}_{\text{m2}}-\Delta\overline{m}_{\text{f2}})=\omega_{\text{m2}}s\Delta\overline{m}_{\text{m2}}-\frac{M_{\text{g}}q_{\text{sc}}p_{\text{sc}}T}{\rho_{\text{sc}}k_{\text{f2}}hT_{\text{sc}}}\frac{\beta_2}{\lambda}\frac{\partial\overline{C}_{\text{mD}}}{\partial r_{\text{m2D}}}\bigg|_{r_{\text{m2D}}=1} \tag{B.129}$$

where:

$$\beta_1=(1-\phi_{\text{f1}}-\phi_{\text{m1}})(1-\omega_{\text{f1}}-\omega_{\text{m1}}) \tag{B.130}$$

$$\beta_2=(1-\phi_{\text{f2}}-\phi_{\text{m2}})(1-\omega_{\text{f2}}-\omega_{\text{m2}}) \tag{B.131}$$

According to Fick's unsteady state diffusion equation, the following equations can be obtained:

$$\frac{\partial\overline{C}_{\text{mD1}}}{\partial r_{\text{mD}}}\bigg|_{r_{\text{mD}}=1}=-\frac{\rho_{\text{sc}}k_{\text{f2}}hT_{\text{sc}}}{M_{\text{g}}q_{\text{sc}}p_{\text{sc}}T}\sigma\left[\sqrt{\lambda s}\coth\left(\sqrt{\lambda s}\right)-1\right]\Delta\overline{m}_{\text{m1}} \tag{B.132}$$

$$\frac{\partial\overline{C}_{\text{mD2}}}{\partial r_{\text{mD}}}\bigg|_{r_{\text{mD}}=1}=-\frac{\rho_{\text{sc}}k_{\text{f2}}hT_{\text{sc}}}{M_{\text{g}}q_{\text{sc}}p_{\text{sc}}T}\sigma\left[\sqrt{\lambda s}\coth\left(\sqrt{\lambda s}\right)-1\right]\Delta\overline{m}_{\text{m2}} \tag{B.133}$$

Inserting Eqs. (B.132) and (B.133) into Eqs. (B.128) and (B.129), we have:

$$-\lambda_1(\Delta\overline{m}_{\text{m1}}-\Delta\overline{m}_{\text{f1}})=g_1(s)\Delta\overline{m}_{\text{m1}} \tag{B.134}$$

$$-\lambda_2(\Delta\overline{m}_{\text{m2}}-\Delta\overline{m}_{\text{f2}})=g_2(s)\Delta\overline{m}_{\text{m2}} \tag{B.135}$$

where:

$$g_1(s)=\frac{\omega_{\text{m1}}}{M_{12}}s+\frac{\beta 1}{\lambda M_{12}}\sigma\left[\sqrt{\lambda s}\coth\left(\sqrt{\lambda s}\right)-1\right] \tag{B.136}$$

$$g_2(s) = \omega_{m2}s + \frac{\beta_2}{\lambda}\sigma\left[\sqrt{\lambda s}\coth\left(\sqrt{\lambda s}\right) - 1\right] \tag{B.137}$$

Combining Eqs. (B.134) and (B.135) with the transient percolation equations Eqs. (B.120) and (B.122), the seepage equations in the inner and outer regions are:

$$\frac{1}{r_D}\frac{\partial}{\partial r_D}\left(r_D\frac{\partial\Delta\overline{m}_{f1}}{\partial r_D}\right) = f_1(s)\Delta\overline{m}_{f1} \tag{B.138}$$

$$\frac{1}{r}\frac{\partial}{\partial r}\left(r\frac{\partial\Delta\overline{m}_{f2}}{\partial r}\right) = f_2(s)\Delta\overline{m}_{f2} \tag{B.139}$$

where:

$$f_1(s) = \frac{\omega_{f1}}{M_{12}}s + \lambda_1\frac{g_1(s)}{g_1(s)+\lambda_1} \tag{B.140}$$

$$f_2(s) = \omega_{f2}s + \lambda_2\frac{g_2(s)}{g_2(s)+\lambda_2} \tag{B.141}$$

Pseudo-steady state diffusion. When gas from the microporosity system into the natural fracture system follows the pseudo-steady state diffusion, we have:

$$-\lambda_1(\Delta\overline{m}_{m1} - \Delta\overline{m}_{f1}) = \frac{\omega_{m1}}{M_{12}}s\Delta\overline{m}_{m1} - \frac{M_g}{\rho_{sc}}\frac{q_{sc}p_{sc}T}{k_{f2}hT_{sc}}\frac{\beta 1}{\lambda M_{12}}\left[\overline{C}_{ED} - \overline{C}_{mD}\right] \tag{B.142}$$

$$-\lambda_2(\Delta\overline{m}_{m2} - \Delta\overline{m}_{f2}) = \omega_{m2}s\Delta\overline{m}_{m2} - \frac{M_g}{\rho_{sc}}\frac{q_{sc}p_{sc}T}{k_{f2}hT_{sc}}\frac{\beta_2}{\lambda}\left[\overline{C}_{ED} - \overline{C}_{mD}\right] \tag{B.143}$$

According to the pseudo-steady state diffusion model described in Eq. (B.87), we have:

$$\overline{C}_{ED} - \overline{C}_{mD1} = -\frac{M_g}{\rho_{sc}}\frac{k_{f2}hT_{sc}}{q_{sc}p_{sc}T}\frac{\sigma\lambda s}{\lambda s+1}\Delta\overline{m}_{m1} \tag{B.144}$$

$$\overline{C}_{ED} - \overline{C}_{mD2} = -\frac{M_g}{\rho_{sc}}\frac{k_{f2}hT_{sc}}{q_{sc}p_{sc}T}\frac{\sigma\lambda s}{\lambda s+1}\Delta\overline{m}_{m2} \tag{B.145}$$

Substituting Eqs. (B.144) and (B.145) into Eqs. (B.142) and (B.143) and arranging them, we have:

$$-\lambda_1(\Delta\overline{m}_{m1} - \Delta\overline{m}_{f1}) = g_1(s)\Delta\overline{m}_{m1} \tag{B.146}$$

$$-\lambda_2(\Delta\overline{m}_{m2} - \Delta\overline{m}_{f2}) = g_2(s)\Delta\overline{m}_{m2} \tag{B.147}$$

where:

$$g_1(s) = \frac{\omega_{m1}}{M_{12}} s + \frac{\beta 1}{M_{12}} \frac{\sigma s}{\lambda s + 1} \tag{B.148}$$

$$g_2(s) = \omega_{m2} s + \beta_2 \frac{\sigma s}{\lambda s + 1} \tag{B.149}$$

Similarity, combining Eqs. (B.146) and (B.147) with Eqs. (B.120) and (B.122), the continuity equations in the inner region are:

$$\frac{1}{r_D} \frac{\partial}{\partial r_D} \left(r_D \frac{\partial \Delta \overline{m}_{f1}}{\partial r_D} \right) = f_1(s) \Delta \overline{m}_{f1} \tag{B.150}$$

$$\frac{1}{r} \frac{\partial}{\partial r} \left(r \frac{\partial \Delta \overline{m}_{f2}}{\partial r} \right) = f_2(s) \Delta \overline{m}_{f2} \tag{B.151}$$

where:

$$f_1(s) = \frac{\omega_{f1}}{M_{12}} s + \lambda_1 \frac{g_1(s)}{g_1(s) + \lambda_1} \tag{B.152}$$

$$f_2(s) = \omega_{f2} s + \lambda_2 \frac{g_2(s)}{g_2(s) + \lambda_2} \tag{B.153}$$

B.5 Microfractures + gas adsorption/desorption + nanopore Knudsen diffusion (model 5)

B.5.1 Transient matrix flow

The continuity equation in the inner natural fracture system is:

$$\frac{1}{r} \frac{\partial}{\partial r} \left(\frac{k_{f1} p_{f1} M_g}{\mu_g} \frac{}{ZRT} r \frac{\partial p_{f1}}{\partial r} \right) = \phi_{f1} c_{fg1} \frac{p_{f1} M_g}{ZRT} \frac{\partial p_{f1}}{\partial t} + \frac{3 \rho_{gm}}{R_m} \frac{k_{app}}{\mu_g} \frac{\partial p_{m1}}{\partial r_m} \bigg|_{r_m = R_m} \tag{B.154}$$

The continuity equation in the inner matrix system is:

$$\frac{1}{r} \frac{\partial}{\partial r} \left(\frac{k_{app} p_{m1} M_g}{\mu_g \, ZRT} r \frac{\partial p_{m1}}{\partial r} \right) = \phi_{m1} c_{mg1} \frac{p_{m1} M_g}{ZRT} \frac{\partial p_{m1}}{\partial t} + q_{des1} \tag{B.155}$$

The desorption flux in the inner region is:

$$q_{des1} = \frac{M_g p_{sc}}{RT_{sc}} \frac{(1 - \phi_{m1} - \phi_{f1}) G_L m(p_L)}{[m(p_L) + m(p_{f1})]^2} \frac{\partial m(p_{f1})}{\partial t} \tag{B.156}$$

The continuity equation in the outer natural fracture system is:

$$\frac{1}{r}\frac{\partial}{\partial r}\left(\frac{k_{\mathrm{f2}}}{\mu_{\mathrm{g}}}\frac{p_{\mathrm{f2}}M_{\mathrm{g}}}{ZRT}r\frac{\partial p_{\mathrm{f2}}}{\partial r}\right)=\phi_{\mathrm{f2}}c_{\mathrm{fg2}}\frac{p_{\mathrm{f2}}M_{\mathrm{g}}}{ZRT}\frac{\partial p_{\mathrm{f2}}}{\partial t}+\frac{3\rho_{\mathrm{gm}}}{R_{\mathrm{m2}}}\frac{k_{\mathrm{app}}}{\mu_{\mathrm{g}}}\frac{\partial p_{\mathrm{m2}}}{\partial r_{\mathrm{m}}}\bigg|_{r_{\mathrm{m}}=R_{\mathrm{m}}} \tag{B.157}$$

The continuity equation in the outer matrix system is:

$$\frac{1}{r}\frac{\partial}{\partial r}\left(\frac{k_{\mathrm{app}}}{\mu_{\mathrm{g}}}\frac{p_{\mathrm{m2}}M_{\mathrm{g}}}{ZRT}r\frac{\partial p_{\mathrm{m2}}}{\partial r}\right)=\phi_{\mathrm{m2}}c_{\mathrm{mg2}}\frac{p_{\mathrm{m2}}M_{\mathrm{g}}}{ZRT}\frac{\partial p_{\mathrm{m2}}}{\partial t}+q_{\mathrm{des2}} \tag{B.158}$$

The desorption item is:

$$q_{\mathrm{des2}}=\frac{M_{\mathrm{g}}p_{\mathrm{sc}}}{RT_{\mathrm{sc}}}\frac{(1-\phi_{\mathrm{m2}}-\phi_{\mathrm{f2}})G_{\mathrm{L}}m(p_{\mathrm{L}})}{[m(p_{\mathrm{L}})+m(p_{\mathrm{f2}})]^2}\frac{\partial m(p_{\mathrm{f2}})}{\partial t} \tag{B.159}$$

Define the following variables:

$$r_{\mathrm{mD}}=\frac{r_{\mathrm{m}}}{R_{\mathrm{m}}},r_{\mathrm{D}}=\frac{r}{L_{\mathrm{ref}}},t_{\mathrm{D}}=\frac{k_{\mathrm{f2}}t}{(\phi_{\mathrm{m}}c_{\mathrm{mgi}}+\phi_{\mathrm{f}}c_{\mathrm{fgi}})_{1+2}\mu_{\mathrm{gi}}L_{\mathrm{ref}}^2},$$

$$\omega_1=\frac{(\phi_{\mathrm{m}}c_{\mathrm{mgi}}+\phi_{\mathrm{f}}c_{\mathrm{fgi}})_1}{(\phi_{\mathrm{m}}c_{\mathrm{mgi}}+\phi_{\mathrm{f}}c_{\mathrm{fgi}})_{1+2}},\alpha=\frac{15}{R_{\mathrm{m}}^2},$$

$$\omega_{\mathrm{f1}}=\frac{(\phi_{\mathrm{f}}c_{\mathrm{fgi}})_1}{(\phi_{\mathrm{m}}c_{\mathrm{mgi}}+\phi_{\mathrm{f}}c_{\mathrm{fgi}})_1},\omega_2=\frac{(\phi_{\mathrm{m}}c_{\mathrm{mgi}}+\phi_{\mathrm{f}}c_{\mathrm{fgi}})_2}{(\phi_{\mathrm{m}}c_{\mathrm{mgi}}+\phi_{\mathrm{f}}c_{\mathrm{fgi}})_{1+2}},$$

$$\omega_{\mathrm{f2}}=\frac{(\phi_{\mathrm{f}}c_{\mathrm{fgi}})_2}{(\phi_{\mathrm{m}}c_{\mathrm{mgi}}+\phi_{\mathrm{f}}c_{\mathrm{fgi}})_2},\alpha_{\mathrm{d1}}=\frac{\phi_{\mathrm{m1}}c_{\mathrm{d1}}}{(\phi_{\mathrm{m}}c_{\mathrm{mgi}}+\phi_{\mathrm{f}}c_{\mathrm{fgi}})_1},$$

$$\alpha_{\mathrm{d2}}=\frac{\phi_{\mathrm{m2}}c_{\mathrm{d2}}}{(\phi_{\mathrm{m}}c_{\mathrm{mgi}}+\phi_{\mathrm{f}}c_{\mathrm{fgi}})_2},\ \lambda_1=\alpha\frac{k_{\mathrm{m}}}{k_{\mathrm{f1}}}L_{\mathrm{ref}}^2,\ \lambda_2=\alpha\frac{k_{\mathrm{m}}}{k_{\mathrm{f2}}}L_{\mathrm{ref}}^2,\ M_{12}=\frac{k_{\mathrm{f1}}/\mu_{\mathrm{gi}}}{k_{\mathrm{f2}}/\mu_{\mathrm{gi}}},$$

$$\theta_{\mathrm{ka\text{-}m}}=\frac{k_{\mathrm{app}}}{k_{\mathrm{m}}}$$

Substituting the above dimensionless variables into the continuity equations yields:

$$\frac{1}{r_{\mathrm{D}}}\frac{\partial}{\partial r_{\mathrm{D}}}\left(r_{\mathrm{D}}\frac{\partial\Delta m_{\mathrm{f1}}}{\partial r_{\mathrm{D}}}\right)=\frac{\omega_1\omega_{\mathrm{f1}}}{M_{12}}\frac{\partial\Delta m_{\mathrm{f1}}}{\partial t_{\mathrm{D}}}+\frac{\lambda_1\theta_{\mathrm{ka\text{-}m}}}{5}\frac{\partial\Delta m_{\mathrm{m1}}}{\partial r_{\mathrm{mD1}}}\bigg|_{r_{\mathrm{mD1}}=1} \tag{B.160}$$

$$\frac{1}{r_{\mathrm{mD1}}^2}\frac{\partial}{\partial r_{\mathrm{mD1}}}\left(r_{\mathrm{mD1}}^2\frac{\partial\Delta m_{\mathrm{m1}}}{\partial r_{\mathrm{mD1}}}\right)=\frac{15\omega_1(1-\omega_{\mathrm{f1}})}{\lambda_1\theta_{\mathrm{ka\text{-}m}}M_{12}}\frac{\partial\Delta m_{\mathrm{m1}}}{\partial t_{\mathrm{D}}}+\frac{15\alpha_{\mathrm{d1}}\omega_1}{\lambda_1\theta_{\mathrm{ka\text{-}m}}M_{12}}\frac{\partial\Delta m_{\mathrm{m1}}}{\partial t_{\mathrm{D}}} \tag{B.161}$$

The continuity equations in the outer region are:

$$\frac{1}{r_D}\frac{\partial}{\partial r_D}\left(r_D\frac{\partial \Delta m_{f2}}{\partial r_D}\right)=\omega_2\omega_{f2}\frac{\partial \Delta m_{f2}}{\partial t_D}+\frac{\lambda_2\theta_{\text{ka-m}}}{5}\frac{\partial \Delta m_{m2}}{\partial r_{mD1}}\bigg|_{r_{mD1}=1} \tag{B.162}$$

$$\frac{1}{r_{mD2}^2}\frac{\partial}{\partial r_{mD2}}\left(r_{mD2}^2\frac{\partial \Delta m_{m2}}{\partial r_{mD2}}\right)=\frac{15\omega_2(1-\omega_{f2})}{\lambda_2\theta_{\text{ka-m}}}\frac{\partial \Delta m_{m2}}{\partial t_D}+\frac{15\alpha_{d2}\omega_2}{\lambda_2\theta_{\text{ka-m}}}\frac{\partial \Delta m_{m2}}{\partial t_D} \tag{B.163}$$

According to the transient interporosity flow model, we have:

$$\frac{\partial \Delta \overline{m}_{m1}}{\partial r_{mD}}\bigg|_{r_{mD}=1}=\Delta \overline{m}_{f1}\left[\sqrt{g_1}\coth\left(\sqrt{g_1}\right)-1\right] \tag{B.164}$$

$$\frac{\partial \Delta \overline{m}_{m2}}{\partial r_{mD}}\bigg|_{r_{mD}=1}=\Delta \overline{m}_{f2}\left[\sqrt{g_2}\coth\left(\sqrt{g_2}\right)-1\right] \tag{B.165}$$

where:

$$g_1=\frac{15\omega_1(1-\omega_{f1}+\alpha_{d1})s}{\lambda_1\theta_{\text{ka-m}}M_{12}} \tag{B.166}$$

$$g_2=\frac{15\omega_2(1-\omega_{f2}+\alpha_{d2})}{\lambda_2\theta_{\text{ka-m}}} \tag{B.167}$$

Substituting Eqs. (B.164) and (B.165) into Eqs. (B.160) and (B.162), the general equation in the natural fracture system can be obtained. In the inner natural fracture system, it is:

$$\frac{1}{r_D}\frac{\partial}{\partial r_D}\left(r_D\frac{\partial \Delta \overline{m}_{f1}}{\partial r_D}\right)=f_1(s)\Delta \overline{m}_{f1} \tag{B.168}$$

In the outer natural fracture system, it is:

$$\frac{1}{r_D}\frac{\partial}{\partial r_D}\left(r_D\frac{\partial \Delta \overline{m}_{f2}}{\partial r_D}\right)=f_2(s)\Delta \overline{m}_{f2} \tag{B.169}$$

where:

$$f_1(s)=\frac{\omega_1\omega_{f1}s}{M_{12}}+\frac{\lambda_1}{5}\left[\sqrt{g_1}\coth\left(\sqrt{g_1}\right)-1\right] \tag{B.170}$$

$$f_2(s)=\omega_2\omega_{f2}s+\frac{\lambda_2}{5}\left[\sqrt{g_2}\coth\left(\sqrt{g_2}\right)-1\right] \tag{B.171}$$

B.5.2 Pseudo-steady state interporosity flow

For pseudo-steady state interporosity flow, the following dimensionless equations in the inner region can be obtained:

$$\frac{1}{r_D}\frac{\partial}{\partial r_D}\left(r_D\frac{\partial \Delta m_{f1}}{\partial r_D}\right)=\frac{\omega_1\omega_{f1}}{M_{12}}\frac{\partial \Delta m_{f1}}{\partial t_D}-\lambda_1\theta_{ka\text{-}m}[\Delta m_{m1}-\Delta m_{f1}] \tag{B.172}$$

$$-\lambda_1\theta_{ka\text{-}m}[\Delta m_{m1}-\Delta m_{f1}]=\frac{\omega_1(1-\omega_{f1})}{M_{12}}\frac{\partial \Delta m_{m1}}{\partial t_D}+\frac{\alpha_{d1}\omega_1}{M_{12}}\frac{\partial \Delta m_{m1}}{\partial t_D} \tag{B.173}$$

Similarity, the dimensionless equations in the outer natural fracture system are:

$$\frac{1}{r_D}\frac{\partial}{\partial r_D}\left(r_D\frac{\partial \Delta m_{f2}}{\partial r_D}\right)=\omega_2\omega_{f2}\frac{\partial \Delta m_{f2}}{\partial t_D}-\lambda_2\theta_{ka\text{-}m}[\Delta m_{m2}-\Delta m_{f2}] \tag{B.174}$$

$$-\lambda_2\theta_{ka\text{-}m}[\Delta m_{m2}-\Delta m_{f2}]=\omega_2(1-\omega_{f2})\frac{\partial \Delta m_{m2}}{\partial t_D}+\alpha_{d2}\omega_2\frac{\partial \Delta m_{m2}}{\partial t_D} \tag{B.175}$$

The expressions of the above dimensionless variables are as follows:

$$r_D=\frac{r}{L_{ref}},\ t_D=\frac{k_{f2}t}{(\phi_m c_{mgi}+\phi_f c_{fgi})_{1+2}\mu_{gi}L_{ref}^2},\ \omega_1=\frac{(\phi_m c_{mgi}+\phi_f c_{fgi})_1}{(\phi_m c_{mgi}+\phi_f c_{fgi})_{1+2}},$$

$$\omega_2=\frac{(\phi_m c_{mgi}+\phi_f c_{fgi})_2}{(\phi_m c_{mgi}+\theta_f c_{fgi})_{1+2}}$$

$$\omega_{f1}=\frac{(\phi_f c_{fgi})_1}{(\phi_m c_{mgi}+\phi_f c_{fgi})_1},\ \omega_{f2}=\frac{(\phi_f c_{fgi})_2}{(\phi_m c_{mgi}+\phi_f c_{fgi})_2},\ \lambda_1=\alpha_1\frac{k_m}{k_{f1}}L_{ref}^2,$$

$$\alpha_{d1}=\frac{\phi_{m1}c_{d1}}{(\phi_m c_{mgi}+\phi_f c_{fgi})_1}$$

$$\alpha_{d2}=\frac{\phi_{m2}c_{d2}}{(\phi_m c_{mgi}+\phi_f c_{fgi})_2},\lambda_2=\alpha_2\frac{k_m}{k_{f2}}L_{ref}^2,M_{12}=\frac{k_{f1}/\mu_{gi}}{k_{f2}/\mu_{gi}},\theta_{ka\text{-}m}=\frac{k_{app}}{k_m}$$

Taking the Laplace transformation for Eqs. (B.172) and (B.175) and then substituting the solutions in the matrix system into them, the following general expressions in the inner and outer natural fracture systems are obtained:

$$\frac{1}{r_D}\frac{\partial}{\partial r_D}\left(r_D\frac{\partial \Delta \overline{m}_{f1}}{\partial r_D}\right)=f_1(s)\Delta \overline{m}_{f1} \tag{B.176}$$

$$\frac{1}{r_D}\frac{\partial}{\partial r_D}\left(r_D\frac{\partial \Delta \overline{m}_{f2}}{\partial r_D}\right)=f_2(s)\Delta \overline{m}_{f2} \tag{B.177}$$

where:

$$f_1(s)=\left[\frac{\omega_{f1}}{M_{12}}+\frac{(1-\omega_{f1}+\alpha_{d1})\lambda_1\theta_{ka\text{-}m}}{\lambda_1\theta_{ka\text{-}m}M_{12}+\omega_1(1-\omega_{f1}+\alpha_{d1})s}\right]\omega_1 s \tag{B.178}$$

$$f_2(s)=\left[\omega_{f2}+\frac{(1-\omega_{f2}+\alpha_{d2})\lambda_2\theta_{ka\text{-}m}}{\lambda_2\theta_{ka\text{-}m}+\omega_2(1-\omega_{f2}+\alpha_{d2})s}\right]\omega_2 s \tag{B.179}$$

Index

Note: Page numbers followed by "*f*" indicate figures, and "*t*" indicate tables.

A

Adsorbed gas, 16–24, 37–39
Adsorption model, 20–24

B

Bessel equation, 153–154
BET adsorption theory, 23
Bottom-hole pressure (BHP), 280–281
  circular gas reservoirs, 74–77
  closed rectangular gas reservoirs, 82–85
  constant, 87–88
Boundary element method (BEM), 192–197, 196*f*, 224
Branch fracture models, 262–264, 262–264*f*, 263*t*

C

Circular composite gas reservoirs
  continuous line source solutions, 155–156
    instantaneous line source solutions, 152–155
  fractured vertical wells
    mathematical solutions, 156–157
    physical model, 156
  pressure and production type curves
    microfractures+gas adsorption/desorption+matrix Fick's diffusion, 164–167
    microfractures+gas adsorption/desorption+nanopore Knudsen diffusion, 171–174
    microfractures+matrix macro pores+gas adsorption/desorption+nanopore Fick's diffusion, 168–170
    microfractures+matrix macro pores +steady state adsorption/desorption and diffusion, 161–163
    microfractures+steady state adsorption/desorption and diffusion, 158–160
Circular gas reservoirs
  bottom-hole pressure, 74–77
  continuous point source solutions, 54–61
    closed outer boundary, 61
    constant pressure outer boundary, 60–61
    infinite outer boundary, 60
    Laplace transformation, 58
    mathematical models, 55–57
    orthogonal transformation, 58–59
  fractured vertical wells, 74–77
    bottom-hole pressure expressions, 75–77
    physical models, 74–75
Clay minerals, 299–300
Clay-rich rocks, 237–238
Closed outer boundary, 61, 154
Composite gas reservoirs
  boundary element models in, 194–197, 196*f*
  Fick's diffusion, 204–206
Composite model, continuous line source in, 346–368
  microfractures+gas adsorption/desorption +matrix
    Fick's diffusion, 352–355
  microfractures+gas adsorption/desorption +nanopore Knudsen diffusion, 364–368
  microfractures+matrix macropores+gas adsorption/desorption+nanopore Fick's diffusion, 355–364
  microfractures+matrix macropores+steady state adsorption/desorption and diffusion, 348–352
  microfractures+steady state adsorption/desorption and diffusion model, 346–347
  pseudo-steady state diffusion, 354–355
  pseudo-steady state interporosity flow, 360–364, 367–368
  pseudo-steady state matrix flow, 351–352
  transient interporosity flow, 355–360
  transient matrix flow, 348–351, 364–366
  unsteady state diffusion, 353–354

Constant bottom-hole pressure, 87–88
Constant pressure outer boundary, 60–61
Continuous line source solutions, 152–156
Continuous point source solutions
  in circular gas reservoirs, 54–61
    closed outer boundary, 61
    constant pressure outer boundary, 60–61
    infinite outer boundary, 60
    Laplace transformation, 58
    mathematical models, 55–57
    orthogonal transformation, 58–59
  in rectangular gas reservoirs, 61–72
    computational remarks, 69–72
    derivation, 62–69
Control volume finite element method (CVFEM), 238, 240
Conventional reservoirs, 237–238
Conventional well model, 263
CVFEM. *See* Control volume finite element method (CVFEM)

D

Darcy's flow, 46
Dissolved gas, 15–16

F

Fick's diffusion, 32–33, 184–187, 216–219, 229–230, 331–336, 352–364
Finite difference method (FDM), 193
Finite element method (FEM), 193, 240
Flow mechanism models, 50
  microfractures+gas adsorption/desorption +Knudsen diffusion model in nanopores, 54
  microfractures+gas adsorption/desorption +matrix Fick's diffusion model, 51–52
  microfractures+matrix macropores+gas adsorption/desorption+Fick's diffusion model in nanopores, 52–53
  microfractures+matrix macropores+steady state adsorption/desorption and diffusion model, 51
  microfractures+steady state adsorption/desorption and diffusion model, 50
Fractured vertical wells
  in circular composite gas reservoirs, 156–157
  in circular gas reservoirs, 74–77
    bottom-hole pressure expressions, 75–77
    physical models, 74–75
  in closed rectangular gas reservoirs, 77–85
    bottom-hole pressure responses, 82–85
    continuous line source solutions, 78–82
  in rectangular composite shale gas reservoirs with SRV, 197–210
    effect of an adsorbed gas volume, 202, 202*f*
    effect of Knudsen diffusion coefficient, 209–210, 209–210*f*
    effect of slippage coefficient, 208, 208–209*f*
    effect of SRV size, 203, 203*f*
    effects of SRV permeability, 201, 201*f*, 205–207, 205*f*, 207–208*f*
    microfractures+gas adsorption/desorption+matrix Fick's diffusion, 204–206, 204*t*, 204*f*
    microfractures+gas adsorption/desorption+nanopore Knudsen diffusion, 206–210, 206*t*, 207–208*f*
    microfractures+steady state adsorption/desorption and diffusion, 198–203, 199*t*, 200*f*
    physical model, 198, 198*f*, 200*f*
    pressure and production type curve analysis, 198–210
Fracture network, 238, 271
Free gas, 15
Free gas molecule flow, 26
Freundlich adsorption theory, 21

G

Gas adsorption/desorption, 142–146
Gas diffusion, 24–25
  in nanometer pores, 32–37
Gas flow, 238
  in fractures and macropore systems, 29
  in nanometer pores, 30–31
Gas reservoir, 177–180, 178*f*. *See also* Shale gas resvoirs
Gas well pressure
  microfractures+gas adsorption/desorption +Fick's diffusion, 102–106
  microfractures+gas adsorption/desorption +nanopore Knudsen diffusion, 111–113

microfractures+matrix macropores+gas adsorption/desorption+nanopore Fick's diffusion, 107–110
microfractures+matrix macropores+steady state adsorption/desorption and diffusion, 96–102
microfractures+steady state adsorption/desorption and diffusion, 90–95
Green formula, 191–193

H

Henry adsorption model, 20
Horizontal well, with multiple fractures, 193
Hydraulic fracture, 11–12, 198, 262, 303–304, 305–306*t*, 307*f*
Hydraulic fracturing operation for well H9-1, 300–303, 304*t*

I

Infinite outer boundary, 60, 154
Inorganic pores, 6–9, 24–25
Instantaneous line source solutions, 152–155
International Union of Pure and Applied Chemistry (IUPAC), 18
Interporosity flow
pseudo-steady state, 329–330, 340–342, 344–346, 360–364, 367–368
transient, 336–340, 342–344, 355–360
Inversion algorithm, 88–89
Isotherm adsorption equation, Langmuir, 323, 326
Isothermal adsorption curves, 18–20
Isothermal Langmuir law, 239
IUPAC *See* International Union of Pure and Applied Chemistry (IUPAC)

K

Knudsen diffusion, 36, 188–189, 206–210, 220–222, 231–236, 342–346, 364–368
Knudsen number, 25–27, 26*f*

L

Langmuir adsorption theory, 21
Langmuir isotherm, 19
Langmuir isotherm adsorption equation, 323, 326
Langmuir law, isothermal, 239
Laplace transformation, 58
Line source solutions, 78–82, 152–156

M

Matrix fick's diffusion model, 43–44, 51–52
Matrix flow
pseudo-steady state, 351–352
transient, 326–329, 348–351
Matrix macropores, 142–146
Microfracture system
effect of, 249, 249*f*
permeability of, 248–249, 248*f*
Multi-stage fractured horizontal wells (MFHWs)
bottom-hole pressure
microfracture+matrix macropores +steady state adsorption/desorption and diffusion, 130–135
microfractures+gas adsorption/desorption+Fick's diffusion, 136–142
microfractures+gas adsorption/desorption+nanopore Knudsen diffusion, 146–149
microfractures+matrix macropores+gas adsorption/desorption+Fick's diffusion in nanopores, 142–146
microfractures+steady state adsorption/desorption and diffusion, 125–130
in circular composite gas reservoirs, 177–180
circular gas reservoirs
mathematical model and solutions, 116–121
physical model, 116
in dual-continuum medium, 249–271
branch fracture models, 262–264, 262–264*f*, 263*t*
effects of fracture network on production, 261–271
local SRV models, 269–271, 270–271*f*, 270*t*
model assumptions and mathematical model, 249–252, 250–252*f*
numerical solution, 252–256
production performance analysis, 257–261, 257–261*f*, 257*t*
rectangular composite models, 264–269, 265–269*f*
physical model of, 178, 178*f*

Multi-stage fractured horizontal wells (MFHWs) (*Continued*)
pressure and production type curves analysis
microfractures+gas adsoption/desorption+nanopore Knudsen diffusion, 188–189, 188–190*f*, 188*t*
microfractures+gas adsorption/desorption+matrix Fick's diffusion, 184–187, 185*t*, 185–187*f*
microfractures+steady state ad-/desorption and diffusion, 180–184, 180*t*, 181–184*f*
in rectangular composite shale gas reservoirs with global SRV, 210–222
microfractures+gas adsorption/desorption+matrix Fick's diffusion, 216–219, 216*t*, 216–219*f*
microfractures+gas adsorption/desorption+nanopore Knudsen diffusion, 220–222, 220*t*, 220–222*f*
microfractures+steady state adsorption/desorption and diffusion, 212–216, 213*t*, 213*f*
physic model, 210–212, 211–212*f*
pressure and production type curve analysis, 212–222
rectangular gas reservoirs
mathematical model and solutions, 122–125
physical model, 121–122
in shale gas reservoirs with local SRV, 222–236
microfractures+gas adsorption/desorption+matrix Fick's diffusion, 229–230, 230–231*f*
microfractures+gas adsorption/desorption+nanopore Knudsen diffusion, 231–236, 232*t*, 232–236*f*
microfractures+steady state adsorption/desorption and diffusion, 224–228
physical model, 223
pressure and production curve analysis, 224–236
in single porosity medium, 238–249
model assumptions and mathematical model, 238–240, 239*f*, 240*t*
numerical solutions, 240–244, 241*f*
physical parameters of gas reservoir, 244–245, 245*t*
production performance analysis, 244–249, 245–249*f*, 245*t*
transient pressure response of, 193

N

Natural microfractures, 10
Non-Darcy's effects, 238
Numerical inversion algorithm, 88–89
Numerical simulation application
hydraulic fracture simulation results, 303–304, 305–306*t*, 307*f*
hydraulic fracturing operation for well H9-1, 300–303, 304*t*
numerical simulation study, 304–306, 308–310*f*, 310*t*
target regions and well, 299–300, 301–302*t*, 303*f*

O

Organic intragranularpores, 10
Original gas in place (OGIP), 101–102, 244–245, 249, 264
Orthogonal transformation, 58–59

P

Partial pressure (Pa), 15–16, 20–24
Poisson's ratio, 4
Pressure buildup test analysis
well H2-2, 287–288, 288–290*f*, 290*t*
well H6-4, 289–290, 291–292*f*, 293*t*
well H6-6, 290–294, 293–296*f*, 296*t*
well H9-6, 294–298, 297–298*f*, 299*t*
Pressure of matrix (Pa), 36–37
Production performance, 125–149
Production type curves, 158–174
Pseudo-steady state diffusion, 33, 335–336, 354–355
Pseudo-steady state flow, 42–43
Pseudo-steady state interporosity flow, 329–330, 340–342, 344–346, 360–364, 367–368
Pseudo-steady state matrix flow, 351–352

R

Rectangular composite shale gas reservoir
fractured vertical wells, 197–210

inner and outer boundary discretization and numbering of, 196*f*
MFHW in, 210–222
Rectangular gas reservoirs, 77–85
bottom-hole pressure responses, 82–85
closed, 82–85
continuous line source solutions, 78–82
continuous point source solutions, 61–72
computational remarks, 69–72
derivation, 62–69
multi-stage fractured horizontal wells
mathematical model and solutions, 122–125
physical model, 121–122
Regime classification, 25–28
Reservoir boundary conditions, 191–192
Reservoir rocks, 240*t*

S

Sedimentary stratum, 299
Shale gas, 1–2, 300
Shale gas production, 238
Shale gas reservoir
abnormal high pressure, 3
accumulation mechanisms, 12–24
brittleness, 4
characteristics, 2–5
cleavage fractures, 13*f*
cutoff values, 5*t*
under different transport mechanisms
microfractures+gas adsorption/desorption+matrix Fick's diffusion model, 331–336
microfractures+gas dsorption/desorption +nanopore Knudsen diffusion model, 342–346
microfractures+matrix macropores+gas adsorption/desorption+nanopore Fick's diffusion model, 336–342
microfractures+matrix macropores +steady state adsorption/desorption and diffusion model, 326–330
microfractures+steady state adsorption/desorption and diffusion model, 323–325
pseudo-steady state diffusion, 335–336
pseudo-steady state interporosity flow, 329–330, 340–342, 344–346
transient interporosity flow, 336–340, 342–344
transient matrix flow, 326–329
unsteady state diffusion, 332–335
evaluation standards, 5*t*
fractured vertical wells in rectangular composite, 197–210
free gas, 3
gas existence status, 3
gas migration, 25–28
generation and accumulation, 2
hydraulic fractures, 14*f*
intragranular fractures, 13*f*
mathematical models
microfractures+gas adsorption/desorption+Knudsen diffusion model in nanopores, 45–47
microfractures+gas adsorption/desorption+matrix Fick's diffusion model, 43–44
microfractures+matrix macropores+gas adsorption/desorption+Fick's diffusion model in nanopores, 44–45
microfractures+matrix macropores +steady state adsorption/desorption and diffusion model, 41–43
microfractures+steady state adsorption/desorption and diffusion model, 39–41
MFHW in a single porosity medium in, 238–249
model assumptions and mathematical model, 238–240, 239*f*, 240*t*
numerical solutions, 240–244, 241*f*
production performance analysis, 244–249, 245*t*, 245–249*f*
MFHW in dual-continuum medium, 249–271
branch fracture models, 262–264, 262–264*f*, 263*t*
effects of fracture network on production, 261–271
local SRV models, 269–271, 270–271*f*, 270*t*
model assumptions and mathematical model, 249–252, 250–252*f*
numerical solution, 252–256
production performance analysis, 257–261, 257*t*, 257–258*f*

Shale gas reservoir *(Continued)*
rectangular composite models, 264–269, 265–269*f*
multiscale flow mechanisms, 24–39
pores, 3
hydraulic fractures, 11–12
inorganic pores, 6–9
natural microfractures, 10
organic intragranular pores, 10
traps, 3
tri-porosity models in, 271–285, 272*f*
model assumptions and mathematical model, 272–273
numerical solutions, 273–281, 273–274*f*, 277*f*
production performance analysis, 281–285, 281*t*, 282–286*f*
Sichuan basin, 5*t*
Single porosity fracture system, 239
Slippage flow, 26
SRV *See* Stimulated reservoir volume (SRV)
S-shape isotherm, 19
State diffusion
pseudo-steady, 335–336, 354–355
unsteady, 332–335, 353–354
Stehfest inversion algorithm, 88–89
Stimulated reservoir volume (SRV), 177, 178*f*, 179, 191–193, 306
boundary element models in composite gas reservoirs with, 194–197, 196*f*
different boundary conditions, 180–182, 181*f*
effect of SRV size, 183–184, 183–184*f*
effects of SRV permeability, 182–183, 182–183*f*
fractured vertical and horizontal wells with, 192*f*
fractured vertical wells in rectangular composite shale gas reservoirs with, 197–210
local SRV models, 269–271, 270–271*f*, 270*t*
MFHW in rectangular composite shale gas reservoirs with global, 210–222
MFHW in shale gas reservoirs with local, 222–236

T

Transient interporosity flow, 41–42, 336–340, 342–344, 355–360
Transient matrix flow, 326–329, 348–351
Transition flow, 26
Tri-porosity models, 271–285, 272*f*
model assumptions and mathematical model, 272–273
numerical solutions, 273–281, 274*f*, 277*f*
production performance analysis, 281–285, 281*t*, 282–286*f*

U

Unsteady state diffusion, 34, 332–335, 353–354

V

Viscous flow, 26

W

Wellbore storage/skin effects, 86–87
Well test analysis model application, 287–298
pressure buildup test analysis
well H2-2, 287–288, 288–290*f*, 290*t*
well H6-4, 289–290, 291–292*f*, 293*t*
well H6-6, 290–294, 293–296*f*, 296*t*
well H9-6, 294–298, 297–298*f*, 299*t*
Wufeng–Longmaxi formation, 299

Y

Young's modulus, 4

CPI Antony Rowe
Eastbourne, UK
June 18, 2019